G. Bauer · W. Richter

Optical Characterization of Epitaxial Semiconductor Layers

Springer

Berlin
Heidelberg
New York
Barcelona
Budapest
Hong Kong
London
Milan
Paris
Santa Clara
Singapore
Tokyo

Günther Bauer · Wolfgang Richter (Eds.)

Optical Characterization of Epitaxial Semiconductor Layers

With 271 Figures

Springer

Professor Dr. Günther Bauer

Institut für Halbleiterphysik
Universität Linz
Altenbergstraße 69
A-4040 Linz, Austria

Professor Dr. Wolfgang Richter

Institut für Festkörperphysik
Technische Universität Berlin
Hardenbergstraße 36
D-10623 Berlin, Germany

ISBN-13:978-3-642-79680-7   e-ISBN-13:978-3-642-79678-4
DOI: 10.1007/978-3-642-79678-4

Library of Congress Cataloging-in-Publication Data
Optical characterization of epitaxial semiconductor layers/Günther Bauer, Wolfgang Richter,
ed. - Berlin; Heidelberg; New York; London; Paris; Tokyo; Hong Kong; Barcelona; Budapest:
Springer, 1996
 ISBN-13:978-3-642-79680-7
NE: Bauer, Günther [Hrsg.]

Production: PRODUserv Springer Produktions-Gesellschaft, Berlin
Cover: Lewis + Leins GmbH; Berlin; Typesetting: Camera-ready by authors
SPIN: 10062779   54/3020 - 5 4 3 2 1 0 - Printed on acid-free paper

DEDICATED

TO

PETER GROSSE

Preface

The characterization of epitaxial layers and their surfaces has benefitted a lot from the enormous progress of optical analysis techniques during the last decade. In particular, the dramatic improvement of the structural quality of semiconductor epilayers and heterostructures results to a great deal from the level of sophistication achieved with such analysis techniques. First of all, optical techniques are nondestructive and their sensitivity has been improved to such an extent that nowadays the epilayer analysis can be performed on layers with thicknesses on the atomic scale.

Furthermore, the spatial and temporal resolution have been pushed to such limits that real time observation of surface processes during epitaxial growth is possible with techniques like reflectance difference spectroscopy.

Of course, optical spectroscopies complement techniques based on the interaction of electrons with matter, but whereas the latter usually require high or ultrahigh vacuum conditions, the former ones can be applied in different environments as well. This advantage could turn out extremely important for a rather technological point of view, i.e. for the surveillance of modern semiconductor processes. Despite the large potential of techniques based on the interaction of electromagnetic waves with surfaces and epilayers, optical techniques are apparently moving only slowly into this area of technology. One reason for this might be that some prejudices still exist regarding their sensitivity.

This book is aimed at filling this gap of knowledge. It is mainly written for graduate students or researchers active in the field of solid state physics or semiconductor engineering. It intends to give an overview of the present status of optical techniques with respect to the analysis of thin semiconductor layers. Whenever possible, the focus has been on in-situ analysis which is of course the most desirable mode of analysis. Since layer properties must be discussed always in correlation with growth we have also included a chapter dealing with analysis during growth, which concentrates on gasphase epitaxial growth, since excellent reviews and analysis tools (RHEED) exist for MBE. The structure of this book is discussed in more detail in Chapt. 1. Not all "optical" techniques – in a general sense – are treated. We have tried to focus on methods with only photons as probes, thus excluding, for example, photo-electron spectroscopy, a technique which would require also a vacuum environment to the sample. Other areas (photoluminescence, modulation spectroscopy, time resolved spectroscopy) were excluded because excellent reviews already exist and, in addition, their inclusion would have increased this book considerably.

In the course of writing and editing we have experienced numerous useful advice and helpful discussions from many colleagues. We would like to specifically mention M. Helm, V. Holý, G. Hughes, O. Hunderi, E. Koppensteiner, H. Krenn, L. Mantese and T. Ryan for critical reading certain chapters of this book. One of us (B.H.) is especially grateful to G. Bachmann and K. P. Hanke, Flachglas AG. It would have been impossible to finish the manuscript whithout the technical help of U. Frotscher, Mrs D. Knoll, M. Keuter, Mrs P. Marsiske, and T. Trepk. We are also grateful to A. Haase, J. Wagner and D. Aspnes for providing unpublished data. The support of the Springer Verlag by K.H. Lotsch and the pleasant collaboration with Mrs G. Maas in the production department should be mentioned especially.

The authors and editors   Summer 1995

Contents

List of Contributors

Günther Bauer
Institut für Halbleiterphysik
Universität Linz
Altenbergstraße 69
A-4040 Linz, Austria

Norbert Esser
Institut für Festkörperphysik
Technische Universität Berlin
Hardenbergstraße 36
D-10623 Berlin, Germany

Jean Geurts
I. Physikalisches Institut
RWTH Aachen
Sommerfeldstraße
D-52074 Aachen, Germany

Bernd Harbecke
Flachglas AG
Auf der Reihe 2
D-45884 Gelsenkirchen, Germany

Bernhard Heinz
I. Physikalisches Institut
RWTH Aachen
Sommerfeldstraße
D-52074 Aachen, Germany

Alois Krost
Institut für Festkörperphysik
Technische Universität Berlin
Hardenbergstraße 36
D-10623 Berlin, Germany

Volkmar Offermann
I. Physikalisches Institut
RWTH Aachen
Sommerfeldstraße
D-52074 Aachen, Germany

Wolfgang Richter
Institut für Festkörperphysik
Technische Universität Berlin
Hardenbergstraße 36
D-10623 Berlin, Germany

Uwe Rossow
Institut für Festkörperphysik
Technische Universität Berlin
Hardenbergstraße 36
D-10623 Berlin, Germany

Wolfgang Theiß
I. Physikalisches Institut
RWTH Aachen
Sommerfeldstraße
D-52074 Aachen, Germany

Joachim Woitok
I. Physikalisches Institut
RWTH Aachen
Sommerfeldstraße
D-52074 Aachen, Germany

Dietrich Zahn
Fachbereich Physik
Technische Universität Chemnitz Zwickau
Reichenhainer Straße 70
D-09126 Chemnitz, Germany

1. Introduction

Günther Bauer, Wolfgang Richter

The last decade has witnessed an enormous progress in semiconductor growth technologies for the oriented deposition of single crystalline layers. Growth methods like Molecular Beam Epitaxy (MBE) [1.1,2.2,1.2–1.4], Metal-Organic Vapour Phase Epitaxy (MOVPE) [1.5], Hot Wall Epitaxy (HWE), Liquid Phase Epitaxy (LPE) and their variants are nowadays matured and applicable for device fabrication techniques. A large number of different combinations of semiconductors have been successfully deposited onto various substrate materials. For perfect epitaxial growth the match of the lattice constants and the thermal expansion coefficients between the substrate and the deposited layer is required if no thickness constraints are imposed. Otherwise, for non-lattice-matched layers, pseudomorphic growth of strained layers is only possible up to a critical thickness. Beyond the critical thickness, misfit dislocations are generated which relieve the strain. For a lattice mismatch, $(a_{film} - a_{substrate})/a_{substrate}$, of 1 percent typical values for the critical thickness are $10\,nm$ (for $Si_{0.75}Ge_{0.25}$ on Si). In Fig. 1.1 the lattice constants of all relevant semiconductors, namely group IV elements, III-V, II-VI and IV-VI compounds are shown. This figure also shows the fundamental energy gap of these semiconductors which is one of the most important quantities for both electronic and optical applications.

Semiconductor heterostructures, Multi-Quantum Wells (MQWs) and Superlattices (SLs) are not only in the focus of present-day research but some of them form the basis of modern electronic and optoelectronic devices. Examples are the High Electron Mobility Transistors (HEMT), the Hetero-Bipolar Transistors (HBT) [1.6, 1.7] and Quantum Well Lasers. Consequently, the characterisation of epitaxial layers and multilayers is very important for the assessment of their mechanical, electronic, optical, and magnetic properties. In order to study certain physical phenomena, well-defined epitaxial layers are necessary. In addition the design and improvement of electronic and optoelectronic devices depends on the precise knowledge of several parameters which characterise the quality of epitaxial layers. Therefore, experimental techniques are required which determine to a high degree of accuracy quantities like crystalline perfection, strains, chemical composition, carrier concentration, carrier mobilities, and physical and chemical properties at interfaces. It would be desirable to get as much information as possible already during the growth, i.e. *in-situ*. Since layer growth takes place at elevated temperatures and not always under ultrahigh-vacuum conditions the number of *in-situ* methods is limited. Moreover, only such *in-situ* techniques which do not disturb the growth or even lead to a partial destruction of the growing film are preferable.

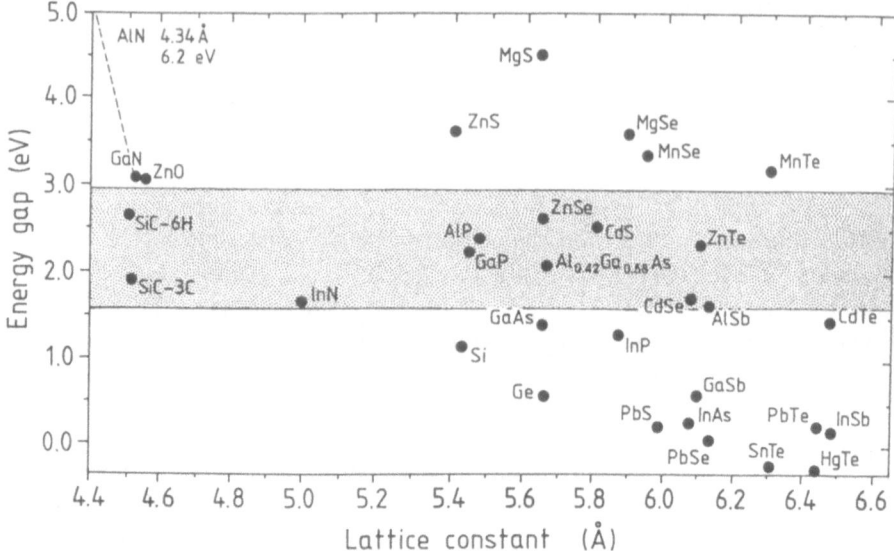

Fig. 1.1. Energy gap vs. cubic lattice constant of elemental group IV semiconductors and of III-V, II-VI and IV-VI compound semiconductors

There are only a few methods which fulfill all these requirements. Among these, the techniques which rely on the interaction of electromagnetic radiation with matter, if properly used, cause rather small perturbations of the growth process. Furthermore, these methods are the only ones which can be applied for *in-situ* control of films grown by gasphase epitaxy, e.g. Metal Organic Vapour Phase Epitaxy (MOVPE).

Apart from the *in-situ* analysis, many properties like carrier concentrations, mobilities and strains are not of interest at growth temperatures but rather at ambient or low temperatures. Therefore, *ex-situ* measurements are equally important as well. For *ex-situ* measurements, nondestructive methods are also preferred. For specific purposes, destructive techniques are unavoidable. These techniques are quite often necessary in order to calibrate nondestructive ones or because they offer the highest achievable sensitivity. In the past many reviews have been devoted to the description of destructive techniques such as Auger electron spectroscopy and secondary ion mass spectroscopy and therefore we concentrate in this series of reviews on nondestructive, i.e. optical ones. These include diffraction, reflection, transmission of electromagnetic radiation, inelastic light scattering, ellipsometry and reflection anisotropy.

This book is organised as follows. Chapter 2 is devoted to the analysis of epitaxial growth. Since many excellent reviews on MBE growth have been published within the last couple of years [1.1, 2.2, 1.8], we concentrate here mainly on the analysis of gasphase epitaxial growth. For these growth methods, Reflection High Energy Electron Diffraction (RHEED) [1.10] cannot be used. Optical investigation of the gasphase as well as of the growing surface are the methods of choice. In Fig. 1.2 we schematically show what kind of information can be extracted. Different optical methods using first- and second-order interaction of light with matter turn out to be useful. Gas phase reactions play an important role and thus information on various quantitities is needed:

— Gas velocity is measured by elastic scattering from injected particles (Laser Doppler Anemometry (LDA)).
— Gas temperature can be determined via the population of excited states by Raman scattering or Coherent Anti-Stokes Raman Scattering (CARS).
— Molecules present in the gasphase can be identified and their concentrations can be measured by absorption, Raman scattering or CARS.

Reactions at the surfaces of course constitute the final step in the growth process and are even more important. Rather modest information has been accumulated up to now on surface processes in gasphase epitaxial growth. So far, most of the models put forward have been derived on the basis of data obtained in Gas Source Molecular Beam Epitaxy (GSMBE) (Metal Organic Molecular Beam Epitaxy (MOMBE), Chemical Beam Epitaxy (CBE)) by standard UHV analysis techniques. The use of these data in VPE at the considerably higher ambient pressures is of course doubtful. Both ellipsometry and reflectance anisotropy spectroscopy [1.11, 1.12] are methods which have the potential to yield relevant information on the surface of the growing film. Especially the exploitation of the anisotropic reflectance (Fig. 1.3) has proven already in the last years its capability of real time monitoring in all situations of III-V-semiconductor MOVPE growth [1.14]. This includes the substrate deoxidation process, selection of the proper pregrowth surface reconstruction

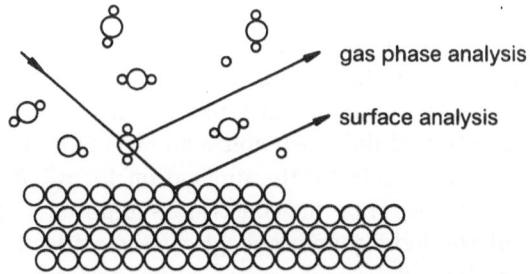

gas phase analysis

surface analysis

Fig. 1.2. Schematic setup of gasphase analysis and surface analysis in gasphase epitaxial growth. *Information needed*: (i) Gasphase: Gas velocity, gas temperature, identification of molecules, concentration. (ii) Surface: structure, composition. *Methods for analysis*: Elastic light scattering, absorption, Raman scattering, Coherent Anti-Stokes Raman Scattering (CARS), Spectroscopic Ellipsometry (SE), Reflectance Anisotropy Spectroscopy (RAS), Second Harmonic Generation (SHG)

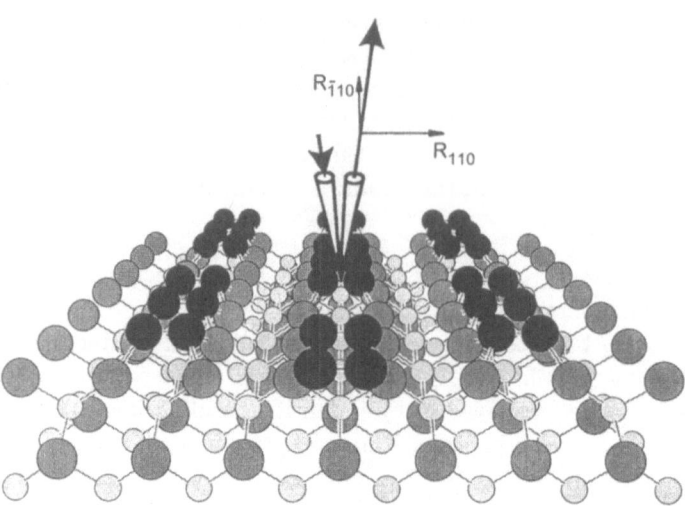

Fig. 1.3. Schematic setup of reflectance anisotropy spectroscopy (RAS) experiment. *Interaction with*: Anisotropic electronic surface states. *Information on*: Surface reconstructions, oxidation/deoxidation, exchange reactions, growth rate, morphology

as well as static and dynamic control during growth. In the latter studies the signal oscillations appearing with monolayer periodicity are especially noteworthy. They give also promise to a deeper understanding of the growth mechanism in MOVPE [1.13]. After this chapter, which focusses on growth aspects, the following are entirely devoted to the different analysis techniques.

The third Chapter deals with ellipsometric studies. Even though this is an old spectroscopic technique, the computerisation of instruments has led to its renaissance in the last decade [1.11,1.12]. In Fig. 1.4, the principles of this measurement technique are sketched and the information which can be extracted is listed. In ellipsometry, the state of polarisation of the reflected light is measured as a function of photon energy. In general, it yields information on the dielectric function in the region of electronic interband transitions. It can therefore be used for the idenfication of unknown epitaxial layers and for the evaluation of structural and chemical defects. Its application for *in-situ* growth control has been somewhat hampered by the constraints imposed due to the geometrical arrangement of the light source and growth chamber as well as the polarisation analysis equipment.

Raman scattering is the subject of the fourth chapter (Fig. 1.5). The experiment determines the frequency of inelastically scattered light, its intensity as well as its polarisation. The majority of investigations deal with lattice vibrations in epitaxial layers and substrates. The modification of phonon branches due to composition and defects, strain and orientation are derived

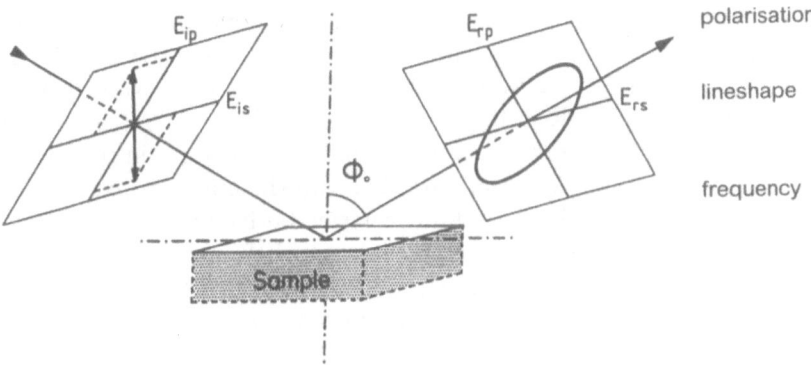

Fig. 1.4. Schematic setup of ellipsometry experiments on epitaxial layers. *Interaction with*: Electronic interband transitions (surface and bulk). *Information on*: Energy gaps versus temperature, strain and chemical composition, layer thickness with submonolayer sensitivity, layer dielectric functions

from frequency positions, spectral widths and polarisation properties. Local strains can be investigated by second-order Raman scattering. Furthermore, Raman scattering by single particle or collective electronic excitations provides information on the properties of carriers in the conduction or valence band. Finally, electronic transitions in impurities are utilised to determine shallow acceptors and donors as well as undesired impurities.

Far Infrared Spectroscopy (FIR) has been a classical tool for the investigation of bulk semiconductors. The amount of information which can be ex-

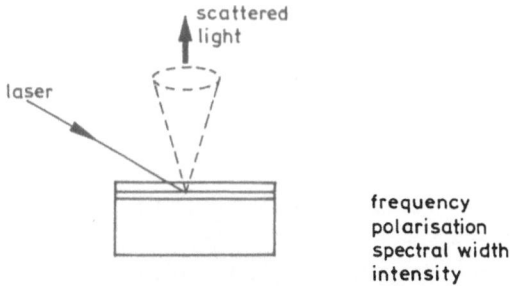

Fig. 1.5. Schematic setup of Raman scattering experiments on epitaxial layers. *Interaction with*: (i) Lattice vibrations: optical and acoustical phonons, local vibrations. (ii) Electronic excitations: plasmons, single particle excitations, impurity states. *Information on*: (i) Structural properties: crystalline quality, orientation, strain. (ii) Chemical properties: composition, reacted phases at interfaces. (iii) Electronic properties: impurities, free carriers, potential barriers

tracted from reflection and transmission experiments on epitaxial layers gives considerable details on their properties (Fig. 1.6). This method is described in detail in Chapt. 5. Reflection and transmission spectra are governed by the frequency-dependent dielectric function. In the far-infrared, mainly optical phonons, free carriers and impurities contribute to the features in the spectra. The information is partly similar and partly complementary to that obtained from Raman spectroscopy. In general, information both on top epitaxial layers as well as on the substrate can be extracted from FIR spectra. The chapter on FIR spectroscopy also contains an elaborate treatment of interaction of light with matter based on Maxwell's equations and stresses the importance of the boundary conditions in the evaluation of spectra of epitaxial layers and layered systems. The spectra contain, due to interference and multiple reflection, information on the geometry, like layer and substrate thicknesses. Even steplike or gradual changes of parameters like carrier concentration or composition can be assessed. The lateral resolution is rather poor (focal diameters are of the order of 3 mm, apart from mid-infrared microscopy techniques where it is about a factor of 100 smaller). Precise determination of the free carrier concentration and their damping parameters is possible. The method is particularly sensitive for the detection of low concentrations of shallow impurities. It has been shown that FIR spectroscopy is well suited to investigate low-dimensional electronic systems like heterostructures, quantum wells, quantum wires and dots.

In Chapt. 6 several techniques based on high-resolution X-Ray diffraction are presented. This method has benefitted from considerable improvements

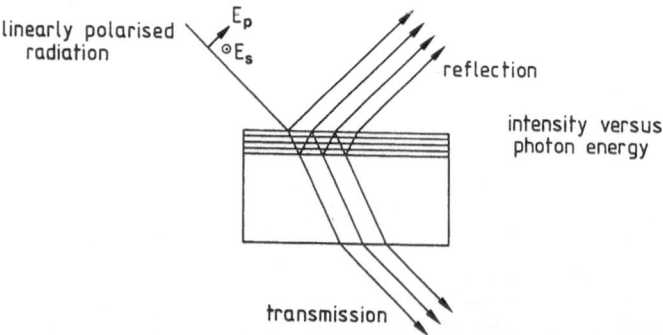

Fig. 1.6. Schematic setup of far infrared spectroscopy experiments on epitaxial layer systems. *Interaction with:* (i) Lattice vibrations: optical phonons, local vibrations. (ii) Electronic excitations: free carriers, impurity states. *Information on:* (i) Structural properties: crystalline quality, layer thickness. (ii) Vibronic properties: frequency and damping of TO phonons and of local modes. (iii) Electronic properties: binding energy and concentration of impurities, free carriers concentration and damping, low dimensional effects like subband energies, minibands, impurities in quantum wells.

of the instrumentation, particularly from high resolution monochromators, position sensitive detectors and sources like rotating anodes or synchrotron radiation [1.15–1.17]. In addition, techniques like grazing incidence diffraction and reflection have been successfully improved for the study of epitaxial layers [1.18]. Using diffraction methods, information on the reciprocal space is obtained. All diffraction methods utilise different scans through the reciprocal lattice (Fig.1.7). Very detailed structural information on epitaxial layers and substrates can be obtained such as film thicknesses, superlattice periods, crystalline quality and mosaic spread. Geometrical distortions like strain and strain relief are measured directly and information on strain gradients can be obtained as well.

Many of these different experimental techniques mentioned above can be applied in order to determine similar physical properties or quantities in epitaxial layers. Therfore, we compare in the following some methods with respect to their sensitivity limits. Structural parameters are obtained by high-resolution X-Ray diffraction. Strains as low as $1 \cdot 10^{-5}$ can be determined with the usual laboratory instrumentation. X-Ray techniques also are the only ones which provide direct information on strains without invoking properties which are related to stress and deformation potentials. As an example we quote the phenomenon of strain relief beyond a certain critical thickness in lattice mismatched layers. It is measured directly by X-Ray diffraction as a change of lattice constant but manifests itself also as a shift of phonon frequencies. Typical minimum stress values evaluated from frequency shifts are in the order of 1 kbar, which corresponds to strains of magnitude $2 \cdot 10^{-3}$. In cases where lateral resolution (1 μm) is required, Raman scattering is advantageous.

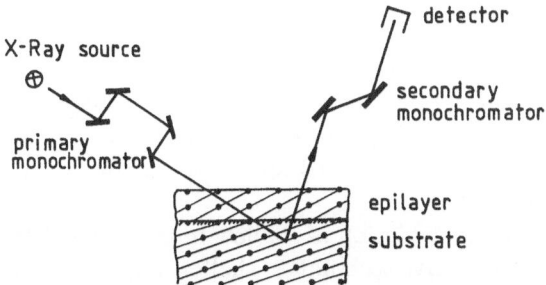

Fig. 1.7. Schematic setup of High Resolution X-Ray Diffraction (HRXRD) of epitaxial layers with primary and secondary monochromators indicated. *Experimental observation*: Bragg intensity for different scans through reciprocal space. *Information on*: (i) Geometry: crystal structure, film thickness, superlattice periods, tilt and terracing. (ii) Strain: elastic strain and relief, distortions, interfacial strain, inhomogenity. (iii) Crystalline perfection: mosaic spread, interface roughness, interdiffusion

Orientation of the epitaxial film with respect to the substrate is conveniently measured using X-Ray diffraction. Under certain circumstances Raman scattering can yield information on orientation as well.

The crystalline quality of epitaxial layers is a very important parameter which is, however, very difficult to quantify. The width of rocking curves is quite often quoted as a measure of crystalline perfection of epitaxial layers. For very thick layers ($> 10\,\mu$m) values as low as 0.16 arcsec have been quoted for silicon, whereas for GaAs this value is about 10 arcsec and for II-VI and IV-VI semiconductors typical values range from 20-100 arcsec. For very thin layers of the order of $1\,\mu$m and below, the method fails due to broadening as a consequence of the finite film thickness unless grazing incidence techniques are used [1.18]. In this thickness range, methods like Raman scattering or ellipsometry are useful by monitoring the influence of imperfections on lattice dynamical or electronic properties.

Layer thicknesses are obtained in FIR spectroscopy from the distance of interference induced extrema in transmission or reflectivity in the frequency region between the plasma edge and the fundamental energy gap. The accuracy is determined by the photon energy and reaches at best a value of about 100 nm. In case the sample consists of multilayers a much better precision is achieved by using X-Ray diffraction. Then a limit of one (embedded) monolayer has been reached. However, the most suitable method for such thin layers is ellipsometry where monolayer resolution on single films can be achieved routinely.

Electronic properties and their changes with chemical composition of the layers are of course the most basic quantities for any kind of investigation directed towards the application in electronic devices. Using conductivity and Hall measurements, carrier concentrations and mobilities are always obtained in combination. These methods are only applicable if the substrate is insulating and the film is doped. Optical methods like infrared spectroscopy and Raman scattering probe the carrier concentration without any contacts and differentiate between the contributions from the substrate and the epilayer. In addition, Raman scattering even provides lateral resolution in the μm scale. The sensitivity limits are about $10^{15}\,\mathrm{cm}^{-3}$ for FIR spectroscopy and about $10^{17}\,\mathrm{cm}^{-3}$ for Raman spectroscopy.

The most sensitive method for the detection of shallow impurities is FIR spectroscopy. In Ge, using photoconductive techniques, shallow impurity concentrations as low as $10^{10}\,\mathrm{cm}^{-3}$ were reported. In Si and III-V compounds the minimum values increase from 10^{11} to $10^{13}\,\mathrm{cm}^{-3}$. Defects like carbon or oxygen impurities can be detected in silicon with concentrations of about 10^{15} to $10^{16}\,\mathrm{cm}^{-3}$ using FIR transmission. Raman scattering generally has a somewhat lower sensitivity but has yielded in certain cases similar sensitivities, i.e. a carbon concentration of $10^{15}\,\mathrm{cm}^{-3}$ can be detected in GaAs.

Quite often also carrier concentration profiles and composition gradients (i.e. in ternary compounds) occur. Those can be evaluated from polarised

FIR transmission and reflection spectra with appropriate model calculations. Gradients in composition and interdiffusion at interfaces can be obtained non-destructively from X-Ray diffraction, but the evaluation of such data is in general tedious and often ambiguous. Consequently such data should be checked using Secondary Ion Mass Spectroscopy (SIMS) with sputtering.

Nearly in all cases of thin film analysis the interface properties to the substrate or to adjacent layers influence more or less the layer properties. Information about the interface contribution is, however, usually extracted only in a very indirect manner from ex-situ studies. In-situ analysis, especially by reflectance anisotropy spectroscopy, has shown the capability to monitor the interface formation processes optically with a resolution in the submonolayer range. Thus very detailed and direct information on the interface may be obtained. This is especially important for non-UHV systems, where RHEED and other techniques cannot be applied.

Table 1.1. Experimental Methods in use for analysis of epitaxial semiconductor layers

Method	ex-situ	in-situ	References
Reflection High Energy Electron Diffraction (RHEED)		×	B. A. Joyce [1.1], P. K. Larsen and P. J. Dobson [1.10]
Low Energy Electron Diffraction (LEED)		×	M. Henzler [1.21] M. P. Seah [1.22]
Auger Electron Spectrocopy (AES)	×	×	M. Grasserbauer and H. W. Werner [1.23]
Secondary Ion Mass Spectroscopy (SIMS)	× destructive		A. Benninghoven et al. [1.24]
Rutherford Back-scattering (RBS)	×		S. T. Picraux et al. [1.25]
Nuclear Reaction Analysis (NRA)	×		W. F. van der Weg and F. H. P. M. Habraken [1.26]
Transmission Electron Microscopy (TEM)	× destructive		H. Cerva, H. Oppholzer [1.27], H. Cerva [1.28], F. A. Ponce et al. [1.29], L. Reimer [1.30]
Scanning Electron Microscopy (SEM)	×	×	D. E. Newbury et al. [1.31]
Scanning Tunneling Microscopy (STM)	×	×	H.-J. Güntherodt and R. Wiesendanger [1.32]

Table 1.1. (continued)

Method	*ex-situ*	*in-situ*	References
Electron Paramagnetic Resonance (EPR)	×		R. C. Newman [1.33]
Electrical Characterisation (Hall effect,DLTS)	× destructive	×	R. A. Stradling and P. C. Klipstein [1.34]
Photo Electron Spectroscopy (PES) (XPS, SXPS, UPS)		×	M. Cardona and L. Ley [1.36], G. V. Marr [1.37]
Extended X-Ray Absorption Fine Structure (EXAFS)	×	×	C. C. Koningsberger and R. Prius [1.38]
Near-Field Optical Microscopy	×		E. Betzig et al. [1.19]
Photoluminescence	×		E. C. Lightowlers [1.39] M. A. Herman et al. [1.40] M. Illegems [1.9]
Electro-Photo-reflectance	×		D. E. Aspnes [1.11] F. H. Pollak [1.41]
Raman spectroscopy	×	×	M. Cardona [1.42]
Ellipsometry	×	×	D. E. Aspnes [1.12] J. B. Theeten [1.43]
Far Infrared Spectroscopy (FIR)	×	×	R. A. Stradling, P. C. Klipstein [1.34] T. Dumelow et al. [1.35]
Reflection Anisotropy Spectroscopy (RAS)	×	×	D. E. Aspnes [1.12]
X-Ray diffraction		×	A. Segmüller et al. [1.16], A. Segmüller [1.18] S. T. Picraux et al. [1.25] P. F. Fewster [1.15] B. K. Tanner [1.17] HRXRD-Conference [1.44]
X-Ray topography	×		Z. J. Radzimski et al. [1.45] R. Köhler [1.46]

All applied techniques bear a spatial resolution given by the beam diameter, which can be reduced down to a size comparable to the wavelength. The standard equipment for e.g. luminescence or Raman spectroscopy provides a spatial resolution of the order of 0.1-1 mm. By using microscope lenses these values can be further diminished to the diffraction limit (micro-luminescene and micro-Raman techniques) which gives resolution in the μm-range. Very recently the development of near-field microscopy has allowed for optical analysis in the nanometer range [1.19, 1.20].

In Table 1.1 we list well-known methods for the characterisation of epitaxial layers according to the following criteria: *ex-situ*, *in-situ*, destructive and nondestructive techniques including those discussed in this volume. References to recent review articles and books are given. Some optical methods like photoluminescence, photoreflectance, X-Ray topography were omitted because excellent review articles exist.

2. Analysis of Epitaxial Growth

Wolfgang Richter, Dietrich Zahn

In the past, crystal growth and crystal characterisation were different topics performed in different laboratories by scientists from different disciplines. With the advent of epitaxial layer growth by Molecular Beam Epitaxy (MBE), Vapour Phase Epitaxy (VPE) and Liquid Phase Epitaxy (LPE) the two areas moved closer together since microscopic knowledge about the growth process turned out to be necessary in order to understand and to control the growth in a reproducible manner.

Especially in the ultrahigh vacuum environment of MBE systems, analytical methods developed in the area of surface science were introduced in order to have *in-situ* control over the growth process and thereby over the properties of the layers grown. Growth of the epitaxial layers, *in-situ* analysis and control of the growth process and epitaxial layer characterisation are today performed in close connection and feedback. This holds also for the MBE variants like gas source MBE (GSMBE), metalorganic MBE (MOMBE) or Chemical Beam Epitaxy (CBE). An excellent description of these MBE or MBE like growth techniques can be found for example in references [2.1–2.3].

In VPE techniques like Hot Wall Epitaxy (HWE) [2.5], Halide-based Chemical Vapour Deposition (HCVD) [2.6] and Metal Organic Vapour Phase Epitaxy (MOVPE) [2.7, 2.8] this close cooperation is much less developed. This is mainly due to the fact that surface science techniques which need high or ultra high vacuum are not applicable, because the growth process requires gas pressures in the range from typically 10^2 to 10^5 Pa. In addition, the gas phase certainly plays an important role in the growth process (transport, chemical reactions). Therefore, not only information about surface processes but also about gas phase reactions is needed in order to understand and control the growth.

Techniques capable of performing *in-situ* diagnostics in and through gas phase environments generally must rely on "optical" methods. Electromagnetic radiation in the near ultraviolet (UV), visible (VIS) and near infrared region (IR) is easily transmitted by the reactor walls generally made of quartz and can easily penetrate through the gas phase to the growing interface. Thereby analysis of the gas phase as well as probing the surface becomes possible.

Experimentally, however, there is the difficulty that not only the epitaxial layer is growing but, in addition, material is also deposited on the reactor walls. Then the walls may become more and more opaque to optical radiation during growth and finally any analysis is no longer possible. By slight modification of typical reactors it is nevertheless practicable to introduce windows (flushed by inert gas) to the reactor which do not disturb the growth process but allow for permanent optical access to the region at or near the growing surface.

Still the combination of realistic growth apparatus with sophisticated optical equipment has not been established in too many laboratories. Progress has been made in the last 20 years first of all with respect to an analysis of the hydro- and thermodynamical aspects of the gas phase in the reactor. Velocity profiles and temperature distributions are nowadays experimentally and theoretically well under control.

The information on reaction kinetics and the evaluation of the most critical reaction steps has made much less progress. Each species to be analysed usually requires a specifically designed experiment and thus progress has been quite slow. However, for the technologically most important semiconductors (Si, III-V) some information about gas phase reactions has been accumulated by now.

The situation is not at all satisfactory with respect to surface processes. Most of the models used today in the description of surface kinetics draw their experimental justification either from gas phase diagnostic results or from hybrid growth methods like Metal Organic MBE (MOMBE) where, at low pressures, standard MBE analysis techniques can be applied. Surface sensitive experiments under gas phase conditions are of course difficult, but recent developments demonstrate that optical techniques might come to a stage where they can be used similarly as standard surface science techniques.

In LPE finally, where the growing surface is surrounded by liquid metallic solutions any *in-situ* diagnostics besides measuring the temperature is hardly possible. However, since LPE is operating near thermal equilibrium the temperature is of course the most important physical quantity. Indeed, it seems that LPE growth can be handled quite well by just controlling the temperature [2.9, 2.10].

This Chapter will therefore be predominantly devoted to vapour phase epitaxial growth. Emphasis in the examples discussed will be on two technologically important systems: MOVPE of III-V-semiconductors and silicon epitaxial growth from Si-hydrides or Si-chlorides. Most of the analytical work published up to date has been concerned with these two areas. The Chapter will be organized according to the preceding discussion. First we will discuss gas phase diagnostics with respect to transport properties (Sect. 2.2), followed by the more difficult task of analysing possible pathways in gas phase reactions (Sect. 2.3). The last two Sections will be devoted to surface analysis (Sect. 2.4.1) and a concluding discussion in terms of growth models (Sect. 2.5).

2.1 Vapour Phase Epitaxy: Basics

The VPE process starts with a gas mixture which contains the precursor molecules necessary for epitaxial growth and a carrier gas (Fig. 2.1). The carrier gas is usually hydrogen with an operating pressure between 10^3 Pa and atmospheric pressure (10^5 Pa). The choice of precursors depends of course on the material to be deposited. In this Chapter we will concentrate on III-V semiconductors for which the standard precursors are metalorganic (mainly alkyl) compounds of the group III elements and hydrides of the group V-elements. The epitaxial growth process is then called Metal Organic Vapour Phase Epitaxy (MOVPE). Because of many undesired properties of the alkyls and hydrides, however, new organic compounds ("alternative precursors") are tested today for replacement. For silicon deposition, on the other hand, mainly silane SiH_4 or chlor-silanes $SiCl_nH_{4-n}$ are the standard precursors. Typical partial pressures for these precursors in the VPE process range from ten to a few hundred Pa. Therefore, the total pressure of the gas mixture and many of its physical properties are essentially determined by the carrier gas.

This gas mixture is fed into a reactor where the heated substrate is placed. Many designs for MOVPE reactors have been proposed, tested, and are in use. The goal of such development work is the homogeneity of properties such as thickness, stoichiometry, and carrier concentration across the whole substrate surface. One common design, which is used in research laboratories as well as in production lines, is the horizontal reactor displayed in Fig. 2.2.

Typical hydrodynamic operating conditions in such a reactor are a laminar flow and gas velocities in the range from 0.1 to 1 m/sec. In growth processes which operate far from thermal equilibrium, as is the case for the decomposition of metalorganic compounds, only the substrate but not the reactor walls are heated (cold wall reactor) in order to avoid decomposition and subsequent deposition onto the walls. As a consequence only the gas in a region close to the substrate is heated (thermal boundary layer) and large

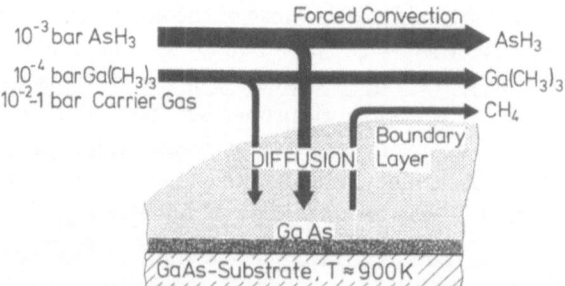

Fig. 2.1. Schematic picture of a gas phase epitaxial growth process using the example of GaAs MOVPE

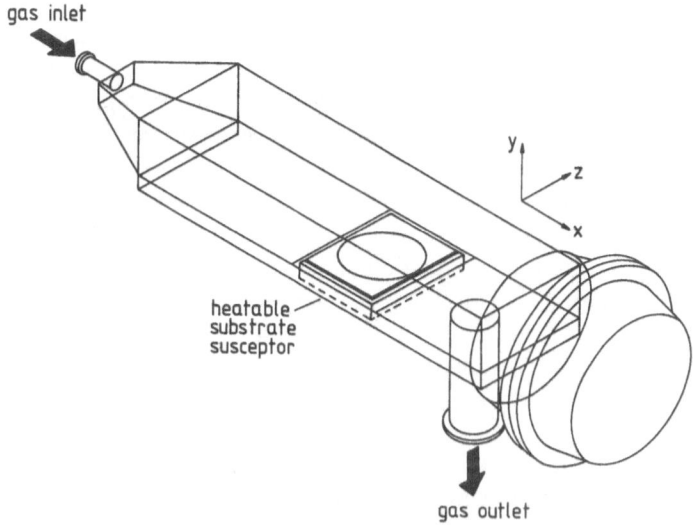

Fig. 2.2. Horizontal reactor for gas phase epitaxial growth

temperature gradients are created in the reactor. Within this region gas phase (homogeneous) reactions and (heterogeneous) reactions at the gas-solid interface take place, leading under appropriate conditions to epitaxial growth of semiconductor layers. Two main factors determine the growth rate, as can be seen from Fig. 2.3. One is given by the gas phase transport of the precursors (or their reaction products containing the group III or V atoms) to the interface.

The other is determined by the reactions near or at the interface converting the precursor molecules into group III or V atoms which are then incorporated into the solid state lattice structure. In the pressure range of 10^3 to 10^5 Pa the transport can be described within the hydrodynamic for-

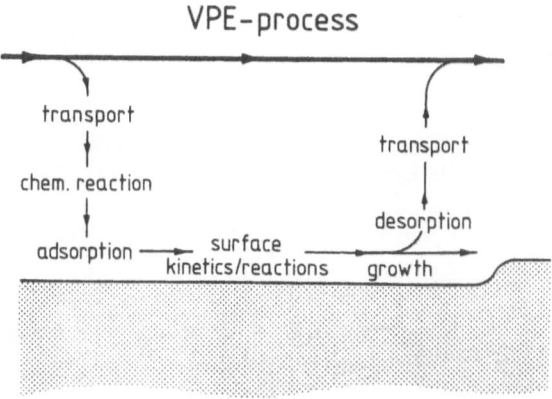

Fig. 2.3. Schematic diagram of single steps relevant in the VPE growth process

malism and is governed by convection and diffusion. Near the substrate the convective flow is essentially zero and diffusion will be dominant. Since the diffusion constant is only weakly temperature dependent in the range of typical growth temperatures (800 to 1100 K) a weak temperature dependence is expected also for the transport contribution to the growth rate. For the chemical reactions taking place near or at the interface the reaction rates with their exponential factors yield a strong temperature dependence. Therefore, the reaction rate contribution to the growth rate will be strongly temperature dependent. These two temperature dependencies are indeed observed in temperature ranges where either one of these factors is the limiting one, i.e. the slowest step (Fig. 2.4). Temperature independence is found at high temperatures and high pressures (transport controlled growth) and exponential behaviour at low temperatures and low pressures (kinetically controlled growth).

Optical techniques useful for analysis in such an environment separate into two groups, namely those methods which are suitable for gas phase analysis and those which provide information from the growing surface. While gas phase diagnostics in principle is an old topic, it has largely benefitted from the advent of the tunable dye laser. Together with properly chosen optical effects it is ideally suited to perform analysis in those quite hostile environments. Considerable progress in the development and application of laser methods has therefore been made in the last decade. The following Sects. 2.2, 2.3 will deal with those techniques. Optical methods for surface analysis, on the other hand, when compared to vacuum compatible electron spectroscopies, are still in a very early stage. This is partly due to the fact that their surface sensitivity was in general not sufficient and, moreover, that it is difficult to extract microscopic information from a macroscopic dielectric function integrating over many microscopic processes. However, progress has been made in the last years and there is hope that these techniques will develop into standard methods. Section 2.4 will deal with this topic.

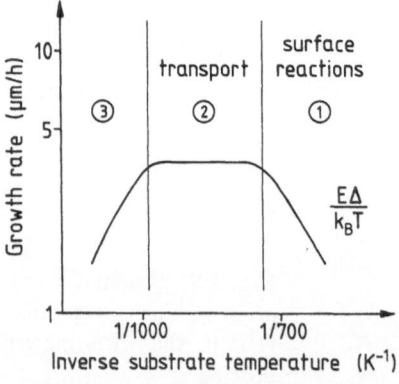

Fig. 2.4. Schematic diagram of growth rate versus temperature

2.2 Gas Phase Diagnostics: Transport

As discussed above, gas phase diagnostics will have to concentrate on the chemical gas phase reactions in order to give information on what species will be present on the growing surface and on the parameters which determine transport of these species to the growing surface. The latter will be especially important in situations of transport limited growth where the homogeneity of growth might be strongly influenced by transport. In this Section we will discuss diagnostics for transport relevant parameters i.e. velocities and temperature profiles in a reactor. Section 2.3 will deal with the analysis of chemical reaction paths in the gas phase.

The mass transport, i.e. the flux $\mathbf{j}_i$ of reactant i to the gas-solid interface is given by convective flow, by diffusion in a concentration gradient and by thermodiffusion (temperature gradient driven). It can be described as follows [2.11]:

$$\mathbf{j}_i = \frac{p_i \mathbf{v}}{kT} - \frac{D_i}{kT}\left[\nabla p_i + \frac{\alpha_i}{T} p_i \nabla T\right] \tag{2.1}$$

where p_i is the partial pressure of the species i, D_i is the diffusion constant of species i within the carrier gas and α_i is the thermodiffusion coefficient. In writing this equation multicomponent diffusion and gas phase reactions have been neglected. The important parameters which determine the flux under these conditions are the velocity $\mathbf{v}$, the temperature T and the partial pressures. While the partial pressures of the growth relevant species will be the result of the chemical gas phase reactions, gas velocity and temperature are essentially a property of the carrier gas only. This is because of the large difference in partial pressures which allow the contribution of the precursor and their reaction products to the hydrodynamical and thermodynamical properties of the gas mixture to be neglected in a good approximation. Thus for the experimental determination of $\mathbf{v}$ and T as well as for theoretical calculations it is sufficient just to consider the carrier gas. This simplifies the experimental studies (no precursors are needed) as well as the theoretical work by reducing the number of necessary equations.

2.2.1 Theoretical Considerations

In order to calculate the temperature and velocity profiles in a reactor, the equations following from the conservation of momentum, mass and energy together with appropriate boundary conditions have to be solved. This is usually performed with finite elements methods. Because of the complex geometries involved it is not really feasible to solve a full 3-dimensional model even with currently available supercomputer capabilities. By considering, however, the reactor to be symmetric in a midplane along the flow direction (like the one in Fig. 2.2) two dimensional solutions in that plane seem to be good approximations and calculations can be performed with realistic boundary conditions on supercomputers. The momentum equation then reads with the coordinate system as defined in Fig. 2.2 for an incompressible gas [2.11]:

$$
\rho \left(v_y \frac{\partial v_y}{\partial y} + v_z \frac{\partial v_y}{\partial z} \right) + \frac{\partial P}{\partial y} - \frac{\partial}{\partial y} \left[\mu \left(\frac{4}{3} \frac{\partial v_y}{\partial y} - \frac{2}{3} \frac{\partial v_z}{\partial z} \right) \right]
$$
$$
- \frac{\partial}{\partial z} \left[\mu \left(\frac{\partial v_y}{\partial z} + \frac{\partial v_z}{\partial y} \right) \right] - \rho g \sin \theta = 0 \tag{2.2}
$$

$$
\rho \left(v_y \frac{\partial v_z}{\partial y} + v_z \frac{\partial v_z}{\partial z} \right) + \frac{\partial P}{\partial z} - \frac{\partial}{\partial y} \left(\frac{\partial v_z}{\partial y} + \frac{\partial v_y}{\partial z} \right)
$$
$$
- \frac{\partial}{\partial z} \left[\mu \left(\frac{4}{3} \frac{\partial v_z}{\partial z} - \frac{2}{3} \frac{\partial v_y}{\partial y} \right) \right] - \rho g \cos \theta = 0 \tag{2.3}
$$

where v_y and v_x are the vertical and axial velocity, respectively. ρ is the density of the carrier gas, μ its viscosity, P the total pressure, g the gravitational constant and θ defines the orientation of the reactor with respect to the horizontal position.

These equations must be combined with the overall continuity balance (mass conservation):

$$
\frac{\partial}{\partial y} (\rho v_y) + \frac{\partial}{\partial z} (\rho v_z) = 0 \tag{2.4}
$$

and the ideal gas law:

$$
\rho = P M_W / RT \tag{2.5}
$$

where R is the ideal gas constant, M_W the molecular weight of the carrier gas and T the temperature. The temperature distribution is then given by the solution of the energy balance equation:

$$
\rho c_p \left(v_y \frac{\partial T}{\partial y} + v_z \frac{\partial T}{\partial z} \right) - \frac{\partial}{\partial y} \left(k \frac{\partial T}{\partial y} \right) - \frac{\partial}{\partial z} \left(k \frac{\partial T}{\partial z} \right) = 0 \tag{2.6}
$$

where k is the thermal conductivity and c_p the heat conductivity of the carrier gas. These equations form a system of coupled partial differential equations for the determination of the four quantities, namely pressure P, temperature T and the two components of the velocity v_x and v_y as a function of the position (x,y) in the reactor. Under the present nonisothermal conditions (2.6) is coupled via the velocity to (2.2 to 2.4) and also via the temperature dependent quantities density ρ, viscosity μ and thermal conductivity k.

The main differences between different calculations are due to the choice of boundary conditions introduced into the calculation. This concerns first of all the walls which quite often have been assumed adiabatic or isothermal. In a realistic reactor, however, they are affected by heat conduction and radiative heat transfer from the susceptor. Another critical choice is constituted by an appropriate input and output velocity distribution.

The outcome of those calculations performed by finite element methods gives two dimensional solutions for the four parameters P, T, v_x and v_y mentioned above. Among these, T is certainly the most important parameter with respect to the thermally driven gas phase reactions in VPE. The velocity profile will deviate of course from the standard parabolic in v_x because of the non-isothermal situations and because natural convection may create (depending on T) a v_y component. As long as no turbulencies are created in the gas flow, which might lead to an irregular flow pattern, transport is in general not too critically affected by this, since the flow velocities anyway have to be zero at the wall or the substrate. The pressure finally is to a good approximation a constant within the reactor and for that reason also not a very crucial quantity. Thus, it appears that temperature and the velocity profiles are the most strongly influenced variables and therefore good test candidates for the verification of experimental data.

2.2.2 Experimental Determination of v and T

2.2.2.1 Measurement of Velocities.
For velocity determination, Doppler frequency shifts (Laser Doppler Anemometry (LDA)) from small particles (TiO_2, diameter 1 μm) moving with the gas stream can be utilised [2.12]. The experiment is sketched schematically in Fig. 2.5.

The moving particle receives the laser light wavelength λ_0 (frequency ν_0) with a frequency ν_p

$$\nu_p = \nu_0 \left(1 - \frac{\mathbf{v} \cdot \mathbf{e}_s}{c}\right) \tag{2.7}$$

where c is the velocity of light. The light scattered from the moving particles (Mie scattering) ν_p is detected with frequency ν_{DE}

$$\nu_{DE} = \nu_p \left(1 - \frac{\mathbf{v} \cdot \mathbf{e}_{DE}}{c}\right)^{-1} \tag{2.8}$$

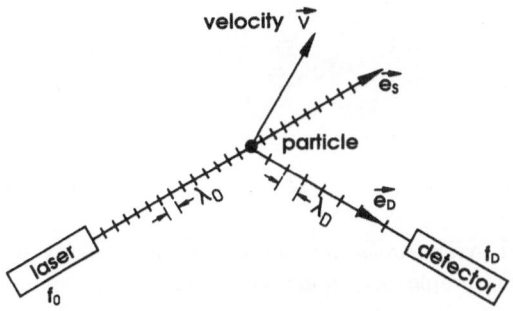

Fig. 2.5. Principle of velocity determination by measuring the Doppler shifted light scattered from the moving particle

and therefore:

$$\nu_{\mathrm{DE}} = \nu_0 \left(1 - \frac{\mathbf{v} \cdot \mathbf{e}_s}{c}\right) \left(1 - \frac{\mathbf{v} \cdot \mathbf{e}_{\mathrm{DE}}}{c}\right)^{-1} \tag{2.9}$$

The Doppler frequency shift:

$$\nu_{\mathrm{D}} = \nu_{\mathrm{DE}} - \nu_0 \tag{2.10}$$

is obtained as the difference frequency from mixing with light unshifted in frequency. By introducing the wavelength $\lambda = c/f$ and for small shifts with:

$$(\mathbf{v} \cdot \mathbf{e}_s)(\mathbf{v} \cdot \mathbf{e}_{\mathrm{DE}})/c^2 \ll 1 \tag{2.11}$$

one obtains:

$$\nu_{\mathrm{D}} = \frac{n}{\lambda_0}\mathbf{v}(\mathbf{e}_{\mathrm{DE}} - \mathbf{e}_s) \tag{2.12}$$

where n is the refractive index and λ_0 the vacuum wavelength. Since the evaluation of (2.12) requires an additional measurement of the angle between the directions of the laser beam and the scattered light, a symmetrical setup as shown in Fig. 2.6 is generally preferred today. There, two beams generated

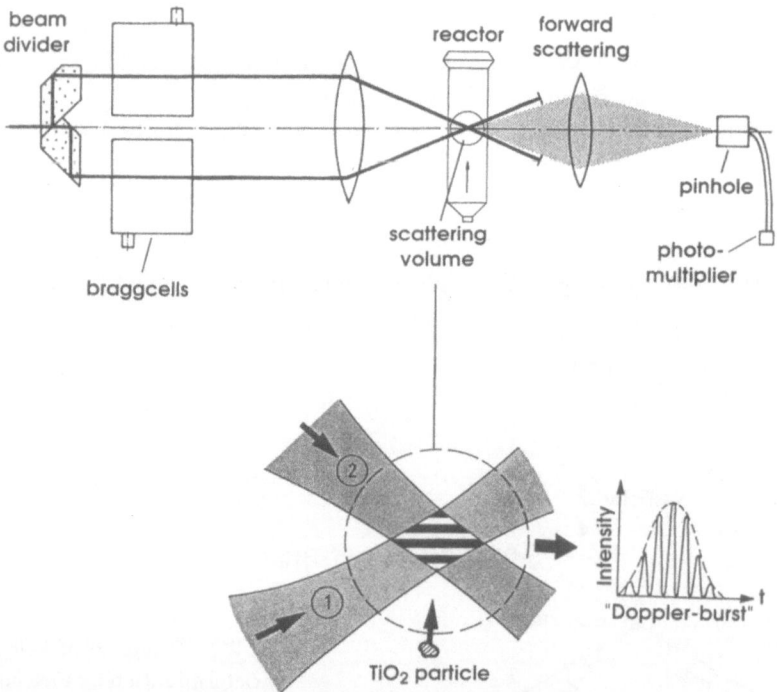

Fig. 2.6. Symmetrical experimental arrangement for measuring the velocity by Doppler shifted scattered light: Laser Doppler Anemometry (LDA)

by a prism arrangement from a single laser (He-Ne) intersect at an angle Θ, which is easily predetermined, and the Doppler frequency is then given by:

$$\nu_D = \frac{2\,v\,n}{\lambda_0}\,\sin\frac{\theta}{2} \qquad (2.13)$$

where v is the velocity component perpendicular to the interference fringes.

This symmetrical arrangement also accounts for the following simple interpretation: the particle with velocity v passing through the interference fringes will strongly emit light at each interference maximum. Neighbouring maxima are apart by the following distance:

$$d = \frac{\lambda_0}{2n\sin\theta/2} \qquad (2.14)$$

and the time of flight needed between them is

$$t = \frac{1}{\nu_D} = \frac{d}{v} \qquad (2.15)$$

Elimination of d from (2.14) and (2.15) yields again (2.13).

The experimental arrangement includes a Bragg cell to shift the Doppler frequency by a known amount for the detection of small velocities with higher precision. Particles are introduced into the gas stream by flowing the carrier gas over a powder of TiO_2 particles. For the measurement of all velocity components either the reactor or the optical arrangement has to be rotated in order to have interference fringes perpendicular to the component to be determined. In the present problem, where two components are to be measured, this can be achieved simply by rotating the optical arrangement by 90° around its optical axis without moving the reactor. The "Doppler burst" generated by the particle when moving through the interference fringes has a typical frequency in the order of a few Hz. After detection by a photomultiplier it can be directly converted to a velocity component.

A different technique of measurement is given by the "Laser 2 Focus" (L2F) arrangement [2.13]. The principle is outlined in Fig. 2.7. A particle passing through both light beams produces two successive pulses of light. The time elapsed between the two pulses yields the component of the particles velocity perpendicular to the optical axis. An optical layout is shown in Fig. 2.8.

Since independent particles may cause uncorrelated pulses, the electronics have to accumulate a number of measurements in the form of a time distribution function. A comparison of both methods is not easy, since both are equally simple (or complex) to apply. In most cases of growth studies, however, LDA has been utilised.

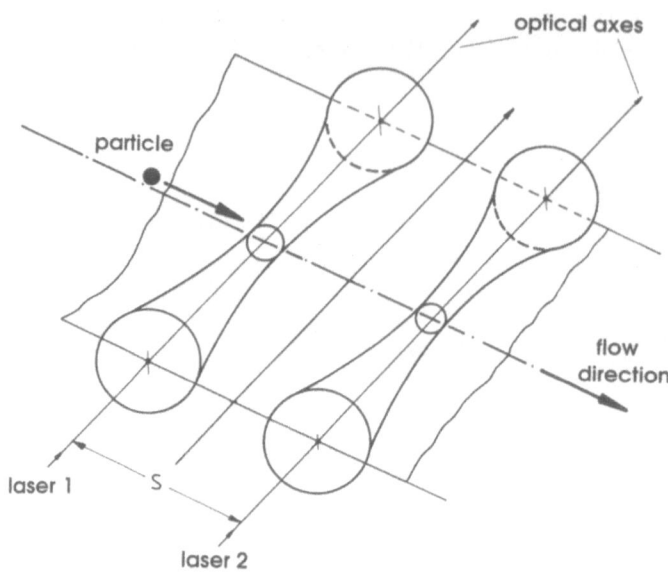

Fig. 2.7. Principle of velocity determination with the Laser 2 Focus (L2F) method. The particle moving with the gas creates a light burst at each focus

In Figs. 2.9 and 2.10 examples of velocity measurements in cold wall reactors taken with the LDA method are shown. The influence of gravity, i.e. horizontal or vertical operation of a reactor, on the velocity profile can be seen in Fig. 2.9. In the horizontal operation the natural convection pushes the hot gas away from the substrate, resulting in less efficient and less homogeneous deposition. The homogeneity of profiles, however, may be simply increased

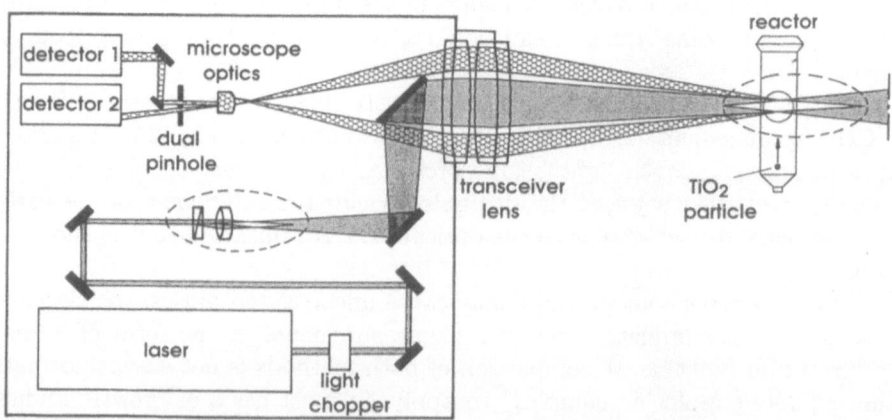

Fig. 2.8. Optical layout for a L2F system creating two laser foci in a reactor [2.13]

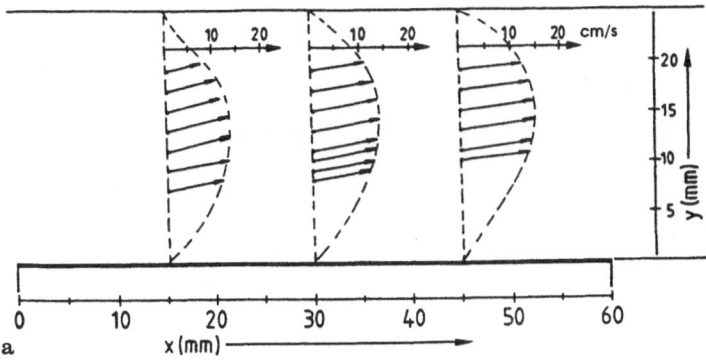

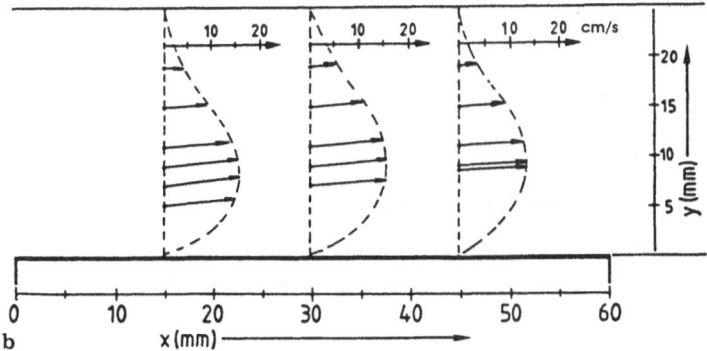

Fig. 2.9. Velocity profiles above the susceptor measured in a reactor midplane (coordinates see Fig. 2.2) by LDA for the same flow rates but for horizontal operation (**a**) and vertical operation (**b**) of the reactor [2.14]

by increasing the flow rate through the reactor [2.15] or by reducing the pressure considerably below atmospheric (Low Pressure VPE) [2.11]. In case the natural convection becomes more dominant than to the forced convection vortex flow (roll cells) may be induced in addition to the laminar flow. Such an example is shown in Fig. 2.10.

2.2.2.2 Measurement of Temperature. Remote temperature measurements of matter can be performed via elementary excitations whose strength of interaction with the optical radiation depends strongly on the thermal distribution functions. Phonons in condensed matter and rotational or vibrational excitations of molecules in gases have typical energies in the range from 10 to 200 meV. This results in strongly varying values for their occupational probability function in the temperature range of relevance for epitaxial growth. Signals depending on the occupational numbers can be most easily

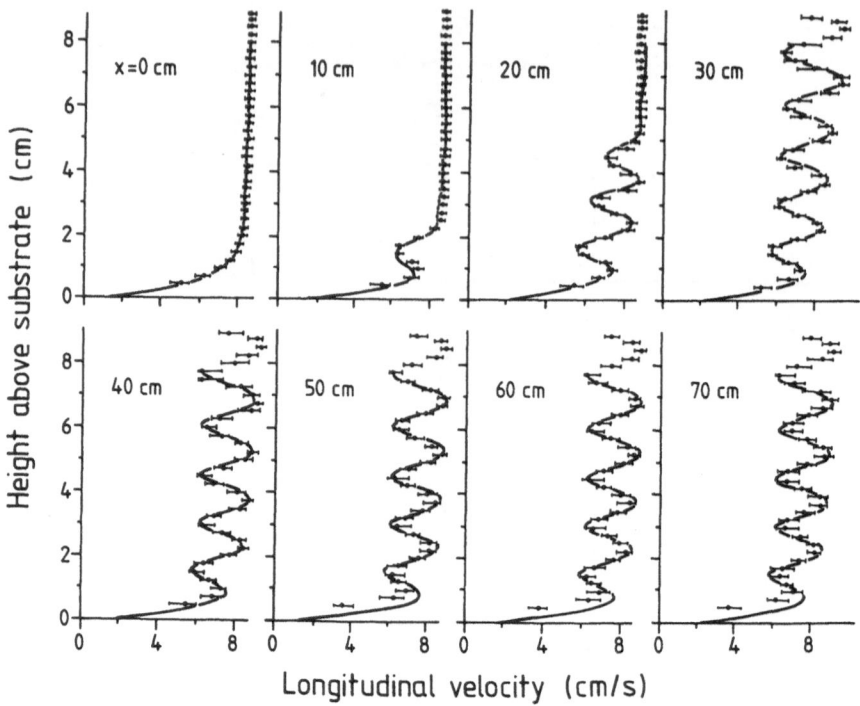

Fig. 2.10. Velocity profiles for a situation where the laminar flow is modulated by the development of roll cells. x denotes the position on the substrate in flow direction (Fig. 2.2) [2.16], data from Chiu and Rosenberger [2.17]

generated by Raman type light interaction processes (Fig. 2.11). This includes spontaneous Raman scattering, which will be discussed here, as well as non-linear Raman type interactions (for example Coherent Anti-Stokes Raman Scattering CARS), which will be discussed in the next Section. Raman type interactions have the advantage that their signal intensities in most cases can be evaluated directly with respect to temperature, without many significant corrections originating from the influence of electronic states.

In solids the intensity ratio of the Stokes and anti-Stokes lines of vibrational modes is usually measured. Because of resonance phenomena, however, the evaluation is rather difficult. Nevertheless, because of its contactless features, it has been intensively exploited for temperature determination in silicon at high temperatures [2.18]. This kind of analysis in principle can be useful for the determination of substrate temperatures.

If the electronic transition matrix elements involved are the same for Stokes and anti-Stokes interaction then the temperature can be determined by

$$\frac{I_S}{I_{AS}} = \frac{A(\omega_S)}{A(\omega_{AS})} \cdot \left(\frac{\omega_i - \Omega_0}{\omega_i + \Omega_0}\right)^3 \cdot \exp\left(\frac{\hbar\Omega_0}{kT}\right) \qquad (2.16)$$

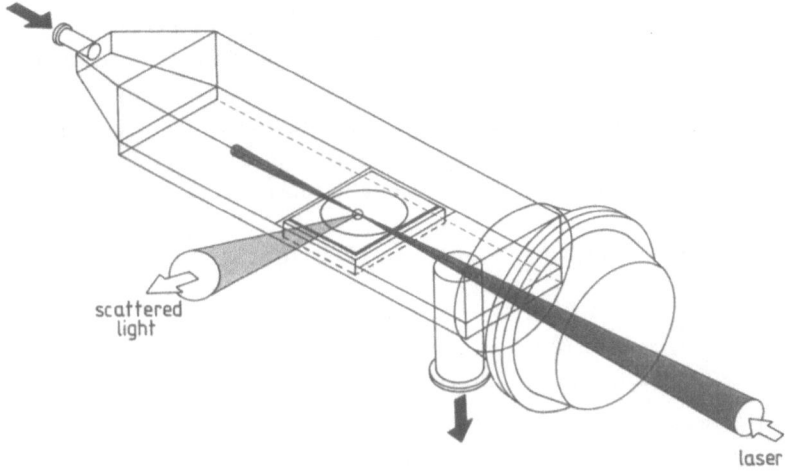

Fig. 2.11. Possible optical geometry to perform a Raman scattering experiment in a reactor

where ω_S, ω_{AS} are Stokes and anti-Stokes frequency and $A(\omega)$ is a correction factor which takes into account the different sensitivity of the analysing spectral apparatus at the Stokes and anti-Stokes frequency. Since both frequencies are quite often several hundreds of wavenumbers apart, this correction may lead to a sizeable effect and has to be performed very carefully.

In molecular gases, however, the large number of rotational or vibrational-rotational transitions may be utilized for an even more accurate and convenient temperature determination. In a typical VPE environment with a carrier gas at higher partial pressures, the measurement of the rotational transitions of the carrier gas (H_2) is the most appropriate choice. Since the concentration is high, large scattering intensities can be expected. In addition the molecular structure is simple and quantitative expressions for the temperature dependence of the intensities can be given [2.19]:

$$
\begin{aligned}
I(J) \;=\; & A(\omega)\frac{N}{Q}g_j(2J+1)\exp\left(-\frac{\hbar\Omega(J)}{kT}\right) \\
\times\; & \gamma_0^2 f(J)\frac{3(J+1)(J+2)}{2(2J+1)(2J+3)}\frac{45(2\pi)^4}{7}\omega^3 I_0,
\end{aligned}
\tag{2.17}
$$

where the factor ω^3 assumes photon counting and $A(\omega)$ is the spectral sensitivity of the spectrometer, which at a known temperature can be easily calibrated with the help of (2.17). N is the number of molecules in the scattering volume, Q is the partition function, $g(J)$ is the nuclear spin degeneracy, J is the rotational quantum number, $\Omega(J)$ are the frequencies of the rotational transitions, γ_0 is the anisotropic matrix element of the Raman tensor, $f(J)$ is a correction term accounting for the anharmonicity and I_0 is the incident laser intensity. Raman scattering measurements can be performed for example in a configuration like the one displayed in Fig. 2.11 using a conventional Raman setup (Figs. 4.4, 4.5). The first measurements have been therefore performed shortly after laser excited Raman spectroscopy became a common experimental technique [2.20–2.22]. A comparison of measured and calculated rotational scattering intensities is shown in Fig. 2.12. Exactly the same temperature dependence of the rotational intensities can be seen in both plots.

For a fast online analysis of the data, (2.17) is rewritten in the form

$$I(J) = m(J) \exp\left[-n(J)\frac{1}{T}\right]. \tag{2.18}$$

This equation clearly reveals the simple dependence of intensity on temperature corresponding to the Boltzmann distribution function which leads to a linear dependence of $\ln I(J)$ versus $n(J)$. This can be used very efficiently to generate a temperature value from intensity data immediately after measurement by linear regression. Typical errors are in the order of a few percent.

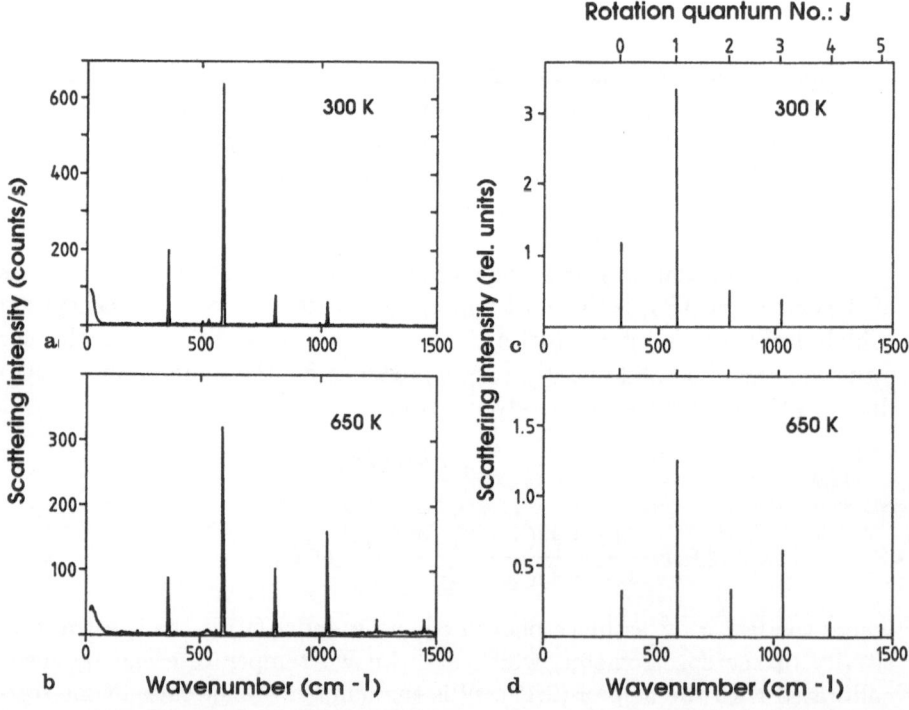

Fig. 2.12. Raman scattering intensities from rotational transitions of H_2 for two temperatures measured (**a,b**) in a configuration as shown in Fig. 2.11 and calculated (**c,d**) by using Eq. 2.17

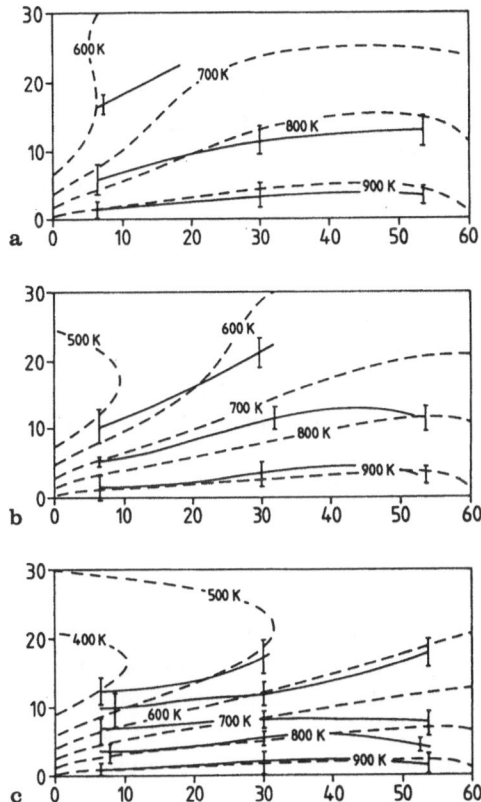

Fig. 2.13. Temperature profiles in a reactor midplane for different flow rates. **a** 2 sl/min, **b** 4 sl/min, **c** 8 sl/min. Solid lines: measured by Raman scattering, dashed lines calculated from (2.2–2.6) with appropriate boundary conditions [2.11]

The result of many of such temperature measurements at different locations within the reactor allows 2- or 3-dimensional temperature profiles to be generated. An example is given in Fig. 2.13 in a reactor midplane for different flow rates of the carrier gas.

Such measurements may also be used to test hydrodynamical calculations. The theoretical results included in Fig. 2.13 were generated by a finite element calculation including detailed boundary conditions for thermal conductivity and also heat transfer by radiation [2.11].

2.3 Gas Phase Diagnostics: Reaction Kinetics

The goal of gas phase diagnostics with respect to reaction kinetics is the identification of the species present and their number densities. In order to get specific "fingerprint"-like information for each molecular species, interaction of the electromagnetic radiation with either electronic, vibrational or rotational excitations may be utilized. Electronic transitions are conveniently observed in the UV/VIS spectral range by absorption spectroscopy or laser induced fluorescence, while vibrational or rotational excitations may be either observed by Raman type interaction conveniently in the VIS, or by direct interaction in the infrared. Methods which can be performed in the VIS or near UV spectral range are usually preferred since the reactor walls, made of quartz, are transparent in this spectral region. Infrared and deep UV measurements require, in contrast, special window materials and quite often also the removal of the ambient air, which might absorb light, too. Thus experiments become more complicated and much less data have been published from those spectral regions. In the following we will describe the optical techniques with respect to their abilities in detecting and determining species in the gas phase quantitatively.

2.3.1 Optical Techniques

2.3.1.1 Absorption Spectroscopy. Different setups have been used for absorption spectroscopy in the gas phase depending on whether electronic transitions with high absorption constants in the UV or vibrational transitions with usually lower absorption constants in the IR are analysed. For large absorption constants single pass configurations (Fig. 2.14a), for small absorption constants multipass configurations (Fig. 2.14b) are in use.

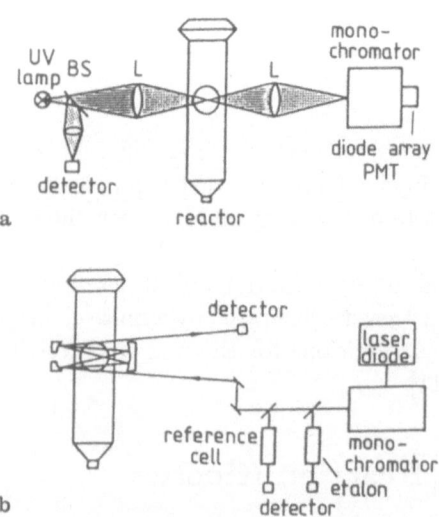

Fig. 2.14. Schematic arrangements for absorption measurements in a reactor with single-pass configuration (**a**) and multi-pass configuration (**b**)

As light sources in the UV, xenon high pressure lamps or deuterium lamps are utilized. UV sensitive intensified diode array detectors or photomultiplier tubes with lock-in amplifiers can be used for detection. The latter is preferred if long term stability and sensitivity are required. In vibrational IR spectroscopy, tunable diode laser and multipass configuration allow for very high resolution spectra and, therefore, identification of species is simplified. FTIR spectrometers may also be employed, however, in comparison to diode lasers their resolution is smaller and the FTIR instrumentation is more difficult to adapt to a growth reactor.

Beer's law can be used to determine number densities:

$$\frac{I}{I_0} = \exp(-SNL) \tag{2.19}$$

where S is the absorption cross section, N the number density and L the length of the light path through the reactor. I, I_0 are the light fluxes with and without absorber. If the absorption cross section is known, it is of course possible to determine the number density. Even then only average values are obtained in the non-homogeneous situation of a cold wall reactor with a heated substrate, because the spatial resolution of an absorption experiment is poor and absorption originates from the whole reactor region traversed by the light beam. In addition, if the reactor windows (walls) acquire some deposits during the measurement, accurate values for I_0 are difficult to obtain. Thus quantitative evaluation of absorption experiments is complicated. However, for monitoring and for the detection of atoms and radicals, absorption spectroscopy is well suited. In the following, a few experimental examples will be discussed. Absorption spectra taken at room temperature in the UV have been published for many metalorganic compounds [2.23–2.25]. They exhibit strong but very broad bands which are not well suited for identification but might well be employed for monitoring purposes.

Figure 2.15 gives an example for these aspects. The ON and OFF switching procedures, as well as the constancy of the precursor partial pressure, can be observed. In addition, from the level of constant absorption, a measure for the precursor concentration can be derived which may well serve for the purpose of controlling the growth process.

In the regions of higher temperatures, new structures originating from precursor decomposition become observable. Examples in Figs. 2.16, 2.17 show spectral features caused by products originating from the decomposition of TMGa (trimethylgallium) and arsine (AsH_3). High resolution detection of methyl radicals (CH_3) by laser diode spectroscopy has been reported for the first time in [2.27]. This work is an excellent example for the potential of laser spectroscopy in gasphase diagnostics.

2.3.1.2 Laser Induced Fluorescence.
In Laser Induced Fluorescence (LIF) a tunable laser is used to selectively excite electronic transitions of an atom or

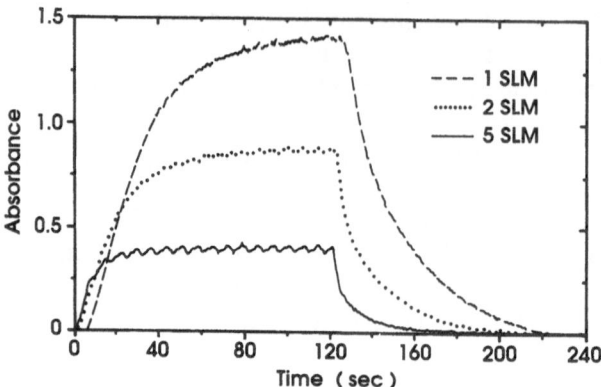

Fig. 2.15. Absorbance of trimethylantimony (TMSb) diluted with hydrogen introduced into the reactor at $t = 0$ s and switched off at $t = 120$ s at different flow rates. SLM: standard liter per minute [2.23]

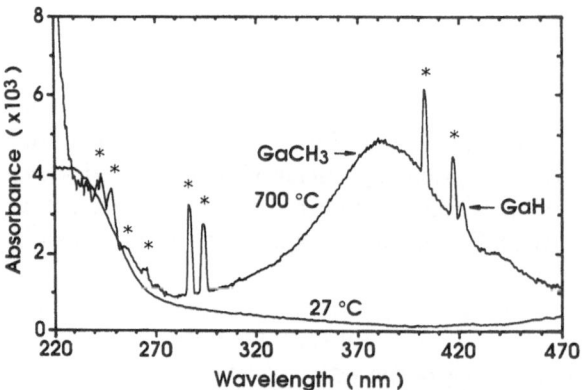

Fig. 2.16. Absorbance of trimethylgallium (TMGa) at room temperature and 970 K. At the latter temperature structure due to gallium radicals (GaH, GaH₃) and Ga atoms (*) is observed [2.26]

molecule to be studied and the resulting fluorescence is analysed spectrally. Corresponding electronic transitions for atoms and molecules are indicated in Fig. 2.18.

The experimental setup is equivalent to that for Raman scattering shown schematically in Fig. 2.11. In contrast to Raman scattering, however, a tunable laser is needed for selective excitation. This laser may operate in cw as well as in pulsed mode. In the latter case it provides also temporal resolution. The LIF technique consequently combines the high spatial resolution of Raman scattering with the temporal and spectral resolution of a laser technique. Moreover, whenever fluorescent transitions exist a high sensitivity can

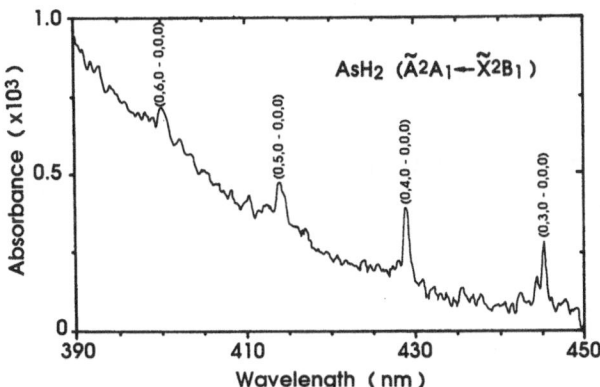

Fig. 2.17. Absorption spectrum of arsine (AsH₃) in the presence of TMGa at an elevated temperature. A structure due to the subhydride AsH₂ appears in the spectra [2.26]

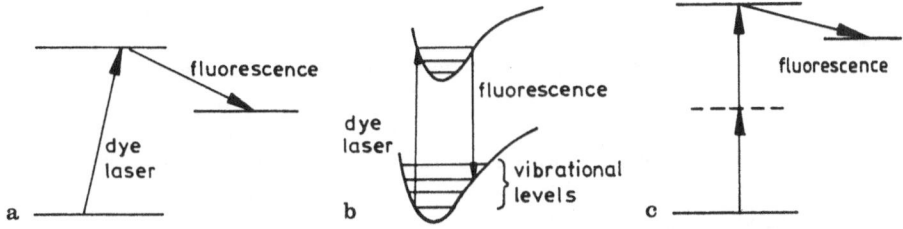

Fig. 2.18. Electronic transitions responsible for laser induced fluorescence (LIF) for the case of a 1-photon excitation process of an atom (**a**) or a simple molecule (**b**) and for a 2-photon excitation process (**c**)

be obtained. Simple biatomic molecules and atoms quite often produce characteristic, fingerprint like structures in the fluorescence spectra with a high quantum yield close to one. The sensitivity of LIF in such a case is in the range of 10^8 molecules/cm³. For larger polyatomic molecules the quantum yield is usually low because these molecules can either return to the ground state by nonradiative processes or possibly photodissociate from the excited state. In such cases the detection sensitivity is substantially lower and other techniques than LIF may be preferred e.g. spontaneous or coherent Raman scattering.

Absolute determination of number densities in LIF is, however, difficult since usually no simple calibration procedure exists. LIF is especially useful for nonstable species. It has been mostly applied to gas phase diagnostics in silicon chemical vapour phase epitaxy [2.28]. Examples for LIF spectra from Si-atoms, Si₂-molecules, and dichloro-silane are presented in Fig. 2.19.

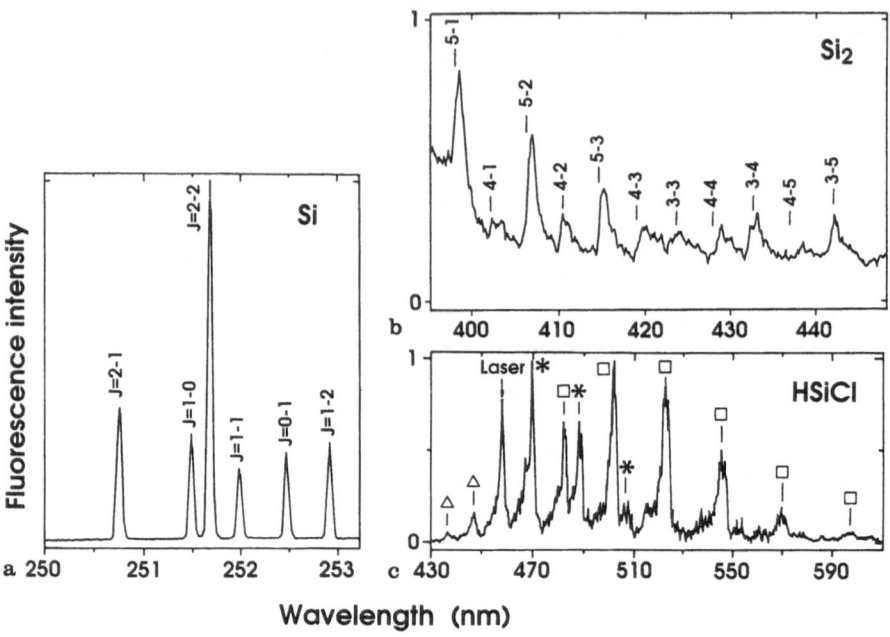

Fig. 2.19. LIF spectrum obtained from: **a** Si atoms during deposition in a reactor. Excitation was at 250.7 nm. The labels denote the total angular momentum of the states involved in the transition [2.29, 2.30]; **b** from Si₂ molecules excited at 390.83 nm. Labels mark the calculated vibrational origin of the transitions involved [2.31]; **c** of HSiCl (excitation: 457.5 nm) observed in Si deposition from dichlorosilane (SiCl₂H₂). Equal labels (□, ∗, △) mark the rotational subband structure of three vibrational bands [2.32]

Clearly, the decrease in spectral sharpness and signal to noise ratio from the simple atom to the more complicated molecule can be noticed.

The results of similar experiments for the hydrides of group V elements have been published, too [2.24]. On the other hand, surprisingly few investigations have been reported for metalorganic precursors. But results have been quoted for example for dimethylzinc [2.33] or trimethylgallium [2.34].

2.3.1.3 Spontaneous Raman Scattering. Spontaneous Raman Scattering (RS) has been applied already quite early to gas phase diagnostics in order to determine temperatures by using rotational transitions in the carrier gas molecules (Sect. 2.2.2.2). Experimentally this provides no problem signalwise since the concentration of carrier gas molecules is high. The situation is, however, different for the detection of reactive species with concentrations usually several orders of magnitude lower. Detection limits for species at partial pressures between 1 to 10 mbar have been reported [2.15, 2.35].

Figure 2.20 gives an example for arsine (AsH₃). A partial pressure in the mbar range in general equals, however, the input partial pressure of precursors

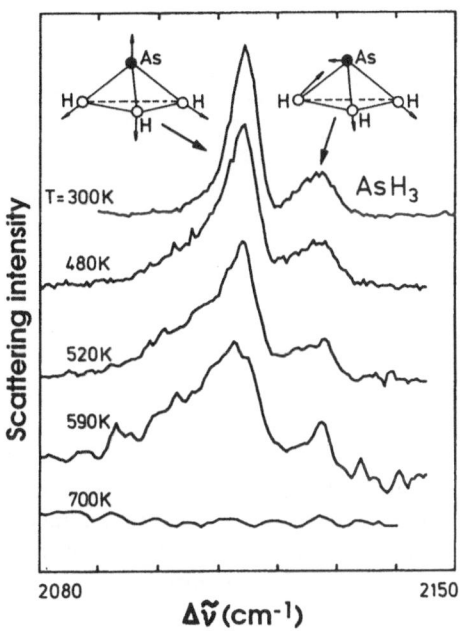

Fig. 2.20. Raman spectrum of the A_1 vibration of AsH_3 at different temperatures. At the highest temperature AsH_3 has been decomposed. Input partial pressure to the reactor was around 20 mbar [2.15]

to the growth reactor. Therefore, the range of useful applications of Raman scattering to gas phase diagnostics with respect to detection of reaction products in VPE is rather limited. Nevertheless, a few results concerning reaction mechanism have been reported [2.35–2.37].

2.3.1.4 Coherent Anti-Stokes Raman Scattering. CARS is a third order nonlinear optical technique. It is described as a four wave mixing process (Fig. 2.21) where the frequency difference of two of the input laser beams matches a vibrational-rotational transition frequency of the molecules under study.

This creates a coherent set of molecules for which the corresponding vibrational-rotational states are occupied. The CARS process may then be thought of as an anti-Stokes Raman Scattering process of the photons in the third input laser beam caused by the coherent molecular excitations.

Not only energy has to be conserved in the CARS process

$$2\omega_L - \omega_S - \omega_{AS} = 0 \tag{2.20}$$

but, since the excitations are coherent, the wavevectors of the light beams, too:

$$2\mathbf{k}_L - \mathbf{k}_S - \mathbf{k}_{AS} = 0 \tag{2.21}$$

The anti-Stokes beam therefore emerges in a well defined direction (Fig. 2.22), which makes its separation from the other laser beams and the corresponding intensity measurements relatively simple.

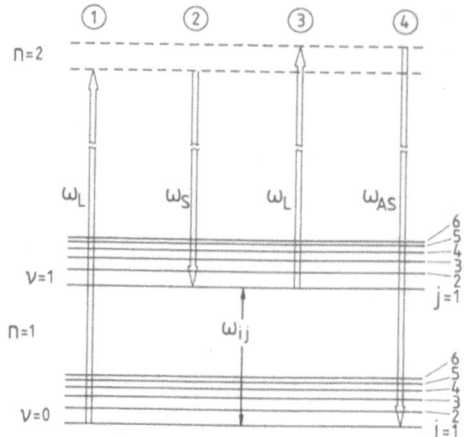

Fig. 2.21. Schematic energy diagram of transitions between molecular levels for a CARS process. Rotational (J), vibrational (v) and electronic (n) quantum numbers are indicated. Transitions 1 and 2 can be thought to be responsible for coherent excitation of molecules and steps 3 and 4 may be viewed as an anti-Stokes Raman process

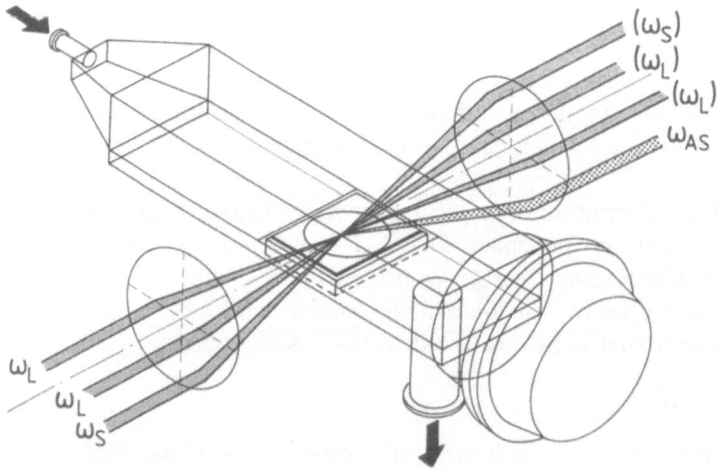

Fig. 2.22. Experimental arrangement for CARS in a VPE reactor. Wavevector conservation, (2.21), requires the anti-Stokes beam to appear under a different angle than the other beams in a so-called folded BOXCARS configuration (all 3 input beams not in one plane). This allows for easy rejection of input laser light

The total power in the CARS beam may be written as [2.38]

$$P_{\text{CARS}} \sim \left| N \sum_{i,j} \left[\frac{d\sigma}{d\Omega} \right]_{ij} \frac{\Delta\rho_{ij}}{\omega_{ij} - \omega_L + \omega_S - i\Gamma_{ij}} + \chi_{\text{nr}}^{(3)} \right|^2 P_L^2 P_S \qquad (2.22)$$

where the matrix elements in the resonant part of the third order nonlinear susceptibility $\chi^{(3)}$ have been expressed by the spontaneous Raman cross section $(d\sigma/d\Omega)_{ij}$ and $\chi_{\text{nr}}^{(3)}$ is the nonresonant part containing the contributions from all transitions which do not fulfill energy conservation as expressed by (2.20). $\Delta\rho_{ij}$ is the difference in occupation between states i and j, Γ_{ij} the transitional linewidth and N the number density. By increasing the laser powers (P_L, P_S) the CARS signal may be enhanced considerably, however, the background ($\chi_{\text{nr}}^{(3)}$) is enhanced, too. Background discrimination may be achieved by exploiting the different polarisation properties of the nonresonant and resonant part of the CARS signal [2.39]. A typical experimental setup for CARS measurements is shown in Fig. 2.23. There, a reference cell (argon,

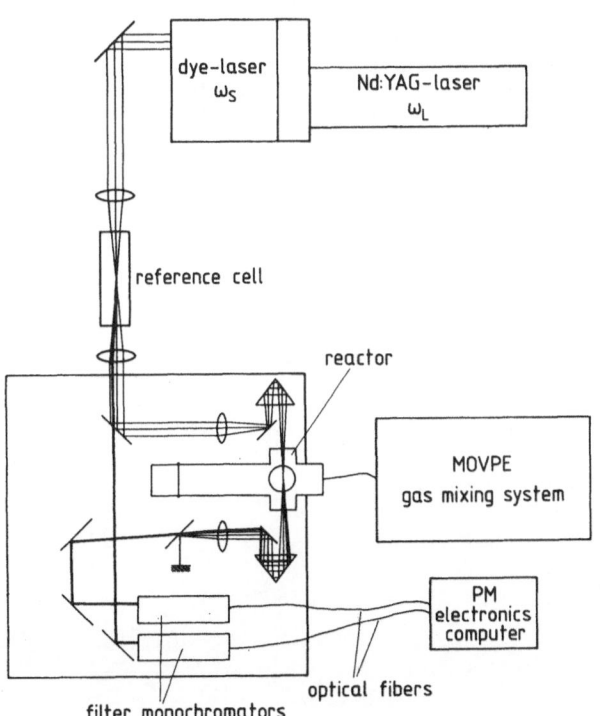

Fig. 2.23. Experimental setup for CARS measurements. Two detectors monitor the reference signal (argon cell) and the signal from the reactor. By moving the cube corners laterally (and synchronously) in front of the windows, different locations inside the reactor can be analysed without disturbing the alignment before and after the reactor

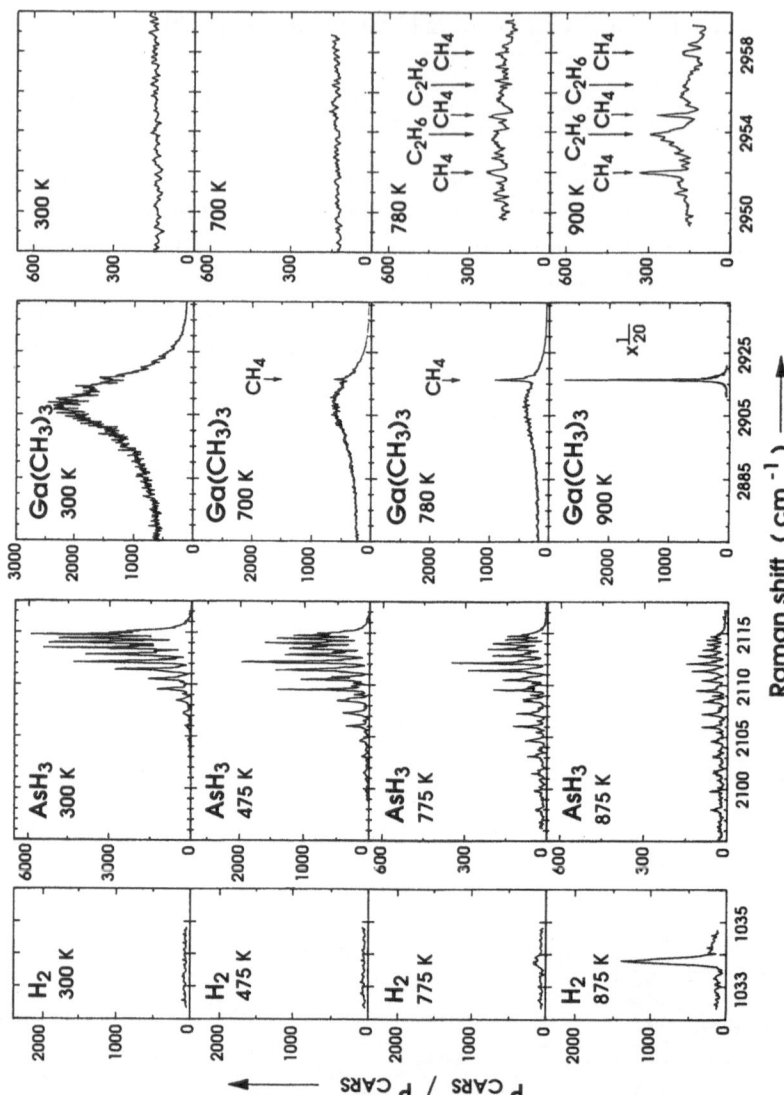

Fig. 2.24. CARS spectra taken at different temperatures in a MOVPE reactor for the growth of GaAs from arsine (AsH₃) and trimethylgallium (Ga(CH₃)₃). The spectra demonstrate with increasing temperature and from left to right: The production of molecular hydrogen, the decomposition of arsine and trimethylgallium and the production of hydrocarbons

1 bar) is introduced in order to normalize for the shot to shot variation in laser pulse power.

In contrast to the incoherent methods described before, the number density enters the CARS signal squared and, therefore, the square root of the normalized signal is proportional to the partial pressure or number density of molecules as shown by the example in Fig. 2.25.

The detectivity of the CARS method in general is limited by laser noise (especially in the case of multimode lasing) and by the background signal originating from all other molecules present. The main background signal contribution naturally is caused by the carrier gas molecules being present in large number. Detectivity levels as low as 10^{-3} mbar may be achieved for biatomic molecules with sharp spectral features, but typical values for larger metalorganic molecules used as reactants in a MOVPE environment are around 10^{-1} mbar. Spatial resolution for a crossed beam arrangement ("Folded BOXCARS") like that in Fig. 2.22 is of the order of a few mm along the beam direction and much better (μm) perpendicular to it.

By tuning the dye laser frequency ω_S, the denominator in (2.22) becomes resonant with the different frequencies ω_{ij}, and spectra such as those in Fig. 2.24 are obtained at different temperatures.

2.3.1.5 Other Methods. The previously described methods have been actively employed in gas phase diagnostics of VPE processes. Other methods, however, which have been mainly used in distinct circumstances (glow discharges, flames) might be useful, too. These include two- or three-photon processes (Fig. 2.18) used to excite certain species to electronic states from which fluorescence may then be observed. An interesting example of this

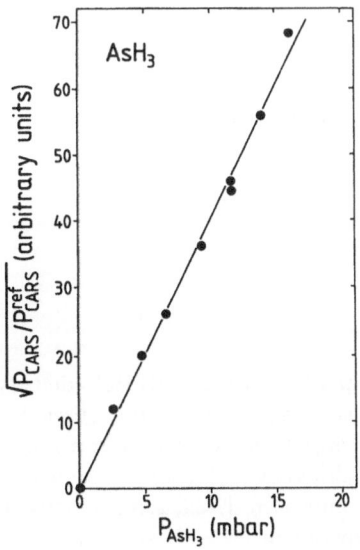

Fig. 2.25. Square root of the normalised CARS signal of AsH_3 versus partial pressure. The linear slope proves the N^2 dependence of the CARS signal predicted by (2.22)

multiphoton fluorescence with respect to MOVPE is the possibility of detecting atomic hydrogen [2.41]. Since atomic hydrogen is supposed to play a major role in the removal of carbon containing radicals in MOVPE, such a diagnostic technique would turn out to be quite useful. Another interesting technique is the opto-galvanic spectroscopy, which is based on the photoionisation of molecular species. The appearance of ions is then detected by a current between two electrodes in the photoexcited volume [2.42]. While this method obviously has a high sensitivity, the presence of electrodes in a VPE environment might disturb the growth process.

2.3.2 Experimental Results

In this Section experimental results concerning the thermal decomposition of precursors and the analysis of products resulting from such a decomposition will be discussed.

2.3.2.1 Thermal Decomposition of Precursors. The thermal decomposition of precursors has always been the primary interest in VPE studies since by such an analysis a lower limit for the growth temperature is determined. Measurements have been either performed as a function of height above the susceptor [2.15, 2.29, 2.30, 2.36] or as a function of temperature [2.28, 2.36, 2.43, 2.44].

Figure 2.26 shows an example obtained with Raman scattering from Si-VPE by using silane (SiH_4) [2.29, 2.30]. The silane concentration is plotted for different susceptor temperatures as a function of height above the susceptor.

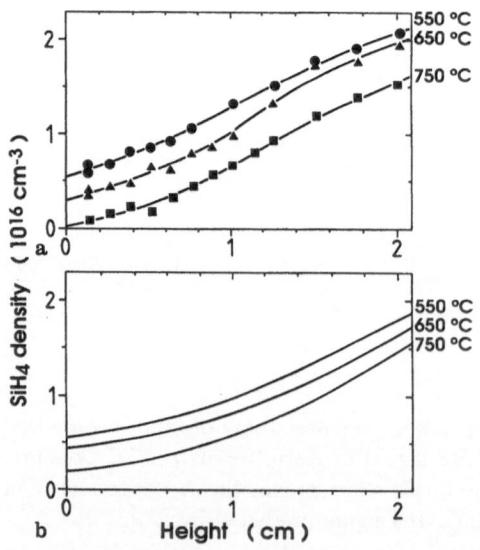

Fig. 2.26. Silane (SiH_4) density versus height above the susceptor during silicon deposition from silane for different susceptor temperatures as indicated [2.29, 2.30] **a** experiment; **b** theory

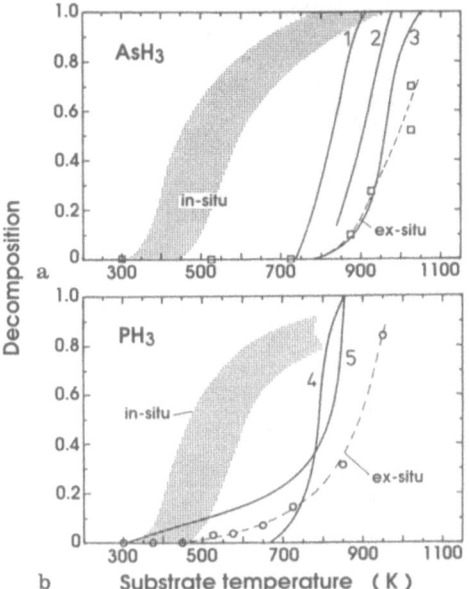

Fig. 2.27. Thermal decomposition of AsH_3 and PH_3 obtained *in-situ* (shaded areas which include error bars) and *ex-situ* by CARS. The other curves (1 to 5) are obtained through *ex-situ* measurements by sampling the reactor gas with a capillary from a region above the substrate to an outside analysis equipment [2.46]

In order to model such a dependence, the gas phase hydrodynamics, diffusion and thermal decomposition had to be involved [2.45]. The calculated results shown also in Fig. 2.26 compare quite well with the experiment. This agreement proves that such measurements can be used to test the quality and validity of growth models in a more direct manner than growth rate data. However, data which can be interpreted more directly are obtained if the precursor concentration is measured as a function of temperature. This can be done with Raman scattering or CARS by measuring simultaneously temperature and concentration. An example is shown in Fig. 2.27 for the decomposition of AsH_3 and PH_3 obtained by CARS.

From such data, activation energies for the decomposition reaction can be obtained. Finally, an example, where the influence of two different surfaces on the decomposition of AsH_3 can be seen, is displayed in Fig. 2.28. This example clearly shows that surface reactions play a critical role in the decomposition of AsH_3 (heterogeneous reaction). Gas phase (homogeneous) reaction alone obviously cannot describe the thermal decomposition in a MOVPE reactor. This must be also the reason why in Fig. 2.27 the data for *in-situ* and *ex-situ* measurements are significantly different. Similar observations have been made for PH_3 or tertiary butylphosphine ($H_2P(CH_3)_3$). In contrast, the decomposition of alkyls with no metal-hydrogen bonds like $Ga(CH_3)_3$ or corresponding ethyls seem to occur via homogeneous reactions because no surface influence was detected.

2.3.2.2 Decomposition Products. In understanding and modelling the growth process not only the decomposition of the precursors but also the

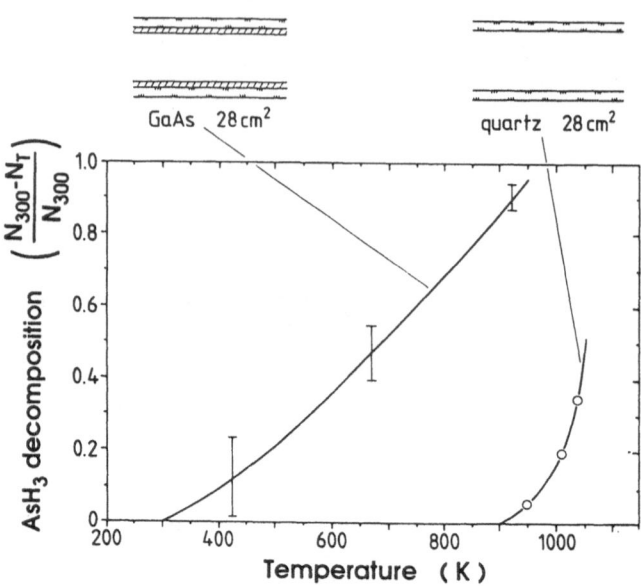

Fig. 2.28. Decomposition of AsH3 versus temperature for two different surfaces (quartz and GaAs) as indicated [2.40]

corresponding reactions and their products should be known. Gas phase diagnostics can at least help to pin point some of the products forming and to give indications which of the many reactions possible are the important ones. In silicon VPE the relative density of Si-atoms and Si_2-molecules above the susceptor was interpreted in favour of dominating gas phase reactions in VPE with silane. Si and Si_2 may be produced by the following reversible reactions:

$$SiH_4 \rightarrow SiH_2 + H_2 \quad \rightarrow \quad Si + 2H_2 \tag{2.23}$$
$$SiH_2 + SiH_4 \quad \rightarrow \quad Si_2H_4 + H_2 \tag{2.24}$$
$$Si_2H_4 \quad \rightarrow \quad Si_2H_2 + H_2 \tag{2.25}$$
$$Si_2H_2 \quad \rightarrow \quad Si_2 + H_2 \tag{2.26}$$

Their relative concentrations were measured by laser induced fluorescence [2.29, 2.30]. The results displayed in Figs. 2.29, 2.30 show maxima at some distance from the susceptor.

This already indicates that the production of these species must be in the gas phase and underlines the importance of gas phase chemistry in Si-VPE from silane.

In MOVPE the detailed decomposition steps of the hydrocarbons containing precursors need to be fully understood because a direct consequence of such reactions is the incorporation of carbon into the epitaxial layers. The pyrolysis of $Ga(CH_3)_3$, for example, should be governed by reactions like

$$Ga(CH_3)_3 \rightarrow Ga(CH_3)_2 + CH_3 \tag{2.27}$$
$$Ga(CH_3)_2 \rightarrow GaCH_3 + CH_3 \tag{2.28}$$
$$CH_3 + H_2 \rightarrow CH_4 + H \tag{2.29}$$
$$GaCH_3 \rightarrow Ga + CH_3 \tag{2.30}$$
$$Ga + H \rightarrow GaH \tag{2.31}$$
$$GaH + H \rightarrow Ga + H_2 \tag{2.32}$$
$$GaH + CH_3 \rightarrow Ga + CH_4 \tag{2.33}$$

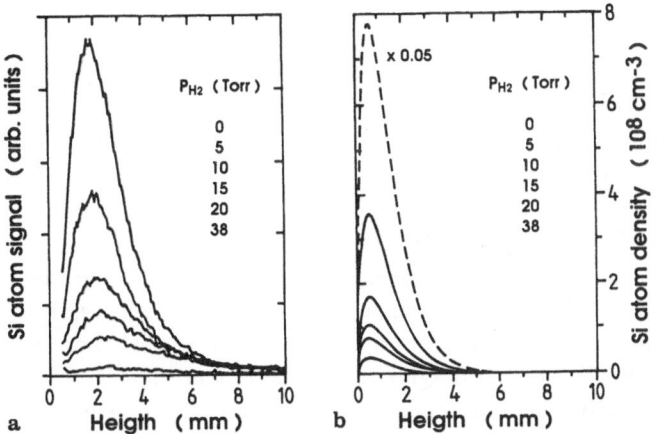

Fig. 2.29. Atomic silicon concentration versus height above the substrate in Si deposition from SiH_4 [2.29, 2.30]. **a** Experiment, **b** theory

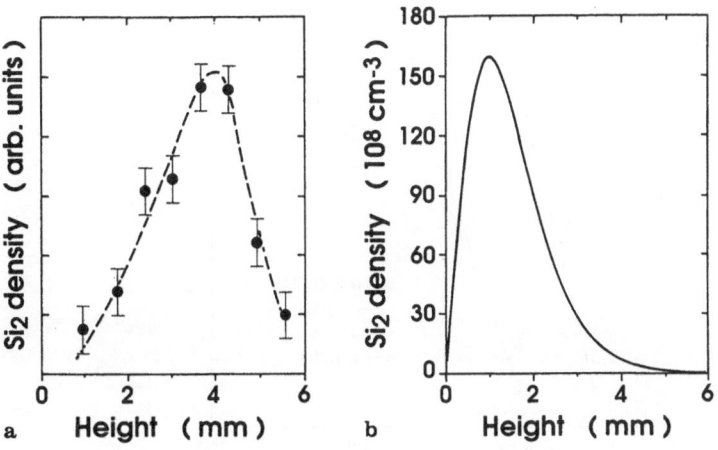

Fig. 2.30. Same as in Fig. 2.29 for molecular Si_2. **a** Experiment, **b** theory

There are many indications for the validity and importance of such a reaction. CH_3 was detected by IR absorption spectroscopy [2.27], CH_4 by CARS [2.44]. CARS also indicated a higher stability of the monomethylgallium, which probably places reaction (2.30) under normal MOVPE conditions at the surface. $GaCH_3$, Ga and GaH were recently directly observed by absorption spectroscopy [2.26]. Figure 2.31 shows the results derived from spectra such as those shown in Fig. 2.17.

Carbon incorporation into the epitaxial layers has been suggested [2.47] to be caused by carbenes generated in reactions like

$$GaCH_3 + CH_3 \rightarrow GaCH_2 + CH_4 \tag{2.34}$$

should there be a deficiency of hydrogen atoms. However, no observation of $GaCH_2$ proving such reactions has been made up to date.

Absorption spectroscopy was also able to detect the arsine subhydride AsH_2 whose existence had been inferred previously from the amount of hydrogen produced in the arsine decomposition [2.44] and from the formation of adducts like CH_3AsH_2 and $(CH_3)_2AsH$.

The examples presented in this Section show that a number of data concerning reaction mechanism can be obtained by *in-situ* gas phase analysis. The necessary experimental techniques are available. They consist, however, in most cases of not easy to bolt-on diagnostics tools, but are mostly highly sophisticated and expensive optical machinery, which prevents their usage as standard monitoring devices.

One of the main purposes of the two Sects. 2.2 and 2.3 on gas phase diagnostics was to show that gas phase processes constitute a complex but important step in gasphase epitaxial growth. This step has to be understood

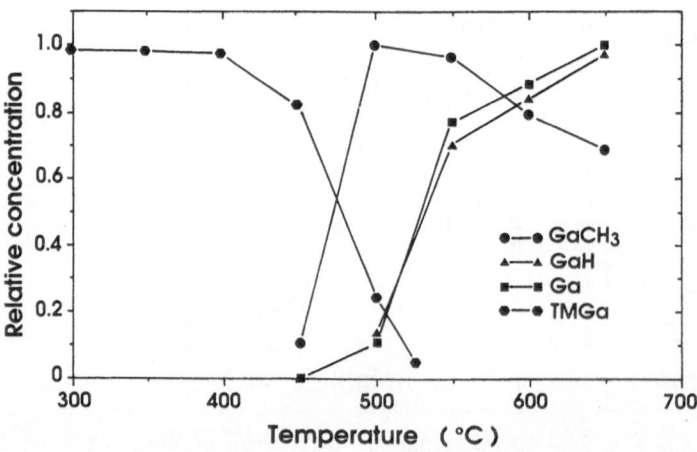

Fig. 2.31. Concentration of several species produced during TMGa decomposition versus temperature [2.26]

in order to obtain the necessary input quantities for a description of the surface kinetics during growth. Gasphase diagnostics by means of optical techniques has contributed strongly towards this goal. The possibilities for surface analysis during gas phase epitaxial growth will be discussed in the following Section.

2.4 Surface Diagnostics

Surface processes of course constitute the most relevant step in epitaxial growth. The information available on surfaces during gasphase epitaxial growth, however, is rather modest. This is certainly due to the fact that standard surface science techniques such as RHEED cannot be applied in the environment of gasphase epitaxy.

The obvious solution seems to be the utilisation of optical techniques in the visible spectral range since these photons with their energies between the electronic ($> 5\,eV$) and the vibronic ($< 0.5\,eV$) molecular absorption bands can easily penetrate the gas phase and probe the growing samples. However, they also penetrate considerably into the substrate itself and thus are in general not surface probes as compared to low energy electrons. This is clearly demonstrated by a plot of the penetration depth d_p (inverse of the optical absorption coefficient) of light with photon energies from 2 to $6\,eV$ (Fig. 2.32). For the three representative materials shown the optical penetration depth d_p at its lowest value still exceeds $5\,nm$ and thus is much

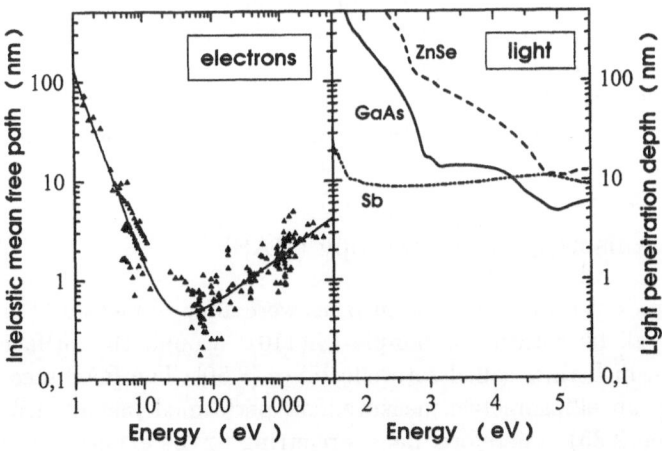

Fig. 2.32. Penetration depth of light d_p in two semiconductors (GaAs and ZnSe) and a metal (Sb) as a function of photon energy compared to the inelastic mean free path of electrons which is nearly material independent (electron data from [2.48], GaAs data derived from [2.49])

larger than the mean free path of low energy electrons. In any ordinary optical measurement, for instance of the reflectance, the contribution of the bulk is therefore dominant. The surface contribution, i.e. that from the upper fraction of a nanometer, on the other hand, is in general negligible except for the near ultraviolet range where it may reach percentage levels.

A couple of recently developed experimental methods have overcome this dilemma. Most of them have in common that the unwanted bulk contribution is suppressed either by analysing surface specific properties or by using p-polarized light under a large angle of incidence (preferably Brewster angle) which in general enhances the sensitivity to surface conditions. Reflectance Anisotropy Spectroscopy (RAS or RDS), Laser Light Scattering (LLS), Infrared Reflection Absorption Spectroscopy (IRRAS) and Second Harmonic Generation (SHG) are representatives of the first group while Spectroscopic Ellipsometry (SE) utilizes the latter aspects. Surface Photo Absorption (SPA), finally, derives its sensitivity from both a specific surface sensitivity together with a Brewster angle arrangement. With the exception of ellipsometry, which is treated in a separate Chapter of this book, all these techniques will be discussed here. The applicability of these methods, mostly discussed here for gasphase environments, is of course in general not limited to such situations but can be applied under UHV conditions (MBE, MOMBE) as well, i.e. what works in a MOVPE setup is in principle also attachable to an UHV chamber for growth. This might be of advantage in certain cases, since the optical methods provide complementary information with respect to electron analysis techniques and in some cases are quite simple to install. From the latter point of view LLS (Sect. 2.4.5) is by far the most simplest method and is already in use nowadays in many growth equipments. SPA (Sect. 2.4.2), IRRAS (Sect. 2.4.3) and SHG (Sect. 2.4.4) represent more elaborate experimental setups and require also two windows with oblique but direct view to the growing surface. With respect to simplicity, RAS is probably one of the most easily attachable optical surface tools since it needs only one window for normal incidence and, moreover, can be desigened in a relatively simple and compact manner.

2.4.1 Reflectance Anisotropy Spectroscopy (RAS)

Reflectance anisotropies at semiconductor surfaces were already measured in 1966 by Cardona et al. by rotating a sample (Si(110)) around the surface normal, a technique which they called rotoreflectance [2.50]. The RAS measurement is basically an ellipsometric measurement performed just at near normal incidence (Fig. 2.33). Therefore, names referring to the experimental arrangement like Perpendicular Incidence Ellipsometry (PIE) or Normal Incidence Ellipsometry (NIE) have been also used for this technique [2.51,2.52].

The experimental setup described here (Fig. 2.33), was basically developed by Aspnes and coworkers [2.54], who were the first ones also to recog-

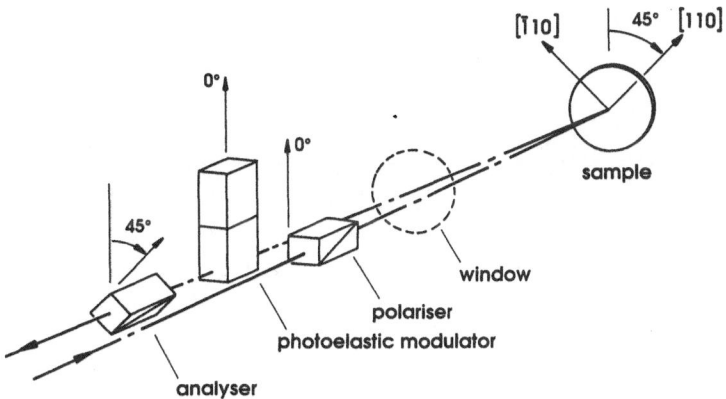

Fig. 2.33. Principal diagram of a RAS setup utilising a photo elastic modulator [2.53]

nize the potential of the anisotropic reflectance with respect to surface science. They named the method Reflectance Difference Spectroscopy (RDS). This name refering to the experimental technique, however, does not reflect the physical origin of the signal. It is sometimes also confused with similar names and acronyms given for a technique which modulates the reflectance by forming an adsorbate, quite often an oxide, on the surface and measuring the difference between the clean and covered surface. This technique has been named Surface Differential Reflectivity (SDR) [2.55] or Differential Reflectance spectroscopy (DRS) [2.56] and is sometimes also abbreviated as RDS [2.57].

The principles of an RAS experiment are most easily understood by considering the best studied example namely the GaAs(001) surface. Figure 2.34 shows the two for epitaxial growth most important surface reconstructions namely the GaAs(001)-(2×4) and the GaAs(001)-c(4×4) [2.58]. Both surfaces are arsenic rich, the former containing one single As layer at the surface the latter a double layer. Both surfaces appear clearly to be anisotropic. They should, moreover, respond also differently to light since the arsenic dimers in the top layers of both reconstructions are oriented perpendicular to each other. Therefore, the optical response from both surfaces, as far as the contribution from the arsenic dimers is concerned, is expected to be anisotropic and also different from each other. The reflectance contribution from such a surface with light polarized along the [110] and the [$\bar{1}$10] directions therefore should be different. Since the bulk is optically isotropic in lowest order for cubic semiconductor materials the complex index of reflection r for the bulk (unreconstructed surface) is in general the same for the [110] and [$\bar{1}$10] directions, i.e. $r_{110} = r_{\bar{1}10}$. As a result, when measuring only the difference between the two directions the bulk contributions will cancel by subtraction and only contributions from the surface with its lower symmetry survive. The

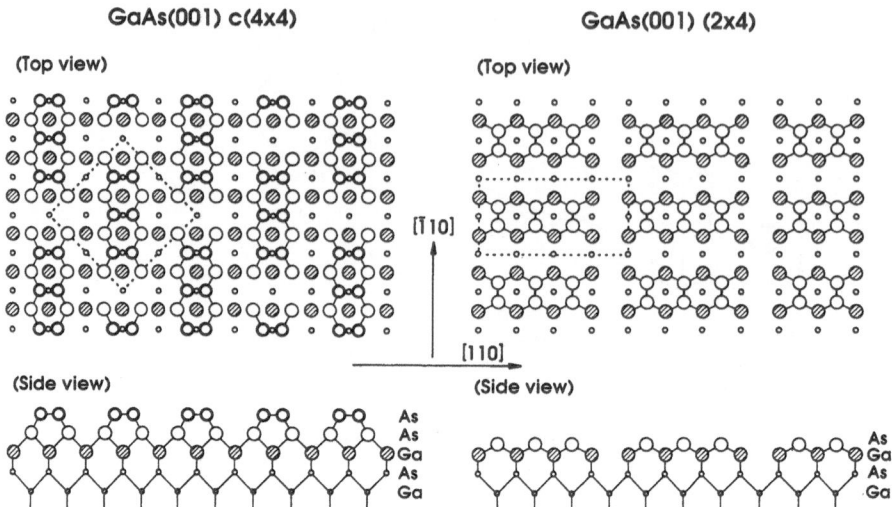

Fig. 2.34. Surface structure of GaAs(001)-c(4×4) and GaAs(001)-(2×4) taken from [2.58]

RAS signal may be written as:

$$\frac{\Delta r}{r} = \frac{r_{\bar{1}10} - r_{110}}{\frac{1}{2}(r_{\bar{1}10} + r_{110})} \tag{2.35}$$

or if the reflectances $R = r \cdot r^*$ are measured, then:

$$\frac{\Delta R}{R} = \frac{R_{\bar{1}10} - R_{110}}{\frac{1}{2}(R_{\bar{1}10} + R_{110})} \tag{2.36}$$

A possible realisation of the experiment which allows the real as well as the imaginary part of the RAS signal to be evaluated is sketched in Fig. 2.33. This setup or very similar versions are now in use in several research laboratories around the world. A detailed discussion and a comparison of this configuration with those using rotating samples or analysers is given in [2.53].

In detail, the equipment of the photoelastic modulator (PEM) configuration consists of a 75 W Xe short-arc lamp with an accessible photon energy range from 1.5 to 6 eV, front surface focusing optics, quartz (or MgF₂) Rochon prisms, a PEM modulator operating [2.59] at 50 kHz, a short focal length grating monochromator to keep the setup small, and an extended S20 photomultiplier (Fig. 2.35). The RAS signal is then extracted using a phase sensitive lock-in amplifier. A typical RAS spectrum can be recorded in a few minutes. Thus, at present real-time monitoring of growth is hardly feasible utilising the full spectral range. This problem of course can be overcome in principle by the use of optical multichannel analyzers. Presently, the detection of changes in the RAS signal at a fixed photon energy can be performed

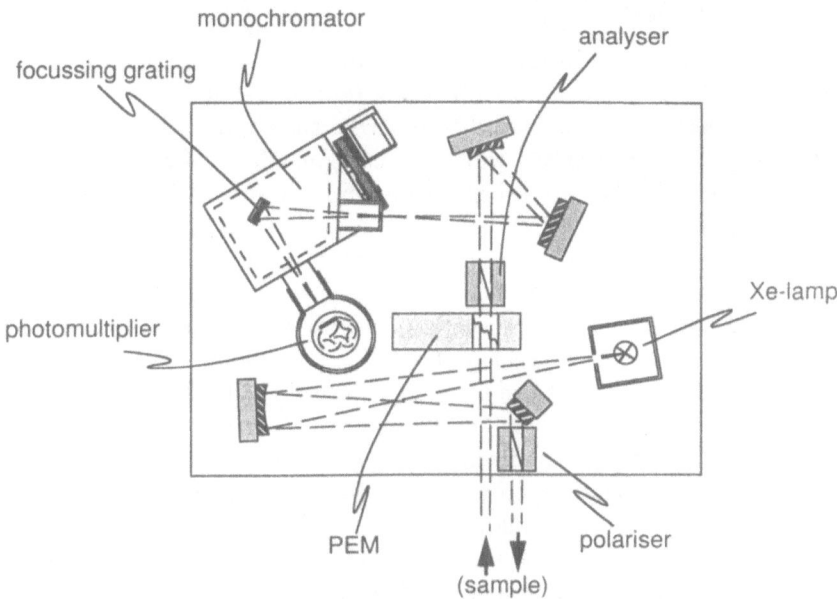

Fig. 2.35. Optical layout of an RAS spectrometer

within 100 ms or less. Consequently, RAS spectra taken over the entire photon energy range can be employed in order to identify the photon energies for which maximum changes occur during growth. Then these photon energies are taken preferably for real time *in-situ* monitoring of growth at fixed photon energies.

As can be seen from the schematic in Fig. 2.33 the RAS optical setup works as a near-null optical bridge. The light coming from the source and passing through the polariser can be divided up in equal parts as light polarised along [110] and [$\bar{1}$10]. In the absence of any anisotropy $r_{110} = r_{\bar{1}10}$, the linear polarisation of the incident beam is simply restored in the reflected beam, i.e. linearly polarised along the principal axes of the modulator. Consequently the output is not modulated and the RAS signal equals zero.

A proper calculation of the output when anisotropy is present, however, yields a rather complicated expression since misalignment of the optical components as well as strain in the window have to be included [2.53]. The normalized voltage output reads to first order:

$$\frac{\Delta V(t)}{V} = 2\left[\text{Im}\left(\frac{\Delta r}{r}\right) + \delta_1 \cos 2\Theta_1 + \delta_2 \cos 2\Theta_2 - 2a_p\right] \cdot J_1(\delta_c) \cdot \sin \omega t$$
$$+ 2\left[\text{Re}\left(\frac{\Delta r}{r}\right) + 2\Delta P + 2\Delta C\right] \cdot J_2(\delta_c) \cdot \cos 2\omega t \qquad (2.37)$$

where ω is the modulation frequency, θ_1, θ_2, $(-45° + \Delta P)$, and $(+45° + \Delta C)$ are the reference azimuth angles of the incident beam window strain, the reflected-beam window strain, the polariser, and the compensator (modulator) respectively, δ_1, δ_2 and δ_c are the retardations due to the incident-beam window strain, reflected-beam window strain, and the modulator, respectively, and J_1 and J_2 are Bessel functions (angles defined as in Fig. 2.33). Two sets of parameters are needed to describe the window strain because it is usually not uniform over the window area. The effects of non-ideal prisms including optical activity are taken into account to first order by representing the transmitted mode for the polariser as $(x + i a_p y)$ [2.53]. The real and imaginary part of the RAS signal have different modulation frequencies and can therefore easily be distinguished by frequency sensitive detection (lock-in). The polariser and modulator azimuthal misalignments (ΔP and ΔC) give rise to an offset in the real part whilst the window birefringence and polariser imperfections affect the imaginary part of the RAS signal. This may prevent absolute measurements. The sensitivity obtained with such a system is found to be about 10^{-5} in $\mathrm{Re}(\Delta r/r)$, but about an order of magnitude poorer in $\mathrm{Im}(\Delta r/r)$. The reason for this is basically due to the extra terms in the imaginary part which allow for example the coupling of mechanical vibrations to $\mathrm{Im}(\Delta r/r)$ via non-uniform window strain. Part of this problem can be removed by using a proper window design with very weak residual strain (so-called strain-free windows) a requirement also demanded in ellipsometry [2.60].

The technique can be further improved [2.61] by using different sample alignments (i.e. angles unequal 45°, see Fig. 2.33) in order to eliminate the offsets. If possible in the experimental apparatus, an elegant way of doing this is by introducing an additional modulation through a slow rotation of the sample at approximately 0.1 Hz. These approaches allow systematic errors in the determination of the RAS signal to be eliminated. Accurate values of $\Delta r/r$ may then be converted into the surface dielectric anisotropy $\Delta(\epsilon \cdot d)$ with d being the thickness of the surface layer using [2.61]:

$$\Delta(\epsilon \cdot d) = (\epsilon_{\bar{1}10} - \epsilon_{110}) \cdot d = i \frac{\lambda}{4\pi}(\epsilon_s - 1) \cdot \frac{\Delta r}{r} \tag{2.38}$$

where ϵ_s is the dielectric function of the bulk material and λ is the wavelength of light. For this conversion precise data for the bulk dielectric function ϵ_s are essential.

A more complicated situation arises in the analysis of RAS if layer systems are under consideration. In this case multiple reflections have to be taken into account as discussed in several parts of this book (see for example Chap. 5, Sect. 5.1.6). However, in the case of RAS they have to be calculated separately for the two polarisations along the surface eigenvectors. Anisotropic properties may occur in such a multilayer system not only at the surface but also at all interfaces. A two layer system (InGaAs/InP) was studied in [2.63]. Anisotropies at the interface as well as on the surface were found. In the case of the Si/SiO$_2$ layer system in contrast Yasuda et al. found that only anisotropy at the Si-SiO$_2$ interface has to be considered [2.64]. The analysis requires in any case RAS spectra taken at several layer thicknesses. This of course can be done easily during real-time studies. In *ex-situ* studies a number of samples with different layer thicknesses should be available.

2.4.1.1 Surfaces Under Pregrowth Conditions. It is appropriate to start this section with those examples where growth studies were performed in an UHV environment since there the RAS spectra can be correlated to RHEED data. Most of the microscopic interpretation of RAS spectra originates up to now from this correlation. Considering the GaAs(001) surfaces (Fig. 2.34) different reconstructions can be observed by RHEED depending on the Ga and As fluxes and the substrate temperature [2.62]. At typical MBE growth temperatures one usually obtains the (2×4)- or at more arsenic rich conditions the c(4×4) reconstruction. RAS spectra for these surfaces and for an even more arsenic rich surface "d(4×4)" are shown in Fig. 2.36. From such spectra it is possible to correlate certain reconstructions with certain RAS spectra.

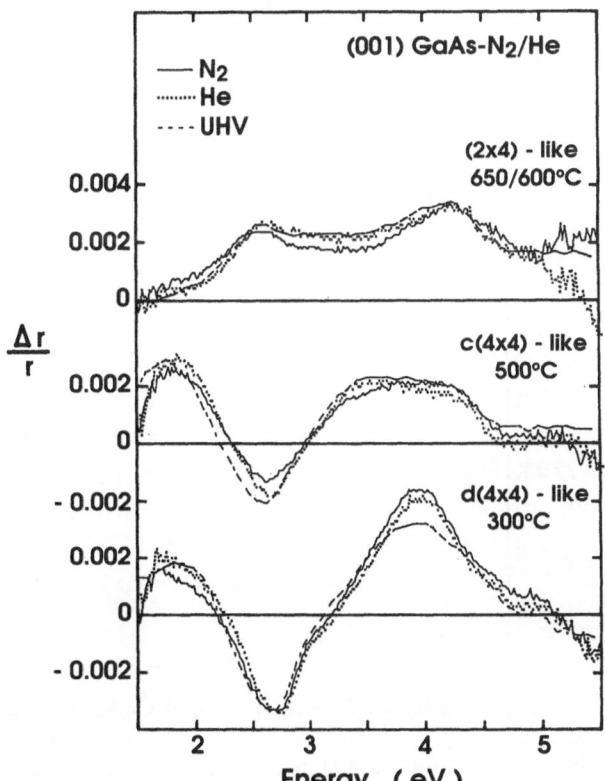

Fig. 2.36. RAS spectra from different reconstructions on GaAs (001) surfaces prepared in UHV and measured either under UHV conditions (dashed line) or under different atmospheric ambients [2.73]

The GaAs(001)-(2×4)- surface shows always up in the RAS spectra with a typical camel back like shape, while the GaAs(001)-c(4×4) one is characterized by a negative minimum at 2.6 eV. Tight binding calculations with an approximated surface structure have indicated that the RAS structure at 2.6 eV is due to electronic transitions from occupied lone pair bands into unoccupied dimer bands [2.74] and is therefore characteristic for As dimers. If the dimers redirect themselves when the surface transforms from one reconstruction to another the RAS signal correspondingly should change at this spectral position. For the case of the two GaAs(001) reconstructions the dimers switch from the [110] to the [$\bar{1}$10] direction and the RAS signal (Eq. 2.35) should change even its sign. This is what is exactly observed in Fig. 2.36. This figure includes also data taken on the same surfaces but in a gaseous ambient of molecular nitrogen or helium. The spectra under the different ambients appear identical. This shows that under these conditions the surfaces are, at least from their optical response, the same as under UHV conditions. This might possibly exclude long range order which is of course not detected by the more local optical response. However, X-ray diffraction experiments at grazing incidence in a MOVPE environment have indeed observed the c(4×4) reconstruction under As stabilized conditions [6.251].

The RAS spectra of GaAs(001) in the pregrowth arsenic stabilized condition of a MOVPE environment can then be described in terms of these surface reconstructions. Figures 2.37, 2.38 give examples for three different situations with H_2 only flowing through the reactor (Fig. 2.37), with additional AsH_3 (Fig. 2.38a) and with the alternative precursor $tBAsH_2$ (Fig. 2.38b). Using

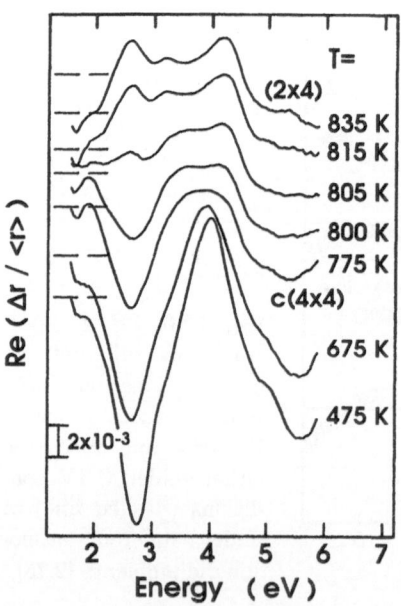

Fig. 2.37. RAS spectra of GaAs (001) in a MOVPE reactor at different temperatures and H_2 partial pressure of 100 mbar [2.83]

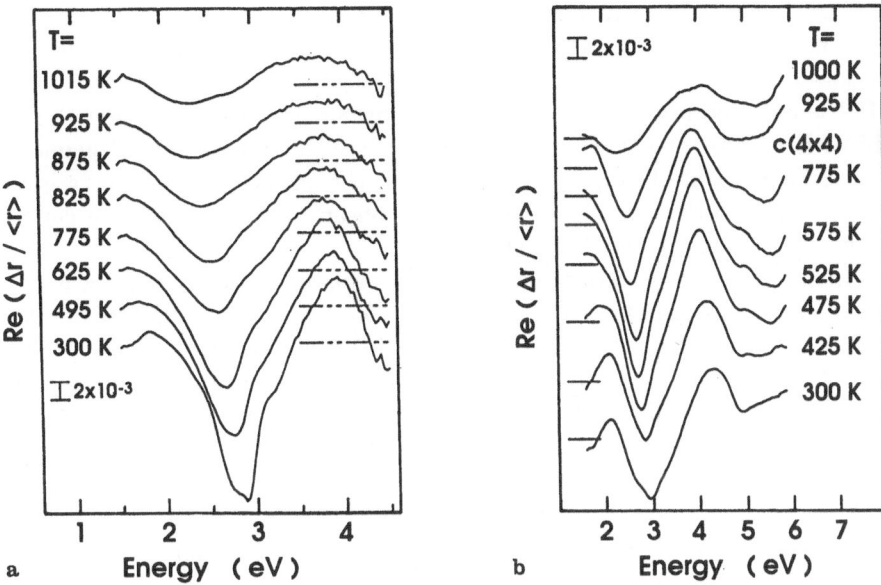

Fig. 2.38. Same as in Fig. 2.37, but under As stabilization with AsH_3 partial pressure of 0.7 mbar (**a**) and $tBAsH_2$ partial pressure of 0.11 mbar (**b**) [2.83]

the RAS-RHEED correlation described above, the reconstructions can clearly be identified as indicated in Fig. 2.37 by c(4×4), and (2×4) reconstructions. In the first case only the arsenic, which is always present from the MOVPE reactor walls, stabilizes the surface. This leads to a (2×4) surface at high temperatures. At intermediate temperatures the RAS spectra can be convoluted by linear combinations of the c(4×4) and the (2×4) RAS spectra [2.83]. If the additional decomposition of arsine delivers more arsenic to the surface than just the reactor walls the c(4×4) configuration is preserved at all temperatures (Fig. 2.38a). This result can be also obtained at even much lower partial pressures with the more efficient arsenic precursor $tBAsH_2$ [2.84]. An exact preparation of the pregrowth surface state with the help of RAS spectra is therefore possible. It turns out that in MOVPE, standard pregrowth conditions are those of a c(4×4) surface while in MBE the surfaces are more at the transition from c(4×4) to (2×4), but typically already within the (2×4) range. MBE growth consequently takes place under less arsenic rich conditions.

Similar results, i.e. the equivalence of RAS spectra under UHV conditions (MOMBE) and a gaseous environment, have been obtained also for InP [2.76]. The spectral shape for the InP(001)-(2×4) and the InP(001)-(4×2) reconstructions again showed characteristic features. In addition RAS spectra have been obtained also with either one of the growth methods: for AlAs(001) [2.77] and InAs(001) [2.88] surfaces under UHV conditions and for InGaAs and GaAsP [2.76] in a MOVPE setup. All these results show that (i) different surfaces of the same material can be identified, (ii) that the spectral features give a fingerprint-like signature of the chemical nature of the surface allowing to differentiate between different materials and surface stoichiometries and (iii) that such surface information may be obtained not only under UHV conditions but also in a gaseous environment as it is present during MOVPE growth or other surface modification techniques. The "chemical" aspects of RAS in obtaining information on the surface stoichiometry is especially noteworthy also for MBE growth, because such information cannot be extracted easily from RHEED data.

In the previous examples only the real part of the RAS signal was presented. If the imaginary part is measured too, for example by utilizing the sample rotation facility usually present in a MBE apparatus, then the surface dielectric anisotropy may be obtained. Figure 2.39 gives examples for different GaAs(001) surfaces [2.61]. It should be noted that the Ga rich (4×2) surface with a strong peak below 2 eV is also included. This peak was attributed to anisotropy contributions from Ga-dimers.

2.4.1.2 Surfaces During Growth. Under growth conditions, i.e. by adding the group III element to the group V stabilized surfaces, the spectra change especially in the case of MOVPE drastically. The magnitude of the change depends on TMGa partial pressure as well as on temperature. Figure 2.40 gives examples for different temperatures and different trimethylgallium partial pressures. These spectra are taken with time constants much larger than the time needed to grow a monolayer of GaAs and thus represent a time av-

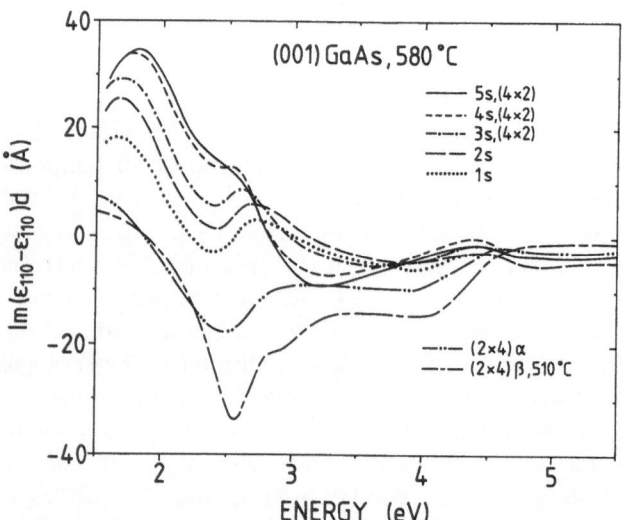

Fig. 2.39. Surface dielectric anisotropy spectra $\Delta(\mathrm{Im}(\epsilon) \cdot d)$ of various reconstructions on GaAs(001). Spectra obtained partly after As-flux interruption at certain times (as indicated) [2.61]

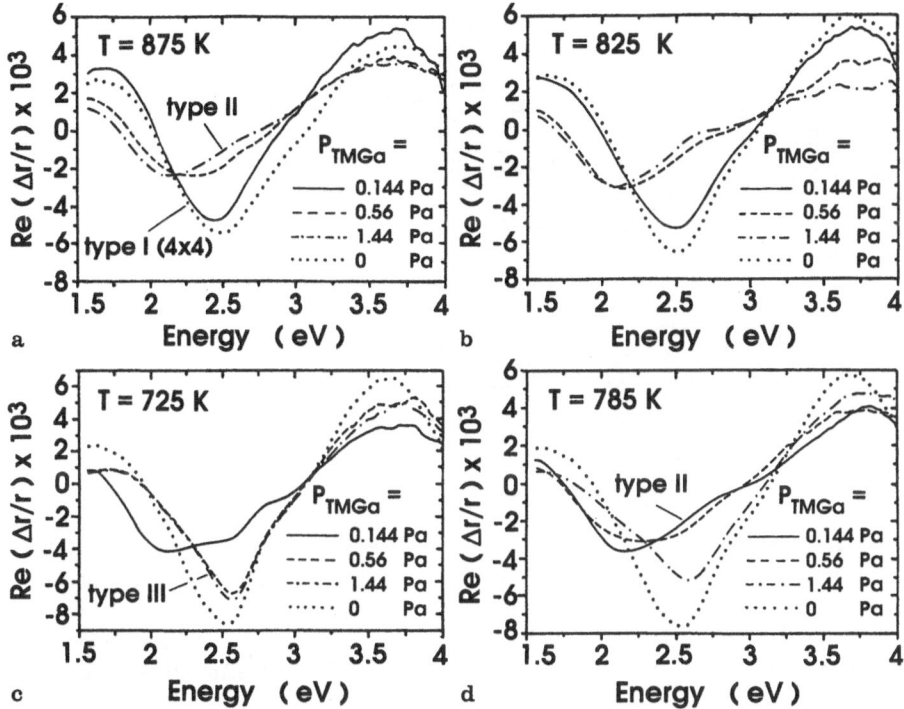

Fig. 2.40. RAS spectra of GaAs(001) in a MOVPE reactor at four different temperatures (a to d) and for different TMGa partial pressures. The partial pressures of H_2 and AsH_3 were 100 mbar (10^4 Pa) and 0.31 mbar (31 Pa), respectively [2.87]

eraged response. Three different types of spectra can be recognized. At high temperatures and low TMGa partial pressures the spectra (type I) are similar to that of the arsenic stabilized c(4×4) with a minimum at 2.5 eV. At intermediate temperatures and partial pressures the spectra (type II) exhibit now a minimum at 2 eV and have negligible anisotropy at 2.5 eV. Finally at low temperatures and large partial pressures of TMGa the spectra (type III) look actually quite similar to the c(4×4) but the minimum appears at 2.6 eV.

These results can be summarized in a p(TMGa)-T diagram (Fig. 2.41) which indicates by the appropriate symbol the type of spectrum obtained at the corresponding temperature and partial pressure of TMGa. While in the type I range the surface must be still covered essentially with a double layer of As, in the type III area as a result of the low temperature and the high TMGa flux the surface is probably covered with TMGa or TMGa fragments like $GaCH_3$. In the intermediate range (type II) a partial coverage with Ga or $GaCH_3$ has to be assumed. This is also in agreement with the growth rate which is linear with TMGa flux in area I and II (transport limited) but saturates in area III (kinetically limited) [2.87].

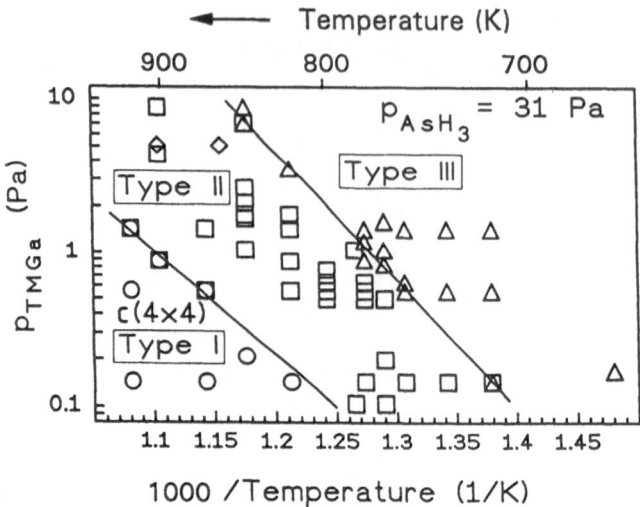

Fig. 2.41. $p(\text{TMGa})$-T phase diagram of GaAs(001) surfaces during MOVPE growth classified according to the type (I, II, III) of RAS spectrum measured (Fig. 2.40). The straight line separating the three regions indicate activation energies around 1.4 eV (I,II) and 1.7 eV (II,III). A partial pressure of 1 Pa corresponds to a flux of $3 \cdot 10^{14}$ cm^{-2}s^{-1} of Ga containing species to the GaAs(001) surface [2.87]

The difference in the RAS signals between the pregrowth and stationary growth situations already suggests that time resolved studies should be quite informative and possibly could allow to observe growth oscillations similar to those observed with RHEED in MBE growth. Indeed the analogy has been first observed by Harbison et al. during MBE growth of AlAs on AlAs(001) as shown in Fig. 2.42 [2.66]. The fixed photon energy (3.44 eV) in these experiments was chosen such that the difference between pregrowth and growth spectra was maximized. The RAS oscillation period observed is identical to that of RHEED and thus proves that RAS is also sampling monolayer growth. There is, however, a phase shift between both oscillations. Similar shifts have been also observed during atomic layer epitaxy [2.69]. This is not surprising since RHEED samples the variation in surface roughness during island formation, while RAS measures the corresponding variation in the anisotropic dielectric response. The latter exhibits dispersion and consequently the phase shift is experimentally observed to change with wavelength, too.

In a non-UHV environment such oscillations were observed for the first time by Samuelson et al. in a metal organic growth technique at very low pressures which they termed CVE (Chemical Vapour Epitaxy) [2.79, 2.80]. For the MOVPE growth under standard total pressure oscillations with monolayer periodicity were first observed by Reinhardt et al. [2.82]. It is evident from Fig. 2.40 that the time variation (increase or decrease) of the signal will strongly vary depending on the choice of photon energy. The time dependence monitored at a photon energy of 2.6 eV where a large difference occurs

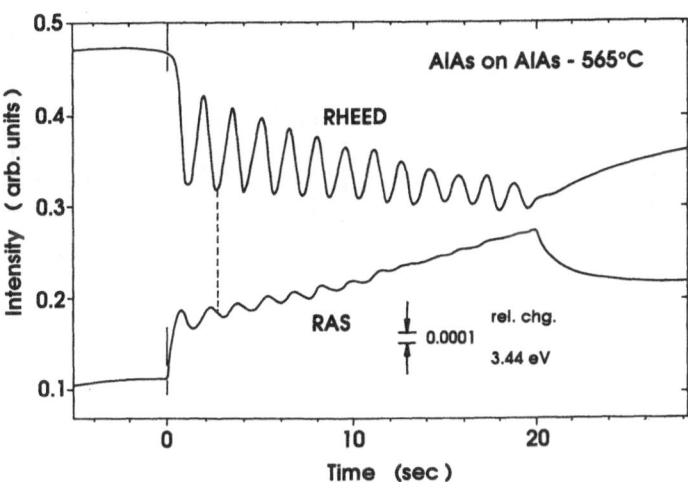

Fig. 2.42. Averages of nine RHEED and RAS traces upon initiation of AlAs growth on an As-stabilised (2×4) AlAs surface. The time needed to grow one momolayer is 1.5 s [2.66]

between the pregrowth and growth RAS signal is displayed in Fig. 2.43 for a number of different TMGa partial pressures [2.82, 2.83]. At $t = 0$ s the TMGa flux is switched on and coincides with the arrival of TMGa at the surface (it is switched off at $t = 35$ to 40 s). The magnitude of the rise at $t = 0$ s corresponds to the difference between two stationary RAS spectra as shown in Fig. 2.40. The most prominent features in Fig. 2.43a which occur after the first rise certainly are the oscillations. Their frequencies increase with increasing TMGa partial pressure. The time of one period corresponds to that needed for the growth of one monolayer of GaAs at these conditions where growth is transport limited and depends linearly on TMGa partial pressure. Indeed the growth rate derived from the oscillation periods increases linearly upon TMGa partial pressure and excellent agreement is found with *ex-situ* growth rate evaluation [2.82, 2.83]. Similar results are obtained for the temperature dependence of the oscillations (Fig. 2.43b) [2.86, 2.87]. The disappearance of the growth oscillations at low temperatures has been attributed to the vanishing growth rate due to limited precursor decomposition. In contrast, at high temperatures the data indicate a transition to a step flow growth mode [2.87].

These RAS oscillations, similar as the RHEED oscillations in MBE, may be employed to grow multilayer structures in MOVPE with an accuracy of parts of a monolayer [2.85]. Figure 2.44 gives an example in form of a RAS time protocol of a superlattice: 30×(5 ML InGaAs/10 ML GaAs). While the 30 periods of the SL are represented by the strong modulation in the top of the figure, the transient of an individual period (No. 26) is shown on an expanded time scale in the lower part. Each monolayer of InGaAs and GaAs can be exactly counted in real time and utilized for an online gas flux control.

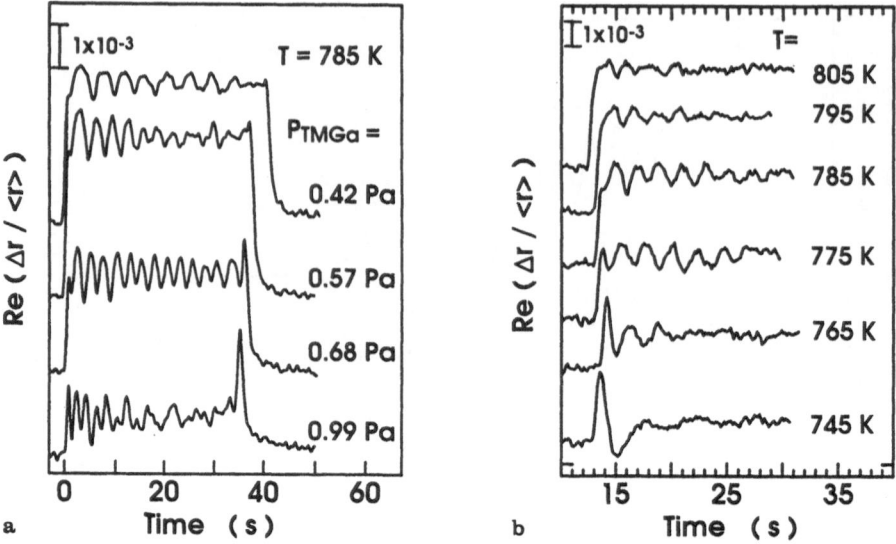

Fig. 2.43a. RAS signal at a fixed photon energy of 2.6 eV on GaAs(001) and H_2 partial pressure of 100 mbar (10^4 Pa) and AsH_3 partial pressure of 0.7 mbar (70 Pa) when starting growth by adding TMGa with different partial pressures between $t = 0$ s and $t = 35$ to 40 s, **b** similar as (**a**) but for different temperatures. The TMGa flux is switched on at $t = 13$ s [2.82, 2.87]

An interesting aspect of RAS is also its sensitivity to anisotropic surface morphologies. This was for example utilized for monitoring the three-dimensional growth mode of lattice-mismatched materials (see e.g. [2.81]). The corresponding effects of roughness observed in the RAS spectra were explained using effective medium theories. A similar approach was recently used in order to describe RAS data obtained for the MBE growth of InAs on GaAs(001) [2.88].

Even many features in the RAS spectra are not understood at present especially of more complicated material systems the future of RAS as a diagnostic tool in gasphase epitaxy looks rather promising. More theoretical work is presently underway and a number of groups have started experimental studies or are planning to do so. The increasing database will fill in the existing gaps of understanding and will increase the potential of RAS beyond just a monitoring device.

2.4.2 Surface Photo Absorption (SPA)

SPA is a technique where the sensitivity originates from a modulation of the surface structure similarly as in the Differential Reflectivity Spectroscopy (DRS) [2.56] but under oblique angles near the Brewster case. This method was developed by Kobayashi and Horikoshi and named surface photoabsorp-

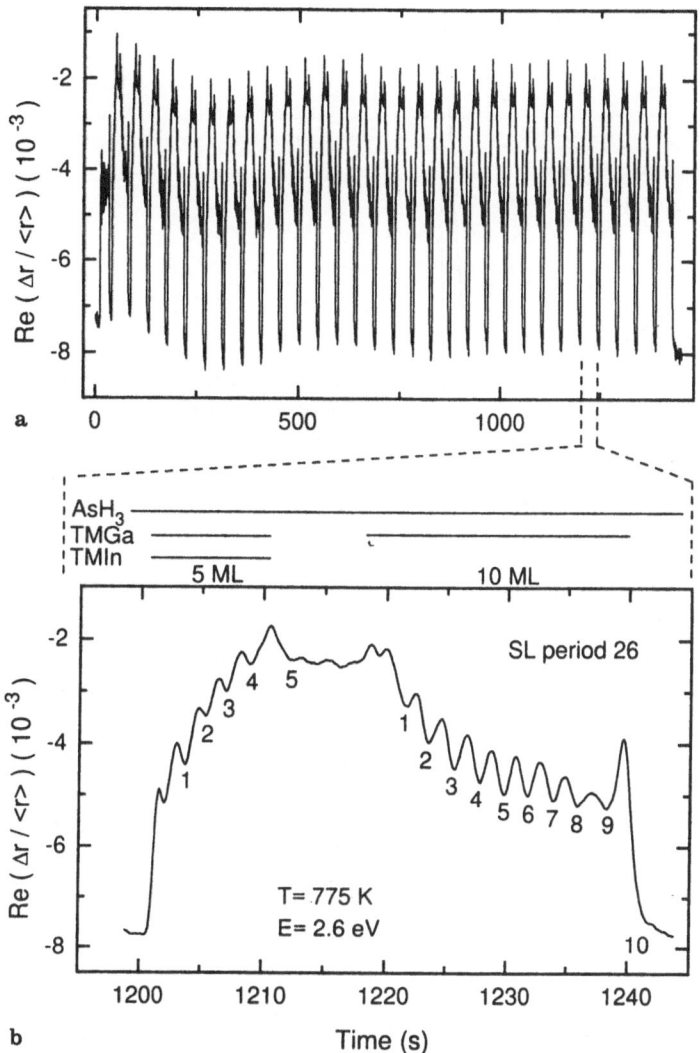

Fig. 2.44a. RAS response for the MOVPE growth of a InGaAs(5 ML)/GaAs(10 ML) superlattice with 30 periods monitored at 2.6 eV, **b** magnification of the RAS response for the growth of one period (number 26) of the superlattice. Monolayer oscillations are seen for each individual layer in the superlattice and are indicated by numbers [2.85]

tion (SPA) [2.89]. Since it derives its sensitivity from the different optical response of different surfaces it is not able to characterize a certain surface but is sensitive to situations where surface changes occur. Its main application lies therefore in non-stationary growth techniques like Atomic Layer Epitaxy.

In SPA the light does not hit the sample surface at normal incidence as in RAS, but rather at a very shallow angle of typically 70° with respect to

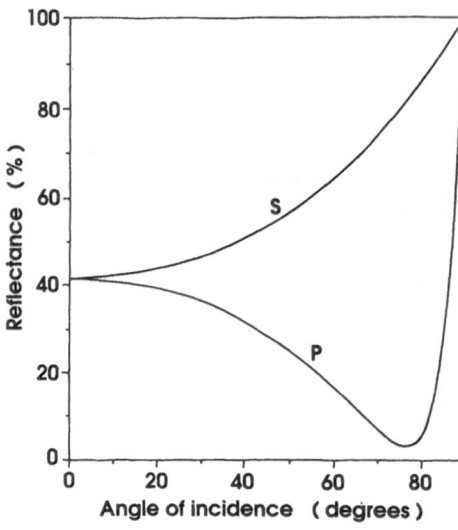

Fig. 2.45. Calculated variation of the reflectance of GaAs upon angle of incidence Θ for p and s-polarised light at $h \cdot \nu = 3.814\,\text{eV}$ using $n = 4 + 2\text{i}$ [2.93]

the surface normal. For most semiconductors this is close to the Brewster angle at which the bulk contribution to the reflected intensity is minimal for light polarised parallel to the plane of incidence. The dependence of the reflectance for light polarised parallel (p) and perpendicular (s) to the plane of incidence on the angle of incidence is sketched in Fig. 2.45 for a GaAs substrate and a wavelength of 400 nm. It can be seen that the reflectance of p-polarised light drops to a few percent near the Brewster angle of 76°. One may therefore expect that any modification of the surface which alters the surface reflectivity will reveal itself most markedly in a change of reflectance for p-polarised light.

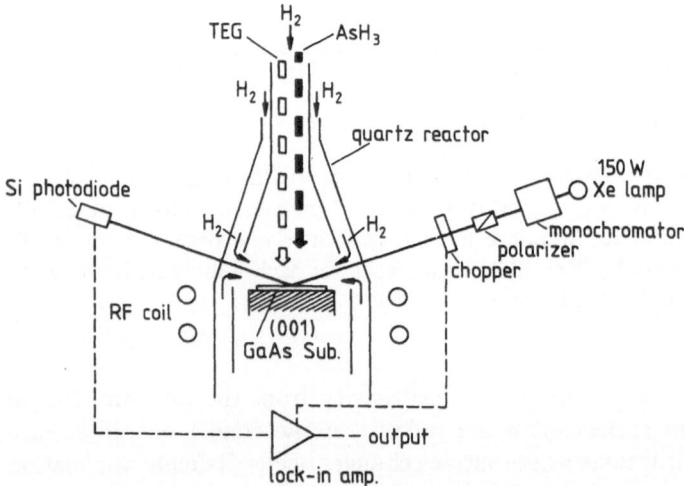

Fig. 2.46. Schematic diagram of a possible SPA experimental setup [2.92]

The experimental setup is sketched in Fig. 2.46 and consists of a light source which can be a laser providing polarised light or a Xe lamp with monochromator and polariser. The intensity of the reflected light is detected by a Si p-i-n photodiode using a lock-in amplifier. The SPA technique has been applied to growth methods like Atomic Layer Epitaxy (ALE) which alternately supply the group III and V elements in quantities which allow one atomic layer to be formed upon deposition. Such an approach is also feasible in MBE e.g. by supplying elemental Ga and As_4 molecules alternately as well as in MOVPE when the constituents are supplied in their gaseous form as e.g. TMGa and AsH_3. The former method is also called Migration-Enhanced Epitaxy (MEE) while the latter in a more general sense has been also named Flow-rate Modulation Epitaxy (FME). According to this growth procedure the surface is alternately terminated by either Ga or As, respectively. Consequently, the authors define their SPA signal as:

$$SPA = \frac{R_{Ga} - R_{As}}{N} \tag{2.39}$$

where R_{Ga} and R_{As} are the absolute reflectance intensities of the Ga- and As-terminated surfaces, respectively, and N is a normalisation quantity which varies in the different publications of the group as R_{As} [2.89–2.91], R the average intensity of R_{As} and R_{Ga} [2.92], or I the intensity of the incident light [2.93].

Considering now the MEE technique it is possible to compare the SPA data with simultaneously recorded RHEED intensities. Figure 2.47 shows the

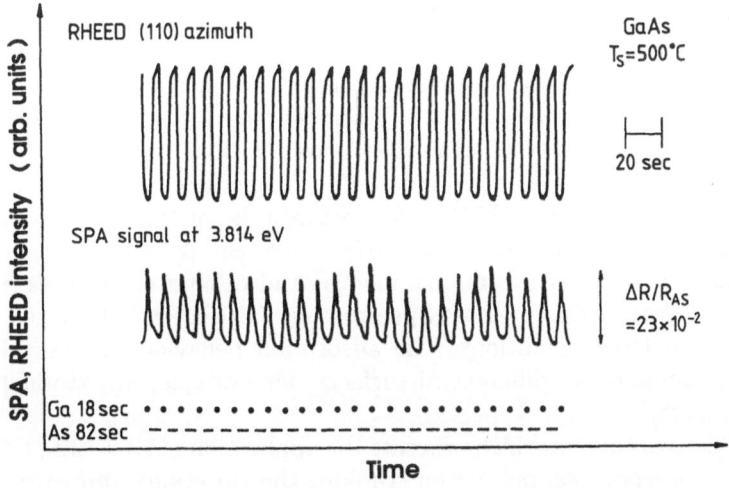

Fig. 2.47. RHEED and SPA intensities for MEE-growth of GaAs. Ga and As MBE cell shutters are opened alternately 1.8 s and 8.2 s, respectively. 1.8 s corresponds to a number of Ga atoms equal to the number of surface sites. A He-Cd laser ($h \cdot \nu = 3.814 \, eV$) was used. The azimuth of incidence is [110] for both techniques [2.89]

result [2.89]. It is apparent that persistent periodic changes not only occur in the RHEED intensity but also in the SPA signal with an excellent signal-to-noise ratio. The difference between the peaks and valleys corresponding to the Ga- and As-terminated surfaces, respectively, is more than 2% of the reflected intensity. It should be noted again, however, that this oscillatory like behaviour only originates from the gas switching procedure and is not at all related to the monolayer oscillations observed during stationary growth by RHEED or RAS as described in the previous Section.

Measurements of the spectral dependence of $\Delta R/R_{As}$ have also shown a resonance enhancement of the SPA signal absorption actually near energies where such effects would be expected from RAS data near 2.6 and 3.8 eV [2.93].

Recently an analytic connection among RAS and SPA has been established on the basis of Fresnel equations [2.94]. The authors show that the results obtained with RAS for different surfaces can be used to interpret the SPA results.

In spite of its high sensitivity a drawback of SPA compared to RAS is of course that SPA measures only the difference between different surfaces but is not capable of characterizing a certain surface. Its main application lies therefore in epitaxial growth techniques like ALE, FME or MEE where the precursor do not flow continuously through the reactor but instead a gas switching procedure is employed.

2.4.3 Infrared Reflection Absorption Spectroscopy (IRRAS)

Considering any growth method using metal-organic compounds as precursors it is of particular interest to identify exactly how the metal-organic species such as TMGa or TEG adsorb and decompose on the growing surface. In principle that can be done by looking at the vibrational modes of the molecules on the surface in combination with symmetry considerations in order to determine the adsorption site (see e.g. [2.95]). The frequencies or energies of the various vibrational modes of TMGa, for instance, lie in the range from approximately 500 to 3000 cm^{-1} or 60 to 370 meV (see e.g. [2.96]). The conventional method for detecting vibrational modes of adsorbates on surfaces is High Resolution Electron Energy Loss Spectroscopy (HREELS) which necessarily depends on UHV conditions. The adsorption behaviour of several metal-organic compounds on different Si surfaces, for example, was studied using HREELS [2.96].

In a MOVPE environment which prevents the application of HREELS, IR optical techniques are good candidates for studying the vibrational properties of the adsorbed species. Infrared light of frequencies that matches those of the vibrational modes interacts with electric dipole moments caused by molecular vibrations. The principles of infrared (IR) spectroscopy are outlined in Chapter 5 of this book. However, when IR measurements are performed to study adsorbates on surfaces in the monolayer coverage regime, the signals are usually very weak and often difficult to detect against the background. Furthermore, the only experimental geometries which seem to be compatible with *in-situ* growth studies are those which utilise external reflection from the front surface for two reasons. Firstly conventional growth reactors usually do not allow access to both substrate surfaces. This would be required for both attenuated total (internal) reflection or single pass transmission measurements. Secondly, these measurements are anyway hampered by the increasing free carrier absorption in the semiconductor substrates at the elevated growth temperatures.

External reflection has already been successfully applied to study adsorbates on metallic surfaces (see e.g. [2.97–2.99]). It was found that in particular for p-polarised light close to grazing incidence, typically 85° with respect to the surface normal, can provide sensitivities which allow 1/1000 of a monolayer of adsorbed molecules to be detected in favourable cases. The molecules absorb some of the light at their vibrational frequencies and absorption peaks occur in the spectrum of the reflected light. This technique is then called infrared reflection-absorption spectroscopy (IRRAS) or reflection-absorption infrared spectroscopy (RAIRS). The resolution of IRRAS (typically around 1 meV) is superior to that of conventional HREELS (typically 2–10 meV) [2.95]. The general advantage of vibrational spectroscopy is that it can provide direct information on the chemical state of the adsorbed molecules from the analysis of the vibrational frequencies observed.

Chesters et al. [2.100] calculated the relative change in the reflectance $\Delta R/R$ at 2000 cm^{-1} for a moderately absorbing film ($n = 1.3 + 0.1\mathrm{i}$) of thickness 0.5 nm on both a metal ($n = 3.0 + 30.0\mathrm{i}$) as well as a Si ($n = 3.44 + 0\mathrm{i}$) substrates using the equation given by McIntyre and Aspnes [2.101] for p-polarised light. They found that the optimum angle for an IRRAS measurement on the Si substrate is 85° which is similar to that on the metal surface. However, the absorbing film on Si leads to an increase in $\Delta R/R$ rather than a decrease observed on a metal surface. The magnitude of the change is approximately 20 times smaller for the film absorbed on Si compared to that on a metal. As a result reflectance changes of the order of 0.01–0.1% can be expected for species adsorbed on a semiconductor surface which is usually only just above the inherent noise level. This low sensitivity may impose a limit to the applicability of IRRAS as a growth monitoring technique, however, because of its molecular fingerprint character it is still a very useful technique for research purposes.

IRRAS was also applied to study the adsorption of TMGa on GaAs(001) surfaces at 300 K [2.103]. The sample was placed in a reflection cell connected to a gas handling system. Figure 2.48 shows a spectrum of the transmittance defined as the reflectance of the adsorbate covered surface divided by the reflectance of the clean surface. The spectrum of the clean surface was obtained from the GaAs(001) surface after annealing the substrate in 10^{-3} mbar H$_2$

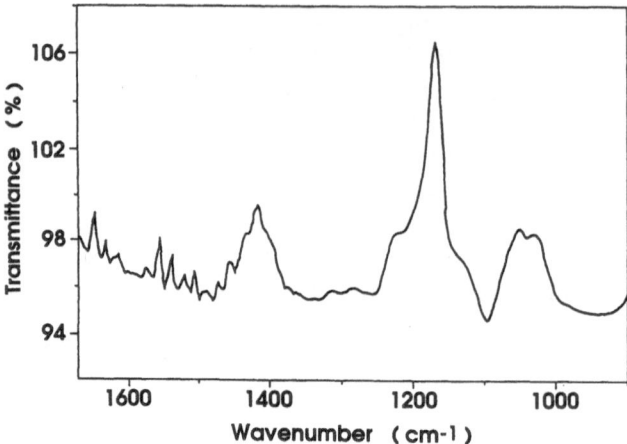

Fig. 2.48. Transmittance ($R_{\text{covered}} - R_{\text{clean}}$) IRRAS spectrum recorded from a GaAs(001) surface exposed to TMGa ratioed to a clean surface spectrum. Taken with a Biorad FTS-40v FTIR spectrometer fitted with a narrow band (800–3000 cm^{-1}), liquid nitrogen cooled MCT detector [2.103]

at 770 K for about 10 min which is believed to result in a clean, nominally Ga rich surface [2.102]. Thereafter the surface was exposed to 10^5 Langmuir TMGa at 300 K. The spectrum was obtained by averaging over 1000 scans within a period of approximately 5 min.

The spectral region shown covers the range where C-H deformation modes usually occur. Clearly three bands are visible at 1420, 1174, and 1050 cm^{-1} with an unexpectedly high change in transmittance around 4–10%. This was explained by the assumption of multilayer molecular physisorption of TMGa. The higher frequency bands were assigned to symmetric and antisymmetric CH$_3$ deformations while the origin of the low frequency one remained unclear.

This work shows that IRRAS is in principle applicable in growth-like situations. Further development is needed in order to explore the application of IRRAS during real growth conditions. For the moment it seems questionable whether monolayer or even submonolayer sensitivity will be achievable at elevated temperatures.

2.4.4 Second Harmonic Generation (SHG)

When two strong electromagnetic fields of frequency ω interact with matter they can combine to produce a field of frequency 2ω. This non-linear process is called Second Harmonic Generation (SHG). It is coherent and the radiating fields have well-defined directions. For example, SHG in the usual reflection geometry emerges along the path of the primary reflected beam and it can easily be separated from the light with frequency ω by the use of appropriate filters. Since SHG is of higher order, however, the cross section for such three-wave mixing is low, with typically one signal photon per 10^{13}–10^{16} incident photons. In a macroscopic picture SHG can be described by

$$\mathbf{P}(2\omega) = \epsilon_0 \chi^{(2)} \mathbf{E}(\omega)\mathbf{E}(\omega) \tag{2.40}$$

where $\chi^{(2)}$ is a second rank non-linear susceptibility tensor. There are bulk as well as surface contributions to this tensor. In the electric dipole approximation this susceptibility vanishes for the bulk of materials such as Si or Ge because of their inversion symmetry. There may, however, be bulk contributions of higher order like magnetic dipole and electric quadrupole terms but these are typically six orders of magnitude smaller than the electric dipole contribution [2.104]. The reduced symmetry at a surface or interface can then lead to a considerable surface (interface) SHG signal which is often comparable to the higher order bulk contribution [2.105]. Information on the surface structure can be deduced by an appropriate choice of the sample geometry and the polarisation vectors [2.106–2.110].

SHG has been widely used to study surfaces and interfaces of materials with inversion symmetry such as centrosymmetric materials. The potential of the technique has recently been reviewed by McGilp [2.111].

Compound semiconductors like GaAs, on the other hand, do not have inversion symmetry in the bulk. As a result the bulk contribution to the SHG signal is usually dominant and exceeds the surface contribution by several orders of magnitude [2.112]. For that reason experimental results for compound semiconductors are very limited. However, it has been suggested by Stehlin et al. [2.113] that the surface sensitivity of the SHG experiment can be restored by a polarisation discrimination of the bulk signal. The authors have exploited this approach in a study of Sn adsorption on GaAs(001). They also provide a list of suitable geometries for the different surfaces which allow the surface contribution of the SHG signal to be separated from that of the bulk.

Very few SHG experiments related to semiconductor growth have been performed so far. Growth studies have been performed for example for the deposition of Si and Ge on Si(001) by MBE [2.114] or for GaAs growth [2.115]. In the latter case the adsorption dynamics of triethylgallium (TEG) on As-rich GaAs(001) at room temperature (RT: 300 K) utilising surface sensitive geometries was studied. The data were obtained using a Nd:YAG laser producing 10 ns pulses at 10 Hz at an energy of 30 mJ/pulse. The 1064 nm light in the reflected beam was separated from the frequency doubled component at 532 nm by a dichroic mirror. The SHG signal was then detected by a photomultiplier tube equipped with a suitable filter.

The As-rich GaAs(001) surface was exposed to $5 \cdot 10^{-9}$ mbar of TEG at RT. The change in the surface SHG signal relative to the clean surface signal is shown in Fig. 2.49 as a function of TEG exposure. The authors have also included the best fit to their data assuming that the system follows a non-dissociative Langmuir adsorption kinetics. This fit describes the rising portion

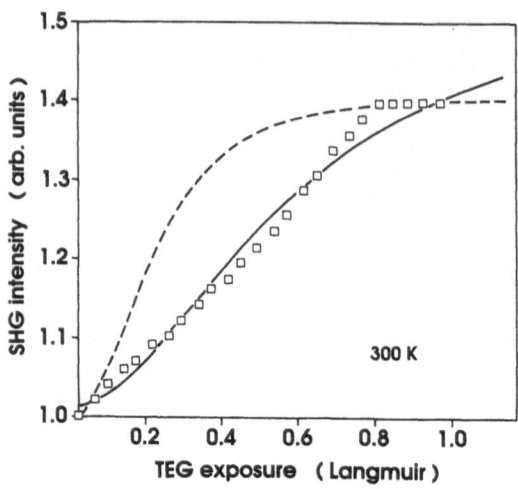

Fig. 2.49. The open squares depict the relative SHG intensity recorded from a GaAs(001) surface as a function of TEG exposure at 300 K. The solid line represents the best fit to the experimental data obtained using the assumption that the system adopts Langmuir adsorption kinetics. The dashed line represents a Langmuir system constrained to saturate at an exposure of 0.8 L [2.115]

of the experimental data but the agreement at the saturation level is rather poor. The agreement is even poorer when Langmuir adsorption behaviour constrained to fit the saturation exposure at 0.8 L is suggested. The authors claim that in analogy to other systems studied by SHG the positive change in the signal is indicative of an electron-rich adsorbate layer and that, considering the structure of the TEG molecule, this is consistent with a dissociative adsorption process.

This example implies that SHG could be a useful tool for the *in-situ* investigation of growth. However, as pointed out above the SHG signal intensity is fairly weak and the high laser powers involved may influence the growth. Thus it seems questionable whether it is a suitable method for real-time growth monitoring at elevated temperatures. Nevertheless, the development of SHG in this area has just begun and, in particular, tunable light sources may lead to a further enhancement of the surface SHG signal when the photon energy is tuned into a resonance with surface electronic states.

2.4.5 Laser Light Scattering (LLS)

When laser light illuminates a sample surface it is not only reflected but also scattered at irregularities at the surface. This elastically scattered light can be detected and used as a measure for the roughness of the surface. This simple diagnostic technique called laser light scattering (LLS) was applied *in-situ* in order to study the MBE and chemical vapour deposition growth of Si [2.116–2.119].

Pidduck et al. have correlated their *in-situ* LLS data with *ex-situ* angle resolved light scattering measurements [2.120]. They find that their results are in qualitative agreement with a "smooth surface" scattering theory [2.121]. From that they concluded that LLS is primarily sensitive to those components of the surface roughness with correlation lengths between 0.5 and 10 wavelengths, depending on the fixed scattering geometry employed. The vertical sensitivity of the method depends on the roughness type: for uniform texture, ready detection of periodic undulations of just few atomic steps mean height was demonstrated, while for monodisperse scattering centres at a density of about 10^6 cm^{-2} a sensitivity to near-nanometer depth was found. In addition, LLS can be strongly anisotropic, for instance in the presence of misorientation steps. The anisotropies in light scattering might be also partly responsible for the RAS signals observed in cases of non-isotropic surface morphologies (Sect. 2.4.1).

Smith et al. have applied LLS in order to monitor surface topography changes during the MBE growth of GaAs [2.122]. The 488 nm line of an Ar$^+$ ion laser was used for illumination. The unfocused beam was impinging on the centre of a GaAs(001) substrate at an angle of 33° to the surface normal fixed by the reactor configuration. The light scattered normal to the substrate was detected using a pulse counting photomultiplier tube equipped with a 488 nm pass-band filter. The background count rates obtained with the laser switched off and the substrate at a temperature of 600 °C were typically < 50 counts per seconds. In the geometry used the magnitude of the scattering vector is $K = 2\pi/\lambda \cdot \sin 33° \approx 7\mu$m^{-1}. Consequently the high sensitivity is achieved for components of the surface roughness with a period of about $2\pi/K \approx 0.9\,\mu$m [2.120]. A typical LLS trace of a growth cycle is shown in Fig. 2.50. The wafer was at $T = 500$ °C at $t = 0$s and was then ramped through the oxide desorption temperature of 620 °C to 630 °C. As can be seen in Fig. 2.50 there initially is very little scattering at $t = 0$s. The rapid increase during the heating period peaking at $t \approx 18$ min coincides with the formation of a reconstructed surface as observed simultaneously by RHEED and therefore indicates the completion of the oxide desorption. The scattering intensity remains almost constant while the substrate is at 630 °C and during cooling to 600 °C in preparation for growth. After opening the Ga shutter for GaAs growth, an initial decrease is observed followed by a smooth increase until the growth is terminated.

In order to interpret such data additional information is required. The authors interrupted the growth process at various stages and quenched cooled the samples in order to preserve as much as possible of the dynamical topography and investigated such samples by atomic force microscopy (AFM). They found that for the LLS maximum at the oxide desorption temperature the surface topography is dominated by 10^9–10^{10} pits per cm^2 of typical lateral size 20–200 nm and depth 5–10 nm. The authors argue that the pits could be produced by an oxide induced etching at exposed weaknesses on the pol-

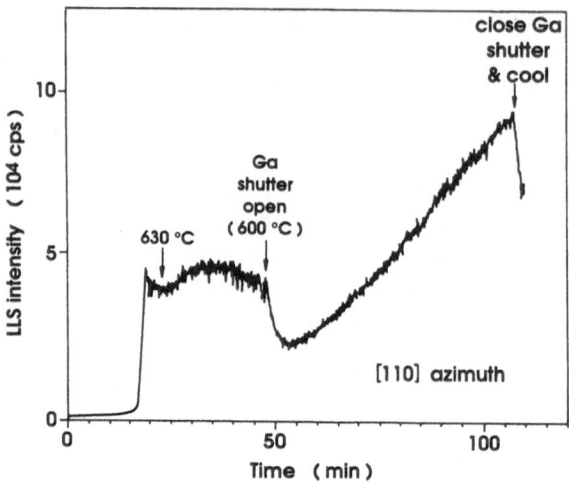

Fig. 2.50. LLS trace of a typical MBE growth cycle. The sample is at 500 °C at $t = 0$, the temperature is increased to 630°C from $t = 0$ to $t = 20$ min, further kept at 630 °C until $t = 46$ min, growth takes place from $t = 47$ min to $t = 106$ min at $T = 600$ °C followed by quench cooling (≈ 200 °C/min) for $t = 106$ min [2.122]

ished GaAs surface. The initial decrease when starting the GaAs growth is believed to be due to an infill of the pits. The following steady increase was found to be associated with the development of a micron-scale undulating surface structure. It is interesting to note that the LLS signal became strongly anisotropic during growth and most intense with the laser beam aligned along the [110] azimuth of the sample. This observation is consistent with the highly anisotropic surface ripple proposed by Briones et al. [2.67]. The decrease in scattering intensity after growth is explained in terms of rapid smoothing due to surface diffusion.

A very recent study performed during GaAs growth has shown also that oscillations periodic with the growth of a GaAs monolayer can be observed [2.68]. The existence of such oscillation opens, similar as in RAS, new possibilities for growth control.

The LLS technique which is simple in its setup certainly requires additional information as for instance supplied by STM/AFM pictures in order to explore the origin of the scattering intensity changes. It it basically limited to the investigation of surface roughness. This, however, affects the signals provided by RHEED and RAS. It therefore seems that this easy to use technique is a very useful complementary diagnostic tool for *in-situ* growth studies. It would certainly help to obtain a better understanding of the damping observed in RHEED oscillations and the roughness induced changes in RAS spectroscopy.

2.5 Conclusions

The experimental methods described in this Chapter have contributed to a large extent to the understanding of gasphase epitaxial growth. The knowledge about gasphase reactions for the most important epitaxial materials systems has made large progress through *in-situ* diagnostic studies. The progress in understanding of surface processes has been much less, but probably will increase strongly within the next years by means of the newly developed or newly applied optical techniques. In general the information available is still not sufficient to serve as sufficient input for complete growth models.

Such models must combine the complex transport phenomena together with the chemical reactions in the gasphase as well as at the surface. While transport processes with the help of modern super computer technology can be handled quantitatively quite satisfactorily for realistic boundary conditions, the situation is still quite different for the chemical reaction path ways. Kinetic data for reactions in VPE processes are very rare. Exceptions for which some data exist are the hydrocarbon reactions [2.123] and also the reactions occurring in silicon deposition from silane or chlorosilanes [2.124]. Therefore, most of the reaction rates still have to be estimated. This concerns especially surface reactions where, as described, less information than in the gas phase is available. Growth models are therefore presently not in a situation where a proof of certain mechanisms can be given, but more in a position where possibly certain critical reaction steps might be identified. This becomes even more clear if one considers the large number of reactions (20–250) [2.45, 2.125, 2.47] which have been partly taken into account in such models.

It thus becomes clear that much more data are necessary in order to describe growth. Especially, among the necessary experimental data, information on surface processes is mostly needed. This should be the main task in the future research of gas phase epitaxial growth. Optical tools for this purpose like Reflectance Anisotropy, Laser Light Scattering, Second Harmonic Generation and IR Absorption are available. They have to be, however, much more developed and especially their interpretation, with the help of theoretical work, has to be put on a more solid microscopic basis. Because of this lack of deeper understanding all the mentioned surface optical techniques qualify already as highly sensitive monitoring devices but still have their limitations as analysing tools with respect to microscopic properties. Fundamental experimental work to solve these questions might be performed in UHV-growth environments (MBE, MOMBE) where additionally classical surface science techniques (LEED, AES, PES, STM) can be applied. However, in addition theoretical work on surface optical properties is strongly needed.

3. Spectroscopic Ellipsometry

Uwe Rossow, Wolfgang Richter

Ellipsometry is an experimental method which dates back to the middle of the last century [3.1, 3.2]. Since then, it has been used to determine the optical properties of all kinds of solids [3.3] including metals, semiconductors and insulators. Ellipsometry has been extensively employed in semiconductor characterisation and has the potential for *in-situ* diagnostics of surfaces. The principle of ellipsometry is based on the fact that the polarisation state of light is changed upon reflection from a surface. This change can be related to the optical properties of the reflecting material. In the most common experimental configuration for ellipsometry, linearly polarised light is incident on a surface and the polarisation state of the reflected light, which is in general elliptically polarised (Fig. 3.1) is analysed. From a knowledge of the polarisation states the ratio of the Fresnel coefficients $\rho = r_p/r_s$ can be derived, since the polarisation is determined by the ratio of the components of the electric field vector.

Using the Fresnel equations, ρ can be directly translated into a complex dielectric function $\epsilon(\hbar\omega)$. Depending on the photon energy $E = \hbar\omega$ of the incident light, $\epsilon(\hbar\omega)$ is the bulk dielectric function of the material under study provided the sample is homogeneous. However, the presence of surface layers or roughness results in inaccurate values for the bulk dielectric function and care has to be taken in the preparation of the samples [3.3]. For most semiconductors, the dielectric function $\epsilon(\hbar\omega)$ has been determined for the

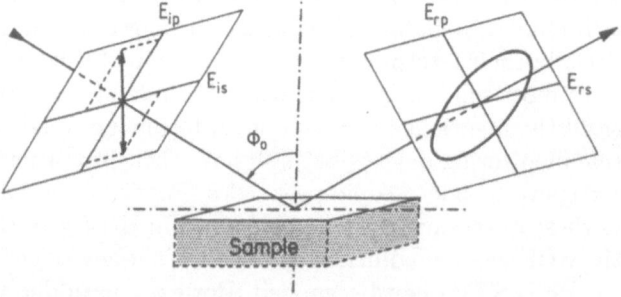

Fig. 3.1. Principle of ellipsometric measurements. Linear polarised light is incident on the sample under an angle ϕ_0 and the polarisation state of the reflected light, which is in general elliptically polarised, is measured

visible and UV spectral range (1.5 to 6 eV photon energy), whereby many contributions of Aspnes [3.3] should be noted.

In most cases the sample under study is not homogeneous but consists of layers or is otherwise structured. In such a case $\epsilon(\hbar\omega)$ is an average over the region penetrated by the incident light and is then a so-called pseudodielectric function or effective dielectric function, written as $<\epsilon(\hbar\omega)>$. From this effective dielectric function $<\epsilon(\hbar\omega)>$ and appropriate models layer properties, such as thickness or layer dielectric functions, can be derived. Using this analysis two experimental parameters – the real and imaginary part of $<\epsilon>$ – (per photon energy) are determined.

Due to the complex structures which can be fabricated by state of the art microelectronics technology, such calculations might be rather time consuming since the effective dielectric function depends on many parameters. However, with the advent of modern computer technology the aspect of time consuming calculations has become less important. As a consequence many quite sophisticated analytical procedures have been developed in order to extract detailed knowledge of the sample beyond an effective dielectric function $<\epsilon(\hbar\omega)>$.

Quite often, however, it is still not possible to separate layer and substrate properties because neither the structure of the sample nor the dielectric function(s) of the layer(s) is known. In such a case at least the existence of surface layers might be clearly observed in $<\epsilon(\hbar\omega)>$ and the modifications of the surface layer can be monitored very sensitively with submonolayer resolution. This is one of the main features for the application of ellipsometry in real time semiconductor growth monitoring and control.

This Chapter deals with all these aspects of ellipsometry in the visible and UV spectral range. A review of ellipsometry in the IR spectral range has recently been published by Röseler [3.4]. The Chapter is organised as follows. In Sect. 3.1, the derivation of $<\epsilon(\hbar\omega)>$ from the knowledge of the polarisation state of the incident and reflected light is presented. The most common ellipsometric setups used for the measurement of the polarisation state of the reflected light are briefly discussed in the second Section. In Sect. 3.3 the emphasis is on the interpretation of $<\epsilon(\hbar\omega)>$ which is explained in detail for some selected examples in Sect. 3.4. The Chapter ends with a discussion of the inherent limitations of ellipsometry.

3.1 Principle of Measurement

The principle of ellipsometry is based on the change in polarisation of light induced by the reflection from a sample surface. As stated above and discussed quantitatively in Sect. 3.1.4, the dielectric function of the sample can be derived from a knowledge of the change of the polarisation state. Therefore,

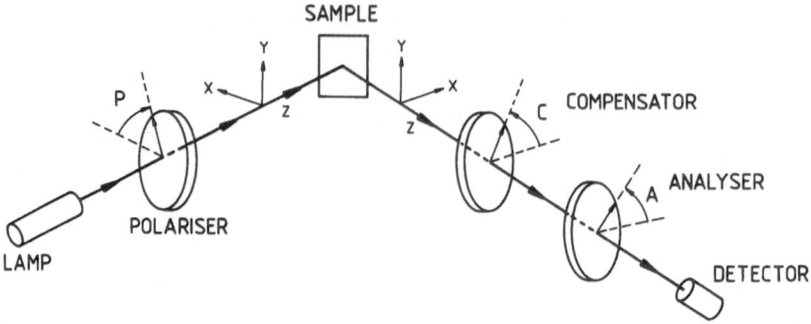

Fig. 3.2. A general scheme of an ellipsometer consisting of a lamp, polariser, sample, compensator (optional), analyser, and detector. The angles of transmission axis of polariser P and analyser A as well as that of the fast axis of the compensator C are relative to the plane of incidence with the x-axis chosen to be in this plane. The positive angles are measured counterclockwise looking against the beam direction

the incident light is often linearly polarised and the polarisation state of the reflected light is measured.

Many experimental configurations have been designed, which distinguish themselves mainly by the way the polarisation of the reflected light is measured. In the older designs ("Null-ellipsometer"), the angles of the polarisation elements, namely the polarisers and compensators (see Fig. 3.2), are arranged in such a way that the detected intensity is minimised. The newer designs (photometric ellipsometers), measure the intensity of the reflected light while modulating the polarisation of the incident or the reflected light. A schematic setup of an ellipsometer, as shown in Fig. 3.2 consists of a light source, polariser, sample, optional compensator, analyser and detector. Following the notation of Azzam and Bashara [3.5] P, C, and A are the angles of the optical elements relative to the plane of incidence.

The compensator is a phase shifting element, usually a quarter-wavelength plate, and is a necessary requirement for null-ellipsometers. This configuration is called a PSCA (Polariser-Sample-Compensator-Analyser) which is equivalent to a PCSA arrangement.

3.1.1 Null-Ellipsometry

In null-ellipsometry the polarisation state of the incident and reflected light is determined by varying the angles of two of the three elements P, C, and A until the intensity measured by the detector is minimised (null-intensity condition, for details see [3.5]). One way of achieving this is to fix the compensator and rotate the polariser until the light which passes through the compensator is linearly polarised again. The linear polarisation state can be verified by the analyser. The intensity at the detector vanishes when the transmission axis of the analyser is orthogonal to the linear polarisation created by the compensator. The name compensator just expresses that this element should compensate the phase shift between the x- and y-component of the electric field vector induced by the (absorbing) sample. In practice more complicated procedures have been used to find the angular positions of P, C, and A which fulfill the "null-intensity" condition [3.6].

Early null-ellipsometers were manually operated with the disadvantage of being slow and user-unfriendly. Automation of the null-ellipsometer is now possible by attaching stepper-motors to the components and by changing their angular positions via a feedback loop of the signal [3.5]. However this procedure is not very fast. Another difficulty is that the phase shift of the compensator varies with wavelength and has to be corrected during spectroscopic measurements. Furthermore, the "null-intensity" condition can only be precisely found if the light intensity is high enough to guarantee a sufficiently high signal-to-noise ratio.

3.1.2 Photometric Ellipsometers

Disadvantages of the null-ellipsometer as outlined above made ellipsometry a very unattractive technique. With the availability of inexpensive computers and Analog-to-Digital-Converters (ADC), photometric ellipsometry became possible [3.7–3.11]. These kinds of ellipsometers can be easily computer controlled and consist (Sect. 3.2.1) of simple elements. All spectra presented in this Chapter were taken with spectroscopic photometric ellipsometers. Moreover, these types of ellipsometers are commercially available.

The most widely used type of photometric spectroscopic ellipsometer is the so-called rotating analyser ellipsometer consisting of a mechanical rotating analyser driven by a DC-motor, a fixed polariser, and an optional (rarely used) compensator [3.8, 3.3]. For this type of ellipsometer, the incident light is linearly polarised and the reflected light, which is (nearly always) elliptically polarised, is modulated by the rotating analyser (Fig. 3.3).

Another configuration which involves modulation of the polarisation is the use of a compensator with time-varying phase shift $\delta(t) = \delta_0 \sin(\Omega t)$, a so-called photoelastic modulator (PEM). Here an AC-voltage is applied to a quartz crystal connected to a piece of quartz glass. The quartz crystal produces stress induced birefringence in the quartz glass via the piezo-effect [3.12]. The use of a PEM is becoming more popular because the setup needs no special mechanical equipment like the rotating analyser [3.7]. It is inherently faster than the rotating analyser ellipsometer since the modulating frequency of the photoelastic modulator is typically 50 kHz compared with 25 to 60 rotations per seconds of the rotating analyser. However, the main disadvantages of using a PEM are:

1) the amplitude of the driving AC-voltage must be regulated to produce a wavelength-independent and known amplitude of the phase shift δ_0.

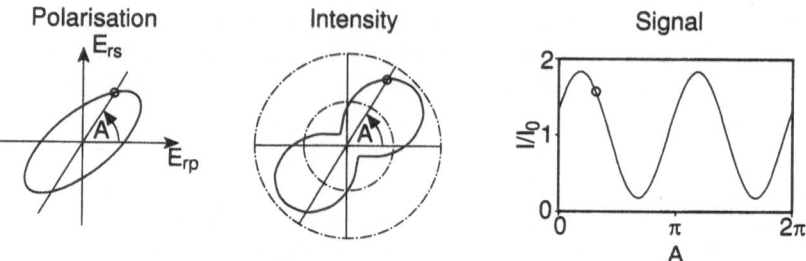

Fig. 3.3. At a fixed location before the analyser, the endpoint of the electric field vector of the reflected light describes an ellipse. This ellipse is examined by the analyser. The intensity of the light determined by the detector is a function of the relative orientation of the analyser transmission axis and the ellipse. Consequently, the signal created by the detector varies as function of the rotating analyser angle. The signal has its minimum (maximum) when the analyser is parallel to the semi-minor (semi-major) axis of the ellipse

2) due to mechanical imperfections of the PEM the amplitude δ_0 may fluctuate with time and temperature.

3) a more sophisticated electrical signal processing is needed because of the high modulation frequency (Sect. 3.2.2).

3.1.3 Description of Light Polarisation

For a quantitative understanding of the operating principle of an ellipsometer, a mathematical formalism which describes how an optical element changes the state of polarisation must be developed.

3.1.3.1 The Jones Formalism. In order to apply the Jones formalism [3.5] it must be assumed that the light is totally polarised meaning that the polarisation state does not fluctuate.

The light is described by plane waves

$$\mathbf{E} = \mathbf{E_0}\exp[\mathrm{i}\,(\mathbf{kr} - \omega t)] = \begin{bmatrix} E_x \\ E_y \\ E_z \end{bmatrix} \qquad (3.1)$$

and the sample, for simplicity, is assumed to be optically isotropic. Thus the dielectric tensor ϵ which is defined by:

$$\begin{bmatrix} D_x \\ D_y \\ D_z \end{bmatrix} = \epsilon_0 \begin{bmatrix} \epsilon_{xx} & \epsilon_{xy} & \epsilon_{xz} \\ \epsilon_{yx} & \epsilon_{yy} & \epsilon_{yz} \\ \epsilon_{zx} & \epsilon_{zy} & \epsilon_{zz} \end{bmatrix} \begin{bmatrix} E_x \\ E_y \\ E_z \end{bmatrix} \qquad (3.2)$$

reduces to [3.15]:

$$\epsilon = \begin{bmatrix} \epsilon & 0 & 0 \\ 0 & \epsilon & 0 \\ 0 & 0 & \epsilon \end{bmatrix} \tag{3.3}$$

and can be replaced by a scalar ϵ

$$\mathbf{D} = \epsilon_0 \epsilon \mathbf{E} \tag{3.4}$$

The z-axis is chosen along the direction of lightpropagation as indicated in Fig. 3.2. Then

$$E_x = E_{0x} \exp[\mathrm{i}\,(kz - \omega t)] \tag{3.5}$$
$$E_y = E_{0y} \exp[\mathrm{i}\,(kz - \omega t)] \tag{3.6}$$
$$E_z = 0 \tag{3.7}$$

and $\mathbf{E}$ can be represented as a column vector with 2 elements. In describing polarisation, the common factor $\exp[\mathrm{i}\,(kz - \omega t)]$ of E_x and E_y is not of interest. Hence, all information about the polarisation is given by the Jones vector defined by:

$$\mathbf{E}^{\mathrm{Jones}} = \begin{bmatrix} E_{0x} \\ E_{0y} \end{bmatrix} \tag{3.8}$$

Light polarised linearly in the x-direction has a Jones vector of the form:

$$\mathbf{E}_x^{\mathrm{Jones}} = E_{0x} \begin{bmatrix} 1 \\ 0 \end{bmatrix} \tag{3.9}$$

For an arbitrary linearly polarised light wave, the electric field vector oscillates along a certain direction in the wave front inclined to the x-axis by an azimuthal angle α. For such a wave, the Jones vector is given by

$$\mathbf{E}_\alpha^{\mathrm{Jones}} = E_0 \begin{bmatrix} \cos\alpha \\ \sin\alpha \end{bmatrix} \tag{3.10}$$

Another simple case is the Jones vector

$$\mathbf{E}_\alpha^{\mathrm{Jones}} = E_0 \begin{bmatrix} 1 \\ \mathrm{i} \end{bmatrix} \tag{3.11}$$

of right-handed circular polarised waves. The imaginary unit i produces the 90° phase shift between the x- and y-component of the electric field vector necessary for circular polarisation.

The matrices J^{Jones} describing the effect of each optical element on the Jones-vectors are defined by $\mathbf{E}_{\mathrm{out}}^{\mathrm{Jones}} = J^{\mathrm{Jones}}\,\mathbf{E}_{\mathrm{in}}^{\mathrm{Jones}}$ and can in general be written as

$$J^{\text{Jones}} = \begin{bmatrix} J_{11} & J_{12} \\ J_{21} & J_{22} \end{bmatrix} \tag{3.12}$$

A simple example for such a Jones matrix is that for an ideal polariser with transmission axis parallel to the x-axis:

$$J^{\text{Jones}}_{\text{ideal polariser}} = \begin{bmatrix} 1 & 0 \\ 0 & 0 \end{bmatrix} \tag{3.13}$$

If the matrix $J^{\text{Jones}}_{\text{ideal polariser}}$ is applied to a general Jones vector then a Jones vector of the form $\mathbf{E}^{\text{Jones}}_x$ (3.9) results.

For a non-ideal polariser with a small ellipticity γ of the transmitted light, the Jones matrix is given by [3.5, 3.7]:

$$J^{\text{Jones}}_{\text{n.-i. polariser}} = \begin{bmatrix} 1 & -i\gamma \\ i\gamma & 0 \end{bmatrix} \tag{3.14}$$

The origin of this ellipiticity may be the optical activity of quartz as discussed in Sect. 3.2.3. The Jones matrix of the sample is:

$$J^{\text{Jones}}_{\text{sample}} = \begin{bmatrix} r_p & 0 \\ 0 & r_s \end{bmatrix} \tag{3.15}$$

Here the Fresnel coefficients are denoted as $r_p = E_{rp}/E_{ip}$ for light polarised in the plane of incidence ($r =$ reflection, $i =$ incidence) and $r_s = E_{rs}/E_{is}$ for light polarised perpendicular to it. The Fresnel coefficients are functions of the dielectric function of the sample and the angle of incidence (Sect. 3.3.4 and [3.5]).

In general, the Jones matrices for optical elements have a simple form if appropriate coordinate systems are chosen. They must be consistent with the symmetry of each element. In (3.9) for example, the x-axis was chosen to be parallel to the transmission axis of the polariser. With this choice the matrices are simple, however, each element has its own coordinate system and one has to change the coordinate system between the optical elements. The change of the coordinate system of one element rotated by an angle α relative to the coordinate system of another element is achieved by using rotation matrices of the form:

$$J^{\text{Jones}}_{\text{rotate}}(\alpha) = \begin{bmatrix} \cos(\alpha) & \sin(\alpha) \\ -\sin(\alpha) & \cos(\alpha) \end{bmatrix} \tag{3.16}$$

The effect of any optical configuration, excluding non-linearities, can be described by a product of Jones matrices.

3.1.3.2 Stokes Vectors and Mueller Matrices.
For partially polarised light, i.e. the degree of polarisation (3.22) is less than one, the Jones formalism

can not be applied. This is the case for polarisers typically used in the infrared (IR) and UV spectral region or when the optical quality of the sample due to rough interfaces is poor (Sect. 3.5.6). In all these cases the more general approach of Stokes-vectors and Mueller matrices must be used.

The Stokes vector consists of 4 elements S_i with $i = 0 \ldots 3$. The components S_i can be expressed by the intensities: (Eq. 1.118 in [3.5]):

$$S_0 = I_0 \tag{3.17}$$

$$S_1 = I_x - I_y \tag{3.18}$$

$$S_2 = I_{+\pi/4} - I_{-\pi/4} \tag{3.19}$$

$$S_3 = I_r - I_l \tag{3.20}$$

where I_0 is the total intensity of the light beam. The parameters I_x and I_y describe the intensities transmitted by a linear polariser with its transmission axis in x- or y-direction. The corresponding transmitted intensities for a linear polariser rotated by $\pi/4$ clockwise and counterclockwise (from the x-axis) are denoted as $I_{+\pi/4}$ and $I_{-\pi/4}$, respectively. Finally, I_r and I_l are the intensities transmitted by a polariser which transmits right-handed (r) and left-handed (l) circularly polarised light. In general one obtains the following relationship:

$$S_0^2 \geq S_1^2 + S_2^2 + S_3^2 \tag{3.21}$$

For totally polarised light the equality is valid, however, for partially polarised light S_0^2 is larger than the rhs term. The degree of polarisation P can be defined by:

$$P = \frac{\sqrt{S_1^2 + S_2^2 + S_3^2}}{S_0} \tag{3.22}$$

For unpolarised light one gets $P = 0$, for totally polarised light $P = 1$.

Similar to the Jones matrices the (4x4) Mueller matrices describe the effect of an optical element on the polarisation state of the wave. Listed here are the Mueller matrices for ideal optical elements neglecting constant prefactors. Others may be found in [3.5] and in Sect. 3.5.6.

Mueller matrix for an ideal polariser or analyser:

$$P^M = A^M = \begin{bmatrix} 1 & 1 & 0 & 0 \\ 1 & 1 & 0 & 0 \\ 0 & 0 & 0 & 0 \\ 0 & 0 & 0 & 0 \end{bmatrix} \tag{3.23}$$

Mueller matrix for a sample:

$$S^M = \begin{bmatrix} 1 & -\cos 2\psi & 0 & 0 \\ -\cos 2\psi & 1 & 0 & 0 \\ 0 & 0 & \sin 2\psi \cos \Delta & \sin 2\psi \sin \Delta \\ 0 & 0 & -\sin 2\psi \sin \Delta & \sin 2\psi \cos \Delta \end{bmatrix} \tag{3.24}$$

ψ and Δ are defined by $\rho = \tan \psi \exp(i \Delta)$.

Mueller matrix for a rotation of the coordinate system:

$$R^M(\alpha) = \begin{bmatrix} 1 & 0 & 0 & 0 \\ 0 & \cos 2\alpha & \sin 2\alpha & 0 \\ 0 & -\sin 2\alpha & \cos 2\alpha & 0 \\ 0 & 0 & 0 & 1 \end{bmatrix} \tag{3.25}$$

3.1.4 Rotating Analyser Ellipsometer in the Jones Formalism

The application of the Jones formalism to describe the operation of a rotating analyser ellipsometer requires, as mentioned above, that the light be totally polarised. The light emitted by the light source, however, is in general unpolarised. Therefore, the Jones-formalism may be used to describe the transformation of the polarisation state only after the light has passed the polariser and has a well defined polarisation state. The polariser is considered to be non-ideal with a small ellipticity γ_p where in the ideal case $\gamma_p = 0$. The Jones vector $\mathbf{E}_{\text{in}}^{\text{Jones}}$ of the polarisation state produced by the polariser is [3.7]:

$$\mathbf{E}_{\text{in}}^{\text{Jones}} = \begin{bmatrix} 1 \\ i\gamma_p \end{bmatrix} E_0 \tag{3.26}$$

where E_0 is a constant.

The rotating analyser ellipsometer (Figs. 3.2, 3.3) is then described by the following expression [3.7]:

$$\begin{aligned} \mathbf{E}_{\text{out}}^{\text{Jones}} &= J_{\text{n.-i. polariser}}^{\text{Jones}} J_{\text{rotate}}^{\text{Jones}}(A - A_s) J_{\text{sample}}^{\text{Jones}} \\ &\times J_{\text{rotate}}^{\text{Jones}}(-(P - P_s)) \begin{bmatrix} 1 \\ i\gamma_p \end{bmatrix} E_0 \end{aligned} \tag{3.27}$$

Substitution of the matrices for the elements gives:

$$\begin{aligned} \mathbf{E}_{\text{out}}^{\text{Jones}} &= \begin{bmatrix} 1 & -i\gamma_a \\ i\gamma_a & 0 \end{bmatrix} \begin{bmatrix} \cos(A - A_s) & \sin(A - A_s) \\ \sin(A - A_s) & \cos(A - A_s) \end{bmatrix} \begin{bmatrix} r_p & 0 \\ 0 & r_s \end{bmatrix} \\ &\times \begin{bmatrix} \cos(P - P_s) & -\sin(P - P_s) \\ \sin(P - P_s) & \cos(P - P_s) \end{bmatrix} \begin{bmatrix} 1 \\ i\gamma_p \end{bmatrix} E_0 \end{aligned} \tag{3.28}$$

A small ellipticity of the analyser, analogous to that of the polariser, is represented by γ_a. The angle P of the polariser relative to the plane of incidence is chosen arbitrary and the angle A is the actual position of the analyser (Fig. 3.2). The polariser and analyser have the angular positions P_s and A_s, respectively, when their transmission axis are in the plane of incidence.

The intensity of the light is proportional to $E_x E_x^* + E_y E_y^*$ and thus, the intensity $I(A)$ at the detector as a function of the analyser angle A can be calculated from (3.28). One gets:

$$I(A) = I_0[1 + a_2 \cos 2(A - A_s) + b_2 \sin 2(A - A_s)] \tag{3.29}$$

with

$$
\begin{aligned}
a_2 &= [\tan^2 \psi \cos^2(P - P_s) - \sin^2(P - P_s)]/D & (3.30)\\
b_2 &= [\tan \psi \cos \Delta \sin 2(P - P_s) + 2\gamma_p \tan \psi \sin \Delta]/D & (3.31)\\
D &= \tan^2 \psi \cos^2(P - P_s) + \sin^2(P - P_s)\\
&\quad - 2\gamma_a \tan \psi \sin \Delta \sin 2(P - P_s) & (3.32)
\end{aligned}
$$

The coefficients a_2 and b_2 are obtained by a Fourier transformation of the signal delivered by the detector which is proportional to $I(A)$. It is important to note that the Fourier coefficients a_2 and b_2 (3.29) are calculated from relative intensities $I(A)/I_0$ and therefore, ellipsometry is in principle insensitive to variations of the absolute intensity of the light source.

The following describes how the dielectric function ϵ is calculated from the Fourier coefficients a_2 and b_2. Details may be found in [3.7].
The azimuth angle Q of the elliptically polarised light is given by:

$$Q + A_F = 0.5 \arctan\left(\frac{b_2}{a_2}\right) + \frac{\pi}{2} u(-a_2) \operatorname{sign}(b_2) \tag{3.33}$$

The parameters A_F and η (3.36) represent the modification of the signal due to the detector as well as the electronic amplifier. They are obtained by a calibration procedure described in [3.7] and Sect. 3.2.4. The functions $u(x)$ and $\operatorname{sign}(x)$ are defined by:

$$u(x) = \begin{cases} 1 & : \quad \text{for } x > 0 \\ 0 & : \quad \text{else} \end{cases} \tag{3.34}$$

and

$$\operatorname{sign}(x) = \begin{cases} 1 & : \quad \text{for } x \geq 0 \\ -1 & : \quad \text{else} \end{cases} \tag{3.35}$$

The modulation of the signal is given by (3.29):

$$\zeta = \eta\sqrt{a_2^2 + b_2^2} \tag{3.36}$$

From this the semi-minor/semi-major axis-ratio r_{ab} of the ellipse can be calculated:

$$r_{ab} = \frac{-2\gamma_a\zeta \pm (1 - \gamma_a^2)\sqrt{1 - \zeta^2}}{(1 + \zeta) - \gamma_a^2(1 - \zeta)} \tag{3.37}$$

The ratio of the Fresnel coefficients ρ introduced above is defined by

$$\rho = \frac{r_p}{r_s} = \tan \psi \exp(\mathrm{i}\,\Delta) \tag{3.38}$$

where ρ can be calculated using:

$$\rho = \frac{\cot[(Q + A_F) - (A_s + A_F)] - \mathrm{i}\,r_{ab}}{1 + \mathrm{i}\,r_{ab}\cot[(Q + A_F) - (A_s + A_F)]}\,\frac{\tan(P - P_s) + \mathrm{i}\,\gamma_p}{1 - \mathrm{i}\,\gamma_p\tan(P - P_s)} \tag{3.39}$$

by substituting Q and r_{ab}.

Finally, the dielectric function of the system can be related to ρ and the angle of incidence ϕ_0 by [3.5]:

$$\epsilon = \sin^2\phi_0 + \sin^2\phi_0\tan^2\phi_0\left[\frac{1-\rho}{1+\rho}\right]^2 \tag{3.40}$$

and hence, ϵ can be calculated by inserting (3.39) into (3.40).

The sign in (3.37) is undetermined. This results in an uncertainty of the sign of r_{ab} and consequently of the imaginary part of ρ if we neglect the small correction due to the optical activity (γ_a and γ_p set to zero). The reason for this uncertainty is that the handiness of the ellipse is not determined [3.5, 3.7].

For bulk material it is sufficient to know the absolute value $|\Delta|$ because $\mathrm{Im}\,\epsilon$ must be positive. Replacing ρ by its complex conjugate which is the same as replacing Δ by $-\Delta$ replaces ϵ by its complex conjugate and hence inverts the sign of $\mathrm{Im}\,\epsilon$. In most other cases the sign of Δ may also be reconstructed. If this is not possible then a compensator is necessary [3.5].

Conclusively, if the Fourier coefficients a_2 and b_2 are evaluated, the dielectric function ϵ can be calculated.

Note: The Fourier coefficients a_2 and b_2 expressed in (3.30, 3.31) can also be obtained using the Stokes vector and Mueller matrices as outlined in the following. If S_o and S_i are the Stokes vectors at the detector, and polariser, respectively, then the effect of the rotating analyser ellipsometer (RAE) can be described by:

$$S_o = R^M(-A)A^M R^M(A)S^M R^M(-P)P^M R^M(P)S_i \tag{3.41}$$

using the Mueller matrices given in (3.23, 3.24, 3.25). Since the light before passing the polariser is unpolarised S_i can be expressed by:

$$S_i = \begin{bmatrix} S_{i0} \\ 0 \\ 0 \\ 0 \end{bmatrix} \tag{3.42}$$

By inserting the Mueller matrices and S_i into (3.41) one obtains the Stokes vector of the light at the detector. For a detector insensitive to the polarisation state only the total intensity S_{o0} is relevant and therefore, neglecting a constant prefactor:

$$\begin{aligned} S_{o0} \sim\ & (1 - \cos 2\psi\cos 2P) + \cos 2A(-\cos 2\psi + \cos 2P) \\ & + \sin 2A(\sin 2\psi\cos\Delta\sin 2P) \end{aligned} \tag{3.43}$$

The Fourier coefficients are finally given by: (see also Eq. (3.29)):

$$a_2^M = \frac{-\cos 2\psi + \cos 2P}{D - \cos 2\psi \cos 2P} \tag{3.44}$$

$$b_2^M = \frac{\sin 2\psi \cos \Delta \sin 2P}{D - \cos 2\psi \cos 2P} \tag{3.45}$$

Using the identities:

$$\cos \psi = \frac{1}{\sqrt{1 + \tan^2 \psi}} \tag{3.46}$$

$$\sin \psi = \frac{\tan \psi}{\sqrt{1 + \tan^2 \psi}} \tag{3.47}$$

$$\sin 2\psi = \frac{2 \tan \psi}{1 + \tan^2 \psi} \tag{3.48}$$

$$\cos 2\psi = \frac{1 - \tan^2 \psi}{1 + \tan^2 \psi} \tag{3.49}$$

it can be shown that the Fourier coefficients a_2^M and b_2^M are equivalent to the Fourier coefficients calculated within the Jones formalism (3.30–3.32) for ideal polarisers ($\gamma_a = \gamma_p = 0$).

3.1.5 The Effective Dielectric Function $<\epsilon>$

The procedure described above for the calculation of the dielectric function from the Fourier coefficients can also be used in case the sample consists of several layers on a substrate or is otherwise structured. In this case the dielectric function obtained by (3.39) and (3.40) is an average over the region penetrated by the incident light and then ϵ is called a pseudodielectric function [3.3] or effective dielectric function and is denoted as $<\epsilon>$. Only if a semi-infinite, isotropic and homogeneous sample is studied then the measured $<\epsilon>$ is the dielectric function of the material to be characterised. But even for the simplest case of a substrate with cubic symmetry (optically isotropic) the surface reduces the symmetry. Moreover the surface can be reconstructed, contaminated and rough. Therefore, the measured $<\epsilon>$ is an approximation to the bulk response of the substrate. However, if the substrate is prepared carefully this is a very good approximation [3.3, 3.14].

Since ϵ is a complex quantity two unknowns of the sample which affect the optical response can be determined for each photon energy (wavelength). For a layer on a substrate, the unknowns could be the dielectric function of the layer (especially the real part for non-absorbing layers) or the thickness of the layer. Often, it is possible to take the slope and curvature of $<\epsilon>$ into account and then more than two parameters can be determined as will be discussed later in Sect. 3.3.4.2.

For anisotropic materials the situation is much more complicated and one must deal with the dielectric tensor ϵ instead of a scalar quantity. This tensor can always be diagonalised by choosing an appropriate coordinate system [3.15]. In general none of the axes of this coordinate system is parallel to the sample surface. This case will not be treated here, however, a discussion of this general situation can be found in [3.5, 3.16, 3.15]. If instead the material is uniaxial or biaxial with the symmetry axis orientated along the surface normal in the plane of incidence or perpendicular to it, then the dielectric tensor is diagonal and can be written as

$$\epsilon = \begin{bmatrix} \epsilon_{xx} & 0 & 0 \\ 0 & \epsilon_{yy} & 0 \\ 0 & 0 & \epsilon_{zz} \end{bmatrix} \tag{3.50}$$

The effect of the sample on the polarisation may still be represented by the Jones matrix

$$J_{\text{sample}}^{\text{Jones}} = \begin{bmatrix} r_p & 0 \\ 0 & r_s \end{bmatrix} \tag{3.51}$$

where r_p and r_s now are functions of ϵ_{xx}, ϵ_{yy}, and ϵ_{zz}. If all three components of the diagonal dielectric tensor (6 unknowns) have to be determined, one has to perform at least three ellipsometric measurements at different angles of incidence since, as mentioned above, one measurement gives only two parameters of the sample.

In the special case of uniaxial materials, the dielectric tensor is represented by a dielectric function for light polarised parallel ($\epsilon_{\parallel}$) and perpendicular ($\epsilon_{\perp}$) to this axis. If this axis is perpendicular to the surface as is the case, e.g., for cleaved Sb, and the refractive index is high, then the electric field vector is nearly perpendicular to that axis and $<\epsilon>$ is nearly equal to $\epsilon_{\perp}$.

3.2 Experimental Details

3.2.1 Rotating Analyser Ellipsometer

The most widely used spectroscopic ellipsometer is of the rotating analyser type. Figure 3.4 shows a typical setup. Common light sources are xenon high pressure arc-lamps, halogen lamps, and less commonly used deuterium lamps (in combination with halogen lamps). A xenon high pressure arc-lamp has a continuous radiation spectrum between 1.5 eV ($\approx$ 800 nm) and 6.2 eV ($\approx$ 200 nm), while the useful spectral range of a halogen lamp is the near infrared region ($\sim$ 1 eV) up to approximately 4.1 eV ($\approx$ 300 nm). Deuterium lamps

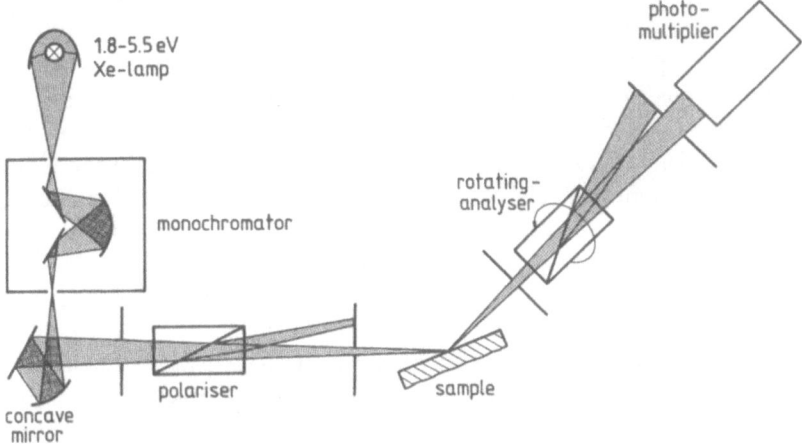

Fig. 3.4. Typical setup of a rotating analyser ellipsometer (RAE) with a Xe-lamp as the light source, Rochon prisms as polariser and a photomultiplier as detector. The arc of the xenon lamp is imaged on the entrance slit of a monochromator. The monochromised light passes through the polariser and is focused onto the sample. The reflected light is then analysed by a second polariser (the rotating analyser) and finally detected by a photomultiplier

are well suited for the UV spectral range [3.17]. Below 1.5 eV, xenon has spectral lines with very high intensity [3.17]. In spite of the high pressure these spectral lines are not broad. Therefore, in this spectral region, the xenon lamps are not useful because the dynamic range of the detector and electronic equipment is limited. A major problem with arc lamps is that the intensity fluctuates with time due to instabilities of the arc [3.17]. Another problem is that the intensity drops nearly exponentially (with increasing photon energy) in the (near) UV region, which is the most interesting spectral range for semiconductors, because the optical gaps with the highest values of Im$<\epsilon>$ are in the (near) UV.

Halogen lamps, on the other hand, are limited in intensity in the UV region. Compared to xenon-lamps, halogen lamps have not only a smooth spectral characteristic, but also more intensity in the near IR. In principle, deuterium lamps could be used in the UV region, where their intensity is higher than that of xenon lamps but they have low intensity in the visible region. A combination of a deuterium lamp and a halogen lamp might be used in the spectrometer instead of a xenon-lamp. Whether this combination is advantageous has yet to be proven in practical applications. So far, xenon arc lamps are most frequently used for investigations in the visible and UV spectral region.

The key elements of all ellipsometers are the polarisers since their imperfections limit the accuracy of the data obtained. The most commonly used polarisers are crystalline quartz Rochon prisms and calcite Glan prisms since only these kinds of polarisers have a sufficient high extinction ratio (10^{-5} or better) [3.18]. The polarisers will be described in detail in Sect. 3.2.3.

The angular position of the polariser is usually fixed when taking a spectrum. The value of the angle P (angle between the polariser transmission axis and plane of incidence) is not critical and should be chosen such that $\mathbf{E}$ has nearly equal components parallel and perpendicular to the plane of incidence. The reflectivity is higher for light polarised perpendicular to the plane of incidence than for light polarised in the plane of incidence. Thus, values for P lower than 45° are appropriate. A value of 30° for P was found to be nearly optimal for a wide range of substrate materials [3.19, 3.7].

The most widely used detectors are photomultipliers (PMT). The advantages of a PMT are: i) a linear relationship between intensity and signal for intensities varying over several orders of magnitude and ii) current amplifications with gains typically in the order of 10^6 to 10^8 and high signal-to-noise ratios. As an inexpensive alternative silicon photo-diodes can be used as detectors for spectroscopy in the near infrared and visible spectral region and also for single wavelength applications. Recently Optical-Multichannel-Analyser (OMA) systems, based on Si diode arrays have been introduced [3.20].

In an RAE the analyser is rotated at a constant frequency of 25 to 60 Hz where the rotation frequency is limited by mechanical vibrations. The angle of the transmission axis of the analyser is measured by an angle encoder (otherwise known as a resolver) mounted on the analyser holder. The accuracy of the angular position is mainly limited by the alignment of the angle encoder and the electronic signal processing and is typically less than 1%.

An angle of incidence, ϕ_0 between 60° and 80° yields highest sensitivity for layer properties in semiconductor substrates [3.5, 3.8]. In silicon, for example, the pseudo-Brewster condition for which the reflectance of light polarised in the plane of incidence has its minimum is nearly fulfilled at 75.6° at a wavelength of 632.8 nm.

The detecting photomultiplier provides a current which is proportional to the intensity of the light at the photocathode. This current is transformed to a voltage, low-pass filtered, and mostly digitised by Analog-to-Digital Converters (ADC, typically with 16 Bit resolution). The ADC is triggered each time an angle encoder placed on the analyser holder gives a pulse when reaching a new angular position. When the digitisation procedure is completed, the ADC often provides an End-Of-Conversion signal (EOC) by which the interface of a computer is informed that valid data can be taken. The angle resolver also delivers a reference signal (NULL) at a certain angular position of the analyser. The computer calculates the Fourier coefficients a_2, and b_2 (Sect. 3.1), the ratio of the Fresnel coefficients ρ and finally the dielectric function (Sect. 3.1).

An alternative to the AD-converter are Lock-In Amplifiers (LIA) with amplitude and phase output. In this case, the reference signal of the LIA can be obtained for example by a generator attached to the analyser driving DC motor creating a sinusoidal varying voltage. The outputs of the LIA at twice the reference frequency are the Fourier-coefficients. The DC component of the signal has to be measured separately.

One advantage of using AD-converters is that an analysis of the high frequency contributions to the signal can easily be done. Typical LIAs don't have this feature and are also more expensive. Furthermore, the separate determination of the DC component may cause inaccuracies.

3.2.2 Photoelastic Modulator Ellipsometer

Photoelastic modulators (PEM) were first introduced into ellipsometry in 1969 [3.10, 3.11]. Recently interest in the use of these devices has reemerged. The useful spectral range of the PEM is limited by the transparency of the materials from which it is fabricated (typically quartz).

A PEM consists of two parts (Fig. 3.5). The first part is made of crystalline quartz. The second part through which the light propagates is quartz glass. The two parts are glued together. If an AC voltage of frequency Ω is applied to the crystalline quartz the piezo-effect causes thickness vibrations in the crystalline quartz in the direction of the applied electric field and perpendicular to it with the same frequency. The variation of thickness results in strain in the quartz glass and therefore the quartz glass becomes birefringent. The difference in the phase shift δ between light polarised parallel and perpendicular to the connecting axis of the two halves is proportional to the stress and the inverse of the wavelength of the light. The phase shift has the same frequency as the applied stress and therefore as the AC-voltage. Hence, it can be written as $\delta(t) = \delta_0 \sin(\Omega t + \phi_0)$, where δ_0 depends on the wavelength of the light, the magnitude of the stress and the thickness of the unstrained quartz glass. For any application, the phase shift of the modulator needs to be known, but it is sufficient to determine δ_0 and perform the measurement such that $\phi_0 = 0$. The amplitude δ_0 can be determined as a function of wavelength in a calibration procedure without any sample. In most cases the magnitude of the applied voltage and hence of the stress induced, is chosen such that δ_0 is independent of the wavelength of the light.

Drevillon has improved the accuracy and flexibility of this kind of ellipsometer [3.21]. The resulting set up is shown in Fig. 3.5. The components, apart from the PEM, are the same as for the rotating analyser ellipsometer. The use of optical fibres, which allow the ellipsometer to be easily adopted to growth chambers, is a new feature.

This set-up, however, differs from the rotating analyser ellipsometer (RAE) in that the modulating element is placed in front of the sample, in which case this configuration resembles more a rotating polariser ellipsometer. In contrast to Fig. 3.1 the incident light is in this case elliptically polarised.

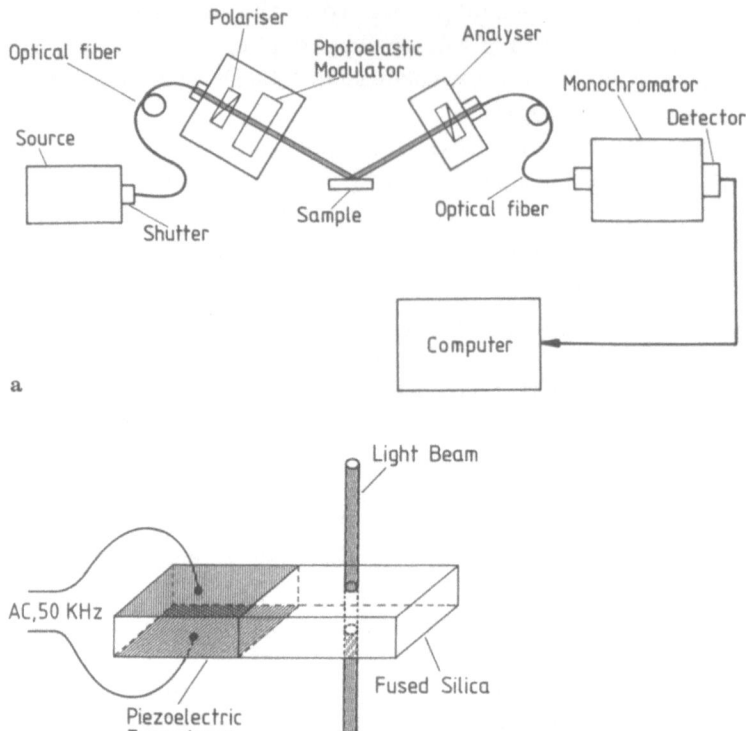

Fig. 3.5a. A schematic optical setup of an ellipsometer using a photoelastic modulator. The optical fibres allow an adaption to various growth chambers, **b** sketch of a photoelastic modulator. It consists of two parts one is built from crystalline quartz the other from quartz glass. The electrical contacts to the crystalline quartz are indicated

But, as stated already in the introduction, the only requirement to the incident light is that its polarisation state is known. However, equivalent results can be obtained with the PEM modulating the reflected light instead.

The detector signal has a form similar to that of the rotating analyser [compare (3.29)]:

$$I(t) = I_0[a_0 + a_2 \sin \delta(t) + b_2 \cos \delta(t)] \tag{3.52}$$

where a_0, a_2, and b_2 are functions of the polariser angle P, analyser angle A, the angular position M of the PEM and the ratio of the Fresnel coefficients $\rho = \tan \psi \, \exp(\mathrm{i}\,\Delta)$.

If, for example, the angles A and P are chosen to be 45° and $M = 0$ then

$$a_0 = 1 \tag{3.53}$$

$$a_2 = \sin(2\psi)\sin\Delta \qquad\qquad (3.54)$$

$$b_2 = \sin(2\psi)\cos\Delta \qquad\qquad (3.55)$$

Therefore, from the determination of a_0, a_2, and b_2, ρ can be calculated. The terms $\sin\delta(t)$ and $\cos\delta(t)$ can be written as Fourier series in $\cos(n\Omega t)$ and $\sin(n\Omega t)$ with Bessel functions $J_n(\delta_0)$ as coefficients. If δ_0 is chosen such that $J_0(\delta_0) = 0$, then the signal can be approximated by (lowest order):

$$I(t) = I_0[a_0 + a_2 2J_1(\delta_0)\sin(\Omega t) + b_2 2J_2(\delta_0)\cos(2\Omega t)] \qquad (3.56)$$

In principle it is sufficient to determine the Ω and 2Ω components of the signal, where Ω is the AC frequency. The Fourier coefficients of these frequencies can be measured using a lock-in amplifier. However, it was found that the accuracy can drastically be enhanced by taking higher frequency components into account [3.21]. In order to pick up all possible Fourier coefficients, the signal is digitised 16 times for a complete cycle of the sine function by a fast AD-converter and processed by a Digital Signal Processor (DSP). Both are necessary because the sampling rate is rather high, namely 16 times 50 kHz (the modulation frequency of the PEM). Care has to be taken that the sampling rate is a multiple of the modulation frequency of the PEM in order to guarantee that the sampling is done at the same values of δ for all cycles of the vibration of the quartz crystal. The high frequencies (the 3Ω coefficient corresponds to a variation of the signal with 150 kHz) pose a significant problem for the electronic circuitry. The bandwidth of the amplifier must be quite high leading as a consequence also to a large noise contribution. Furthermore, the amplification of the signal has to be highly linear with a very low distortion.

3.2.3 Polarisers

The most commonly used polarisers are crystalline quartz Rochon prisms and calcite Glan prisms, since only these kinds of polarisers have a sufficiently high extinction ratio (10^{-5} or better) [3.18].

The quartz Rochon prisms are built from two wedges with the optical axes of the quartz perpendicular to each other. The incident light is split into two components. The component with the electric field vector perpendicular to the optical axis of the quartz in both wedges passes through the Rochon prism unaffected and is linearly polarised. The electric field vector of the other component is parallel to the optical axis in one of the wedges and perpendicular to it in the other. Hence, this component is deviated and linearly polarised perpendicular relative to the first component [3.18].

Glan prisms are also built from two parts separated by air (for UV applications) or a glue with a lower refractive index than the birefringent material used, usually calcite. In contrast to the Rochon prisms the optical axes in both parts are parallel to each other. For the Glan prism the wedge angle γ

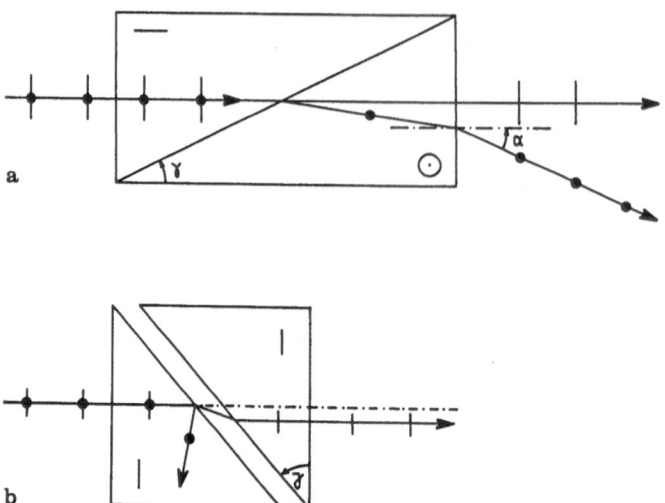

Fig. 3.6. A sketch of a Rochon prism (**a**) and a Glan(-air) prism (**b**). The directions of the optical axis are shown. The main difference between these two types of polarisers is that the optical axes in the two halves of the Rochon prisms are perpendicular to each other, while they are parallel in the Glan prism

(see Fig. 3.6b) is chosen such that the component with the higher refractive index is totally internal reflected and the other one is transmitted.

In choosing the prism type – Rochon or Glan – the material is also fixed. Calcite cannot be used for Rochon prisms because it is not possible to produce smooth surfaces for the optical axis in the surface and perpendicular to it. The smoothness of the surfaces is essential for the performance of the prism. Glan prisms, on the other hand, cannot be built from crystalline quartz since the loss of intensity of the transmitted component due to the reflections at the surfaces would be much too high.

The major disadvantage of using quartz is its high optical activity especially along the optical axis. The **E** vector of linearly polarised light with, e.g., $\lambda = 589.3\,$nm is rotated by $21.7°\,$mm^{-1}. Since the interface of the two wedges of the Rochon prism is not perpendicular to the direction of propagation, in the first half of the prism the rays of the light beam travel different distances depending on the location in the cross section of the beam. Therefore, the rotation of **E** varies over the cross section of the light beam mixing a large number of different polarisation states and the first half of the prism is said to be depolarising. In the second half of the prism the light propagates perpendicular to the optical axis, therefore the optical activity has little effect and the second half of the prism is polarising [3.18]. Due to the combined effect of optical activity and birefringence the unaffected component is not exactly linearly polarised but has instead a small ellipticity [3.7, 3.15]. This has to be taken into account in evaluating the data (Sect. 3.1).

Calcite, on the other hand, does not exhibit optical activity, but absorbs light in the UV region. The onset of the absorption edge is strongly dependent on the quality of the calcite. The useful spectral range typically ends at about 5 eV, which is for example the spectral position of the E_1-gap in cubic CdS. For (near) UV applications, quartz is better suited than calcite. Nowadays MgF_2 is used also as material for building Rochon-polarisers. This material is especially suited for VUV applications transmitting light with photon energies higher than 9 eV. The disadvantage of these polarisers is, however, that the technology of growing and polishing this material has not reached the same degree of sophistication compared with quartz and thus polarisers built with MgF_2 are much more expensive.

3.2.4 Calibration Procedures

Prior to all measurements, the angular positions of the polariser and analyser have to be calibrated, since the absolute angular positions of the transmission axis relative to the plane of incidence has to be known. In a procedure developed by Aspnes [3.7], the values P_s and A_s can be determined, when the transmission axis of the polariser and analyser are in the plane of incidence. Furthermore, for a calculation of the normalised Fourier coefficients from (3.29), the DC component a_0 must be determined. Unfortunately, all low-pass filter elements and amplifiers induce a phase shift and may reduce the high-order Fourier coefficients like a_2, b_2 relative to a_0. Aspnes introduced the parameter η which should compensate for the amplitude reduction of the high-order Fourier coefficients relative to the DC-component and an angle A_F (3.33), which represents the phase shift of the signal due to the electronics [3.8].

The unknown parameters P_s, A_s, A_F, and η can all be determined in one calibration procedure described in detail in [3.7]. The procedure is based on the fact that light linearly polarised in the plane of incidence is reflected as linearly polarised light in the same plane, if the sample is not optically active.

3.2.5 Experimental Limits

There are several parameters which can limit the accuracy of the data. So far as the ellipsometer itself is concerned, there are two main origins of systematic errors. One is the angle of incidence and the other is the influence of windows in applications, where the sample must be kept in a certain ambient, such as an UHV chamber.

3.2.5.1 Angle of Incidence. The angle of incidence ϕ_0 is a parameter, which strongly affects the accuracy of the data as shown in Fig. 3.7. Using the same Fourier-coefficients, but different angles of incidence leads to large differences in the dielectric function calculated using (3.40).

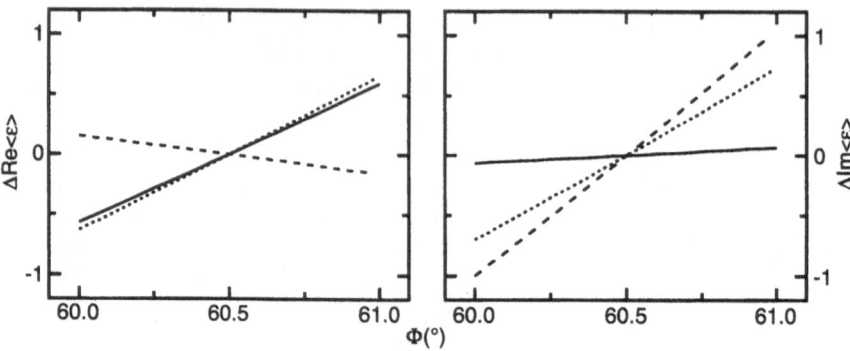

Fig. 3.7. The change of Re<ϵ> and Im<ϵ> for GaAs at 2.0 (solid line), 3.0 (small dashes), and 4.8 eV (long dashes) when using wrong values for the angle of incidence. The values ΔRe<ϵ> and ΔIm<ϵ> were calculated with the dielectric function of GaAs (taken from [3.3]) as input using (3.40) and a true angle of incidence of 60.5°

The problem is that the angle of incidence is very difficult to determine with an accuracy better than the resolution in <ϵ>. For SiO_2 on Si with known dielectric function for layer and substrate, some authors fit the measured effective dielectric function by varying ϕ_0 as well as the layer thickness. For non-transparent layers this procedure obviously does not work. One solution to this problem is that one measures instead the angle between the optical benches on which the elements of the ellipsometer are mounted. If one assumes that the light propagates in the middle axis of the benches or parallel to that axis, the angle between the benches is twice the angle of incidence. This is only a good assumption if the alignment is made accurately, the light path is long, the diameter of the light beam is small, and the light is nearly non divergent. One can estimate that all the methods to determine ϕ_0 have an accuracy of approximately 0.05°. This error in ϕ_0 results in a significant inaccuracy of the dielectric function as is demonstrated in Fig. 3.7. At 4.8 eV, for example, the true value of Im<ϵ> is 25.18 [3.3]. An uncertainty of 0.05°corresponds to an error of 0.1 (0.4%) in Im<ϵ>. This is much larger than the resolution in the values of the dielectric function measured, which is limited by the AD-converter used (normally 16 Bit 4.5 valid digits, 12 Bit 3.5 valid digits). The reproducibility of the spectra, however, is much better. Two measurements of the same sample lead to spectra, which plotted together cannot be distinguished. Hence, if the alignment is performed properly the reproducibility of ϕ_0 is much better than 0.05°.

3.2.5.2 Influence of the Windows. Windows in a setup may affect the polarisation by several mechanisms. First, there may be residual strains in the window from e.g. the glass blowing process or the polishing. These strains might vary across the window. In UHV applications, there are additional sources for strain. Every time the chamber is baked new strains due to temperature gradients in the window can build up. The largest effect, however, arises from the pressure gradient across the window caused by air on one side and UHV on the other. This results in a strain, which should be nearly centrosymmetric, but tensile on one side and compressive on the other side.

The former effects should be negligible when all elements are properly aligned and the baking process is done carefully. But the effect of the pressure gradient can be significant, especially in spectral regions where the samples are hardly absorbing e.g. silicon below the E_1-gap.

3.2.6 Trends and New Developments

Many attempts have been made to improve the experimental setups and the data evaluation of ellipsometric spectra. One important result is the extension of the accessible spectral range to the IR [3.4,3.22–3.24] and far-IR [3.25] in order to determine the lattice contribution to the dielectric function. Another important development was the extension to the vacuum-UV spectral region [3.26] in order to characterise wide band gap materials like Si_3N_4 or diamond by ellipsometry.

New types of ellipsometers were developed in the last years which are reviewed in several papers [3.27–3.29]. Using a special setup [3.28], the Mueller matrix of a sample can be measured, yielding information e.g. about the depolarisation effect of a sample (compare Sect. 3.5.6). Woollam combined the possibility of varying the angle of incidence and a spectroscopic feature (VASE) [3.30]. For anisotropic samples, the ability to vary the angle of incidence is essential. For multilayer structures, the variation of the angle of incidence provides a data set necessary for a multilayer analysis (Sect. 3.4.6). The disadvantage of varying the angle of incidence is that the spot size and for non-ideal alignment, the location of the spot can vary, therefore, the samples should be laterally homogeneous.

Fast ellipsometers have been constructed for applications, where short response times are needed, e.g., spontaneous crystallisation or initial stages of oxidation. The best time resolution reached so far is 25 nsec (single wavelength) [3.31].

3.3 Interpretation of the Effective Dielectric Function

In general the sample under investigation is not homogeneous, nor optically isotropic with a sharp interface to the ambient. While this is obvious for layered materials, as pointed out earlier, substrate materials may also be contaminated and oxidised or may have a reconstructed surface. Therefore, the notation $<\epsilon>$ was introduced [3.3] for an averaged or effective dielectric function. In this Section, the basics for the evaluation and interpretation of $<\epsilon>$ are presented and procedures are described for the derivation of layer dielectric functions ϵ_{layer} from the measured $<\epsilon>$. Application to some selected examples is given in the next Section.

3.3.1 Examples of Dielectric Functions

Figure 3.8 shows the dielectric function for (001) orientated GaAs prepared by chemically etching and measured in a chemically inert nitrogen atmosphere [3.3]. Four major features are visible in $Im<\epsilon>$ for GaAs termed E_1 (2.9 eV), $E_1 + \Delta_1$ (3.1 eV), E_0' (4.6 eV), and E_2 (5.0 eV). They are caused by van-Hove singularities of the Joint Density of States (JDOS) often referred as interband critical points or briefly as optical gaps [3.32–3.34]. The interband critical points are also indicated in the bandstructure shown in Fig. 3.9.

For comparison Fig. 3.10 shows the dielectric function of silicon. For silicon the E_0'-gap (3.3 eV) is very close to the E_1-gap (3.4 eV) and only a single feature is visible in ϵ_{Si}. The spin orbit splitting is small and thus the $E_1 + \Delta_1$-gap can also not be resolved at room temperature. The second peak corresponds to the E_2-gap (4.2 eV) of silicon and occurs at lower energies compared to GaAs.

In Fig. 3.11 the imaginary part of the dielectric functions of α-tin, InSb [3.3], and CdTe [3.36] is shown. All the chemical elements — Sn, In, Sb, Cd, and Te — are in the same period of the periodic table and the materials are (nearly) latticed matched to each other. The lineshapes of the dielectric functions show remarkable similarity, but the gap positions are shifted to higher energies and the absolute values of $Im\epsilon$ is decreasing from α-tin to CdTe corresponding to their ionicity. Such similarity of dielectric functions can serve as a guideline for the interpretation of measured dielectric functions of ternary materials the dielectric functions of which are not reported in the literature.

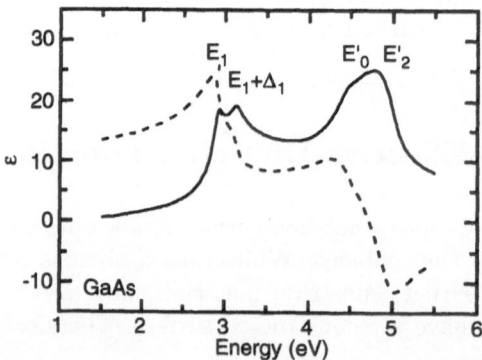

Fig. 3.8. The complex dielectric function of GaAs after [3.3]. (solid line: imaginary part, dashed line: real part) Labeled are E_1, $E_1 + \Delta_1$, E_0', and E_2, which are the critical points of GaAs in the visible and UV spectral region

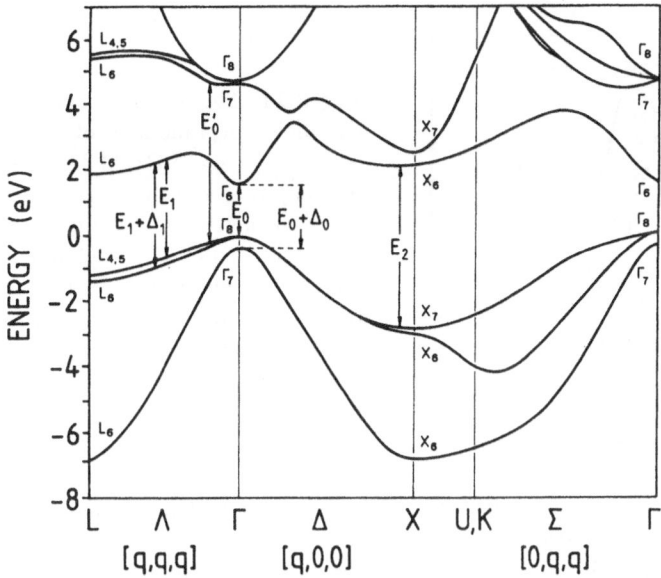

Fig. 3.9. The bandstructure of GaAs [3.35]. The critical points E_1, $E_1 + \Delta_1$, E_0', and E_2 are highlighted. The E_2-gap has at least two contributions of transitions near X and Σ [3.33]

The dielectric functions of a number of other materials have been reported. In Table 3.1 the most important semiconductors are listed with the corresponding reference of the dielectric function. Dielectric functions of other materials can be found in [3.37–3.39].

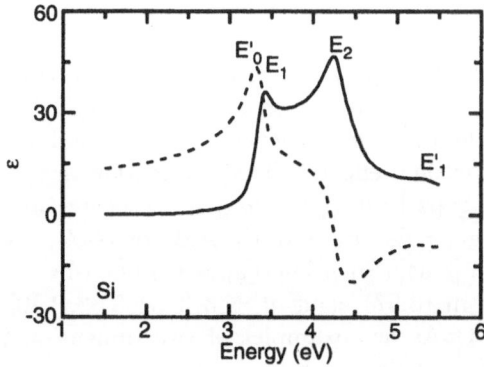

Fig. 3.10. The complex dielectric function of Si. (solid line: imaginary part, dashed line: real part) Labeled are the features E_0', E_1, E_2, and E_1' which are the interband critical points of Si in the visible and near UV spectral region [3.3]

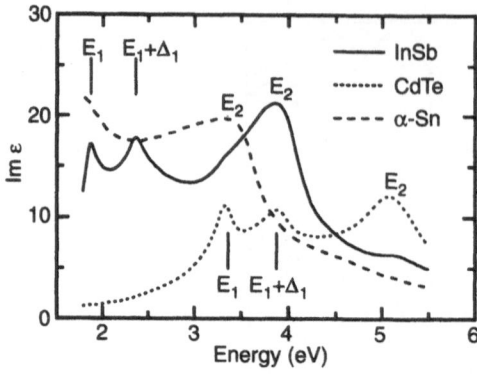

Fig. 3.11. The imaginary part of the dielectric functions of α-tin, InSb [3.3], and CdTe [3.36]. The lineshapes of the dielectric functions are similar, but the gap positions are shifted to higher energies and the absolute values of Imε are decreasing from α-tin to CdTe with increasing ionicity

3.3.2 Lineshape Analysis of Optical Gaps

Since the 1960's, the spectral positions and broadening of the optical gaps were determined by modulation techniques like photoreflectance and electroreflectance [3.40, 3.33]. It was empirically found, that the lineshape of the optical transitions can be described using functions of the form [3.34, 3.40, 3.33]:

$$\epsilon'' = A e^{i\phi} (E - E_g + i\Gamma)^{-n} \tag{3.57}$$

with ϕ, n listed in Table 3.2. The second derivative of the dielectric function ϵ'' with respect to energy was chosen instead of ϵ for the fitting procedure in order to reduce the influence of artefacts like oxide overlayers (Sect. 3.4.1.3).

In (3.57), A is the amplitude, E_g the gap energy, and Γ the broadening parameter of the optical transition. The exponent n depends on the dimensionality of the optical transition. The angle ϕ was introduced, such that all functions describing the lineshape for the various kinds of optical gaps can be written in one form [3.34].

For values of ϕ differing from 0°, 90°, 180°, and 270° a mixing between Reε and Imε takes place indicating that there are many particle contributions to ϵ [3.34, 3.42]. For GaAs and many other semiconductors ϵ is modified by the electron hole interaction (excitonic effects). This interaction drastically increases the JDOS near the gaps leading to sharper structures and higher values of Imε. The optical gaps E_0, E_1 and $E_1 + \Delta_1$ of GaAs are dominated by these excitonic effects [3.34]. Such lineshapes modified by excitonic effects can also be included in (3.57) using $n = 3$ [3.34, 3.33, 3.40]. The so called E_0'- and E_2-gaps of GaAs are examples of two dimensional (2D) interband critical points. A different approach to the lineshape analysis is discussed in [3.41] where the E_1 interband critical point is considered to be composed of a 2-dimensional critical point and a 3-dimensional critical point.

Table 3.1. List of some materials for which the dielectric functions can be found in the corresponding references

Material	Reference	Material	Reference
Si	[3.3, 3.44]	WSi_2	[3.87, 3.88]
Si (heavily doped)	[3.45, 3.46]	$MoSi_2$	[3.87]
μc-Si	[3.47, 3.48]	$CoSi_2$	[3.89, 3.90]
a-Si	[3.45, 3.48]	$NiSi_2$	[3.91, 3.90]
Ge	[3.3, 3.49]	Al_2O_3	[3.92]
a-Ge	[3.50]	SiO_2 (fused)	[3.94]
μc-Ge	[3.22]	Si_3N_4	[3.39]
a-Sb	[3.51]	GaN	[3.68]
p-Sb	[3.51–3.54]	GeO_2	[3.95]
c-As	[3.55]	GaAs oxide	[3.93, 3.45]
a-As	[3.55]	GaSb oxide	[3.93]
Bi	[3.56]	GaP oxide	[3.93]
Ga	[3.57]	InSb oxide	[3.96]
Al	[3.58]	AlF_3	[3.92]
Graphit	[3.59]	CaF_2	[3.97]
a-Te	[3.60]	CeF_3	[3.92]
a-C	[3.61]	LaF_3	[3.92]
SiGe	[3.62]	HfF_4	[3.92]
SiC	[3.63]	ScF_3	[3.92]
AlSb	[3.64]	ThF_4	[3.98]
AlAs	[3.65]	YF_3	[3.92]
GaAs	[3.3, 3.66, 3.67]	HfO_2	[3.92]
GaP	[3.3, 3.67]	ZrO_2	[3.92]
GaSb	[3.3]	Sc_2O_3	[3.92]
Ga_2Se_3	[3.69, 3.70]	Y_2O_3	[3.92]
InP	[3.3, 3.67]	ThO_2	[3.92]
InAs	[3.3]	H_2O	[3.99]
InSb	[3.3, 3.71]	$Al_xGa_{1-x}As$	[3.100, 3.66]
ZnSe	[3.72–3.74]	$GaAs_xP_{1-x}$	[3.100]
ZnTe	[3.78]	$In_{0.53}Ga_{0.47}As$	[3.101–3.103]
ZnS	[3.74, 3.75]	$In_{0.52}Al_{0.48}As$	[3.103]
CdS	[3.76, 3.74]	$In_{1-x}Ga_xSb$	[3.104]
CdTe	[3.36, 3.80]	ZnS_xSe_{1-x}	[3.105]
CdSe	[3.77, 3.79, 3.74]	$ZnSe_xTe_{1-x}$	[3.106]
PdO	[3.81]	$Zn_xCd_{1-x}Te$	[3.107]
PbTe	[3.82]	$Hg_xCd_{1-x}Te$	[3.36, 3.108]
ZnO	[3.83, 3.74]	$Cd_xMn_{1-x}Te$	[3.109]
CdO	[3.84, 3.74]	InGaAsP	[3.110, 3.102, 3.67]
GeS	[3.85]	$(Al_xGa_{1-x})_{0.5}In_{0.5}P$	[3.112]
a-GeSe	[3.86]	$Bi_2Sr_2CaCu_2O_8$	[3.111]

Table 3.2. All possible combinations of ϕ and n defined by (3.57) for the van-Hove singularities of the Joint Density Of States (JDOS)

dimensionality	kind of singularity	ϕ	n
3D	minimum	0°	3/2
	saddle point	90°	
	saddle point	180°	
	maximum	270°	
2D	minimum	0°	2
	saddle point	90°	
	maximum	180°	
1D	minimum	270°	5/2
	maximum	0°	5/2
excitonic			3

The spectral positions of the optical gaps depend on the bandstructure of the material and thus are characteristic for the material. Therefore, the chemical composition of a substrate or thick layer can be determined from the spectral position of the gaps (Sects. 3.4.1, 3.4.2.1).

From the values of Γ, defect concentrations in the material can be estimated as will be discussed in Sect. 3.4.1.2. The smallest values of Γ should be found for the material displaying the best crystalline quality.

3.3.3 Direct Inspection of $<\epsilon>$

In many cases, a direct interpretation of the measured $<\epsilon>$ is possible without carrying out complicated evaluation procedures like single or multilayer analysis. Oxidation, for example, leads to a spetral shift in Im$<\epsilon>$ near the E_2-gap of Si or GaAs as can be seen in the simulation shown in Fig. 3.12.

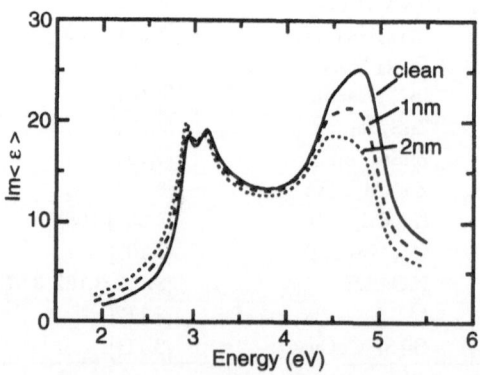

Fig. 3.12. Calculated Im$<\epsilon>$ for GaAs with oxide layers on top with thicknesses 1 nm (long dashes) and 2 nm (short dashes). The calculation was performed using a 3-phase model (see below) vacuum-oxide-bulk from the data of [3.3, 3.113]

The same is observed in the presence of other transparent materials like polymers. Knowing the decrease of Im$<\epsilon>$ near E_2, therefore, allows an estimate on the film thickness. For thicknesses beyond a few nm a decrease of the gap positions occurs. Therefore, care has to be taken in analysing the spectral positions of the optical gaps using the measured $<\epsilon>$.

Regarding the crystallinity of the material, a rough estimate can also be derived from direct inspection of $<\epsilon>$. It can be stated as a rule of thumb that the sample of highest crystalline quality of a given material should show the highest values of Im$<\epsilon>$ near the gaps [3.43]. This can be understood by analogy to the harmonic oscillator. With increasing damping, the height of Im$<\epsilon>$ at the eigenfrequency of the oscillator decreases. The electronic system can loose energy to the crystal e.g. via non-radiative recombination involving defect states or lattice vibrations. For example, in the case of ion implantation, where defects are created, the height of Im$<\epsilon>$ near the gaps decreases (Sect. 3.4.1.2). For high defect concentrations the translational symmetry is broken and finally amorphous material is created. Furthermore, surface roughness also leads to a reduction of the observed Im$<\epsilon>$ near the gaps.

3.3.4 Single Layers on a Substrate

The most common case studied is the characterisation of single layers with either unknown ϵ_{layer} or thicknesses d_{layer}, where the incident light penetrates into the substrate underneath. This analysis can be extended quite easily to the more general case of a layered structure (Chap. 5 about FIR spectroscopy of this book or [3.5]).

3.3.4.1 The 3-Phase Model.
The ratio ρ for a layer (1) on a substrate (2) in vacuum (0) for an angle of incidence ϕ_0 assuming sharp interfaces (the so called 3-phase model) is given by [3.5]:

$$\rho = \frac{r_{01p} + r_{12p}e^{i\,2\beta}}{1 + r_{01p}r_{12p}e^{i\,2\beta}} \frac{1 + r_{01s}r_{12s}e^{i\,2\beta}}{r_{01s} + r_{12s}e^{i\,2\beta}} \tag{3.58}$$

In this equation r_{ijp}, r_{ijs} are the Fresnel coefficients for the reflection of p- and s-polarised light at the interface between phase $i = 0, 1$ to phase $j = 1, 2$. The Fresnel coefficients can be expressed by:

$$r_{ijp} = \frac{\epsilon_j\sqrt{\epsilon_i - \epsilon_0 \sin^2 \phi_0} - \epsilon_i\sqrt{\epsilon_j - \epsilon_0 \sin^2 \phi_0}}{\epsilon_j\sqrt{\epsilon_i - \epsilon_0 \sin^2 \phi_0} + \epsilon_i\sqrt{\epsilon_j - \epsilon_0 \sin^2 \phi_0}} \tag{3.59}$$

and

$$r_{ijs} = \frac{\sqrt{\epsilon_i - \epsilon_0 \sin^2 \phi_0} - \sqrt{\epsilon_j - \epsilon_0 \sin^2 \phi_0}}{\sqrt{\epsilon_i - \epsilon_0 \sin^2 \phi_0} + \sqrt{\epsilon_j - \epsilon_0 \sin^2 \phi_0}} \tag{3.60}$$

where ϵ_0, ϵ_1, and ϵ_2 are the complex dielectric functions of 0(vacuum), 1(layer), and 2(substrate). The phase shift β is given by

$$\beta = 2\pi \frac{d_{\text{layer}}}{\lambda} \sqrt{\epsilon_{\text{layer}} - \epsilon_0 \sin^2 \phi_0} \tag{3.61}$$

where λ is the wavelength of light in vacuum. From ρ calculated with (3.58) an effective dielectric function $<\epsilon>$ is calculated by using (3.40). Equation (3.58) is also obtained from the more general expressions for multilayers on a substrate given in Chapt. 5 of this book about Far-infrared spectroscopy by using the following replacements:

$$r_{01} = \rho_{a2} \tag{3.62}$$
$$r_{12} = r_{2b} \tag{3.63}$$
$$\phi_2^2 = \exp(i\, 2\beta) \tag{3.64}$$

3.3.4.2 Determination of Layer Properties. Different situations may be considered in the analysis of a layer on a substrate. The easiest task is to determine the unknown thickness d_{layer} of the layer with known dielectric function ϵ_S for substrate and layer ϵ_{layer}. One has to calculate an effective dielectric function $<\epsilon>_{\text{calc}}$ by a 3-phase-model (vacuum-layer-substrate) using (3.58) and (3.40) with the layer thickness d_{layer} as a parameter and vary d_{layer} until the function Θ defined by

$$\Theta = |<\epsilon> - <\epsilon>_{\text{calc}}|^2 \tag{3.65}$$

is minimised for all energies. For this purpose well known routines like the Newton or Levenberg-Marquardt one are applied [3.114]. In the case of known layer thickness d_{layer} and dielectric function ϵ_S of the substrate, but unknown ϵ_{layer} of the layer, the analysis is analogous.

In the general case with unknown d_{layer} and unknown ϵ_{layer} the determination of these parameters is much more difficult, because only two parameters are measured for each wavelength and three have to be determined. Several guidelines can help to treat this problem: (i) ϵ_{layer} for layers with similar thickness should be similar, (ii) Im ϵ_{layer} below the fundamental gap should be small, (iii) structures from the substrate optical gaps should not appear in ϵ_{layer}. The last condition requires further explanation. The optical gaps are specific for the material and differ in the spectral position. If d_{layer} is chosen too large, the calculated layer ϵ_{layer} "incorporates" the structures of the substrate. If a too small d_{layer} is chosen, ϵ_{layer} has to compensate this and the resulting structures will also be visible in ϵ_{layer}. However, the last point is very stringent and it is not always possible to eliminate the substrate contribution by variation of the thickness. For example, if the ϵ_S used does not represent well the underlying substrate, or if an interlayer is formed during growth, the elimination of the substrate structures in ϵ_{layer} is usually not successful (Sect. 3.4.2.3).

3.3.4.3 Ultrathin Layers. For very thin layers the effective dielectric function can be approximated by [3.51]:

$$<\epsilon> = \epsilon_{sub} + \frac{4\pi i d}{\lambda} \sqrt{\epsilon_{sub} - \sin^2\phi_0} \; \frac{\epsilon_{sub}(\epsilon_{sub} - \epsilon_{layer})(\epsilon_{layer} - 1)}{\epsilon_{layer}(\epsilon_{sub} - 1)} \qquad (3.66)$$

This formula can be easily solved for ϵ_{layer} and thus allows the determination of the dielectric function of very thin layers. The upper limit for the layer thickness in order to guarantee the validity of the formula depends on the system and is in the order of magnitude of 1 nm.

3.3.5 Inhomogeneous Layers

The effective dielectric function of lateral inhomogeneous layers composed of different materials or phases depends on the dielectric functions of all constituents and their geometrical arrangement which is the topology of the layer. Theories like Maxwell-Garnett [3.115], Looyenga [3.116] or Bruggeman [3.117] allow a calculation of the effective dielectric function $<\epsilon>$ from known dielectric functions of the constituents for a particular topology (Maxwell-Garnett: isolated spheres in a matrix). In order to apply these effective dielectric functions it is necessary that the dimensions for all constituents (besides the matrix material) are small compared to the wavelength of light, but large enough to have the same dielectric function as the bulk material. However, for microscopic rough surfaces some authors succeeded in fitting their spectra with the Bruggeman effective medium theory [3.118].

For the situation of particles with dielectric function ϵ_p, embedded with a volume fraction f, in a matrix M with dielectric function ϵ_M, these theories allow $<\epsilon>$ to be calculated using:

Looyenga [3.116]:
$$\sqrt[3]{<\epsilon>} = f\sqrt[3]{\epsilon_p} + (1-f)\sqrt[3]{\epsilon_M} \qquad (3.67)$$

Maxwell-Garnett [3.115]:
$$\frac{<\epsilon> - \epsilon_M}{<\epsilon> + 2\epsilon_M} = f\frac{\epsilon_p - \epsilon_M}{\epsilon_p + 2\epsilon_M} \qquad (3.68)$$

Bruggeman [3.117]:
$$f\frac{\epsilon_p - <\epsilon>}{\epsilon_p + 2<\epsilon>} + (1-f)\frac{\epsilon_M - <\epsilon>}{\epsilon_M + 2<\epsilon>} = 0 \qquad (3.69)$$

Whether the assumption that the particles (and the matrix) have the bulk dielectric function is fulfilled depends on the material of the particles and their size. In the case, where the diameter of the particles is very small size effects may occur, and the dielectric function may differ significantly from that of the corresponding bulk value (Sects. 3.4.1.4, 3.4.4).

In order to determine whether the above simple theories can be used, one has to apply a more sophisticated theory like the Bergman theorem. In this theory, the effective dielectric function $<\epsilon>$ can be calculated from the dielectric function of the particles ϵ_p embedded in the matrix with dielectric function ϵ_M [3.119, 3.121]:

$$<\epsilon> = \epsilon_M \left(1 - f \int_0^1 \frac{g(n)}{t - n} \, dn \right) \tag{3.70}$$

with the definition

$$t = \frac{\epsilon_M}{\epsilon_M - \epsilon_p} \tag{3.71}$$

where f is the filling factor. The function $g(n)$ describes the topology of the system and is independent of the dielectric functions of the constituents. If the absolute value of t is much greater than 1, the denominator in the integral of (3.70) does not become resonant and in this case $<\epsilon>$ is not sensitive to $g(n)$, i.e. to the geometry of the system and the simpler theories are applicable. For details see [3.121] or Chapt. 5 of this book about Far-infrared spectroscopy.

If the wavelength of light is in the order of the dimensions of the constituents, the situation is much more complicated. Then the samples produce stray light, which is often unpolarised or mixed polarised. There are no general theories available for this case. The other limiting case, where the wavelength is shorter than the dimensions of the particles, is investigated extensively for instance by Grosse et al. in the IR spectral region [3.121].

3.4 Characteristic Experimental Examples

In this Section examples are presented to highlight the capability of ellipsometry and to clarify the procedures described in Sect. 3.1. Of course, no complete overview on ellipsometry can be given, instead the examples presented demonstrate the wide range of problems which can be treated with spectroscopic ellipsometry.

First, the influence of several parameters like strain, sample temperature, or defects on the spectral positions and broadening of the interband critical points is discussed. At present ellipsometric investigations deal mostly with semiconductor heterostructures of which a few examples are discussed. Thereafter, an application of an effective medium theory is given. These investigations were all performed *ex-situ*, i.e. after growth of layers.

There are several problems which may occur in the interpretation of *ex-situ* measurements. Quite often interfaces are not abrupt and/or interlayers might have formed. The surface may become rougher with increasing layer thickness and the layers may grow inhomogenously in depth. For an analysis of such samples one often needs more information on the sample than achievable by *ex-situ* investigations. Furthermore, the sample may change due to oxidation after growth and before measurement.

These problems can be partially circumvented by performing *in-situ* measurements, where the dielectric function of the substrate used can be measured and the growth of the layers can be monitored. Furthermore, processes that effect the sample like contamination during or after growth can be observed. Three examples of *in-situ* studies and a multilayer analysis will end this Section.

3.4.1 Interband Critical Points

The electronic structure and consequently the dielectric function is influenced by several internal and external parameters like defects, strain or temperature. In this Subsection, the changes induced in the spectral positions and broadenings of the optical gaps are discussed. This knowledge is important for the evaluation of the experimental spectra and is used to distinguish, for example, between (quantum) size effects and strain [3.122].

3.4.1.1 Influence of Temperature. Semiconductor epitaxial layers are usually grown at elevated temperatures. The growth temperatures often vary during the growth process, e.g., buffer layers are grown at lower temperatures. The dielectric function depends strongly on the sample temperature and for *in-situ* characterisation this dependence as well as the sample temperature has to be known. On the other hand, if this dependence is known, the sample temperature can be determined from the spectral position and broadening of the gaps as is now discussed.

For semiconductors with diamond or zincblende structure, an increase in temperature is accompanied by a red shift of the spectral positions of the optical gaps. There are two principle reasons for the change of the optical gaps: the thermal expansion of the crystal and the electron-phonon interaction. The thermal expansion of the sample changes the bandstructure and consequently the transition energies. The second effect, the renormalisation of the states caused by the electron-phonon interaction, was found to be much more important for semiconductors with diamond or zincblende structure [3.34].

The change of the dielectric functions with temperature for Si, GaAs, and InP in the range from 22 K to 793 K is reported in [3.34]. Typical spectra for Si at various temperatures are shown in Fig. 3.13. With lower sample temperature the features in the dielectric function become much sharper, the height of Im$<\epsilon>$ near E_0', E_1, E_2 increases, and the gaps shift to higher energies.

The evaluation of the gap parameters versus temperature as shown in Fig. 3.14 for Si gives a quantitative picture of these trends. This dependence of the gap parameters on temperature was used to measure the temperature of sample surfaces in UHV chambers. The accurate determination of the sample temperature is always a problem, since thermocouples cannot have a direct contact to the sample surface. Otherwise the surface could be contaminated by the thermocouple. Pyrometers, on the other hand, give the temperature accurately above approximately 600 °C and only if calibrated for the specific

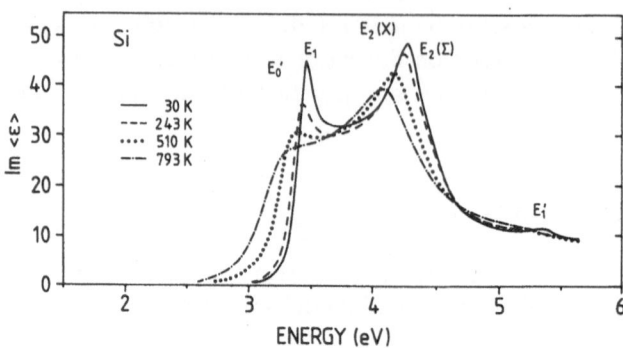

Fig. 3.13. The dielectric function of Si for various temperatures. Lowering the temperatures results in sharper features in the dielectric function and consequently, the values of the imaginary part of the dielectric function are strongly increasing [3.125]

material. This procedure to derive the temperature from the measured dielectric function was applied in [3.123] and good agreement was found between the temperature obtained by the gap parameters with those measured by a thermocouple close to the sample surface.

3.4.1.2 Influence of Defects: Si Implanted GaAs.

Ionimplantation is a well known technique for the doping of semiconductors or the introduction of recombination centres in order to achieve highly ohmic or even insulating material. For a certain implanted material the projected range which is the average penetration depth depends on both the mass and kinetic energy of the ions. Therefore, the doping depth can be controlled by choosing the kinetic energy of the ions. In general, the ions create defects (point defects like vacancies or interstitials and dislocations) in the implanted material. The defect density depends mainly on the dose of the ions for a fixed target material, while the kind of defects depends on the mass ratio of the ions and the atoms of the target material. Details can be found in [3.124].

The damage due to the implanted ions and the crystalline perfection after annealing has to be assessed by the investigation of ion-implanted semiconductors. From the ellipsometric point of view, it is interesting to see which parts of the spectra of $<\epsilon>$ are changed by defects, how sensitive ellipsometry is to changes caused by a small dose of implanted material and whether one can distinguish between severely damaged crystalline and amorphous material. If the relationship between the implanted dose and the change of the measured dielectric function is known as well as the defect concentration created by the ions, it would be possible to quantify the defect concentration of epitaxial layers.

In Fig. 3.15, $\mathrm{Im}<\epsilon>$ is plotted for $100\,\mathrm{KeV}$ Si^+_{29} as-implanted GaAs with doses ranging from $1\cdot10^{12}$ to $1\cdot10^{16}\mathrm{cm}^{-2}$ compared to an unimplanted sample

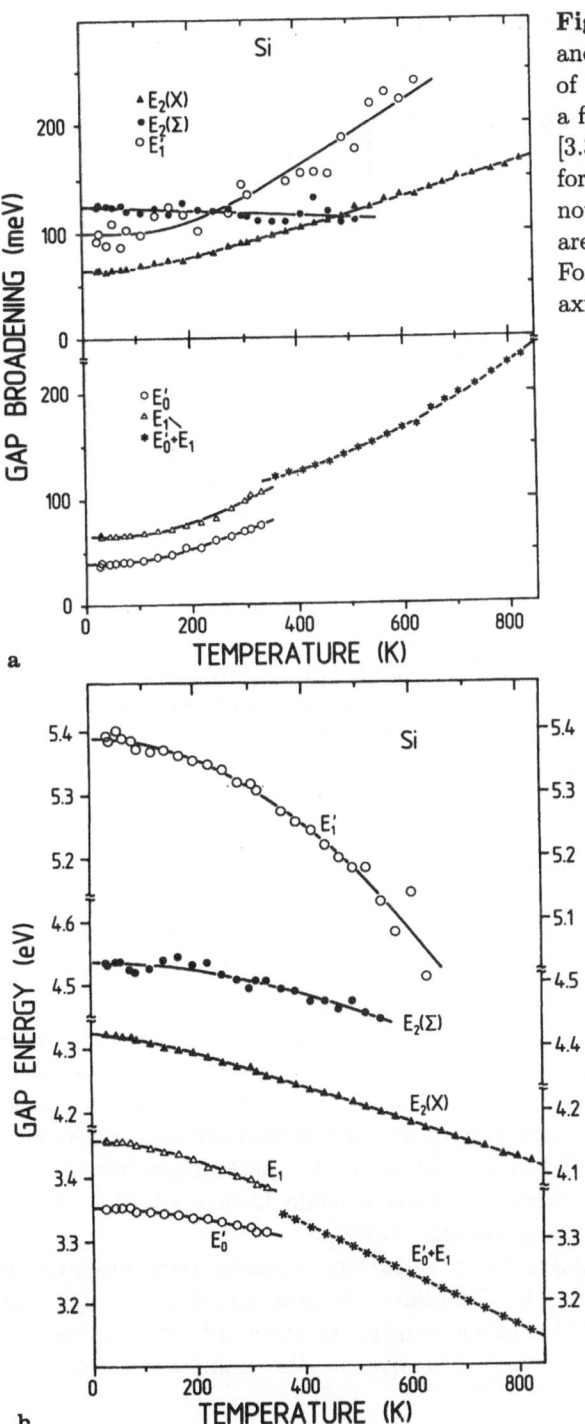

Fig. 3.14. Spectral position (**b**) and broadening (**a**) of the gaps of Si versus temperature using a fitting procedure described in [3.34]. The gaps E_0' and E_1 for elevated temperatures cannot be resolved and therefore are fitted as one feature $E_0' + E_1$. For simplicity, the gap energy axis is not continuous [3.125]

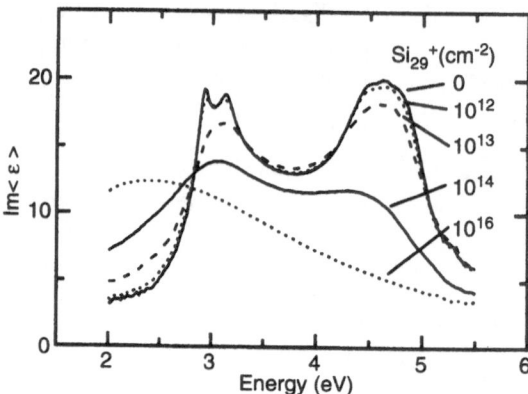

Fig. 3.15. Im$<\epsilon>$ for Si$_{29}^+$ implanted GaAs with 100 KeV for a dose ranging from $1 \cdot 10^{12}$ to $1 \cdot 10^{16}$cm^{-2} compared to an unimplanted reference sample. All samples were wet etched by a standard method before implantation. The imaginary part of the dielectric function at the gaps is reduced with increasing dose and becomes smoother [3.126]

as reference. Increasing the dose reduces the height of Im$<\epsilon>$ at the optical gaps and broadens them. Moreover, below the E_1-gap, between 2 and 2.5 eV, Im$<\epsilon>$ increases with implantation dose. Therefore Im$<\epsilon>$ in this spectral region seems to be most sensitive to defects.

For the highest dose $(1 \cdot 10^{16}$cm$^{-2})$, all original features in Im$<\epsilon>$ of crystalline GaAs have vanished leaving a broad maximum at about 2.5 eV. A smooth $<\epsilon>$ is always an indication for amorphous material (Sect. 3.4.1.4, [3.127]). The spectral position of the maximum, however, does not correspond to that expected for amorphous GaAs (about 3.5 eV) [3.128] or that found for high dose arsenic implanted GaAs [3.129]. This might be explained by considering the density of Si atoms in the implanted layer. The projected range for Si$_{29}^+$ in GaAs is about 65 nm [3.130]. For a dose of $1 \cdot 10^{16}$cm^{-2} one gets a volume density of approximately $1 \cdot 10^{21}$cm^{-3} Si atoms in the implanted region, which can be regarded as an alloy of Si and GaAs. This might cause a totally different electronic structure or at least heavily strained GaAs and may be the reason for the discrepancy between the spectral positions of the two maxima in Im$<\epsilon>$. Another explanation could be a weakening of the Ga-As-bonds due to voids in the material [3.127].

Implanted GaAs with doses up to $1 \cdot 10^{13}$cm^{-2} can be considered to be equivalent to GaAs with defects. Therefore, the evaluation of the spectral positions and broadening of the optical gaps for these samples is possible and a comparison of the values obtained with those for the unimplanted reference samples is reasonable.

Table 3.3. The spectral positions and broadening of the gaps E_1 and $E_1 + \Delta_1$ for the implanted GaAs compared to two unimplanted reference samples

dose[cm^{-2}]	E_1		$E_1 + \Delta_1$	
	E_g(eV)	Γ(eV)	E_g(eV)	Γ(eV)
reference	2.908	0.101	3.135	0.138
reference	2.907	0.103	3.134	0.139
$5 \cdot 10^{11}$	2.905	0.115	3.124	0.151
$1 \cdot 10^{12}$	2.906	0.122	3.125	0.161
$2 \cdot 10^{12}$	2.912	0.132	3.124	0.177
$3 \cdot 10^{12}$	2.916	0.143	3.123	0.193
$5 \cdot 10^{12}$	2.924	0.158	3.120	0.223
$1 \cdot 10^{13}$	2.957	0.192	3.107	0.258

Despite a scattering in the data detailed in Table 3.3 the general trend is, that the spectral separation of the two gaps E_1 and $E_1 + \Delta_1$ is reduced and that the broadening parameter Γ increases with defect density as expected.

3.4.1.3 Oxide Overlayers. For oxidised or contaminated samples, the spectral positions and broadening of the optical gaps calculated from the measured $<\epsilon>$ differ slightly from those for clean bulk material. This can be demonstrated by a simulation using an effective dielectric function calculated for a 1.5 nm thick oxide layer on top of crystalline GaAs following the procedure explained in Sect. 3.3.4.

Figure 3.16 shows the real part of the second derivative of $<\epsilon>$ with respect to the energy denoted as $<\epsilon>''$ for oxidised and clean (non-oxidised) GaAs. The values obtained by a fit using (3.57) to the gaps are listed in Table 3.4. The difference in the values of the spectral positions and the broadening of the E_1- and $E_1 + \Delta_1$-gaps calculated directly from $<\epsilon>''$ of clean and oxidised GaAs is small, although the measured $<\epsilon>$ is strongly influenced by oxide overlayers (Sect. 3.3.3).

3.4.1.4 Size Effects: Microcrystalline Si. Due to the decreasing dimensions used in modern semiconductor device fabrication, size effects become more and more important. These size effects influence the dieletric function and a qualitative understanding is therefor necessary. The characteristic length below which the dielectric function changes has to be determined. This length depends on the material under investigation.

There are very few studies about this subject. For GaAs it is reported that the optical gaps shift to higher energies with decreasing layer thickness while

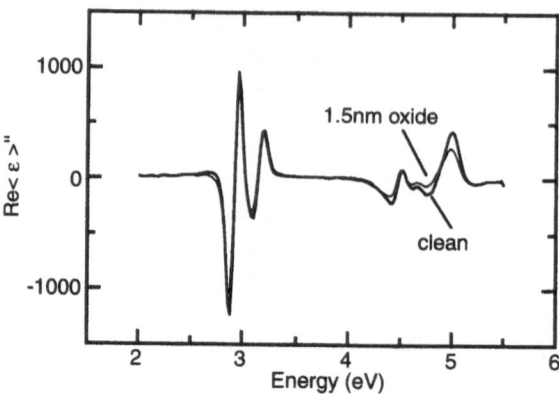

Fig. 3.16. Calculated real part of $<\epsilon>''$ for oxidised and clean (non-oxidised) GaAs. The values from Aspnes for ϵ for GaAs [3.3] and its oxide [3.113, 3.45] are used. The spectral position and the broadening of the gaps E_1 and $E_1 + \Delta_1$ calculated from the spectra of the "oxidised" GaAs are hardly distinguishable from that of clean GaAs

Table 3.4. Spectral positions and broadening of E_1 and $E_1 + \Delta_1$ from a fit using (3.57) to the spectra for clean and the simulated oxidised GaAs

	E_1		$E_1 + \Delta_1$	
	$E_g(eV)$	$\Gamma(eV)$	$E_g(eV)$	$\Gamma(eV)$
clean	2.910	0.102	3.132	0.140
oxidised	2.910	0.104	3.129	0.144

the shape of the spectra and consequently the broadening remains constant [3.131].

The spectral position and broadening of the optical gaps were investigated for poly- and microcrystalline (μc-Si) silicon layers as a function of the average crystallite diameter. The diameters were determined with convential and cross sectional Transmission Electron Microscopy (TEM) [3.122, 3.48]. Figure 3.17 shows the dependence of the spectral positions E_g and the broadening parameter Γ of the gaps on the crystallite diameter.

Two regimes can be distinguished. For crystals larger than 40 nm, neither the spectral position nor the broadening of the gaps were dependent on the crystal diameter. The discrepancies between the values obtained for E_g and Γ of the layers with large crystallites and bulk Si are attributed to strain in the layers. On the other hand, for diameters smaller than 40 nm, a blue shift of the $E_0' + E_1$ and a red shift of E_2 is observed. An increase of Γ for all gaps with decreasing crystallite size is found.

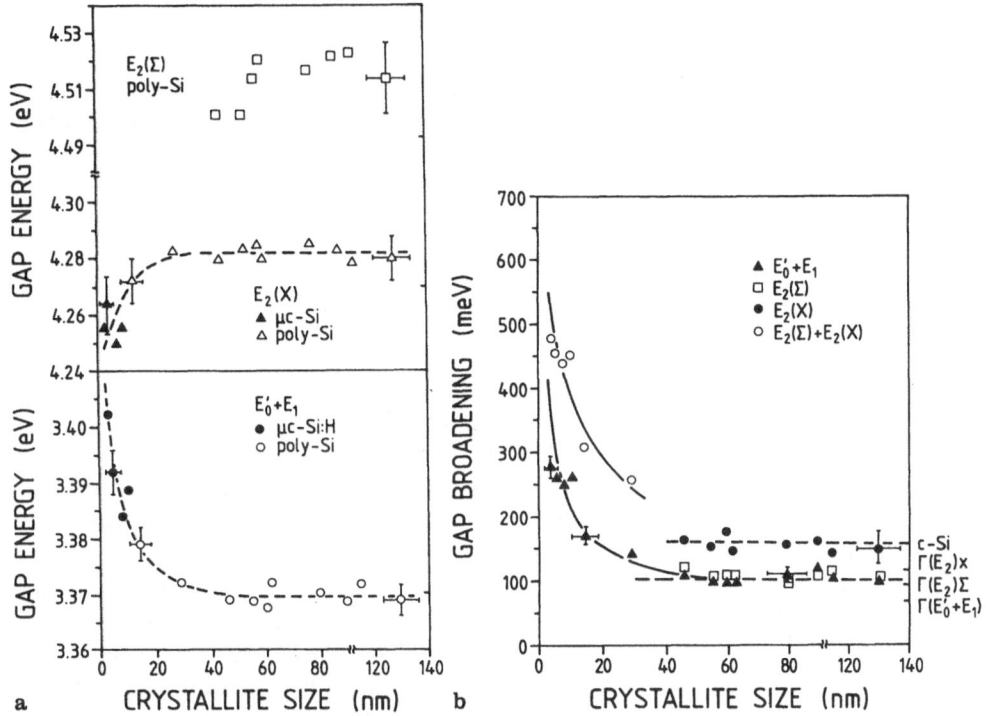

Fig. 3.17. Dependence of the spectral positions (**a**) and broadening (**b**) of the $E_0' + E_1$-, $E_2(X)$- and $E_2(\Sigma)$-gaps on the average crystallite size. Solid symbols correspond to μc-Si samples. Dotted lines are drawn to guide the eye, arrows indicate the corresponding volume reference values [3.122]

3.4.2 Semiconductor Heterostructures

Optoelectronic devices mainly employ semiconductor heterostructures. The optimisation of the growth conditions and the knowledge of the properties of semiconductor heterostructures is essential in producing high quality devices. Therefore numerous publications deal with these systems and in the following it will be shown that ellipsometry is a powerful tool for characterising semiconductor heterostructures.

The semiconductors most widely used for optoelectronic applications are compounds from elements of the third group (In, Ga, Al) and the fifth group (As, Sb, P) with direct gaps. The stoichiometry is chosen either that the layer and substrate are lattice matched e.g. $In_{0.53}Ga_{0.47}As$ on InP or that the overlayer is (biaxially) strained e.g. InGaAs on GaAs. The capability of choosing the stoichiometry is based on the high degree of perfection of the growth methods like MOVPE and MBE as discussed in Chapt. 2 about Analysis of Epitaxial growth.

The problem of the characterisation of heterostructures by ellipsometry is, that the dielectric function of only some of these materials are (well) known [3.102]. Consequently, there are no standards for their dielectric functions, except for a few like $Al_xGa_{1-x}As$ or $GaAs_{1-x}P_x$ [3.100] as described in the next Section.

3.4.2.1 AlGaAs, GaAsP.

For $Al_xGa_{1-x}As$ and $GaAs_{1-x}P_x$ layers grown on GaAs the dielectric functions given in [3.100] can serve as a standard. Their dielectric functions and the optical gaps depend on the composition parameter x. For instance, $Al_xGa_{1-x}As$ has a direct gap for $x \leq 0.4$ and an indirect gap if x is greater than 0.4. It is therefore possible to derive the composition from the spectral position of the optical gaps [3.3, 3.66].

In Fig. 3.18 the dielectric functions of $Al_xGa_{1-x}As$ are shown for various compositions x. It is obvious that the optical gaps E_1 and $E_1 + \Delta_1$ shift to higher spectral positions when x increases. In extracting the spectral positions from the spectra by a lineshape analysis (Sect. 3.3.2) one obtains the dependence between spectral position and composition x (see Fig. 3.19).

From this dependence plotted in Fig. 3.19 it is possible to get the composition x if the spectral positions of E_1 or $E_1 + \Delta_1$ are determined.

However, this requires that x is known for these reference spectra and therefore, the accuracy of an unknown x is limited by the knowledge of x for these standard samples. In [3.100], the value of x for these standard samples was determined by Auger electron spectroscopy.

3.4.2.2 InP on InGaAs.

The procedure to obtain layer dielectric functions as described in Sect. 3.3.4.2 was applied to the simple case of a very thin (top) layer supposed to be InP. As a result the thickness of the layer was found to be 2 nm and the corresponding layer dielectric function (imaginary part) is shown in Fig. 3.20.

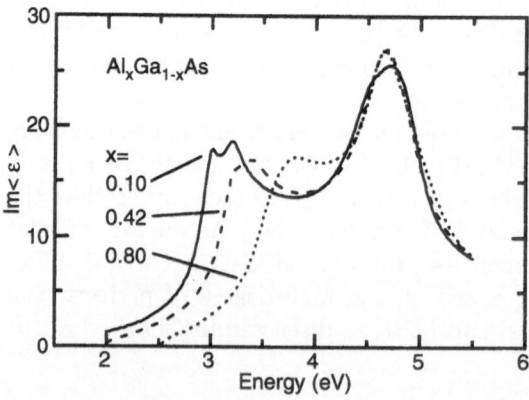

Fig. 3.18. Imaginary part of $<\epsilon>$ for $Al_xGa_{1-x}As$ for various compositions x

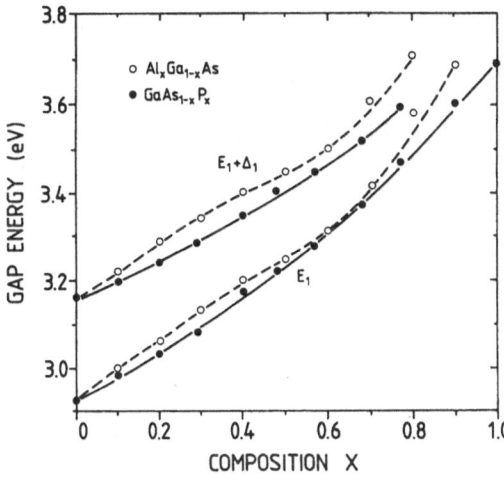

Fig. 3.19. Variation of the threshold energies of the E_1- and $E_1 + \Delta_1$-gaps with composition for $Al_x Ga_{1-x}As$ and $GaAs_{1-x}P_x$. The lines represent guidelines to the eye [3.100]

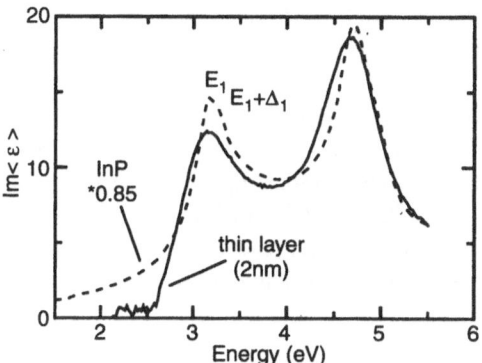

Fig. 3.20. Im ϵ_{layer} for a thin layer of supposedly InP grown on a $1.5\,\mu m$ thick InGaAs layer on a InP substrate. Since the InGaAs is thick it can be regarded as substrate for photon energies above 2.5 eV [3.101]. From the similarity of the shape of the thin layer dielectric function and that of InP [3.3] it can be concluded that the top layer is mainly InP

For comparison, the dielectric function of bulk InP is also shown. The similarity of these dielectric functions is obvious. However, the gaps of the thin layer are somewhat broader, most likely due to size effects [3.122] and slightly shifted to lower energies. These shifts can be explained by assuming that the layer is not only composed of In and P, but contains also some arsenic.

The spectral positions of the E_1- and $E_1 + \Delta_1$-gaps of a ternary material of the form $InAs_xP_{1-x}$ are between those of InAs (2.50 eV) and InP (3.15 eV). The incorporation of arsenic into the InP layer could arise during growth due to the so called carry over effect [3.132] from the arsenic precursor used for the growth of the underlying InGaAs layer.

3.4.2.3 CdS on InP. Using wide gap II-VI compound semiconductor layers on III-V substrates, it was recently possible to grow blue light emitting diodes [3.133, 3.134]. Their application as insulators in MIS-structures might also be possible. One example of a lattice matched II-VI compound on a III-V substrate is CdS on InP. Whereas bulk CdS has hexagonal wurtzite structure, epitaxially grown CdS on InP has cubic symmetry (zinc blende structure) induced by the substrate ([3.135, 3.136] and Chapt. 4 about Raman spectroscopy). Therefore, the dielectric function of the layer differs from that of bulk CdS. Thus, before an analysis of the interface properties can be performed, the dielectric function of the cubic CdS modification has to be determined. Care has to be taken that no interfacial layer formation occurs in the sample(s) from which the layer dielectric function is determined. Otherwise the data are not very accurate especially below the direct gap.

The measured $<\epsilon>$ for two CdS layers grown on InP without the presence of an interlayer are shown in Fig. 3.21. For this heterostructure the method of determining ϵ_{layer} and d_{layer} as described in Sect. 3.3.4.2 is demonstrated. Layer dielectric functions $\epsilon_{\text{layer}}^*$ for different layer thicknesses d_{layer}^* were calculated from one of the CdS/InP (110) spectra in Fig. 3.22 (solid line). For test layer thicknesses $d_{\text{layer}}^* = 20\,\text{nm}$ and $d_{\text{layer}}^* = 25\,\text{nm}$, the E_1 and $E_1 + \Delta_1$ features of the InP substrate (around 3 eV) appear as edges in the calculated $\epsilon_{\text{layer}}^*$. The

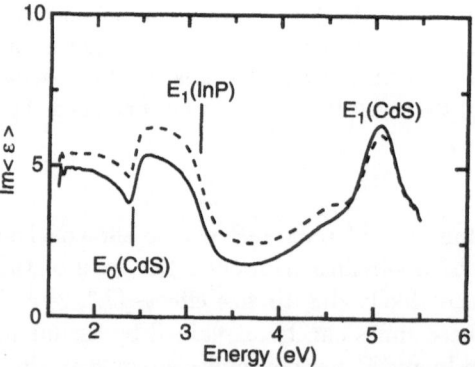

Fig. 3.21. Im$<\epsilon>$ for two typical CdS layers on InP(110) with thicknesses 23 nm (solid line) and 20 nm (dashed line). Indicated are the spectral position of the direct gap of CdS and the E_1-gaps of InP and CdS, respectively

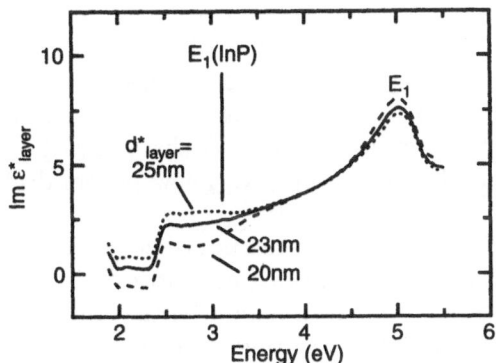

Fig. 3.22. Resulting $\epsilon^*_{\text{layer}}$ for the cubic CdS layer for test thicknesses $d^*_{\text{layer}} = 20$, 23, and 25 nm. The "best" ϵ_{layer} with negligible contribution of the substrate (InP-feature) is that with the layer thickness of 23 nm

correct value of d_{layer} must be in between these two values. For $d_{\text{layer}} = 23$ nm no substrate related features are visible in $\varepsilon^*_{\text{layer}}$. Consequently this is the "true" thickness d^*_{layer} of the CdS layer. The corresponding best layer dielectric function $\varepsilon^*_{\text{layer}}$ is shown in Fig. 3.23. This demonstrates the sensitivity of spectroscopic ellipsometry for the determination of layer thicknesses d_{layer}.

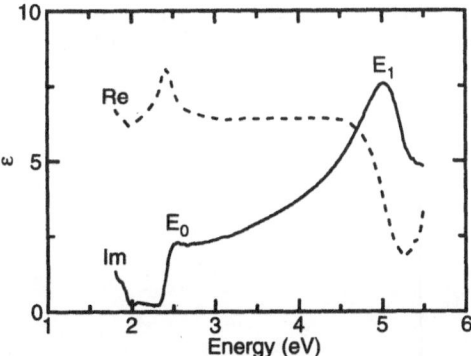

Fig. 3.23. The dielectric function of the cubic CdS layer corresponding to the best thickness of 23 nm in Fig. 3.22. The ϵ_{layer} has negligible contributions of the substrate dielectric function around the spectral position of the E_1-gap of InP. A lineshape analyses of ϵ_{layer} gives values of 2.41 eV for the direct gap E_0 and 5.16 eV for E_1

3.4.3 Strained Layers of InGaAs

Strained layers are important for many applications. The highest electron mobility for InGaAs has been reported for a strained layer [3.137] and strain can improve the efficiency of semiconductor lasers [3.138]. A characterisation of strain by ellipsometry is still a challenge and only very few papers deal with this problem.

The influence of strain on the dielectric function can be studied either by applying external stress or in heteroepitaxy by growing non-lattice matched material. If in the former case, the stress is applied in an arbitrary direction the degeneracy of the dielectric tensor is lifted resulting in three (different) dielectric tensor components, a case which is very difficult to analyse. If instead hydrostatic stress or uniaxial stress in the symmetry directions of the crystal e.g. along [001] is applied, then the problem is much easier. For hydrostatic stress the main effect is an energy shift of the optical gaps while the lineshape of the dielectric function is not or hardly affected. For uniaxial stress the lineshape is changed as well [3.139]. For heterostructures, however, the effect of biaxial strain is the most interesting case. The biaxial strain can be regarded as composed of hydrostatic and uniaxial strain. Therefore, the lineshape of the dielectric function and the spectral positions of the optical gaps change. This is shown in Fig. 3.24 for InGaAs on GaAs.

The energy shift of the optical gaps complicates the determination of the stoichiometry of the strained layers [3.142]. Figure 3.25 shows how the spectral positions of the E_1- and $E_1 + \Delta_1$-gaps depend on the indium concentration. In thick InGaAs layers, the strain relaxes via dislocations. Consequently the relationship between the spectral position of the gaps and the indium concentration differs in the cases of a strained or a completely relaxed layer.

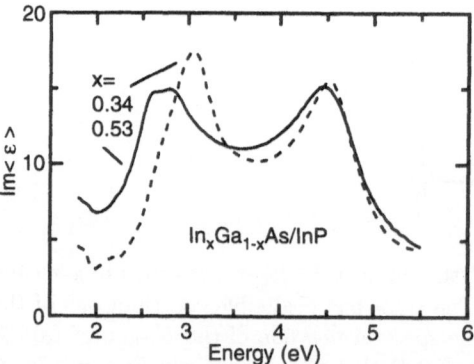

Fig. 3.24. The Im$<\epsilon>$ of lattice matched (53% In) and non-lattice matched (66% In) InGaAs on InP. Both the lineshape of the dielectric function and the spectral positions of the optical gaps differ

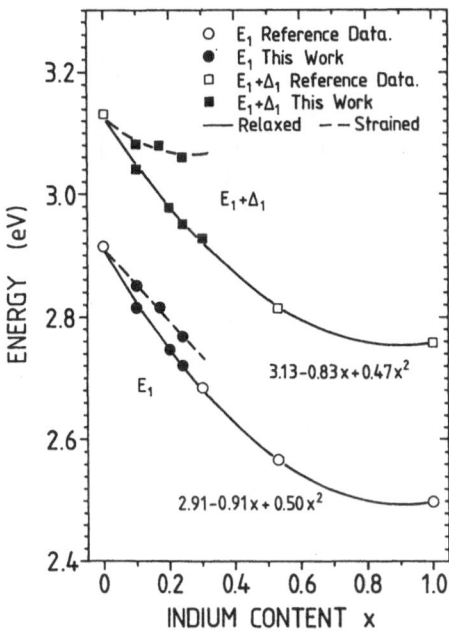

Fig. 3.25. The variation of the E_1- and $E_1 + \Delta_1$-gaps of InGaAs on GaAs as function of the composition of the layers. Shown are two dependencies one for strained InGaAs layers and one for relaxed layers [3.142]

3.4.4 Inhomogeneous Systems: Porous Silicon Layers

Porous silicon layers are of increasing interest, because they can be used for fabrication of light emitting diodes or possibly as substrate material for epitaxy of III-V compounds on Si [3.143]. Further aspects are the enormously enlarged surface area and the possibility to test the validity of effective medium theories [3.144]. The porous layers are produced via a HF (hydrogen fluoride) based electrochemical process ([3.144] and references therein for details). Depending on the current density and doping of the layers the structure and porosity (volume fraction of the pores) can be varied. Since p-type material is easier to fabricate, most papers about porous silicon layers deal with this type of samples. Using lowly p-type doped material results in the fabrication of extremely small pores separated by crystallites both with diameters in the nm range. Increasing the doping concentration to heavily p-type doped material (resistivity $\leq 0.2\,\Omega$cm) increases the diameter of the pores and the silicon skeleton [3.145].

Spectra for typical porous silicon layers with medium porosities of 36%, 48%, and 65% as produced from p-doped material are shown in Fig. 3.26a (non-oxidised). For comparison equivalent samples but oxidised in dry O_2 for 1 h at 300 °C (preoxidised) are shown in Fig. 3.26b. Thin film thickness interferences effects occur in the spectral region 1.7 to 2.5 eV, where the penetration depth in dense, crystalline silicon (c-Si) is rather high ($2\,\mu$m around 2 eV) [3.3]. The interference pattern is not well resolved due to the limited

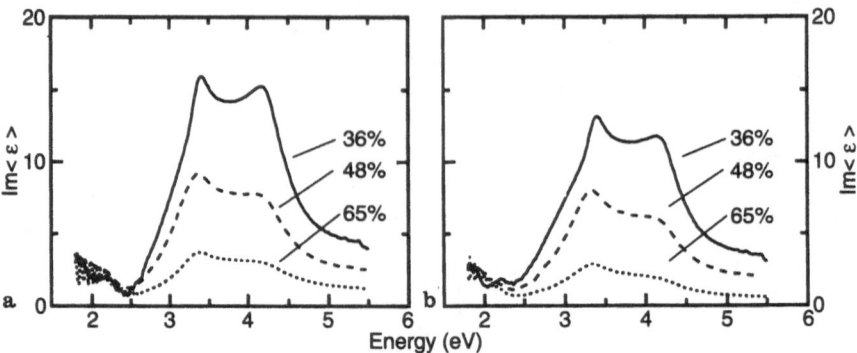

Fig. 3.26. Measured Im$<\epsilon>$ for 5 μm thick porous Si layers with porosities 36%, 48%, and 65% produced from p^+- doped (0.01 Ωcm) Si(100). (**a**) shows the spectra for non-preoxidised samples, (**b**) same as (**a**) but preoxidised. Below 2.5 eV interference fringes are visible, while above 2.7 eV the penetration depth of light is smaller than the layer thickness and only the layer contributes to the measured effective dielectric function

spectral resolution of the monochromator used in this study. Above 2.5 eV the interference effect disappears indicating that above this value only layer properties are probed. At 3.4 eV and 4.2 eV features due to the E_1- and E_2-gaps of bulk crystalline Si are visible in the measured dielectric function (compare Fig. 3.13). Therefore, the silicon skeleton of the porous layer is still crystalline. The values of Im$<\epsilon>$ near the E_1 and E_2 features are reduced compared to that of bulk Si. The oxidation of the layers leads to a further reduction of Im$<\epsilon>$ due to an decrease of the volume fraction of silicon. Obviously, the magnitude of $<\epsilon>$ drops non-linearly with increasing porosity. The decrease of the values is expected, since the density of oscillators is lowered when the porosity increases. The nonlinearity of the decrease indicates that the dielectric function is not simply given by the porosity weighted average of the dielectric function of the Si skeleton and vacuum. Instead effective medium theories have to be applied.

Increasing porosity (or volume fraction of silicon oxide) leads to a drop in the imaginary part of the calculated Bruggeman effective dielectric function Im$<\epsilon>$, but conserves the shape of Im$<\epsilon>$ [3.146]. The spectral positions and broadening of E_0', E_1, and E_2 remain constant and the absorption below E_1 is very low (Im$<\epsilon> \approx 0$) as is the case for dense c-Si. Furthermore, the ratio Im$<\epsilon>$(E near E_1)/Im$<\epsilon>$($E = E_2$) stays approximately constant. This is shown in Fig. 3.27 where an effective dielectric function for pores (vacuum) embedded in crystalline silicon calculated by a Bruggeman effective medium theory (Sect. 3.3.5) with a volume fraction of 48% is compared to dense crystalline Si. From a lineshape analysis the gap parameters are obtained for the calculated $<\epsilon>$ and details are given in Table 3.5. The nearly identical

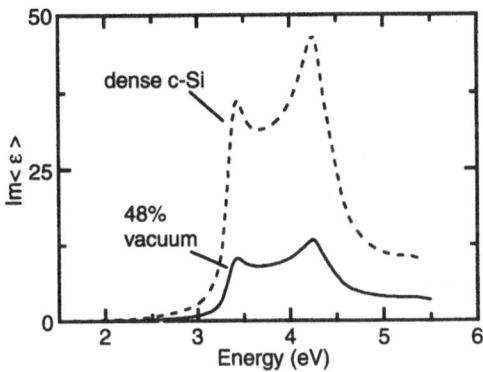

Fig. 3.27. Im$<\epsilon>$ for 48% vacuum in Si calculated with a Bruggeman effective medium dielectric function (3.69) compared to that of dense crystalline Si. It is important to note that the shape of Im$<\epsilon>$ is similar to that of bulk crystalline Si

values demonstrate that the parameters of the optical gaps E_g and Γ can be evaluated directly from the measured spectra. From measured spectra such as those schown in Fig. 3.26 it was found that the gaps are broadened and that the spectra are similar to that for micro-crystalline Si. Using the relationship between the broadening and the average crystallite size, it turns out that the size is about 5 nm or less (see Fig. 3.17b).

A fit of the measured dielectric function for porous silicon layers with a Bruggeman effective medium theory was tried before in [3.147] and also for the spectra of Fig. 3.26 [3.148]. For a fit to the data, one has to take into account the oxide layer on top which should be more important for this porous structure than for dense Si. An additional oxide layer on top just decreases the height of Im$<\epsilon>(E = E_2)$ while Im$<\epsilon>(E$ near $E_1)$ is hardly affected. This gives a guideline to separate the contribution of the oxide layer to $<\epsilon>$. The validity can also be proved comparing the non-preoxidised samples in Fig. 3.26a with the preoxidised ones in Fig. 3.26b.

Figure 3.28 illustrates the result of trying to use a Bruggeman effective dielectric function to fit the spectrum for a sample with 48% porosity

Table 3.5. Spectral positions and broadening of the E_1-gap of dense crystalline silicon (c-Si) compared to the calculated spetrum (Fig. 3.27) of silicon with pores

	E_0'		E_1	
	E_g(eV)	Γ(eV)	E_g(eV)	Γ(eV)
porous Si	3.321	0.092	3.406	0.124
dense c-Si	3.321	0.091	3.406	0.123

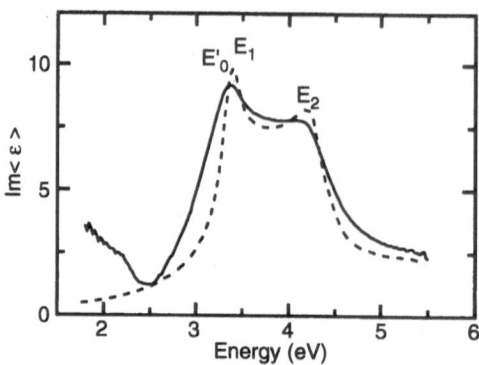

Fig. 3.28. Calculated Bruggeman effective dielectric function (dashed line) for a layer composed of 5% SiO$_2$ in c-Si with 48% vacuum with a 3.5 nm oxide film on top compared to the non oxidised sample with 48% porosity (solid line) shown in Fig. 3.26. Obviously, the gaps are asymmetricly broadened and therefore, the dielectric function of the Si skeleton deviates from that of bulk material

(Fig. 3.26a). Because of the neglectance of size effects this calculated dielectric function is not able to fit the measured $<\epsilon>$ resulting in a large discrepancy between the calculated and the measured dielectric function. While this is typical for dielectric functions measured for porous silicon layers, still general trends can be extracted. The thickness of the oxide film on top seems to be somewhat larger than expected for dense crystalline material which should be 1 to 1.5 nm. By a simple calculation, it can be shown that the SiO$_2$ content of the skeleton in the porous layer is much lower than that which would result if the pores are oxidised to a depth of 1.5 nm (natural oxide). The other interesting fact is that the fit is even less satisfactory below the E_1-gap giving lower values for Im$<\epsilon>$ than the measured data. Moreover, the gaps are broadened compared with the calculation as expected since the bulk dielectric function of silicon was taken for the fit. Consequently, the dielectric function for bulk dense crystalline silicon is not appropriate in a simulation of porous silicon layers as is also discussed in [3.146, 3.149]. This becomes even more apparent when comparing Im$<\epsilon>$ for different doping levels of the substrate (see Fig. 3.29). For low doping levels (high resistivities) a broad structure appears at a spectral position around the E_2-gap of bulk silicon. A comparison shows, however, that the lineshape of $<\epsilon>$ for these porous layers differ from that of amorphous material. The lineshape is also different for porous layers grown on highly doped silicon. A strong feature close to the E_1-gap of bulk silicon is always visible for theses layers.

Beside the problem that the dielectric function of bulk crystalline silicon does not represent the silicon skeleton, the Bruggeman effective medium theory is also not applicable. This is expected, since in the Bruggeman effec-

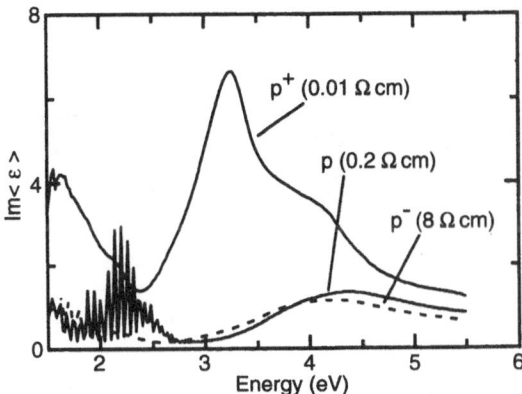

Fig. 3.29. Im$<\epsilon>$ of porous silicon layers formed on lowly and highly doped silicon substrates. While in the lowly doped case only a broad feature at an spectral position of the E_2-gap of bulk silicon remains, porous silicon layers formed on highly doped silicon show also a feature most likely related to E_1-gap of silicon

tive medium theory, the percolation cannot be varied but is fixed for a given porosity. This is discussed in detail in Chapt. 5 about Far-infrared spectroscopy.

3.4.5 *In-Situ* Studies

During recent years ellipsometry has been more widely applied to study growth processes of both epitaxial (Sect. 3.4.5.1) and amorphous semiconductors layers. For amorphous silicon, the influence of the substrate on the quality and surface roughness of the layers [3.127] has been studied, as has the addition of B_2H_6 for doping purposes [3.150] and the interface with different amorphous materials like a-C:H [3.151]. Other examples include the cleaning of semiconductor substrates during reactive ion etching [3.152], the modification of diamond films by ion beams [3.123] or etching of silicon in KOH monitored by ellipsometry [3.153]. Recently, ellipsometry was also used to control the stoichiometry of ternary III-V compounds [3.154, 3.155] during heteroepitaxial growth. In the following some characteristic examples of *in-situ* studies are discussed.

3.4.5.1 Study of GaAs/Al$_x$Ga$_{1-x}$As Interfaces.
In an early study it was shown that *in-situ* applied ellipsometry can detect the sharpness of the interface width of the GaAs $\rightarrow$ Al$_x$Ga$_{1-x}$As and Al$_x$Ga$_{1-x}$As $\rightarrow$ GaAs transitions during MOVPE growth [3.156]. Single wavelength ellipsometry using a He-Ne laser (632.8 nm line) as the light source was utilised in these experi-

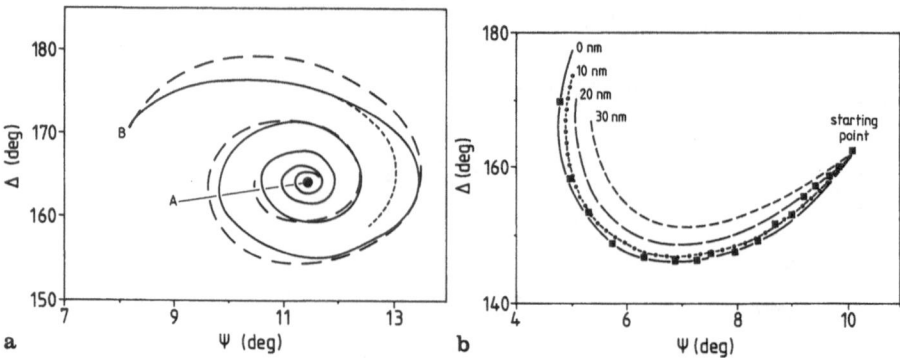

Fig. 3.30a. Real-Time examination of a $Al_xGa_{1-x}As \rightarrow GaAs$ (MOVPE) heteroepitaxy (solid line). Point B is the stable state for $Al_xGa_{1-x}As$ epitaxy ($x = 0.4$). The long dashed line is the (Δ,ψ) locus calculated for an abrupt transition between (Al,Ga)As and GaAs layers. The short dashed line corresponds to the loci calculated by assuming a 20 nm linear transition region between $Al_xGa_{1-x}As$ and GaAs. All lines begin at the point B, **b** Real-Time examination of a $GaAs \rightarrow Al_xGa_{1-x}As$ (MOVPE) heteroepitaxy (solid line). The (Δ,ψ) loci respectively calculated for 10 nm, 20 nm, and 30 nm transition width are also given [3.156]

ments (Fig. 3.30). The laser light of this wavelength penetrates deep (around 300 nm [3.3]) into these materials. In order to interpret the data, one has to refer to the equations in Sect. 3.3.4. If the growing layer was non-absorbing, then β given in (3.61) is real and can be expressed by a real constant times the layer thickness d_{layer}. Since $\rho = \tan\psi \exp(i\,\Delta)$ depends on the thickness only through the term $\exp(-i\,2\beta)$ (3.58) and β is real, ρ is a periodic function of the layer thickness. If instead of ρ, the pair (Δ,ψ) is plotted for increasing thickness, one would get a closed contour with a shape depending on the dielectric function of the layer and substrate as well as the angle of incidence [3.5]. If the layer is weakly absorbing as is the case here, β is complex and ρ is no longer periodic in d_{layer}. Instead, one gets a so-called exponential spiral which, for increasing d_{layer}, converges to a certain point in the (Δ,ψ) plane. This point is given by the dielectric function of the layer (and ambient).

The starting point B in Fig. 3.30a is the stable state when all precursors (AsH_3, $(CH_3)_3Ga$ and $(CH_3)_3Al$) are flowing. When $(CH_3)_3Al$ is switched off, the experimental point (Δ,ψ) starts moving along a spiral shape locus: both GaAs and $Al_xGa_{1-x}As$ are only weakly absorbing at 632.8 nm and hence, Fabry-Perot interferences manifest as a spiral in the (Δ,ψ) plane. When the GaAs layer is thick enough, the (Δ,ψ) point stabilises at point A which is representative of the GaAs layer. The inverse growth sequence of $Al_xGa_{1-x}As$ on GaAs shown in Fig. 3.30b needs not to be equivalent to the former case. Figure 3.30b shows a qualitative agreement between the experimental curve and the calculated one, assuming an abrupt interface. Calculated curves for 10 nm, 20 nm, and 30 nm transition width is also shown. The $Al_xGa_{1-x}As$

growth was deliberately stopped at a thickness of 40 nm. On the other hand, the $Al_xGa_{1-x}As \to GaAs$ transition given in Fig. 3.30a exhibits an experimental locus which is in qualitative agreement with a calculated one, assuming a linear composition profile extending over 30 nm.

Using a light source, however, with a fixed photon energy imposes strong limitations which are overcome by using a broad band light source which allows the wavelength to be chosen with respect to signal optimisation. This was done in the following example.

3.4.5.2 Control of Composition. Aspnes applied ellipsometry to control the composition of $Al_xGa_{1-x}As$ grown with chemical-beam epitaxy [3.157]. This was done by controlling the Al flux depending on the value of the measured dielectric function. If the measured $<\epsilon>$ deviated from the target value, the Al flux was changed until the desired $<\epsilon>$ and thus the Al concentration was reached. In order to achieve this, the dielectric function was measured at a fixed photon energy of 2.6 eV. Figure 3.18 shows that the sensitivity in this spectral region to composition is high and is further increased at the growth temperature of around 600 °C since the gaps shift to lower energies. From values of the dielectric functions obtained for various compositions x at 2.6 eV, a relationship between $Im<\epsilon>$ and x was determined. Inverting this relationship gives the composition x for the measured value of $<\epsilon>$. When the profile in $x(d)$, where d is the increasing layer thickness, is pre-defined, the target values $Im\epsilon_t(d)$ can be calculated.

In Fig. 3.31 target values $Im\epsilon_t(d)$ are shown together with those obtained during the growth of the structure. The excellent agreement between the target and measured values is remarkable. At the start and the end of the growth overshoots occur due to the finite response time of the system. This

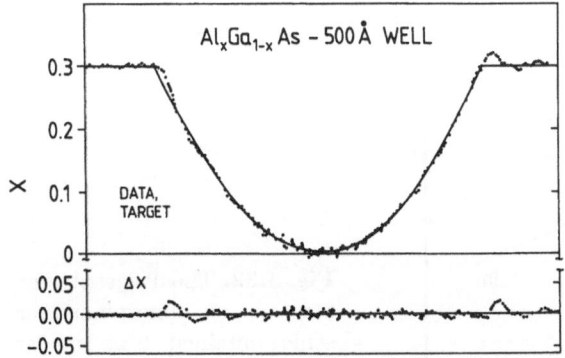

Fig. 3.31. Data for a 50 nm $Al_xGa_{1-x}As$ parabolic quantum well. Top: values of x, determined from $Im<\epsilon>$ for the outer running 1.15 nm of material, compared to target values. Bottom: difference between determined and target values [3.157]

limits the accuracy of the composition to $\Delta x = \pm 0.02$. It is expected that for faster systems the accuracy in x can be further enhanced.

3.4.5.3 Arsenic Layers on Silicon.

The interest in arsenic layers is two-fold. Firstly, arsenic forms ordered and stable monolayers on silicon surfaces [3.158]. Secondly, arsenic layers can be used to protect surfaces (capping) against contaminations [3.159–3.161,3.55,3.162] and thus conserving the surface for a longer period. This method of protection can also be helpful in the growth of II-VI compound semiconductors on III-V substrates. First, a III-V buffer layer is grown and capped by arsenic. The arsenic is then desorbed in a second growth chamber, where the II-VI compound is deposited. Arsenic capped surfaces can also be used in surface studies since after desorption of the arsenic layer, a clean surface is established for further investigations. However, small crystals of arsenic oxides form after some time in the presence of oxygen which finally leads to a destruction of the caps [3.162, 3.163].

Figure 3.32 shows a typical spectrum for a freshly grown, thick arsenic layer on silicon. The broad structure around 3 eV indicates the amorphous structure of the arsenic. The spectral position of the maximum and the shape of ϵ compare well with that of amorphous As [3.55].

After desorbing such arsenic caps at temperatures below 300 °C, one ordered monolayer of arsenic was obtained [3.163]. This monolayer can also be desorbed at temperatures around 700 °C [3.164]. Fig. 3.33 shows a deposition/desorption cycle [3.164]. After thermal cleaning of the surface, As was deposited at room temperature (33 °C). The As deposition and desorption was monitored at a fixed wavelength of 300 nm (4.13 eV). This spectral position is close to the E_2-gap of bulk Si and provides the lowest penetration depth in Si. Every 20 sec a value of $<\epsilon>$ was taken, whereby the time resolution of the ellipsometer for typical values of the analyser rotation frequency and number of averaged cycles is around 10 sec. In the case of the arsenic desorption the changes in the dielectric function are faster than the response time

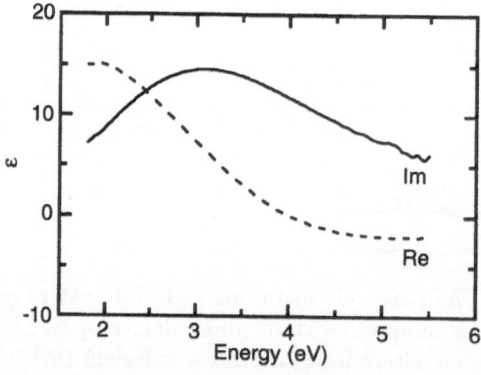

Fig. 3.32. The dielectric function ϵ of amorphous arsenic obtained from an arsenic capped silicon surface. The broad structure in Im$<\epsilon>$ indicates the amorphous character of the layer

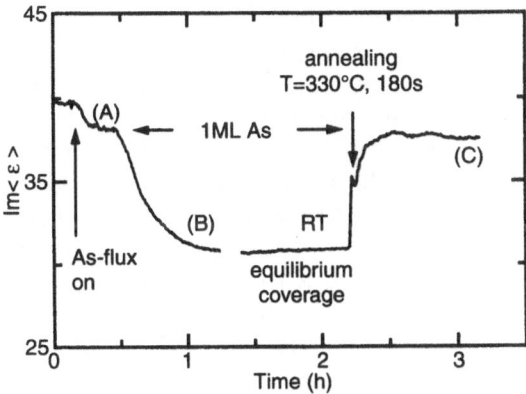

Fig. 3.33. Arsenic adsorption and desorption studied on Si(100) at 4.13eV. First As was deposited onto the silicon surface at a temperature of 33 °C. For 1 ML As coverage a plateau in Im<ϵ> is reached (A), further deposition decreases the magnitude of Im<ϵ> until a saturation after a few nm (B) is achieved. After stopping the As flux no change in Im<ϵ> occurs. The As was desorbed by raising the sample temperature up to 330 °C. The value of Im<ϵ> after cooling the sample to room temperature (C) is nearly at the same level as the plateau found in the adsorption of As, as expected

of the ellipsometer. Thus the measured transients show a slightly too small desorption rate. This is a general problem when monitoring fast processes with ellipsometry.

After a clean silicon surface has been prepared the arsenic filled MBE cell is heated up to approximately 300 °C. The pressure in the chamber rises up to the 10^{-6} mbar range giving rise to a very high arsenic flux i.e. > 1 Langmuir (one monolayer (ML) per second). Then, first a "plateau" in <ϵ> is reached (A), when 1 ML is adsorbed. After some time Im<ϵ> starts to decrease again, but no new plateau is reached. This may be explained by the low sticking coefficient (number of atoms or molecules adsorbed on the surface divided by the number of incoming atoms or molecules) for As_4 on As. Hence, a high flux or a long deposition time is needed to grow a thicker layer of arsenic on silicon. Following further As deposition, an equilibrium coverage (B) is reached, where the number of desorbing arsenic molecules equals the number of adsorbing molecules. The equilibrium coverage is strongly dependent on the sample temperature and arsenic flux, since the desorption rate and the sticking coefficient are functions of temperature. This is also known from the growth of the arsenic caps. At room temperature, the sticking coefficient of arsenic is too low to grow caps. The samples have to be cooled down below 0 °C to achieve layers of several nm thick [3.165].

In Fig. 3.33 also the transient in Im<ϵ> during the desorption of the arsenic layer is shown. After stopping the arsenic deposition and without raising the temperature Im$<\epsilon>$ does not change. Increasing the temperature leads to an immediate increase in Im$<\epsilon>$ accompanied by an increase of the chamber pressure. This demonstrates that most of the arsenic is loosely bound. Further desorption ends at approximately 250 °C in agreement with desorption experiments of the arsenic caps on silicon. The new saturation level in $<\epsilon>$ (C), which is reached when the sample cools down to room temperature, is nearly the same as the plateau (A) corresponding to 1 ML As.

For a quantitative analysis of the arsenic desorption it has to be kept in mind that $<\epsilon>$ is temperature dependent (Sect. 3.4.1.1). The E_2-gap of Si shifts to lower energies and broadens when increasing the sample temperature. The broadening lowers Im$<\epsilon>$, while the shift of the gap position increases the values. Since the broadening is the stronger effect a decrease of Im$<\epsilon>$ with increasing temperature of the sample is expected. The dip in Imϵ during the desorption of the arsenic (near 2.2 h) is probably caused by a slight misalignment of the sample holder during the annealing process of the sample.

The experimental results for the adsorption and desorption of arsenic were reproduced several times with similar results. The first plateau (A) can consequently be identified as that state where one monolayer is adsorbed on the silicon surface.

The dielectric functions of one monolayer of As on Si(100) and that of the corresponding clean surface are shown in Fig. 3.34. The surface of the wafer is slightly misorientated (3° off to [011]). Therefore, the surface is microscopically rough and the height of Im$<\epsilon>$ near the E_2-gap is reduced compared to literature data for bulk silicon [3.3]. The effect of the As-monolayer on the dielectric function is small but significant. The measured difference in Im$<\epsilon>$ agrees well with calculations reported in [3.166].

The significance of the small differences in Fig. 3.34 can easily be judged from the signal to noise ratio in Fig. 3.33. It is also obvious from this figure that ellipsometry is sensitive to *submonolayers* coverages. But on-line moni-

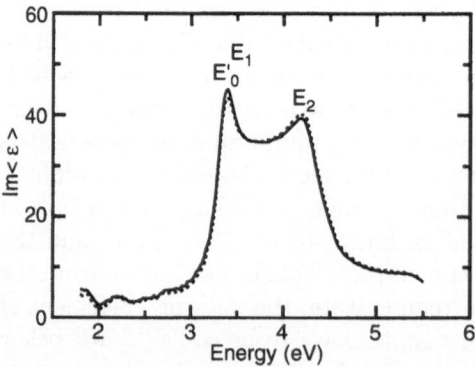

Fig. 3.34. Im$<\epsilon>$ for 1 ML As on Si(100) (solid line) compared to that for the clean sample (dashed line)

toring of the desorption process only becomes possible if the time resolution of the ellipsometer used is much higher than the time needed to desorb this monolayer of arsenic.

3.4.6 Multilayer Analysis

Recently, more complex systems like AlGaAs-GaAs multilayers [3.167–3.169] and the damage profile of silicon implanted silicon [3.172] have been characterised using ellipsometry. In order to demonstrate the capability of spectroscopic ellipsometry in the analysis of such structures, we discuss in detail an investigation of a GaAs/AlGaAs multilayer structure [3.169]. The analysis of this system is typical in the sense that most problems of multilayer analysis occur. The MOVPE growth of this structure was performed such that the thicknesses of the barriers and the quantum wells change as a function of the lateral position on the sample. This structure has the advantage that the energy levels of the optical transitions observed can be related directly to the thickness of the quantum wells and the barriers.

In Fig. 3.35 a sketch of the side view of the sample is given. The total diameter of the sample is 2". Therefore, a conventional ellipsometer with a focus of a few square mm is not suited for the analysis of this kind of sample with high lateral inhomogeneity. In this study, an ellipsometer with a spatial resolution of 100μm was used [3.170]. The analysis starts with the determination of the thicknesses of the oxide layer on top d_0, the top AlGaAs layer d_1, the quantum well d_2, and the barrier d_3.

There are two possible ways of determining these thicknesses. The spectra in Fig. 3.36, illustrate that interference effects occur in the low-energy spectral region for a lateral position, where the layers are thickest and the wells are

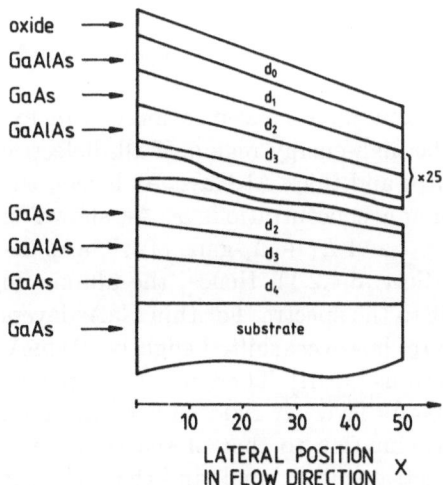

Fig. 3.35. Schematic representation of the multilayer structure of the sample used in Fig. 3.36 an 3.37. It exhibits a thickness gradient along one direction (x-axis) of the 2"wafer [3.170]

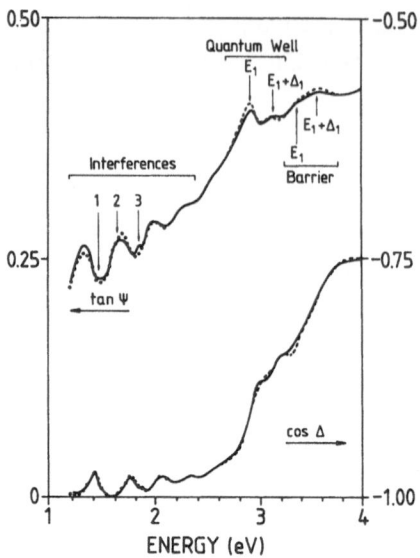

Fig. 3.36. Example of measured ellipsometric spectra tan ψ and cos Δ (full lines). In the low-energy part of the spectra, the interferences due to the total thickness of the MQW can be fitted by an appropriate multilayer model (dotted lines). Weak structures, superimposed on the interference regime, are due to the optical transitions in the MQW. The excitonic lines are labeled 1, 2, and 3 and correspond to the following transitions: 1: e1 → hh1, 2: e2 → hh2, 3: e3 → hh3. The transition e1 → lh1 (1') does not occur. For higher energies, the optical response originates from the top AlGaAs and GaAs layers. Multilayer modelling (dashed lines) of this region allows the determination of the thicknesses of these top layers, including the top oxide layer [3.170]

uncoupled. Between 3 and 4 eV the E_1- and $E_1+\Delta_1$-gaps of GaAs and AlGaAs are separately visible.

In principle, the thicknesses and the aluminum concentration can be extracted by a multilayer analysis from the high-energy region, if all dielectric functions are known. For GaAs substrates and thick $Al_xGa_{1-x}As$ layers, the data are well known for various aluminum concentrations x. As shown in Fig. 3.19, the spectral positions of the E_1- and $E_1 + \Delta_1$-gaps of AlGaAs depend on the aluminum concentration (Sect. 3.4.2.1). Hence, the aluminum concentration can be obtained from a fit to the spectra. For thin GaAs layers in a quantum well structure, the gaps are, however, shifted slightly (24 meV here) due to the confinement of the electrons [3.131]. Therefore, this spectral shift depends on the thickness of the GaAs layer. A rigid shift of the bulk GaAs ϵ value can be taken as an approximation to that of the thin GaAs layer in the quantum well. Since the spectral shift is small and the dielectric functions of such layers are not exactly known, the thicknesses obtained in this way are not accurate. Furthermore, for smaller thicknesses d_2 of the well and d_3 of the barrier, a shifted bulk ϵ value cannot represent the dielectric functions of the well and barrier. Finally, the penetration depth is low in the spectral region of the E_1-gaps and thus only a few layers of the total number are probed.

Another possible way to determine the thicknesses of the multilayer structure is to evaluate the features in the low-energy region of the dielectric function caused by Fabry-Perot interference effects. Because the interference pattern depends mainly on the total thickness of the multilayer stack, all parameters cannot be measured independently and one has to make reasonable assumptions. It is assumed that [3.169]:

1. the aluminum concentration is constant all over the wafer
2. the ratio d_3/d_2 is constant and for this structure equal to 1.025
3. the top AlGaAs layer thickness is related to the barrier thickness by the following relationship: $d_1 = d_3 - 2.5\,\text{nm}$

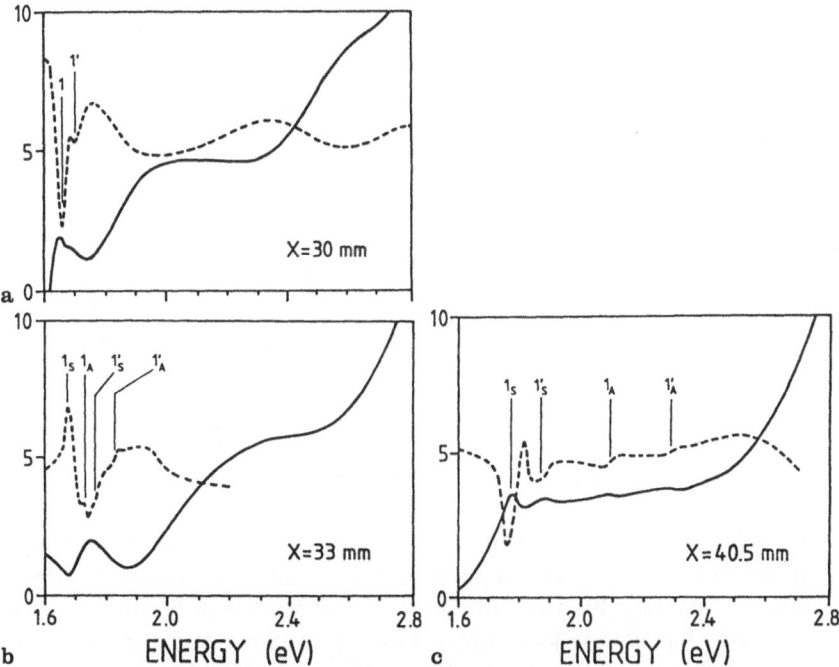

Fig. 3.37. Imaginary part of the effective dielectric function (full line), as well as the first derivative of the real part of the effective dielectric function (dashed line) measured by spectroscopic ellipsometry in the 1.6–2.8 eV range. All spectra were taken at room temperature. **a** $X = 30\,\text{mm}$, $d_2 = 2.9\,\text{nm}$. The QWs are uncoupled. The heavy- and light-hole transitions (1 and 1') are well separated, **b** $X = 33\,\text{mm}$, $d_2 = 2.4\,\text{nm}$. The QWs are weakly coupled, and the splitting of the states 1 and 1' becomes observable, **c** $X = 40.5\,\text{mm}$, $d_2 = 1.4\,\text{nm}$. The system is strongly coupled and both symmetric and antisymmetric levels are clearly seen [3.170]

The third assumption requires that the thickness of the top oxide layer is uniform over the wafer. However, the choice of d_1 is not very critical for the determination of d_2 and d_3 from the interference pattern, since the total MQW (or SL) thickness is $d_1 + 25(d_2 + d_3)$.

The thicknesses determined by these two procedures differ only within the uncertainties of the calculation and hence, both procedures can be used to determine sample thicknesses.

It is found that for d_2, $d_3 < 2.5\,\mathrm{nm}$, the quantum wells become strongly coupled and a transition to superlattice behaviour is observed. In Fig. 3.37 Im$<\epsilon>$ and the real part of the first derivative Re$<\epsilon>'$ with respect to energy are plotted together for different barrier thicknesses and hence, different coupling between the QWs. Due to the coupling of the QWs the transitions $T = 1, 1', 2 \ldots$ split into two transitions T_S and T_A corresponding to transitions between symmetric and antisymmetric states. By taking spectra at different positions on the sample and evaluating the thicknesses of the QWs and the barrier as function of position, one can produce the dispersion curves for the transitions reported in [3.169].

These values may be compared with a calculation solving a Schrödinger effective mass equation for electron and holes in a well. The potential depth of the well is determined by the band lineup of the AlGaAs/GaAs. Using the conduction- and valence-band offsets, the calculated dispersion curves are in good agreement with the measured data [3.169].

3.5 Sample Related Problems

Problems which can arise relating to the nature of the sample in ellipsometric measurements are briefly discussed. In most cases the complex structure of the sample leads to difficulties in the interpretation of the spectra.

3.5.1 Sample Preparation

Since the sensitivity of spectroscopic ellipsometry is in the monolayer range, sample preparation, contamination, and oxidation may significantly influence the measurements. This became clear in the early studies of Drude involving the preparation of antimony [3.2]. Later, more systematic studies of the influence of the sample preparation on the data obtained were performed and it became apparent that only certain preparation procedures lead to reproducible results [3.173]. Until the pioneering work of Aspnes [3.3] it was not clarified which preparation method gives the best results concerning the measured dielectric function. In these studies, it was shown that the correct sample preparation procedure is crucial in order to obtain accurate and reproducible spectra. This work led to a set of data for semiconductors which in most cases today is the basis for quantitative analysis of ellipsometric spectra [3.3].

For samples which are particularly difficult to prepare, sample preparation procedures limit the accuracy of the data. For example, aluminum containing samples have a tendency to form a very stable surface layer of Al_2O_3. The strong influence on sample preparation was one of the reasons why ellipsometry remained unpopular in the past.

3.5.2 Multilayer Structures

Problems in analysing multilayer structures have already been discussed for the GaAs/AlGaAs MQW structure in Sect. 3.4.6. In most cases there are too many unknowns compared to the two values measured with the ellipsometer for each photon energy and angle of incidence. Therefore, certain assumptions have to be made for a quantitative analysis of the spectra. Even in the simple case where all dielectric functions of the layers and the substrate are known and the interfaces are atomically abrupt, it is difficult or impossible to determine all the unknown thicknesses. One of the reasons for this is that frequently deeper lying layers make a small or negligible contribution to the effective dielectric function of the sample. Other reasons include noisy spectra and the correlation of parameters to each other [3.174]. However, there are a few examples in which quite complicated structures have been analysed [3.172].

The situation which arises when the layer thicknesses as well as one of the layer dielectric functions are unknown is at present only solvable in certain cases. The simplest of these cases is when there is only one layer on a substrate, as discussed in Sect. 3.4.2.3. But even this simple problem is quite often too complex to analyse because of non-abrupt interfaces or inhomogeneities of the layer.

3.5.3 Gradually Varying Composition

Layers with gradually varying composition are used as buffer layers in the epitaxial growth of lattice mismatched materials [3.175] or in device applications [3.157]. They may also occur in layered structures where there is interdiffusion at the interfaces. The analysis of graded layers is complicated, however. One approach is a multilayer analysis of this structure. The thicknesses of the single layers used to model the graded layer must be chosen much smaller than the variation of the composition. The problem can be solved by taking the dielectric function for each of these compositions and assuming a certain profile for the variation of the composition. Unfortunately, there are not enough data available to really perform such an analysis. The dielectric function of ternary materials for example are only known for certain values of the composition. Furthermore, this approach can be successful only if the gradient is small. If the gradient is high, many very thin layers are required in order to calculate the effective dielectric function resulting in severe numerical problems [3.176]. Consequently to our knowledge, a successful analysis of ellipsometric spectra obtained for a layered structure with gradual varying composition has not yet been performed.

3.5.4 Anisotropies

As discussed in detail in Chapt. 2, the optical response of surfaces is often anisotropic [3.177,3.178]. The surface may be regarded as an anisotropic layer on an isotropic substrate (to first order [3.179]). This must be taken into account in interpreting the spectra. In this case the spectra depend on the orientation of the sample.

This is shown in the following example for GaAs. GaAs(001) has two different Fresnel coefficients $r_{[110]}$ and $r_{[1\bar{1}0]}$ for the [110] and the [1$\bar{1}$0] directions, respectively [3.178]. The anisotropy of GaAs(001) and most other materials is very small (Chapt. 2). Figure 3.38 shows $<\epsilon>$ for GaAs(001) with the sample orientated with the [100] axis in the plane of incidence and after rotating the sample by 90° around the surface normal. Hence, if two measurements are to be compared one must ensure that the orientation of the samples is the same or that the anisotropy is zero.

3.5.5 Quantification of Defects and Strain

The example of ion implanted GaAs showed that it is possible to detect defects using ellipsometry. A quantification of the defect density, however, is not yet possible. The same is valid for strain effects. There are a few recent pub-

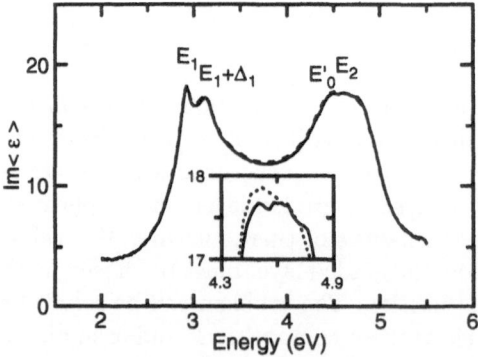

Fig. 3.38. $<\epsilon>$ for GaAs(001) with the [100] axis in the plane of incidence and perpendicular to it. There is a small peak between the E_0'- and E_2-gap which is more pronounced for one orientation (inset). Therefore, the lineshape of Im$<\epsilon>$ between those two gaps can be used to orientate the samples

lications about the influence of strain on the dielectric function as measured by ellipsometry. Further experiments, however, must be carried out in order to be able to quantify the strain from the measured dielectric function.

3.5.6 Depolarisation

Surfaces which are rough tend to depolarise the reflected light or, to reflect light with a mixture of different polarisation states [3.180]. Both situations cannot be distinguished by conventional ellipsometers.

The simplest case is a sample with a wedge-like layer on a flat substrate similar to the one sketched in Fig. 3.35. In this case the layer thickness varies across the light spot on the sample and the polarisation of the reflected light is a function of the location within the light spot on the sample. Therefore, the state of polarisation varies over the cross section of the reflected light and the signal created by the detector is a function of all these polarisation states. Consequently, the signal obtained is equivalent to that of partially polarised light.

The depolarisation caused by the sample can be determined using setups which are capable of measuring the complete Mueller matrix (Sect. 3.1.3) instead of the Jones matrix [3.28,3.4]. In the simplest case the Mueller matrix of a depolarising sample is given by [3.181]:

$$
S_D^M = 1/D \begin{bmatrix} D & -\cos 2\psi & 0 & 0 \\ -\cos 2\psi & 1 & 0 & 0 \\ 0 & 0 & \sin 2\psi \cos \Delta & \sin 2\psi \sin \Delta \\ 0 & 0 & -\sin 2\psi \sin \Delta & \sin 2\psi \cos \Delta \end{bmatrix} \quad (3.72)
$$

where D is the inverse of the polarisation degree of the reflected light. For $D = 1$ the light is totally polarised and the Mueller matrix S_D^M reduces to that of (3.24). For $D = \infty$ the Mueller matrix S_D^M is:

$$
S_D^M = \begin{bmatrix} 1 & 0 & 0 & 0 \\ 0 & 0 & 0 & 0 \\ 0 & 0 & 0 & 0 \\ 0 & 0 & 0 & 0 \end{bmatrix} \quad (3.73)
$$

In the latter case the reflected light is unpolarised independently of the state of polarisation of the incident light. For this simple case of depolarisation the value of D can be determined and the measured dielectric function can be corrected for the influence of depolarisation. A correction of the measured dielectric function for depolarisation using just the ellipsometry data is not simple since the calibration parameters (η in the case of the rotating analyser) also depend on D. Moreover, (3.72) is not valid in the case where the sample itself acts as a (partial) polariser, a case which may occur when there is a grating-like structure on the sample surface.

3.6 Summary

As shown in the last Sections, ellipsometry is suitable for the characterisation of thin layers with known or unknown optical properties. The advantages of ellipsometry are reproducibility, precision, sensitivity to small modifications of the sample and a broad spectral range. The dielectric function measured by ellipsometry contains information about the electronic structure of the grown layer(s), the layer thickness(es), the abruptness of interfaces, interfacial layers, the topology of inhomogeneous layers, and the surface condition after sample preparation. The electronic structure depends on the material and, particularly in the case of compound semiconductors, on the chemical composition. The electronic structure, and consequently the dielectric function, is influenced by strain, sample (growth) temperature, defects and crystallinity of the layers. Conversely, all these parameters which are important for a qualitative and quantitative examination of epitaxially grown layers can be obtained by an analysis of the effective dielectric function $<\epsilon>$. This analysis requires a model for the sample structure. In most cases a 3-phase or multiphase-model is sufficient. In more complicated cases with non-abrupt interfaces, interfacial layers, or layers with graded composition, the analysis of $<\epsilon>$ is very difficult and in certain cases impossible. A further complication is the lack of suitable reference dielectric functions.

Spectroscopic ellipsometry yields more information than single wavelength measurements. In contrast to the single wavelength mode, spectroscopic ellipsometers determine the dielectric function for each wavelength as well as the lineshape or derivatives with respect to energy of the dielectric function.

Further information is obtained from *in-situ* ellipsometric measurements during the growth process. The difference in $<\epsilon>$ before and after a thin layer is deposited depends mainly on the properties of the thin layer [3.176]. This allows ellipsometry to be used for real-time growth control [3.157].

4. Raman Spectroscopy

Norbert Esser, Jean Geurts

For the analysis of semiconductor layers, heterostructures and interfaces, Raman spectroscopy has become a widely used method. Among the main reasons for its wide-spread application is its sensitivity for thin films down to monolayer (ML) thickness, combined with its variable information depth. It allows a nondestructive analysis of the surface regions within some nm below the surface as well as deeper regions up to the μm range, e.g. buried interfaces. As a consequence of the recent evolution of the experimental equipment, nowadays Raman experiments can be performed very efficiently, especially due to the employment of multichannel detector systems. Besides, lateral resolution in the micrometer range can be achieved in so-called micro-Raman spectroscopy. This development has opened the field of laterally structured heterostructure systems and devices for Raman analysis.

The information provided by a Raman-spectroscopic analysis includes lattice dynamics as well as electronic properties. The lattice dynamics reflects structural information such as the identification of materials and compounds, including reacted phases at interfaces, but also aspects such as the composition of mixed compounds, layer orientation, stress, and crystalline perfection. This information is obtained from scattering by phonons through the evaluation of the phonon frequencies, half widths, lineshape and their intensities in the Raman spectrum. Furthermore, the resonance behaviour of the scattering intensity reflects the energies of the critical points in the electronic energy bands.

Electronic characterisation of interfaces and layers can be performed through electric-field induced Raman scattering (EFIRS) from longitudinal optical (LO) phonons and scattering from coupled plasmon-LO-phonon modes. For superlattices, in addition, new phenomena occur in the Raman spectrum, e.g. folded acoustical phonons due to the modified periodicity length, confined optical phonon modes due to the spatial localisation of optical vibrations in each material, and interface modes resulting from coupling of vibrations which are essentially confined at the various interfaces. These new modes can be used to analyse the layer thicknesses and interface sharpness.

Several review articles of Raman spectroscopy have been published. General aspects of light scattering by crystals were treated by Hayes and Loudon [4.1]. Richter emphasized the resonance behaviour of Raman scattering in semiconductors [4.2]. The series "Light scattering in Solids, Volume 1 to 6", edited by M. Cardona and G. Güntherodt, covers a wide variety of aspects of light scattering, such as resonance effects [4.3], scattering by free carriers [4.4], and by vibrations in superlattices [4.5].

Here we will first treat the fundamentals of Raman scattering in Sect. 4.1. Besides the Raman scattering principles, the various mechanisms and the symmetry-imposed selection rules are included. The experimental aspects of Raman scattering are discussed in Sect. 4.2. Subsequently, applications of Raman scattering to heterostructures are presented, illustrating the variety of properties which can be studied. Section 4.4 is concerned with the analysis of structural properties, such as lattice perfection and orientation, strain, composition of mixed compounds, and reactions at interfaces. At the end of this Section a short survey is presented of specific phenomena in the Raman spectra of superlattices: folded acoustic, confined optical, and interface phonons. Since Raman scattering from superlattices was recently reviewed extensively (e.g. [4.5, 4.6]), only the main features will be treated. Section 4.5 deals with electronic characterisation by impurity excitations and collective electronic excitations. Here also the basic aspects of electronic Raman scattering from subbands in superlattices are described. Section 4.6 focuses on band bending at interfaces, which can be analysed from plasmon-LO-phonon modes and from electric-field induced Raman scattering (EFIRS). Finally, a summary is given in Sect. 4.7.

4.1 Theory of Raman Spectroscopy

4.1.1 Principles of Raman Spectroscopy

In Raman spectroscopy, inelastic light scattering processes are analysed, i.e. scattering processes in which energy is transferred between an incident photon with energy $\hbar\omega_i$ (incident) and the sample, resulting in a scattered photon of a different energy $\hbar\omega_s$ (scattered). The amount of transferred energy corresponds to the eigenenergy $\hbar\Omega_j$ of an elementary excitation labelled "j" in the sample, e.g. a phonon, a polariton, a plasmon, a coupled plasmon-phonon mode or a single electron or hole excitation. A Raman setup is shown schematically in Fig. 4.1.

A Raman spectroscopy experiment yields the eigenfrequencies of the elementary excitations through the analysis of the peak frequencies ω_s in the scattered light, since the frequency of the incident light ω_i is well defined by the use of a laser-light source. Energy conservation yields:

$$\hbar\omega_s = \hbar\omega_i \pm \hbar\Omega_j. \tag{4.1}$$

Here the "$-$" sign stands for those Raman processes in which an elementary excitation is generated. These are called Stokes processes. Those which imply the annihilation of an elementary excitation correspond to the "$+$" sign. They are referred to as anti-Stokes processes. Since the efficiency of Stokes and anti-Stokes processes has a characteristic temperature dependence, the intensity ratio between the Stokes and the anti-Stokes peak can be applied for a de-

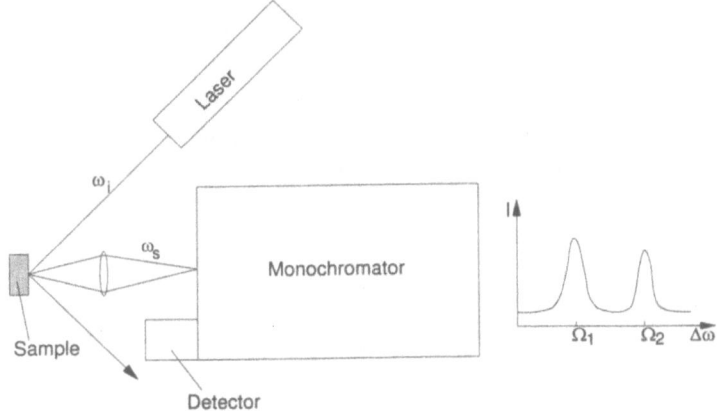

Fig. 4.1. Schematic picture of the main components of an experimental setup for Raman scattering and a Raman spectrum, showing the scattered light intensity I vs. frequency shift $\Delta\omega$. Positive frequency shift $\Delta\omega$ corresponds to a Stokes process (energy loss). ω_i: incident light frequency; ω_s: scattered light frequency; Ω_j: eigenfrequencies of elementary excitations ($j = 1, 2$)

termination of the sample temperature [4.3]. However, in most experimental investigations only Stokes processes are studied. As shown in Fig. 4.1, a plot of the intensity of the scattered light versus frequency difference $\Delta\omega = \omega_i - \omega_s$ yields peaks at the eigenfrequencies Ω_j of the elementary excitations.

In analogy to energy conservation, the quasi-momentum conservation law gives the correlation between the wave vector $\mathbf{k}_i$ of the incident light, $\mathbf{k}_s$ of the scattered light and the excitation wave vector $\mathbf{q}_j$:

$$\mathbf{k}_s = \mathbf{k}_i \pm \mathbf{q}_j. \tag{4.2}$$

Here the wavevectors inside the sample are involved, thus $\mathbf{k}_{i,s} = n(\omega_{i,s})\mathbf{k}_{i,s}^0$, where $n(\omega_{i,s})$ is the index of refraction for the light frequencies $\omega_{i,s}$. Since investigations of thin semiconductor films usually imply a backscattering configuration, (4.2) can be reduced to a scalar form. For the Stokes process q_j is given by

$$q_j = \frac{1}{c_0}\Big(n(\omega_i)\,\omega_i + n(\omega_s)\,\omega_s\Big). \tag{4.3}$$

Here c_0 is the light velocity in vacuum. The combination of energy and momentum conservation implies that in a scattering process only certain combinations of energy and momentum can be transferred to the sample. Thus, well-defined (Ω, q) pairs out of the whole range given by the dispersion relations of the solid's excitations may be involved in the scattering process.

Furthermore, Eq. (4.3) yields a quasi-momentum transfer which is approximately proportional to the laser frequency. A deviation from this proportionality may arise due to the frequency dependence of the refractive index $n(\omega)$. Equation (4.3) yields q-values in the range of $10^6\,\mathrm{cm}^{-1}$. These small q-values result because the light wavelength exceeds the lattice constant by a factor of more than 100. Thus, the excitations involved are within a few percent from the centre of the Brillouin zone. For scattering from phonons the energy transfer $\hbar\Omega_j$ is independent of the incident frequency ω_i, since the phonon dispersion is negligible near the Brillouin zone centre. The plasmon-LO-phonon mode frequencies, however, have a pronounced q-dependence, which can be analysed in Raman scattering by varying the laser frequency.

The conservation of momentum is an idealized picture which may be limited in real experiments. One source of such limitations causing a relaxation of the momentum conservation rule is simply the light absorption in the sample. The ideal case of a plane wave with a well defined k vector corresponds to an infinitely propagating wave. The attenuation of the penetrating wave in the solid (in real space), however, is equivalent to a spectrum of different momentum values (in **k**-space) thus leading to a spectral band of solid state excitations fulfilling the momentum conservation rule. Even more stringent limitations are often imposed by the solid itself since a well defined momentum in the solid requires an ideal homogeneity of the lattice. In real crystals impurities and dislocations are responsible for momentum non-conservation which is reflected by a corresponding broadening of the spectral lines observed. In case of disordered media such as amorphous solids the Raman phonon spectra can be understood as approximately the density of the phonon states by averaging entirely over the whole Brillouin zone. The line shape of phonon lines observed in Raman spectra can thus be used to characterise the degree of disorder in a solid.

In the Raman-scattering process the interaction between the photons and the elementary excitations is indirect. It is mediated by electronic interband transitions, because these transitions define the dielectric susceptibility χ in the visible spectral range, where Raman experiments are usually performed. Raman scattering occurs when the interband transitions are influenced by a phonon excitation for example. The generation of scattered light with frequency ω_s by incident light with frequency ω_i can be described by a generalised dielectric susceptibility tensor $\tilde{\chi}(\omega_i, \omega_s)$:

$$\mathbf{P}(\omega_s) = \epsilon_0 \tilde{\chi}(\omega_i, \omega_s)\mathbf{E}(\omega_i). \tag{4.4}$$

Here $\mathbf{P}(\omega_s)$ is the oscillating polarisation which gives rise to the scattered light wave and $\mathbf{E}(\omega_i)$ is the oscillating electric field of the incident light wave. The Raman scattering intensity can thus be expressed by the dipole radiation intensity using the generalised dielectric susceptibility $\tilde{\chi}(\omega_i, \omega_s)$:

$$I_s = I_i \frac{\omega_s^4 V}{(4\pi\epsilon_0)^2 c^4} \left| e_s \tilde{\chi}(\omega_i, \omega_s) e_i \right|^2 \tag{4.5}$$

Here $I_{i,s}$ and $e_{i,s}$ denote intensity and polarisation unit vector of incident and scattered light, and V is the scattering volume. The Raman scattering efficiency can be defined from Eq. (4.5) by normalizing to the incident power I_i.

The influence of the phonon on the generalised susceptibility tensor can be expressed by a Taylor expansion of the tensor components $\chi_{\alpha,\beta}(\omega_i, \omega_s)$ in terms of the lattice deformation Q_j due to the phonon normal modes j. For phonon excitations of the form $Q_j = Q_j^0 \exp(i\,(\Omega_j t - q_j r))$, one obtains:

$$
\begin{aligned}
\chi_{\alpha,\beta}(\omega_i, \omega_s) \;=\;& \chi_{\alpha,\beta}^0(\omega_i) \\
&+ \sum_j Q_j \cdot \left(\frac{\partial \chi_{\alpha,\beta}(\omega_i)}{\partial Q_j} \right) \\
&+ \sum_j i\, Q_j \cdot q_j \left(\frac{\partial \chi_{\alpha,\beta}(\omega_i)}{\partial \nabla Q_j} \right) \\
&+ \sum_{j,\gamma} Q_j \cdot E_\gamma \cdot \left(\frac{\partial^2 \chi_{\alpha,\beta}(\omega_i)}{\partial Q_j \partial E_\gamma} \right) \\
&+ \sum_{j,j'} Q_j Q_{j'} \cdot \frac{1}{2} \left(\frac{\partial^2 \chi_{\alpha,\beta}(\omega_i)}{\partial Q_j\, \partial Q_{j'}} \right) \\
&+ \dots
\end{aligned}
\tag{4.6}
$$

Here Ω_j is the phonon frequency, Q_j^0 the amplitude, and q_j the wave vector of mode j. α and β are the directions of the scattered and the incident electric field, and E_γ is the γ-component of a possible additional static electric field. Besides the susceptibility $\chi_{\alpha,\beta}^0(\omega_i)$ of the crystal without phonons, the right-hand side of Eq. (4.6) contains three terms due to one-phonon processes, giving rise to scattered photons whose energy shift is equal to Ω_j. The last term in (4.6) originates from two-phonon processes, which lead to scattered photon frequencies $\omega_s = \omega_i \pm (\Omega_j \pm \Omega_{j'})$. Thus, they can cover a wide variety of eigenmode combination frequencies.

The one-phonon terms originate from the susceptibility modulation by a lattice deformation Q (term 2), a deformation gradient ∇Q (term 3), and a lattice deformation in the presence of a static electric field $\mathbf{E}_\gamma$ (term 4). Scattering due to those terms which only involve one or more phonon amplitudes but otherwise leave the crystal symmetry invariant (term 2 and 5), is often called "symmetry-allowed" Raman scattering, while "symmetry-forbidden" scattering requires further symmetry-reducing elements, such as a deformation gradient or a static E-field (terms 3 and 4). In this sense "forbidden" means forbidden in the first order of the selection rules, where only the symmetry properties of the phonon eigenvectors are involved. The partial derivatives in (4.6) constitute the Raman polarisability $\tilde{\chi}$, often termed as Raman tensor $\tilde{R}$.

Alternatively to (4.6), light scattering can also be described in a microscopic quantum mechanical time-dependent perturbation theory, which represents the photon-electron and the electron-lattice interaction by three electronic transitions [4.7]:

- the electronic transition from the ground state $|0\rangle$ to an excited state $|e\rangle$: creation of an electron-hole pair due to the absorption of a photon with the energy $\hbar\omega_i$.

- the electron-lattice interaction, i.e. the electronic transition from $|e\rangle$ to $|e'\rangle$ under creation or annihilation of a phonon with $\hbar\Omega$.

- the transition from $|e'\rangle$ to the ground state $|0\rangle$: recombination of the electron-hole pair under emission of a photon $\hbar\omega_s$.

For the combination of these processes the third-order perturbation theory yields as the dominant term [4.8]:

$$\chi_{\alpha,\beta}(\omega_i, \omega_s) = \frac{e^2}{m_0^2 \cdot \omega_s^2 \cdot V} \sum_{e,e'} \frac{\langle 0|p_\alpha|e'\rangle \langle e'|H_{E-L}|e\rangle \langle e|p_\beta|0\rangle}{(E_{e'} - \hbar\omega_s)(E_e - \hbar\omega_i)} \tag{4.7}$$

Here m_0 is the electron mass and V the scattering volume, p_α and p_β are vector components of the dipole operators of the scattered and incident light, and H_{E-L} is the electron-phonon interaction Hamiltonian. E_e and $E_{e'}$ are the energies of the excited electronic states.

4.1.2 Electron-Phonon Interaction

The electron-phonon interaction processes which give rise to the derivative $\partial\chi_{\alpha,\beta}/\partial Q_j$ in (4.6) and to the Hamiltonian H_{E-L} in (4.7) are classified in terms of the specific properties of the electronic band structure which are influenced by the phonon, or in terms of the phonon properties (lattice deformation or electric field) which are responsible for the interaction.

The electronic band structure properties which may be influenced by a phonon are the electronic transition energies and the eigenfunctions of the electronic states.

The transition energies can be modulated through phonon-induced energy shifts of the valence and/or conduction bands. In terms of (4.7) this process corresponds to a transition between intermediate states $|e\rangle$ and $|e'\rangle$ in the same energy band, thus to intraband electron-phonon interaction. In this case the total Raman process involves only two bands: one valence band and one conduction band. It is called a two-band process.

Phonon-induced modulations of the electronic eigenfunctions lead to a change of the transition probabilities in (4.7). Since an eigenfunction modification may be described as an admixture of states of adjacent bands, it implies according to (4.7) an electronic transition between states $|e\rangle$ and $|e'\rangle$ of different bands (interband electron - phonon interaction). Therefore, the total scattering process is a three-band process: it implies either one valence band and two conduction bands or one conduction band and two valence bands.

As can be seen from (4.7), in most cases (when the spacing between the energy bands is not too small) the scattering efficiency of two-band processes by far dominates the three-band scattering. For two-band processes, both factors of the denominator are approximately zero for a properly chosen photon energy, while for three-band scattering both factors can not vanish simultaneously.

The phonon may influence the electronic band structure by two different properties. The most common influence is due to the lattice deformation itself, giving rise to an additional lattice-periodical potential and thus affecting the electronic band structure. This interaction, called Deformation Potential (DP) scattering, applies for Transverse Optical (TO) as well as Longitudinal Optical (LO) phonons. It is the dynamic analogon to the piezo-modulation of the electronic band structure. The effect of such a strain to the critical points of the band structure was calculated by Kane [4.9]. The influence of the deformation potential on the different energy gaps in III-V semiconductors and its implications for Raman scattering have been treated extensively by Richter [4.2].

For polar phonons, e.g. in III-V and II-VI semiconductors, an additional interaction mechanism occurs for the longitudinal optical mode, since the lattice deformation of the LO mode implies a macroscopic electric field [4.10,4.11]. Besides acting as an additional restoring force and thus leading to an enhanced frequency of the LO phonon with respect to the TO, this field may also give rise to light scattering, since it may influence the electronic system. It leads to a long range interaction mechanism which is called Fröhlich interaction [4.12]. This interaction is crucial for the analysis of electric properties of interfaces by Electric Field Induced Raman Scattering (EFIRS), which will be discussed in Sect. 4.5.2.

4.1.3 Resonance Effects

From the quantum mechanical picture of the scattering process as described by Eq. (4.7) it is evident that the probability of electronic transitions in the spectral regime of the incident and scattered photons on the one hand and the strength of the electron-phonon interaction on the other hand are the physical quantities determining the cross section of the inelastic scattering process. Since the energy conservation law is valid for the whole process only, the electron-hole pair states involved are virtual and must not necessarily coincide with the real electronic band structure. However, if the energy corresponds to real states the excitation or annihilation probability of the electron hole pair increases dramatically. As a consequence, the Raman scattering cross section can vary over several orders of magnitude in dependence of the photon energy used for excitation. Of crucial importance are the so-called critical points of the electronic band structure, where a filled and an empty band develop parallel in k-space giving rise to a large combined density of states. The so-called Resonance Raman scattering takes advantage of the large scattering enhancement which is observed for photon energies of the incident or scattered light close to a critical point of the electronic band structure.

The Raman resonances can also be described within the susceptibility approach. The resonance is reflected in (4.6) by an enhancement of the partial derivatives for appropriate incident light frequencies. According to Eq. (4.4) and (4.6) the scattering efficiency is determined by the generalised susceptibility (also termed Raman susceptibility) which can be expanded in different terms according to the specific scattering process. The Raman susceptibility is described by the time dependent terms, i.e. derivatives of the susceptibility with respect to the phonon deformation. The Raman susceptibility (e.g. first order scattering) can be rewritten as:

$$\frac{\partial \chi}{\partial Q} = \frac{\partial \chi(E_C)}{\partial E} \cdot \frac{\partial E_C}{\partial Q} \tag{4.8}$$

Here $\partial E_C/\partial Q$ is the phonon deformation potential describing the electron-phonon interaction at the optical gap E_C. The validity of the susceptibility model can thus experimentally be verified by studying the Raman efficiency as a function of exciting photon energy and comparing the result to the derivative of the susceptibility. This in fact has been done for several bulk semiconductors such as GaAs [4.55], InP [4.56] and Si [4.57] and will be described in Sect. 4.4.1.

The resonance of Raman scattering is of importance mainly for three reasons. Firstly, the resonance determines the scattering cross section of Raman scattering. If for instance an extremely thin layer is to be analysed the resonance condition is crucial to allow recording any signal at all. Secondly, if higher order scattering processes are of interest the scattering cross section is generally low in off-resonance condition. Therefore, for higher order scattering the resonance is even more pronounced than for first order scattering since additional terms of $(E_e - \hbar\omega_i)$ appear in the denominator of Eq. (4.7) in this case [4.2, 4.3]. This is e.g. important for the electric-field induced scattering which may be used to analyse fields occurring at semiconductor interfaces as described in Chapt. 4.5. Thirdly, the resonance behaviour itself can be used as analytical tool to determine electronic properties of the sample under study. By varying the photon energy of the exciting light and monitoring the Raman efficiency the electronic band structure of a solid can be explored [4.2, 4.3].

Finally, one should keep in mind that the summation over the electron-hole pair states $|e\rangle$, $|e'\rangle$ in Eq. (4.7) samples the joint density of states. Regarding low-dimensional structures the resonances may become of even larger importance due to the more pronounced maxima in the density of states at critical points compared to the three-dimensional case.

4.1.4 Selection Rules

Selection rules are symmetry considerations, based on group theory, which give necessary conditions for a phonon to be observable in Raman spectroscopy, i.e. to be Raman-active. They determine, which components of the Raman tensor $\tilde{\chi}$ may be nonzero; this means the polarisation directions $\mathbf{E_i}$ and $\mathbf{E_s}$ of incident and scattered light, for which a nonvanishing Raman scattering intensity may be expected. Commonly, the experimental configuration is given in a compact notation as $\mathbf{k_i}(\mathbf{E_i};\mathbf{E_s})\mathbf{k_s}$, which is usually called the Porto-notation .

For materials whose crystal lattice has a centre of inversion, group theory predicts that a phonon can be observed exclusively either in Raman spectroscopy or in Infra Red (IR) spectroscopy, depending on its symmetry properties. The reason for this exclusion criterion is that in such crystals the irreducible representations of the phonons can be classified in either even modes (g-modes; "gerade") or odd ones (u-modes; "ungerade"). Even symmetry means invariance of the lattice deformation against inversion while for odd symmetry modes inversion implies a 180° phase shift of the phonon.

Figure 4.2 gives examples for both cases. The first example shows a row of atoms along the [111] direction of Si, whose lattice consists of two face centered cubic (fcc) lattices, shifted by a 1/4 space diagonal (O_h-symmetry: diamond structure), yielding a diatomic base whose atoms are located at (0;0;0) and (1/4;1/4;1/4). This structure has a centre of inversion, located in the centre of the base at $1/2 \cdot (1/4;1/4;1/4)$. Obviously, the Si lattice vibration, whose atomic displacements are indicated by arrows in the figure, is invariant against inversion, thus its symmetry is even. A different situation occurs for NaCl, which consists of two fcc lattices, shifted by 1/2 space diagonal. Here the

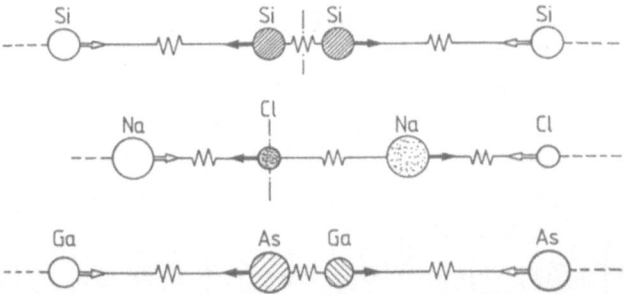

Fig. 4.2. Phonon symmetry for different crystal lattices: The dashed circles are the diatomic base of an elementary cell, open circles are atoms of neighbouring cells. The dash-dotted vertical lines mark the centre of inversion. (a) Si (O_h^7-structure) has a g-mode (even symmetry), which is Raman-active. (b) NaCl (O_h^5-structure) has a u-mode (odd symmetry), which is IR-active. (c) GaAs (T_d-structure) shows both Raman and IR-activity

atoms of the diatomic base are located at $(0,0,0)$ and $(1/2,1/2,1/2)$, yielding an inversion centre at the atoms themselves. This implies that the phase of the lattice vibration is shifted by $180°$ by inversion, thus the phonon mode has odd symmetry.

Due to symmetry arguments, phonons with even symmetry can be Raman active, those with odd symmetry IR-active. The requirement of even symmetry for Raman activity can be understood from (4.7), since only even phonons lead to coupling of electronic states $|e\rangle$ and $|e'\rangle$ of equal symmetry. The equal symmetry of $|e\rangle$ and $|e'\rangle$ is a requirement for coupling both of them to the ground state $|0\rangle$ by vector-like dipole operators p_α and p_β, respectively. The IR-activity of odd phonons means that these vector-like modes can interact directly with the vector-like IR electric field, leading to IR absorption.

It should be mentioned that the exclusion criterion of Raman and IR activity of phonon modes does not apply to crystal lattices like III-V compounds. Their structure is like Si, but without a centre of inversion, because of the nonequivalent sublattices: one consisting of group III atoms, the other of group V (T_d-symmetry: zincblende). This allows the observation of the lattice vibrations in Raman as well as in IR spectroscopy. Moreover, the degeneracy of the TO and LO phonon modes is lifted by the shift of the LO frequency due to its macroscopic electric field.

The deduction of the nonvanishing components of the Raman tensor from group symmetry considerations can be understood from (4.7) by analysis of the symmetry of the electronic states which are coupled by the phonon mode through the Hamiltonian H_{E-L}. Group theory yields for deformation potential scattering, that a lattice deformation in x-direction can couple odd symmetry electronic states of y-like symmetry to z-like states, while y-deformation couples x-states to z-states, and z-deformation couples x-states to y-states [4.13]. Coupling of y-states to z-states by x-deformation implies that for incident light polarised in y-direction the scattered light is polarised along z when the sublattices are displaced along x-direction.

Therefore, for deformation potential scattering from phonons with zero wave vector, the Raman tensors for O_h and T_d symmetry in the main crystal axis system ($x = [100]; y = [010]; z = [001]$) only have off-diagonal elements [4.14]:

$$\tilde{R}(x) = \begin{pmatrix} 0 & 0 & 0 \\ 0 & 0 & a \\ 0 & a & 0 \end{pmatrix}, \quad \tilde{R}(y) = \begin{pmatrix} 0 & 0 & a \\ 0 & 0 & 0 \\ a & 0 & 0 \end{pmatrix}, \quad \tilde{R}(z) = \begin{pmatrix} 0 & a & 0 \\ a & 0 & 0 \\ 0 & 0 & 0 \end{pmatrix}. \quad (4.9)$$

Here the indices x, y and z give the direction of the phonon-induced lattice deformation. Furthermore, as can be seen from Fig. 4.3, for backscattering at a (100) surface, the polarisation of the incident and the scattered light must be in the y,z-plane, since we deal with transverse incident and scattered waves, propagating approximately in x and $-x$ direction, respectively. Therefore only the yy, yz, and zz-components of the Raman tensors can contribute to

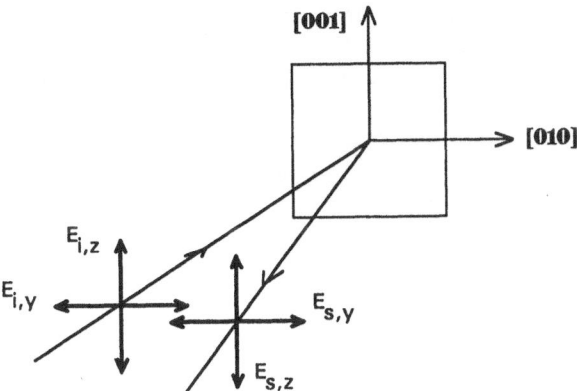

Fig. 4.3. Possible polarisation configurations for backscattering at a (100) surface. Since the polarisation of incident and scattered light must be perpendicular to the [100] direction, according to (4.9) only a lattice deformation in [100] direction, i.e. a longitudinal phonon can be observed

the scattering process and, according to (4.9), a nonvanishing contribution can only originate from $\tilde{R}(x)$. This implies that only a lattice deformation in x-direction can be observed. Since the momentum conservation law (4.3) requires the phonon propagation to be also in x-direction, we deal with a LO phonon. Transverse phonons which propagate in x-direction have a lattice deformation in y- or z-direction, giving rise to tensor components xz and xy. Thus, for backscattering at a (100) surface, only the LO phonon is allowed. For polarisation along the main crystal axes it appears in the off-diagonal scattering configuration, e.g. $100(010; 001)\bar{1}00$, whereas it can be observed in diagonal configuration for polarisation along the surface diagonal axes, e.g. for $100(011; 011)\bar{1}00$.

For misoriented surfaces the Raman selection rules are modified due to polarisation components along other crystal directions. This may lead to the appearance of phonon peaks which were originally forbidden. When switching from (100) to (110) or (111) surfaces, the application of the corresponding rotation matrices to the tensors of (4.9) yields that in backscattering experiments at (110) surfaces, only the TO phonon can be observed, while at (111) surfaces both the TO and the LO phonon may appear in the Raman spectrum. In Table 4.1 these results are summarised. Here it has been assumed that the direction of the incident and the scattered light is exactly normal to the surface. Deviation from this criterion, e.g. by the finite aperture of the collection optics for the scattered light and possibly also by oblique incidence of the laser beam, have the same effect as a misorientation of the surface. In Table 4.1 also the polarisation directions which are required to observe the phonons are listed. Besides the deformation potential scattering, the observation of the LO phonon through the Fröhlich mechanism is also included.

Table 4.1. Observation of TO and LO phonons from Deformation Potential scattering (DP) or Fröhlich scattering (F) on materials with T_d symmetry for various crystal-face directions and polarisation directions of the incident and scattered light E_i and E_s

surface	E_i	E_s	TO	LO
(100)	[010]	[001]	-	DP
	[010]	[010]	-	F
	[001]	[001]	-	F
	[011]	[011]	-	DP + F
	[011]	[0$\bar{1}$1]	-	-
(110)	[1$\bar{1}$0]	[1$\bar{1}$0]	DP	F
	[1$\bar{1}$0]	[001]	DP	-
	[001]	[001]	-	F
(111)	[1$\bar{1}$0]	[1$\bar{1}$0]	DP	DP + F
	[1$\bar{1}$0]	[11$\bar{2}$]	DP	-
	[11$\bar{2}$]	[11$\bar{2}$]	DP	DP + F

For this scattering process, group theory yields only diagonal tensor components [4.2, 4.3] and the LO phonon may be observed not only from (100) and (111) surfaces, but also from (110).

4.2 Experimental Setup for Raman Scattering

The essential parts of a Raman setup are a laser light source, a spectrometer and a very sensitive photon detector. In the following Sections we discuss the constituent parts of the experimental setup in detail. A possible realisation is shown in Fig. 4.4. Different arrangements of the sample in a Raman experiment are possible. While postgrowth analysis can be performed in a cryostat or even under ambient conditions, the Raman study of the growth of semiconductors is performed *in-situ* in an UHV vessel. The experimental aspects of *in-situ* Raman analysis are discussed in Sect. 4.2.5.

4.2.1 Light Source

Raman spectroscopy requires a monochromatic light source which should have a stable and considerably high intensity. Therefore, it has benefitted most by the development of lasers on the visible spectral range. Usually an Ar^+ or a Kr^+ ion laser is applied. These lasers offer a variety of discrete lines. An

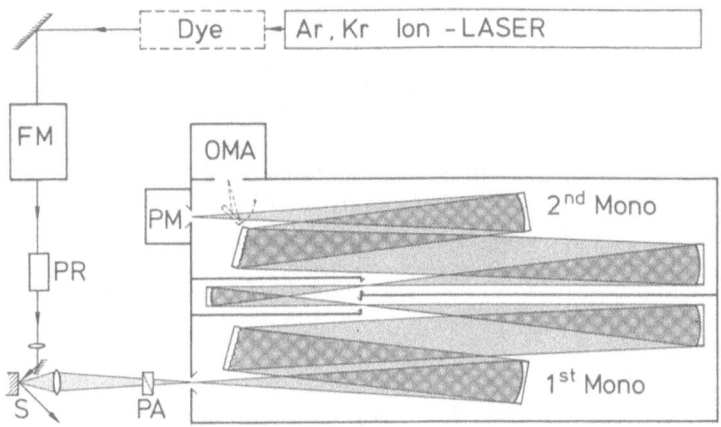

Fig. 4.4. Schematic view of an experimental arrangement for Raman spectroscopy, using an additive double monochromator. For the detection, either a single channel photomultiplier (PM) or a multichannel detector can be used. FM: foremonochromator; PR: polarisation rotator; PA: polarisation analyser; S: sample

Ar^+ laser has up to 10 lines in the spectral range from green to ultraviolet, the strongest at 514 nm, 488 nm and 457 nm. The lines of a Kr^+ laser are in a wider spectral interval, ranging from IR to UV. The strongest lines are in the red spectral region: at 647 nm and 676 nm. Through this variety of laser lines different penetration depths in semiconductors may be chosen and the conditions for resonant Raman scattering can be improved. Since the violet Kr^+-laser lines are near the E_1 gap of GaAs and InP, they are very important for EFIRS experiments. To optimise the resonance conditions, often lasers are applied which allow a quasi-continuous variation of the laser wavelength. These lasers are based on a laser medium which gives stimulated spectral emission in a broad interval in which a laser wavelength can be chosen by variation of the resonator conditions. The required population inversion in the lasing medium is achieved through optical pumping by an Ar^+ or Kr^+ laser. Classically, liquid dyes are applied as a lasing medium. The spectral range of each dye is limited to about 50... 100 nm but, due to the availability of a large number of different dye materials, the whole spectral region between 400 nm and 900 nm can be covered. More recently, a quasi-continuously variable solid state laser whose laser medium is a Ti-sapphire crystal was developed to cover the spectral range between 600 nm and 1 μm. Besides the cleaner operation, an essential advantage of the Ti-sapphire laser is its constant efficiency over a wide power range, in contrast to the dye lasers whose output saturates due to heating effects.

In spite of the monochromatic character of the laser light, a spectral filter has often to be introduced into the laser beam. This filter eliminates nonlasing emission lines from the light source which, due to diffuse reflection from the sample, could otherwise enter into the monochromator and simulate Raman peaks in the spectrum. In most cases an interference filter or a prism monochromator is used for their suppression.

The polarisation of the incident light can be varied by a $\lambda/2$ plate or a Fresnel rhombus [4.15]. The laser beam is focused to the sample by a spherical or a cylindrical lens. The latter offers the advantage of a reduction of the power density in the focus, thus minimising heating effects. If optimum lateral resolution is required, a microscope objective is applied, yielding a focus in the μm range (see Sect. 4.2.4)

4.2.2 Raman Spectrometer

The scattered light is collected through a lens system and focused onto the entrance slit of a spectrometer. A high-quality imaging system is required since the entrance slit must be quite narrow to achieve a sufficient spectral resolution. Especially spherical aberration effects should be minimized, because they considerably reduce the effective aperture angle. Inserting a polarisation analyser in front of the entrance slit allows the selective detection of scattered light with a well-defined polarisation to separate the individual components of the Raman tensor.

For the frequency analysis of the scattered light, a grating monochromator is applied. Because of the extremely low efficiency of the Raman scattering process, the monochromator must give a very high contrast. Therefore a double or triple monochromator is used to separate the very weak Raman light from the by far more intense elastically-scattered light. The mode of operation of the monochromator depends on the choice of the light detector, which can be either a Photo Multiplier (PM) or a multichannel detector. A multichannel detector consists of up to 1000 detector elements in a row of about 2.5 cm, enabling the parallel detection of a specific spectral interval [4.16].

When using a PM, an additive double grating monochromator is usually applied, as is shown in Fig. 4.4. The frequency resolution in the Raman spectrum is determined by the focal length of the monochromators, the line density of the gratings and the slit widths. For example, a double monochromator of a focal length $f = 80$ cm, equipped with 1800 lines/mm gratings, gives for wavelengths near 500 nm a spectral halfwidth of $1.3 \, \text{cm}^{-1}$ when the geometrical slit width is set to 100 μm. The throughput is determined by the relative aperture, which is usually about 1:8. Often a spatial filter between both monochromators serves to improve the stray light reduction. For multichannel detection a broad spectral interval has to be transmitted to the detector, which can be achieved by a considerable enhancement of the slit width between first and second monochromator (beyond 1 cm). However, this procedure may drastically deteriorate the stray light reduction.

To meet the requirement of the transmission of a broad spectral interval, combined with an efficient reduction of the stray light, triple monochromator systems with an optimized geometrical arrangement have been developed. As can be seen from Fig. 4.5, these systems consist of a subtractive double monochromator to filter out the elastically scattered light, followed by a third monochromator to project the Raman spectrum onto the multichannel detector.

The exit slit of the first monochromator is set such that after the geometrical separation of the spectral components of the scattered light the interval which contains the Raman lines is transmitted to the second stage. The laser frequency, in contrast, is blocked by the slit edge. By the subtractive operation of the second monochromator with respect to the first one, the spectral separation is compensated and the whole spectral interval of the Raman lines has a common focus on the slit between the second and the third monochromator. This concept allows a very narrow width of this slit and, therefore, the system gives a very efficient stray light reduction which is even improved when utilising a spatial filter between the first and the second monochromator, although the filtering is in this case less effective than in the case of PM-operation with a double monochromator.

Finally, in the third monochromator, the interval of the Raman lines is spectrally separated and focused to the multichannel detector elements. The spectral width of the detected interval is determined by the focal length and the grating period of the third monochromator. A monochromator with 60 cm focal length, equipped with a grating of 1800 lines/mm, gives for wavelengths in the region of 500 nm an interval of 650 cm^{-1} on a detector of 2.4 cm width, which means about 0.65 cm^{-1} per detector element. Sophisticated versions of triple monochromators allow a fast switching to a mode with an improved spectral resolution, achieving about 0.2 cm^{-1} per detector element. This change can be performed by employing a grating with an increased line density, mounted on the same stage in the third monochromator, or by switching the first and second monochromator from the subtractive to the

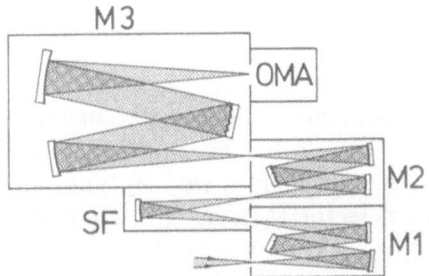

Fig. 4.5. Triple monochromator setup for Raman spectroscopy, consisting of a subtractive double monochromator (M1; M2) with spatial filter (SF) for stray light reduction, followed by a third monochromator (M3) to project the spectrum onto the multichannel detector (OMA)

additive operation mode. In the latter procedure the light path is modified by introducing additional mirrors. As an important consequence, in this operation mode the width of the detected spectral interval is determined by the entrance slit width of the third monochromator, since this slit is no longer a common focus for all frequencies. The required slit widening deteriorates the stray light reduction. Therefore the additive mode is only applicable for samples with very flat surfaces, which lead to a low stray light intensity.

For a very efficient reduction of the stray light at the laser frequency, very recently holographic filters have become commercially available. These so-called super-notch filters have in their transmission spectrum a very steep dip with a transmission of 10^{-6} at the centre and a spectral half width of about $150\,\mathrm{cm}^{-1}$. Because of their interference-based operation principle the reduction of the laser line requires that the light must pass through these filters as a parallel beam. This development has led to a novel concept of Raman spectrometers, which do not contain any grating. Instead, the essential part of the first stage is a notch filter, while in the second stage the wavenumber of the Raman light to be transmitted is tuned by tilting an interference filter. The application of a twodimensional (CCD) multichannel detector (see below) allows an imaging of the sample, using a specific Raman line, e.g. the $520\,\mathrm{cm}^{-1}$ line of silicon. Thus, a material-specific picture can be obtained. Alternatively, conventional Raman spectra of a single spot can be recorded by tilting the interference filter. In this spectral scanning operation mode, the image of the sample spot to be analysed covers only the centre of the CCD detector, thus eliminating the advantages of multichannel detection.

For samples with a strong luminescence in the visible region, Raman spectroscopy in the near infrared can provide an alternative. Here Fourier-transform Raman spectrometers are available. As a light source they use a Nd:YAG laser, operating at a wavelength of $1.03\,\mu\mathrm{m}$. The concept of Fourier-transform Raman spectroscopy could only be realized after the development of high-performance filters for the blocking of the stray light prior to the interference stage.

4.2.3 Multichannel Detector

The development of multichannel detector systems has led to a tremendous increase of the efficiency in detecting Raman light [4.16]. The most commonly used types are the diode array, the CCD (Charge Coupled Device) type, and, to a lesser extend, the position-sensitive detector (Mepsicron). They were compared in a review by Tsang [4.17].

In a diode array detector the spatial distribution of light intensity is stored as a charge distribution pattern on an array of Si diodes. To achieve the required sensitivity for Raman experiments an intensifier stage is placed in front of the diode array. The intensifier contains a large area photocathode which emits electrons due to the Raman light. A microchannel plate, placed directly behind the photocathode, leads to a multiplication of these electrons by a factor of about 10^3. The resulting electron bunches are transformed into light pulses by a luminescence screen at the end of the multichannel plate. The diode array is attached directly to the luminescence screen.

The sensitivity of the system is about 0.2 pulses per photon, a value which is comparable to a PM. Cooling of the detector to $-35\,^\circ$C reduces the electronic background per channel to about three pulses/second.

The fundamental limit for the spectral resolution is the diode width of $25\,\mu$m, while a further reduction is imposed by a slight smearing out of light between the intensifier stage and the diode array. This leads to illumination of neighbouring diodes and results in an effective geometrical width of $75\,\mu$m. Since typical dispersion values for commercially available multichannel / triple monochromator systems are $1\,\mathrm{cm}^{-1}$ per channel in the subtractive mode and $0.3\,\mathrm{cm}^{-1}$ per channel for additive operation at wavelengths near $500\,$nm, the resulting spectral half widths are approximately $3\,\mathrm{cm}^{-1}$ and $1\,\mathrm{cm}^{-1}$, respectively.

Recently, significant progress has been achieved in the development of multichannel detectors based on Charge Coupled Devices (CCD). Here the detector is realised as a silicon integrated circuit (IC) [4.18]. The readout of the information is performed by shifting the accumulated charges across the detector elements towards the preamplifier. For this purpose the voltage levels of the subsequent elements in the IC are varied stepwise. The IC-chip also contains the preamplifier. It can be cooled down to liquid nitrogen temperature. This concept leads to a high sensitivity (quantum efficiency beyond 60 percent), combined with an extremely low background and noise level, thus enabling very long accumulation times. Besides, the two-dimensional array of detector elements on a CCD chip allows the simultaneous recording of Raman spectra from a finite range of the sample when a line focus is applied for the illumination.

A different concept is followed in the so-called position-sensitive detector, called "Mepsicron" (Microchannel Electron Position Sensor with time (ChRONos) resolution). Instead of separate detector elements, here a single large-area photosensitive plate is applied, which has electrical contacts at the corners. The position of the incident photons is evaluated by comparison of the arrival times of the signals at the various contacts. These detectors have a high efficiency, but do not allow the unambiguous detection of quasi-simultaneously impinging photons. Thus, the maximum count rate is limited to about 10^5 counts per second over the whole detector area. This limited dynamical range may hamper the detection of weak signals adjacent to extremely strong ones.

4.2.4 Micro-Raman Spectroscopy

Commercial Raman setups nowadays also offer the capability of performing spatially resolved Raman experiments. This can be achieved using a light microscope as optics for both focusing the incident beam onto the sample and collecting the scattered light. For this purpose, the microscope contains a beam splitter, inserted between the objective and ocular lens system. By this setup a lateral resolution limited by the light diffraction to approximately 0.5 micron for visible light is achieved. The lateral resolution is of great importance for characterising patterned semiconductor samples which is in general the case for the analysis of semiconductor devices.

In comparison to standard optics with lenses of long focal lengths in the centimeter regime the microscope setup collects a much larger solid angle of scattered light. This results in a much higher collection efficiency thus enhancing the sensitivity of the setup but, on the other hand, the propagation direction of the scattered wave is less well defined which is of importance with regard to the selection rules of Raman scattering. Another aspect is that the small spot of the focused light leads to a large power density of the illumination which should be regarded with care. Considerable heating occurs for samples with small heat conductivity such as amorphous materials. Besides, chemical reactions at the surface can be enhanced by illumination, and finally, when opaque samples are studied photoinduced carriers can modify the spectral response of the sample (see e.g. [4.19]). Low-temperature micro-Raman spectroscopy is possible by the use of extremely compact cryostats in combination with objectives with a working distance up to 8 mm.

Figure 4.6 shows schematically three possible realisations of micro-Raman spectroscopy from heterostructures. One important application is the analysis of systems which are laterally structured on the micrometer scale. This analysis may be concerned with e.g. strain effects at the edges of lithographic structures (see Sect. 4.3.2). Furthermore, micro-Raman spectroscopy is applied for focusing on cleaved side-faces of heterostructures, as shown in the middle part of the figure. A scan along such a side-face from the upper surface to the substrate interface corresponds to an in-depth scan of the heterostructure: the individual layers subsequently dominate the spectrum. The picture on the right symbolises an additional method to obtain depth-resolution through micro-Raman spectroscopy: a lateral scan on the front surface after bevel

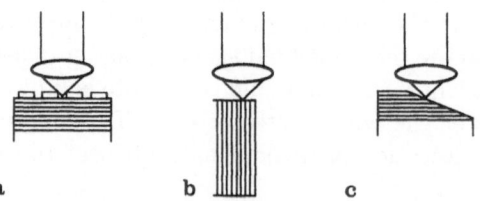

Fig. 4.6. Possible realisations of micro-Raman spectroscopy from heterostructures. **a** Locally resolved analysis of structured surfaces, **b** scan on a cleaved side-face, **c** scan on surface after bevel etching

etching or bevel polishing. Through a small-angle bevel the lateral resolution is converted into a very sensitively tunable depth resolution.

Recently it has been demonstrated that the diffraction limit of the resolution can be overcome by the so-called Scanning Near-Field Optical Microscope (SNOM). Such microscope typically utilizes laser illumination of a sample through an optical fiber which is terminated by an extremely small pin hole. Light propagating through the pin hole illuminates the sample placed within nanometer distance and allows for a spatially resolved characterisation limited by the pin hole diameter rather than the light wavelength [4.21, 4.22]. Future applications of SNOM may also Raman spectroscopy. Recently it has been demonstrated that SNOM can be utilized as spectroscopic probe recording the photoluminescense of semiconductor heterostructures [4.23].

4.2.5 In-Situ Experiments

The Raman scattering analysis can be extended to in-situ monitoring under special gas phase surroundings such as in gas phase epitaxy as well as under ultrahigh vacuum (UHV) conditions. This allows, for instance, to analyse clean surfaces, thin films down to single monolayer thickness and epitaxial growth. Since in Raman spectroscopy the lattice vibrations, whose eigenfrequencies are in the infrared spectral region, are analysed through incident and scattered light in the visible, a chamber with an appropriate arrangement of standard glass or quartz windows allows the performance of in situ analysis.

Figure 4.7 shows a Raman setup attached to an UHV vessel. Requirements are one or two optical viewports in the UHV vessel for the incoming

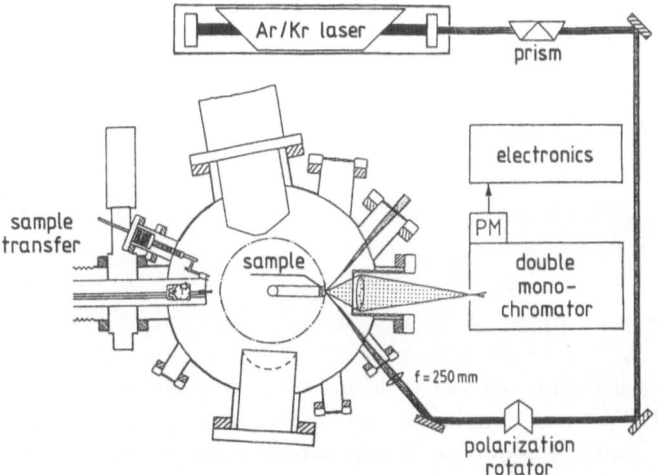

Fig. 4.7. Raman scattering setup for *in-situ* investigations under UHV conditions. The special window arrangement offers a sufficient angle of acceptance for the scattered light

and scattered light. The sample should be rather close to the collecting lens to maintain a large acceptance angle of the collection optics for high sensitivity. This might be a problem in a standard UHV vessel, where the sample manipulator does not allow the movement of the sample close enough to the viewports. A possible solution is a re-entrant window, located some centimeters inside of the UHV vessel, allowing the collecting lens to be placed quite close to the sample.

Especially for *in-situ* experiments, the application of a multichannel detection system offers important advantages, besides the ability to observe very weak structures in the Raman spectrum:

- Since the required registration time for a spectrum is drastically reduced, growth processes can be monitored on-line, and the danger of surface modification by residual gas adsorption during the experiment is diminished.
- Due to the improved detection efficiency the power density of the incident light can be significantly reduced. Hence photoinduced reactions at the surface can essentially be avoided.

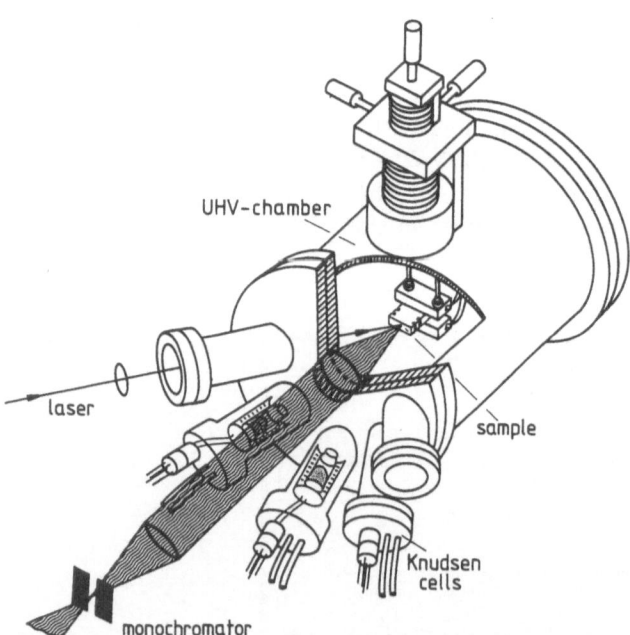

Fig. 4.8. Raman scattering setup for online *in-situ* investigations during MBE growth in UHV. The UHV vessel contains several evaporation sources which are arranged such around a center window that Raman experiments can be performed during continuous evaporation. The monochromator must be equipped with a multichannel analyser if spectrally resolved information is to be extracted during the growth process [4.20]

Using a Raman-UHV setup as shown in Fig. 4.8 with an optimized arrangement of sample, evaporation cells, and windows for sample illumination and Raman-light collection, the growth of overlayers can be observed on line, without moving any part of the equipment or affecting the growth process [4.20]. The latter opens new applications in Raman spectroscopy. In contrast to standard setups where long sampling times for a spectrum prevent the application of Raman spectroscopy for monitoring of dynamical processes, the setup shown in Fig. 4.8 allows real online analysis of semiconductor heterostructure fabrication by MBE.

4.3 Analysis of Lattice Dynamical Properties

Utilizing various inelastic light scattering mechanisms, Raman spectroscopy allows one to determine a variety of sample properties. Informations about the crystal lattice (structure, orientation, composition) as well as electronic properties (impurity incorporation, carrier concentration, band bending) can thus be obtained. In this chapter, representative examples will demonstrate how Raman scattering by lattice vibrational modes can be utilised to characterise semiconductor heterostructures and interfaces.

In comparison to electron spectroscopy methods like Photo Emission Spectroscopy (PES), Raman spectroscopy is advantageous for the investigation of buried interfaces, since the information depth of visible light may clearly exceed that of electrons. In PES experiments the information depth is restricted by the escape depth of electrons. The escape depth varies as a function of the kinetic energy of the electrons from a few monolayers at 40 – 100 eV typically achieved in Ultraviolet- (UPS) or in Soft X-Ray Photoemission Spectroscopy (SXPS) to a few tens of atomic layers above 1000 eV obtained in X-Ray Photoemission Spectroscopy (XPS) [4.24].

The information depth of Raman spectroscopy, on the other hand, is determined by the absorption of the incident and scattered light. For semiconductor overlayers the absorption coefficient of photons may vary over many orders of magnitude, depending on whether the photon energies are below or above the fundamental gap of the overlayer. As an example, the fundamental gap of ZnSe is 2.7 eV at 300 K [4.25], leading to transparency at a laser wavelength of 514.5 nm ($E_{photon} = 2.41$ eV), whereas a penetration depth as low as 100 nm at a wavelength of 406.7 nm ($E_{photon} = 3.05$ eV) is achieved [4.26]. Accordingly, by an appropriate choice of the laser wavelength the region to be analysed can be shifted between the front surface and the underlying interface [4.27].

In comparison to other optical methods, for instance Fourier Transform Infrared Spectroscopy (FTIR), Raman scattering by lattice vibrations can have the advantage of a much higher sensitivity to extremely thin films at interfaces. Whereas FTIR experiments utilise the direct interaction of the light with the lattice vibrations (in the same spectral regime), Raman scattering of visible light is mediated by the electronic band structure of a solid. The latter allows one to obtain resonance between the light probe and critical points of the band structure by tuning the exciting laser light which then gives rise to a enhancement of the scattering intensity of several orders of magnitude [4.2, 4.3]. In Sect. 4.3.1 it will be shown that Raman spectroscopy is able to monitor early stages of epitaxial growth using a standard experimental setup whereas IR absorption spectra require special multiple transmission/reflectance arrangements to investigate thin adsorbate layers [4.28]. For some systems even submonolayer sensitivity has been demonstrated with Raman scattering and small amounts of chemically-reacted phases at interfaces can be detected and identified.

Furthermore, the orientation of crystalline overlayers or interlayers can be determined via the selection rules of Raman scattering, and changes in the eigenfrequency of lattice vibrations can be used for a quantitative analysis of strain in layered systems. All these aspects will be addressed in the following Sections.

4.3.1 Crystalline Order

In this Section the determination of crystalline order during growth of thin overlayers on semiconductors is described. For this purpose examples of highly ideal interfaces are chosen where chemical reactions are excluded. Typical examples for ideally abrupt interfaces are group V elements (As, Sb, Bi) on III-V- and IV-semiconductor surfaces. Detailed *in-situ* Raman investigations were performed for the growth of Sb and Bi on III-V-(110) and for As on Si(111). The structural analysis by Raman spectroscopy may be performed with even submonolayer sensitivity under resonant excitation [4.32–4.34]. These systems have found a great experimental interest since they form not only abrupt but also geometrically well defined interfaces [4.29–4.31]. Due to the different lattice structures, however, an epitaxial growth of layers thicker than a single monolayer cannot be achieved. Instead either metastable crystalline modifications of a few monolayer thickness, amorphous growth or polycrystalline islands are observed. The examples described in here will outline some characteristic results of such *in-situ* Raman experiments.

In more common cases where resonance conditions are not ideally fulfilled or the overlayer growth is not ideally ordered, Raman spectroscopy may require a few monolayers of coverage to give a detectable signal. This has been shown in a wide range of growth studies by Raman spectroscopy for various semiconductor heterostructures. A large number of Raman scattering studies of heterostructures has been performed for II-VI- on III-V-semiconductors. This class of heterostructures is of large technological interest because the combination of II-VI with III-V allows to fabricate wide band gap to narrow gap semiconductor junctions which are feasible for optoelectronic devices [4.67]. Besides, several II-VI/III-V combinations show a good lattice matching which is an important condition for high quality growth of epitaxial layers. On the other hand, chemical exchange reactions occur in general at II-VI/III-V interfaces leading to the formation of III-VI-compounds. Examples of II-VI on III-V growth are thus treated in this Section with regard to structural properties as well as in Sect. 4.3.4 where interfacial reactions are addressed.

4.3.1.1 Vibrational Modes of Monolayers.

Among epitaxial monolayer adsorbates on semiconductor surfaces Sb on III-V(110) represents a widely studied and well characterized example. The overlayer grows in a crystallographic structure determined by the underlying substrate and thus distinct from the bulk structure of the adsorbed material. For Sb deposited on III-V-(110) surfaces the first monolayer adsorbs as zigzag chains along the $[1\bar{1}0]$ direction. Each Sb atom is bonded to one substrate atom and to two neighbouring Sb atoms, thus essentially continuing the III-V bulk structure [4.38, 4.39, 4.40, 4.41].

Such highly-ordered systems are expected to exhibit specific dynamic properties which should be observable by Raman spectroscopy. In fact clear experimental evidence was found for characteristic features due to Sb monolayers on InP-, GaAs- and GaP(110) surfaces [4.33,4.42–4.44]. The strongest signals were reported for Sb on InP(110) as shown in Fig. 4.9. The spectra were recorded between stepwise submonolayer deposition at 300 K on cleaved InP(110) surfaces in UHV using the 2.41 eV laser line of an Ar laser. For the clean surface only the TO and LO phonon modes of InP at 305 cm^{-1} and at 345 cm^{-1} and very weak multiphonon structures are observed. With increasing Sb coverage, six new peaks evolve (marked by the dash-dotted vertical lines). Three of them (96 cm^{-1}, 157 cm^{-1}, 185 cm^{-1}) are in the spectral range of Sb vibrations, the other three (289 cm^{-1}, 321 cm^{-1}, 354 cm^{-1}) are near the InP phonon frequencies. Striking are the high intensities, in the order of the substrate phonon intensities, and the spectral sharpness. Note that the peaks essentially evolve between 0.5 and 1 ML. In this coverage regime the overlayer develops from separate fragments to ordered chains. Consequently, the peaks are not just due to adsorbed Sb atoms, but they originate from interface vibrations of the ordered epitaxial monolayer. The subsequent growth of monolayer patches to a finally complete layer is also reflected by the evolution of the eigenfrequencies of the interface vibrations. Up to 1 ML the peaks show strong shifts of several wave numbers to higher frequencies, whereas after completion of the monolayer they remain essentially constant. The large number of peaks can be explained by assuming that the Sb epilayer and the upper InP layer are involved in the vibrations. This gives rise to a 4 atom base (In, P, and 2 Sb atoms), thus to 9 possible peaks. A strong anisotropy appears in the spectra for different polarisation configurations: by far the most intense peaks appear if the incident and scattered light are both polarised parallel to the

Sb chain direction [1$\bar{1}$0]. This is in accordance with the expected anisotropic electronic polarisability of the chains. Crucial for the observation of these monolayer vibrations is the appropriate choice of the exciting laser line. By using the various lines of an Ar- and a Kr-laser the resonance behaviour was tested yielding maxima between 2.4 eV and 2.7 eV [4.32]. The spectra shown in Fig. 4.9, for instance, were recorded at 2.41 eV (19436 cm^{-1}). The Resonance Raman experiments are discussed in more detail in Sect. 4.4.1.

For Sb on GaAs(110) the interface peaks can also be observed. However, their resonance behaviour has not yet been investigated in detail. For the laser line of 2.41 eV the scattering intensity is much weaker than on InP, furthermore the peaks evolve at different frequencies. Both effects confirm the influence of the substrate material on the electronic structure and the vibration dynamics of the interfaces.

An important consequence of the optical identification of the Sb monolayer is the ability to monitor the vibrational modes at the interface even after further deposition of Sb. In this way it was shown that the interface peaks persist for coverages above 14 ML, where the overlayer undergoes a phase transition from the amorphous to the crystalline state (see following Section). Consequently, the Raman analysis yields that the first Sb ML is essentially unaffected by the rearrangement of the overlying Sb atoms due to its strong bonding to the substrate [4.44, 4.36, 4.42].

Raman signals from monolayers were also reported for arsenic on Si(111) [4.34, 4.36]. For a laser excitation close to the resonance in the UV spectral

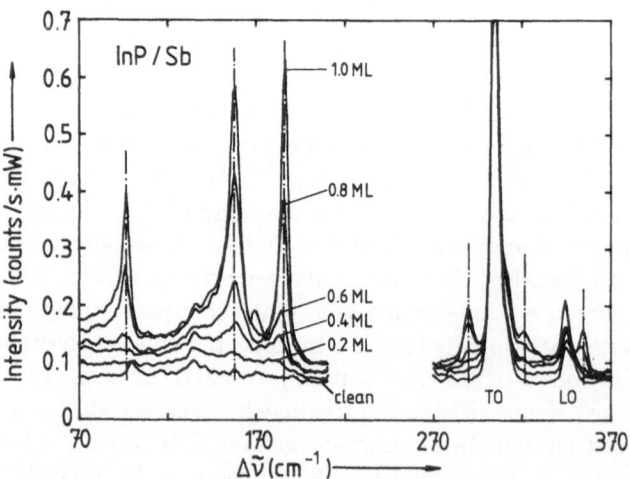

Fig. 4.9. *In-situ* Raman spectra of 0 … 1 ML Sb on InP(110) [4.33]. The interface vibrational modes due to the Sb are marked by dash-dotted lines. Their frequencies are different from bulk Sb modes. The spectra were recorded with the 2.42 eV laser line, the polarisation of incident and scattered light was along the [1$\bar{1}$0] direction

region a peak was observed at $356\,\mathrm{cm}^{-1}$ which was identified as a bending vibrational mode of the first ML of As. Also an indication of a second, stretching vibrational mode was found in agreement to results of Helium atom scattering [4.35]. The corresponding, well ordered As monolayer was produced by thermal desorption of an amorphous As cap which is formed during As deposition at liquid-nitrogen temperature. This procedure utilises the high thermal stability of the As monolayer in comparison to the weakly bonded amorphous As cap.

4.3.1.2 Structure of Thin Overlayers. As mentioned above, Sb films undergo structural phase transitions at a thickness of several monolayers. The type of structure formed depends on overlayer thickness as well as substrate temperature. For room temperature deposition the overlayer structure on top of the ordered monolayer develops from an amorphous phase which is stable up to approximately 15 ML into crystallites of the bulk Sb structure for higher coverages.

The Raman spectra shown in Fig. 4.10 were recorded with the excitation line of $\lambda = 514\,\mathrm{nm}$ (2.41 eV) which is close to the E_1-gap resonance of bulk crystalline Sb. The polarisation of the scattered light was not defined. For coverages below 12 ML they show a broad feature which is typical for amorphous

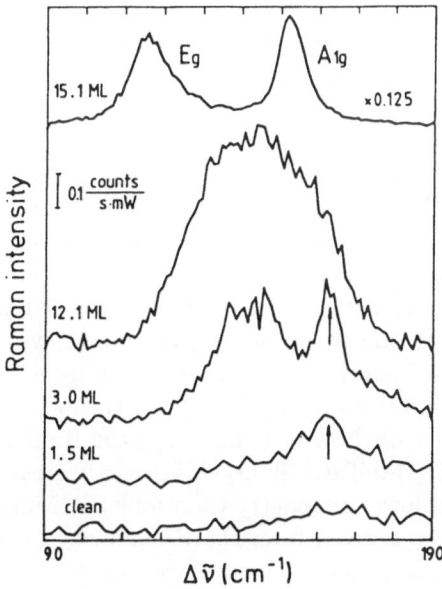

Fig. 4.10. Raman spectra of an Sb overlayer up to 15 ML on GaAs(110) recorded with the 2.41 eV line. The drastic change of the spectral shape between 12 and 15 ML reveals the crystallisation of the overlayer [4.46]

Sb. The spectra of amorphous materials essentially reflect the phonon density of states throughout the Brillouin zone. Accordingly a broad band appears, centered near $135\,\mathrm{cm}^{-1}$ with a halfwidth of about $30\,\mathrm{cm}^{-1}$ [4.45, 4.47].

The sharper peak at $180\,\mathrm{cm}^{-1}$ for 1.5 and 3 ML coverage is due to the epitaxial first ML (note that the substrate is GaAs(110) in this case). Beyond 12 ML the broad band of the amorphous Sb is replaced rather abruptly by the crystalline Sb peaks. The optical phonon modes are split into an A_{1g} mode (deformation along the crystal axis) and a doubly degenerate E_g mode (deformations perpendicular to the axis) with frequencies of $151.5\,\mathrm{cm}^{-1}$ and $113\,\mathrm{cm}^{-1}$, respectively [4.48, 4.49]. For coverages beyond 15 ML the signal disappears due to amorphous Sb, which means that the crystallisation involves the whole amorphous material. A more detailed analysis of the selection rules of the crystallised Sb reveals that it is textured with the c-axis perpendicular to the surface [4.46].

Investigations of the Sb overlayers by Raman spectroscopy were also performed for different substrate temperatures. They prove that different growth modes occur depending on the temperature chosen. If, for instance, the substrate is kept at 90 K during deposition, the crystallisation of the overlayer is suppressed and a macroscopic thick amorphous Sb layer is formed [4.45]. If, on the other hand, thin amorphous layers (between 2 ML and 10 ML) are annealed to temperatures between 370 K and 500 K, metastable crystalline modifications are formed.

By a combined Raman scattering and LEED analysis it has been shown that the crystal lattices of the metastable Sb layers on GaAs(110) and InP(110) are distinct from the bulk structure due to the influence of the underlying substrate [4.36, 4.44, 4.50]. The LEED patterns show rectangular surface meshes with lattice constants $a = 4.82\,\text{Å}$ or $4.89\,\text{Å}$ and $b = 4.44\,\text{Å}$ or $4.15\,\text{Å}$ on GaAs(110) or InP(110), respectively. The lattice constants a, b clearly demonstrate the effect of the substrates. In both cases the lattice constants are very different to the Sb bulk values of $a = 4.51\,\text{Å}$ and $b = 4.31\,\text{Å}$.

Consequently the Raman spectra display characteristic lattice vibrational modes distinct from the bulk modes (Fig. 4.11). The degeneracy of the two modes of E_g symmetry (displacements perpendicular to the threefold axis) is lifted in the substrate-stabilized lattices due to a lowering of crystal symmetry. Three optical modes with eigenfrequencies at $134\,\mathrm{cm}^{-1}$, $146\,\mathrm{cm}^{-1}$, $158\,\mathrm{cm}^{-1}$ and $136\,\mathrm{cm}^{-1}$, $148\,\mathrm{cm}^{-1}$, $162\,\mathrm{cm}^{-1}$ are found for Sb on GaAs(110) or InP(110), respectively. The occurrence of three eigenmodes in the Raman spectra demonstrates that the substrate stabilized Sb crystallises in a structure with two atoms per unit cell but of lower symmetry than bulk Sb. More insight into the crystal symmetry can be derived from the Raman selection rules of the phonon modes. The two high frequency modes are found to occur for parallel, the low frequency mode for crossed polarisations only. These selection rules are in agreement with an orthorhombic or monoclinic lattice, whereas a triclinic structure can be excluded [4.36].

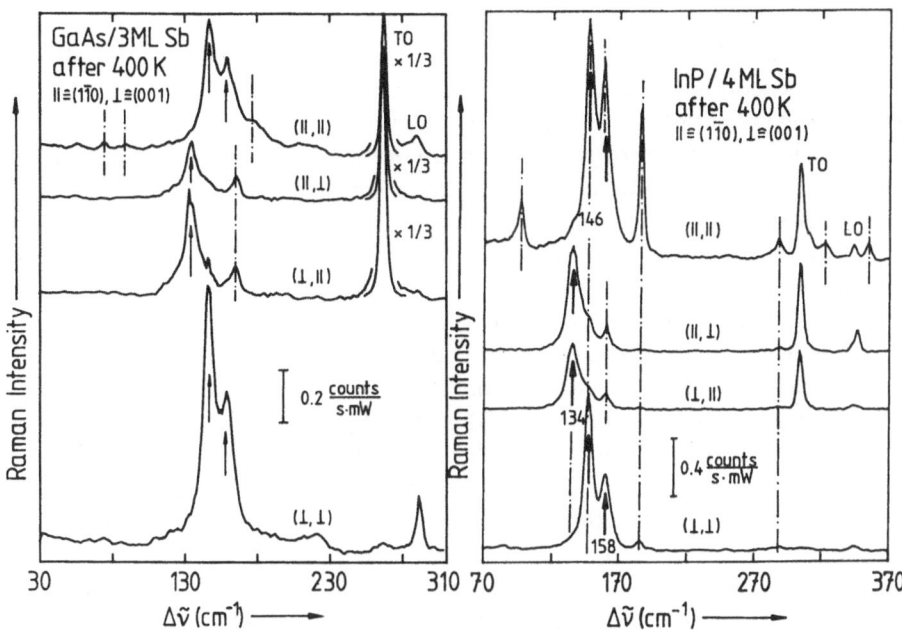

Fig. 4.11. Raman spectra of Sb on GaAs(110) and InP(110) after annealing to 400 K, taken for different polarisation configurations. The arrows mark the peaks due to lattice vibrational modes of the orthorhombic Sb modifications [4.42]

The interface of Bi on III-V-semiconductors is another example of an ideally abrupt interface. However, due to the larger covalent radius of Bi compared to Sb, the Bi monolayers are not perfectly ordered [4.52, 4.53]. This may explain why interface vibrational modes were not detected for Bi on InP(110) and GaAs(110). [4.42]. For room temperature deposition of coverages above 1 ML, two different crystalline Bi modifications are formed below 10 ML and above, respectively. The corresponding Raman spectra reveal phonon modes which are shifted by several wavenumbers for the low coverage modification and coincide with the bulk positions for the larger coverages. LEED patterns show for 4 ML to 10 ML of Bi structures similar to those described above for Sb, whereas for larger coverages trigonal patterns arise [4.51]. In contrast to Sb on GaAs(110), the lattice mismatch of the substrate-stabilized Bi modification is by far lower. The evaluation of the LEED patterns reveals lattice constants of $a = 4.75$ Å(4.75 Å) and $b = 4.48$ Å(4.54 Å) (parenthesis: bulk lattice constants). Accordingly, the phonon spectrum of the substrate-stabilized Bi reveals strain effects rather than a new crystal structure of lower symmetry.

The examples of group V elements on III-V-(110) surfaces illustrate that the combination of LEED and Raman scattering yielding the statical and dynamical lattice properties is extremely suitable for identifying lattice properties of thin ordered films.

As a second type of overlayer-substrate systems, Ge overlayers on GaAs (110) will be discussed where the criterion of an abrupt interface is still fulfilled. Since both Ge and GaAs crystallise in the zincblende structure and exhibit nearly perfect lattice matching (Ge: $a_0 = 5.658$ Å, GaAs: $a_0 = 5.653$ Å [4.54]), the epitaxial growth is not restricted to a single layer [4.60, 4.61]. In this case sensitive Raman scattering investigations can be performed using the 2.41 eV line of an Ar-laser which is close to the $E_1 + \Delta_1$-gap of Ge. Figure 4.12 shows a set of Raman spectra for increasing Ge thickness [4.62]. The transverse optical phonon mode of the Ge is detected from 6 ML on. For the growth of the Ge overlayer, the substrate temperature is the crucial parameter. For temperatures above 600 K as displayed in Fig. 4.12 a single crystalline layer grows. Due to momentum non-conservation the TO line of the Ge layer is broadened and shifted in its frequency position for thicknesses of a few monolayers. Above 25 ML, however, the TO phonon line approaches the frequency position and the width of bulk Ge samples, indicating epitaxial growth. For lower substrate temperatures, polycrystalline or even amorphous growth of the Ge is observed, which can be identified in the Raman spectra by a broadening and a shift of the TO line [4.63].

A new parameter in semiconductor epitaxy can be introduced by extending the growth from Ge to $Si_x Ge_{1-x}$ alloys on GaAs. Since Si exhibits a lattice constant of 5.431 Å, a tunable lattice mismatch can be obtained by varying the silicon content. This example is used in the following Section to demonstrate the analysis of strained layers by Raman spectroscopy.

Many other growth studies of semiconductor heterostructures performed by Raman scattering, especially the growth of II-VI on III-V-semiconductors,

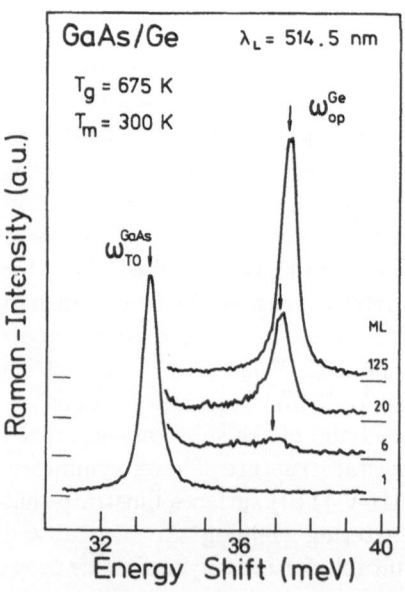

Fig. 4.12. Raman spectra of Ge/GaAs (110) heterostructures for different Ge thicknesses. At the substrate temperature of 675 K the Ge overlayer grows epitaxially. From 25 ML on the TO phonon line of Ge refers to spectra as recorded from bulk samples (not shown here) [4.62]

deal with more complicated systems due to chemical exchange reactions at the semiconductor interface. In these cases, the application of Raman scattering is of interest for both monitoring interfacial reactions as described in Sect. 4.3.5 and structural properties of the II-VI overlayer. A typical example of that type is the growth of CdS on InP(110). By Raman spectroscopy it was demonstrated that epitaxial CdS layers deposited on InP(110) substrates (lattice mismatch of 0.8 percent) grow in the zincblende structure [4.64] whereas the lattice structure of bulk CdS is of wurtzite type [4.66].

From the phonon lines appearing in the Raman spectra both structural modifications can be clearly identified. Figure 4.13 shows for example Raman spectra of a 20 ML thick CdS layer on InP(110) and, for comparison, spectra taken from a (110)-surface of a bulk CdS sample after cooling to 90 K substrate temperature. The spectra were recorded using the 457 nm (2.71 eV) excitation line with parallel polarisations of incident and scattered light. In the case of the (110) surface of the InP substrate the polarisation was along the [001] direction. Due to the near-resonant excitation and the large Fröhlich interaction in CdS strong LO and 2LO phonon lines at $308\,\mathrm{cm}^{-1}$ and at $716\,\mathrm{cm}^{-1}$, respectively, appear in this polarisation configuration, whereas for scattered light polarisation perpendicular to the incident one (not shown here) a weaker, deformation potential induced TO mode at $240\,\mathrm{cm}^{-1}$ can be observed as expected corresponding to zincblende phonon modes. In the hexagonal bulk structure, however, the $A_1(TO)$ at $228\,\mathrm{cm}^{-1}$, the $E_1(TO)$ at $240\,\mathrm{cm}^{-1}$, the E_2 at $257\,\mathrm{cm}^{-1}$, the $E_1(LO)$ at $308\,\mathrm{cm}^{-1}$ and the $2E_1(LO)$ at

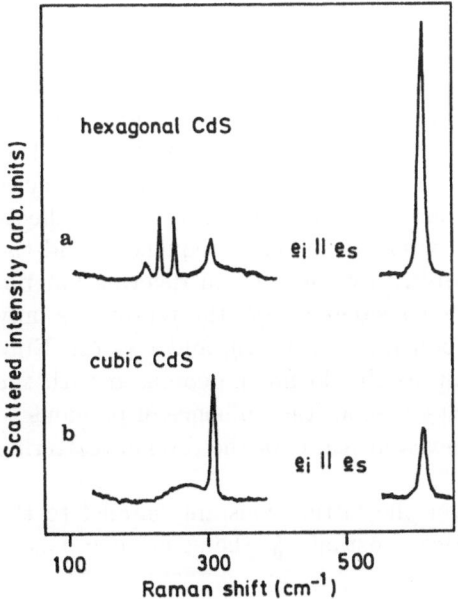

Fig. 4.13. Raman Spectra of a bulk CdS crystal (wurtzite structure) (**a**) and of 20 ML CdS deposited on InP(110) (**b**). Both spectra were recorded using the 457 nm Ar laser line as excitation. The samples were cooled to 90 K, the polarisation of incident and scattered light were chosen parallel [4.65]

$716\,\mathrm{cm}^{-1}$ are observed. This confirms the importance of the substrate lattice for the growth.

Besides, the Raman spectra also reveal the presence of a reacted phase at the interface indicative as a broad spectral feature between 200 and $300\,\mathrm{cm}^{-1}$ which has been assigned to In_2S_3 [4.64]. More examples for the detection of reacted phases by Raman spectroscopy are described in Sect. 4.3.5 in detail.

Raman scattering has also been widely applied for the characterization of crystalline phases at metal-silicon interfaces (Pt/Si, Ti/Si) and for the characterization of structural changes due to semiconductor processing such as etching, ion bombardment, thermal annealing, laser annealing and so forth. These aspects have recently been reviewed by F. H. Pollak [4.68].

4.3.2 Strain

Generally, the elemental or compound materials which constitute the layers of semiconductor heterostructures have different lattice constants. While the relative lattice mismatch is in the range of 0.1 percent when atoms of corresponding rows of the periodic table are involved in the different layers (e.g. Ge/GaAs, ZnSe/GaAs, CdS/InP), layers of atoms from different rows show a relative mismatch distinctly beyond one percent, except for AlAs/GaAs, which is a nearly lattice-matched system. The lattice mismatch causes a biaxial stress, leading to pseudomorphic strained epilayers. While for highly strained systems the strain may be partially relaxed in an early stage of growth by a coherent island growth mode [4.69], in case of smaller lattice mismatch pseudomorphic strained layers may be grown up to a critical thickness, at which dislocations lead to a partial relaxation of the epilayer. The critical thickness, which is essentially determined by the lattice mismatch [4.70–4.72], amounts to about $0.2\,\mu\mathrm{m}$ for a relative mismatch of about 0.3 percent, while a critical thickness of only some ML is predicted for a mismatch of 4 percent.

The strain in pseudomorphic epilayers and its release during the relaxation can be analysed through an exact evaluation of the phonon frequencies in the Raman spectrum, since the stress-induced lattice deformation affects the lattice dynamics. The relation between the phonon frequencies and the strain tensor $\tilde{\epsilon}$ is described quantitatively by the phonon deformation potential tensor. For the cubic materials considered here, the tensor has only three independent nonvanishing components called p, q, and r [4.73]. Note that these phonon deformation potentials should not be confused with the electronic deformation potential which describes the influence of phonons to the electronic band structure and thus is a measure for the Raman scattering efficiency.

Since for a pseudomorphic epilayer the lattice constant parallel to the interface $a^{\parallel}$ equals the substrate lattice constant a_0^{S}, the strain component parallel to the interface amounts to

$$\varepsilon_L^\| = \varepsilon_{xx} = \varepsilon_{yy} = (a_0^S - a_0^L)/a_0^L \tag{4.10}$$

Here a_0^L is the equilibrium lattice constant of the epilayer.

Through the compliance tensor $\tilde{S}$ this leads to a tetragonal distortion:

$$\varepsilon_L^\perp = \frac{2 \cdot S_{12}}{S_{11} + S_{12}} \varepsilon_L^\| \tag{4.11}$$

Through the deformation potentials, the tetragonal distortion induces a splitting of the threefold degenerate phonon oscillator into a singlet mode whose sublattice displacement is perpendicular to the interface, and a doublet mode with a displacement parallel to the interface. For (100) backscattering Raman experiments the singlet mode is the LO phonon, which is symmetry-allowed, while the doublet mode is the symmetry-forbidden TO phonon. These strain effects were treated in detail by Anastassakis [4.74, 4.75].

The strain-induced shift of the phonon frequencies can be approximated by:

$$\Delta\omega_{LO} = \left(\frac{p + 2q}{3\omega_0^2} \cdot \frac{(S_{11} + 2S_{12})}{(S_{11} + S_{12})} + \frac{2}{3}\frac{q - p}{2\omega_0^2} \cdot \frac{(S_{11} - S_{12})}{(S_{11} + S_{12})}\right) \cdot \omega_0 \cdot \varepsilon_L^\| \tag{4.12}$$

$$\Delta\omega_{TO} = \left(\frac{p + 2q}{3\omega_0^2} \cdot \frac{(S_{11} + 2S_{12})}{(S_{11} + S_{12})} - \frac{1}{3}\frac{q - p}{2\omega_0^2} \cdot \frac{(S_{11} - S_{12})}{(S_{11} + S_{12})}\right) \cdot \omega_0 \cdot \varepsilon_L^\| \tag{4.13}$$

The materials to be considered here have compliance tensor elements S_{11} in the range of $2 \cdot 10^{-11}\,\mathrm{Pa}^{-1}$ and S_{12} in the range of $-1 \cdot 10^{-11}\,\mathrm{Pa}^{-1}$ [4.76]. Realistic phonon deformation potential values p and q are $-\omega_0^2$ and $-2\omega_0^2$, respectively [4.76]. For a lattice mismatch of one percent this leads to phonon shifts of about 1 percent for the allowed LO phonon and about 0.5 percent for the TO phonon.

Assuming an inaccuracy of $\pm 0.2\,\mathrm{cm}^{-1}$ in the Raman peak frequency determination, for frequencies of about $300\,\mathrm{cm}^{-1}$ strain effects at pseudomorphic layers can be observed for lattice mismatch beyond about 0.1 percent (see e.g. [4.27]).

Distinct strain effects were observed e.g. from GaAs layers on Si substrates for layer thicknesses far beyond the critical thickness [4.77]. This may seem surprising, since full relaxation of the GaAs layer should be expected. However, due to the different thermal expansion coefficients of Si and GaAs, a thermally induced strain occurs during the cooling from growth temperature after the growth process. The temperature-dependent Raman experiments showed a thermal strain value of about $2 \cdot 10^{-3}$ (tensile) at room temperature continuously decreasing with increasing temperature, and turning into compressive biaxial strain at $500\,\mathrm{K}$.

Although the strain determination with Raman spectroscopy does not achieve the accuracy of X-ray diffraction, it offers some interesting advantages, such as the lateral resolution in the $\mu\mathrm{m}$ range which is obtained in micro-Raman spectroscopy. This technique has been used e.g. for the analysis of the strain in Ge and GaAs/Ge structured layers on oxidised Si substrates [4.78] and of the strain distribution in laterally structured silicon which had been oxidised [4.79, 4.80]. Here a lateral resolution below one micrometer was achieved. The strain was essentially localised at the edge of the oxidised regions, where maximum phonon shifts of $0.4\,cm^{-1}$ were detected (corresponding to a stress of $2\,kbar$), while a relaxation occurred within $2\,\mu m$ from the edge. In contrast to X-ray diffraction, Raman signals can already be obtained from layers as thin as some ML for many materials, and for thicker layers the information depth can be tuned. For a ZnSe layer of $180\,nm$ on GaAs(100) the variation of the laser line from $488\,nm$, for which the ZnSe is transparent, to $413\,nm$, which has a depth of information of only $50\,nm$, results in a phonon red shift of $1\,cm^{-1}$ [4.27]. This shift reveals a reduced strain value in the surface region (below 10^{-3}) as compared to the interface strain of $2.7 \cdot 10^{-3}$ which corresponds to the lattice mismatch.

For opaque layers on transparent substrates the epilayer strain at the surface and at the interface can be determined separately by taking Raman spectra from the front and back of the sample, respectively. This method has been applied e.g. for GaAs on $CaF_2(100)$ [4.85], and for InSb on $BaF_2(111)$ [4.86].

As an example of strain relaxation with increasing layer thickness, the development of the peak frequencies vs epilayer thickness for SiGe on GaAs(110) [4.63] is shown in Fig. 4.14. Here quite abrupt frequency shifts of the epilayer phonon modes (Si-Si, Si-Ge, and Ge-Ge vibrations) are observed at an epilayer thickness of $40\,nm$. These shifts enable the monitoring of the relaxation of the epilayer strain when the layer thickness exceeds the critical thickness. In Sect. 4.5.2 it will be shown that the relaxation leads to additional defects at the interface which can be detected by Electric-Field Induced Raman Scattering.

A depth-resolved analysis of strain in epitaxial layers can be performed when utilising micro-Raman spectroscopy for cleaved side faces of heterostructures, as was schematically shown in Fig. 4.6. This method was applied e.g. to investigate the strain at the ZnSe/GaAs(100) interface [4.81, 4.82]. Thick ZnSe layers (1000 to $4000\,nm$) were grown on GaAs(100) by MOVPE and after growth characterised by micro Raman analysis from the cleaved (110) side face. In these studies the strain in the ZnSe layer was monitored as a function of the distance to the interface. A lateral resolution limited by the focus diameter of $500\,nm$ was achieved, and the cross section was obtained by recording a sequence of Raman spectra shifting the focus in $200\,nm$ steps perpendicular to the interface.

Figure 4.15 shows a corresponding sequence of Raman spectra. The top spectrum is recorded for a focus location mainly on the GaAs side, the bottom spectrum on the ZnSe side of the interface, respectively. The light polarisations were chosen parallel to the $[1\bar{1}0]$ direction giving rise to allowed TO

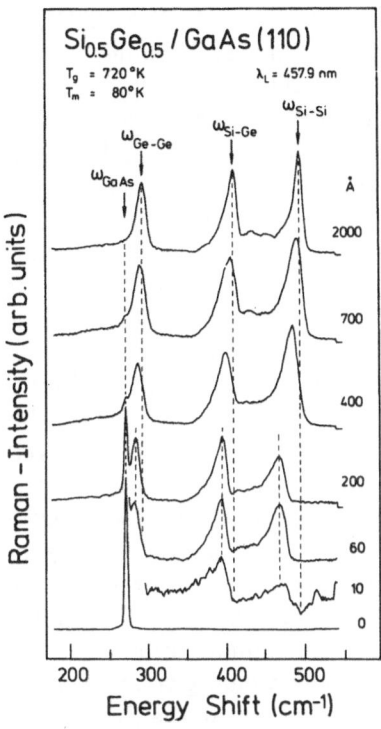

Fig. 4.14. Development of the strain-induced shift of the Raman frequencies vs the layer thickness for Si_xGe_{1-x} on GaAs. The distinct shift of the phonon frequencies at 40 nm is due to the relaxation of the epilayer strain [4.63]

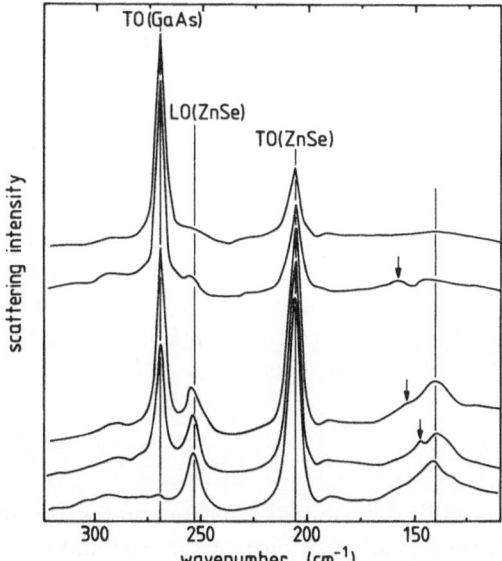

Fig. 4.15. Raman spectra of ZnSe on GaAs(100) microprobed at successive distances from the interface 200 nm apart each, starting at the top. Vertical lines mark the unstrained ZnSe line frequencies, the arrow indicates the TO-TA strain shifted peak positions [4.81]

scattering from ZnSe and GaAs, second order scattering of the ZnSe 2TA and TO-TA lines and weak, symmetry-forbidden scattering (Fröhlich-induced) of the ZnSe LO phonon mode. The TO-TA line is found to exhibit the most pronounced shift (by approximately $18\,\mathrm{cm}^{-1}$) due to the interfacial strain while approaching the interface from a pure ZnSe focal location. Also the LO mode is shifted, although by the much smaller amount of $1\,\mathrm{cm}^{-1}$, whereas the TO line remains at a constant frequency position. The latter result has been interpreted as due to a cancellation of strain-induced shifts since the singlet and doublet TO modes of ZnSe show shifts of opposite sign when applying strain in a [100] direction [4.83]. The large difference of the LO and TO-TA shifts remains a surprising result. Comparing to the hydrostatic pressure dependence of both lines the LO shift would correspond to approximately 3 kbar whereas the TO-TA shift would correspond to 36 kbar [4.81, 4.82]. This remarkable difference has been interpreted in terms of the difference of macroscopic versus microscopic strain at the interface. The first order scattering from optical (LO, TO) modes infers zone center phonon modes with long wavelengths (approximately 500 nm) whereas the TO-TA mode is a combination of zone boundary modes at the X and L point of the Brillouin zone with wavelengths in the order of 1 nm. Thus, the optical modes measure an average strain, the TO-TA modes a local strain, respectively. In bulk samples strained homogeneously, e.g. by hydrostatic pressure, both modes would give an accurate measure of the strain, whereas the large strain gradient at the ZnSe/GaAs interface induced by strain release through stacking faults reduces the average strain by a lot compared to the microstrain at the interface.

Another experimental technique to enable scanning of a strain gradient at the interface of a semiconductor heterostructure is polishing or etching the surface after growth under a small angle to expose deeper lying regions. A micro-Raman spectroscopy scan on the bevel allows for a well controlled, stepwise shift of the region of interest from the upper surface towards the epilayer/substrate interface. This was successfully applied to monitor e.g. the strain relief at the Ge_xSi_{1-x} layer on Si(100) [4.84] from frequency shifts of the Si-Si, Ge-Ge and Si-Ge modes as a function of the focus position on the bevel. The depth resolution is determined by the light penetration depth rather than by the focus diameter. In the latter study a penetration depth below 100 nm was achieved using the 457.9 nm (2.71 eV) line as excitation.

4.3.3 Orientation

The orientation of crystalline layers can be determined from Raman scattering by using the selection rules, which were treated in Sect. 4.1.4. For materials with zincblende structure, the occurrence of either LO or TO phonon scattering or both reveals the orientation of the surface normal of the overlayer: for (100) orientation only the LO phonon is observed, for (110) only the TO, and for (111) both LO and TO. For a proper interpretation of the spectra the polarisation configuration should be chosen such that only deformation-potential scattering is allowed. Furthermore, rotation of the polarisation directions of incident and scattered light leads to a modulation of the phonon peak intensities, which can be used to identify the axes parallel to the surface, e.g. the [010] and [001] axis at (100) surfaces. For (100) and (110) layers this determination is performed straightforwardly in normal incidence. For (111) layers, however, the procedure must be performed under oblique incidence to obtain phonon intensity variations [4.86]. The accuracy obtainable in the determination of the orientation by Raman scattering is about one degree.

For the orientation analysis an advantage of Raman spectroscopy with respect to X-Ray diffraction analysis is its sensitivity for very thin overlayers, for some materials down to some ML. Besides, for those systems where substrate and overlayer phonons can be observed, the relative orientation of the in-plane axes can be determined, since also in the case of perfect lattice matching the Raman peaks of substrate and overlayer are generally well separated, in contrast to the X-Ray peaks, which coincide in this case.

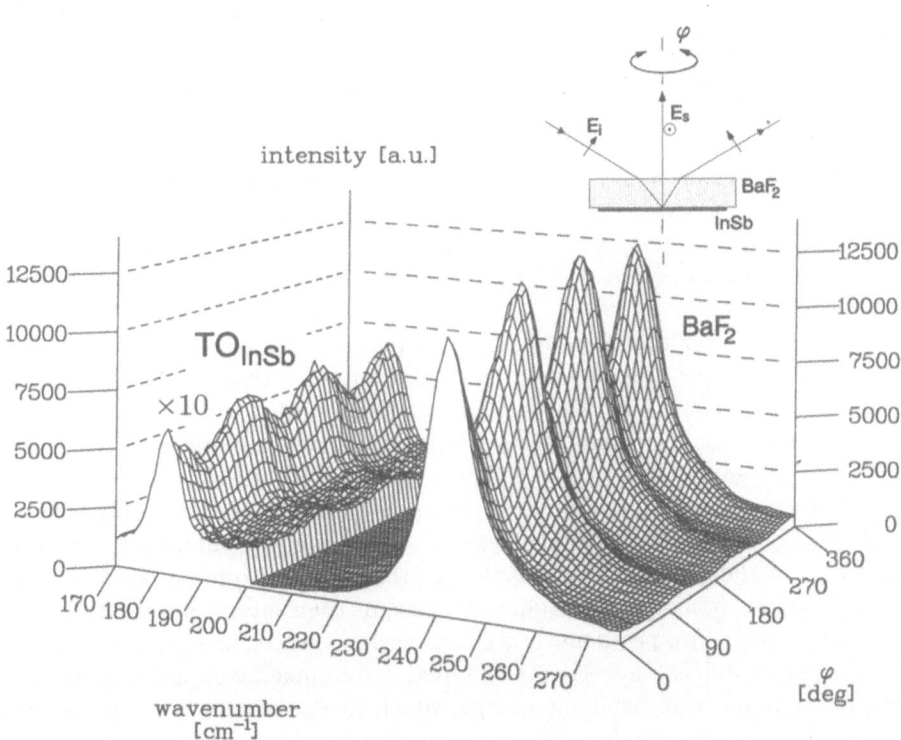

Fig. 4.16. Determination of the orientation of an InSb layer on a BaF$_2$(111) substrate from the phonon-intensity modulation in Raman spectra due to a systematic variation of the angular positions of the sample. The spectra, taken from the back of the transparent substrate show phonons from InSb as well as BaF$_2$ [4.86]

As an example of an orientation analysis by Raman scattering, Fig. 4.16 shows the results from an InSb layer, grown on $BaF_2(111)$ by hot-wall epitaxy. The series of spectra represents different rotation angles ϕ of the sample around the surface normal. In order to determine the relative orientation of substrate and overlayer the spectra were taken from the back of the substrate. Phonon peaks from both the substrate and the InSb overlayer are observed since the BaF_2 substrate is transparent. The intensity modulation which is observed experimentally when rotating the sample indeed corresponds to the symmetry-induced selection rules. From the identical modulation of the overlayer and substrate peak intensities it can be deduced that (i) the InSb layer orientation is also (111), and (ii) the substrate and the overlayer have also their in-plane axes in common, e.g. [$1\bar{1}0$] and [$11\bar{2}$] [4.86]. Several other material combinations, e.g. $EuSe/BaF_2(111)$ epitaxial heterostructures, have only the same surface normal orientation, while the in-plane axes of the epitaxial layer are rotated with respect to those of the substrate [4.87]. Furthermore, Raman scattering from the $InSb/BaF_2(111)$ heterostructure allowed a depth-resolved determination of the strain in the InSb overlayer by comparison of the InSb phonon frequencies measured from the front and the back.

4.3.4 Composition and Ordering of Mixed Compounds

Many applications of heterostructures require ternary or quaternary compounds, either for lattice matching, such as e.g. $In_{.53}Ga_{.47}As$ on InP, or $In_{.49}Ga_{51}P$ on GaAs, or for tuning of the energy gap, e.g. nearly lattice-matched $Al_xGa_{1-x}As$ on GaAs or strained Ge_xSi_{1-x} on Si. In these cases the control of the compound composition and its lateral homogeneity is crucial for device performance.

Usually the composition of mixed compounds is analysed by determination of their lattice constant by high resolution X-ray diffraction. However, it can also be deduced from the fundamental band gap, which is derived from photoluminescence, or from the lattice vibrational eigenfrequencies, which are obtained by Raman spectroscopy. These optical methods offer special advantages. The required minimum layer thickness is lower than for X-Ray diffraction, and for thicker layers even a depth-resolved analysis can be performed by variation of the incident light wavelength. Furthermore, optical methods can be applied for homogeneity studies because of their lateral resolution in the μm range. They are particularly well suited for selectively grown structures.

In photoluminescence the stoichiometry of a mixed compound is derived from its fundamental band gap energy, which varies between the gap energies of the constituent materials. However, indirect-bandgap materials like Si and Ge for example, have an extremely weak photoluminescence, and for some mixed compounds of III-V materials, like e.g. $In_xGa_{1-x}As$, the spectral difference of the gaps of the constituent binaries is so large, that a very high spectral flexibility would be required to cover the necessarily wide compositional range. An additional complication is the strong decrease of the luminescence efficiency due to lattice imperfections when the mixed compounds are not quite close to the lattice-matched composition.

In Raman spectroscopy the composition is deduced from the eigenfrequencies of the lattice vibrations, which depend characteristically on the stoichiometry [4.88–4.90]. In the dynamics of the vibrational modes with varying composition, three types can be distinguished: one-mode, two-mode and three-mode behaviour. In the simplest case, the one-mode type, only a single TO and a single LO phonon mode exist throughout the composition range. The phonon mode frequencies vary approximately linearly with composition between the frequencies of the constituent materials. In the two-mode case, which holds e.g. for $Al_xGa_{1-x}As$ as shown in Fig. 4.17a, separate vibrational modes exist for the two possible pairs of neighbouring atoms, Ga-As and Al-As vibrations for $Al_xGa_{1-x}As$, thus two TO - LO pairs, whose relative intensity and mode frequencies characteristically depend on the composition. The three-mode case applies e.g. for Si_xGe_{1-x}, where mixing takes place in both FCC sublattices. Here Si-Si, Si-Ge, and Ge-Ge-vibrations are observed (Fig. 4.17b), whose intensities and frequencies vary with composition.

The ability of Raman scattering to analyse a wide compositional range with a high spatial resolution was used for the analysis of patterned epitaxial

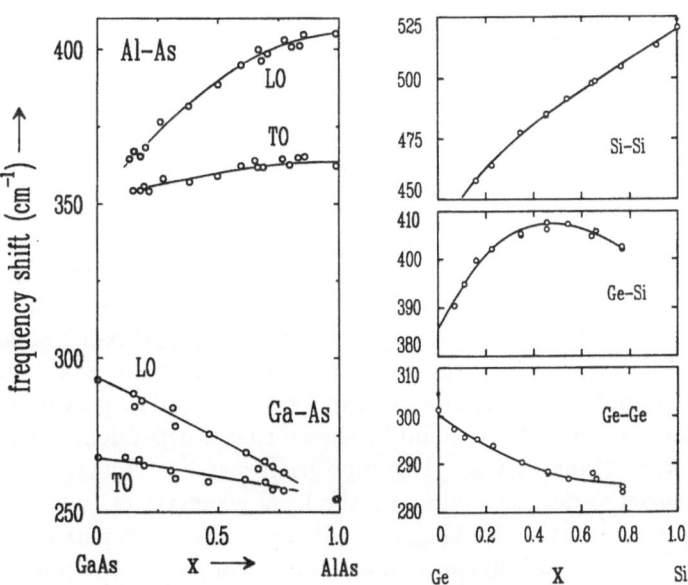

Fig. 4.17. Multi-mode phonon behaviour in mixed crystals. Two-mode phonon behaviour in $Al_xGa_{1-x}As$: Ga-As and Al-As TO and LO modes (**a**) [4.91], three-mode phonon behaviour in Si_xGe_{1-x}: Si-Si, Si-Ge, and Ge-Ge mode (**b**) [4.92]

$In_xGa_{1-x}As$ structures, selectively grown on InP by low pressure MOVPE [4.93]. Here a strong correlation between the local composition and the pattern geometry was observed, which could be attributed to a selective diffusion of the different constituents in the gas phase.

A further example for spatially resolved Raman spectroscopic composition analysis is a study of Ge_xSi_{1-x} strain-relief layers [4.94]. In epitaxial growth of high quality Ge or Ge_xSi_{1-x} on Si substrates strain-relief layers are deposited prior to the growth of the epilayer. They consist of a pile of thin Ge_xSi_{1-x} layers whose Ge-content gradually increases towards the epilayer. Tsang et al. performed a depth-resolved composition analysis of the strain-relief layer by a series of Raman spectra, scanning the surface of a polished small-angle bevel with microscopic lateral resolution. Using a laser line with a penetration depth below 30 nm, the lateral micro-Raman resolution was converted into depth resolution [4.94].

From a detailed evaluation of the peak positions and relative intensities of the three vibration modes in Ge_xSi_{1-x} compounds the composition and strain can be deduced. Such studies were performed by Mooney et al. for uniformity control [4.95] and by Schorer et al. for analysing the diffusion behaviour of Si and Ge during annealing of Si/Ge multilayer systems [4.96].

In the mixed compound $In_xGa_{1-x}P$ the In and Ga atoms tend to a spontaneous atomic ordering in the group III sublattice [4.97]. From lattice images obtained by high-resolution transmission electron microscopy it is known that ordered CuPt-type domains may coexist with disordered domains for $Ga_{0.51}In_{0.49}P$ grown lattice matched on GaAs(100). Two differently oriented ordered phases can exist where either the $\{1\bar{1}0\}$ or $\{111\}$ planes are occupied alternately by P-Ga and by P-In. The degree of order strongly depends on the growth parameters. In principle, this effect might influence the Raman eigenfrequencies and obscure the composition dependence. However, it was shown that the eigenfrequency of the longitudinal optical phonon is not affected by the degree of order [4.98]. On the contrary, electronic band gap shifts up to 90 meV to lower energy due to ordering were reported [4.98, 4.99]. Thus, the combination of Raman spectroscopy and photoluminescence allows the separate determination of composition and degree of order. The experimental realisation of this combination is quite easy, since the $In_xGa_{1-x}P$ bandgap is in the visible spectral region, where the Raman experiments are performed. By such a combined analysis at $In_xGa_{1-x}P$ layers on GaAs(100), grown selectively by low pressure MOVPE, it could be shown that independent of the pattern geometry a constant $In_xGa_{1-x}P$ composition can be achieved, but that the degree of order varies, depending on the local geometry [4.100].

TEM results have shown that for $Ga_{0.51}In_{0.49}P$ layers, grown on patterned GaAs(100) substrates, ordered regions with a geometrical size in the micrometer range can be achieved. For such samples, it should be possible to focus on the ordered and disordered regions separately by micro-Raman spectroscopy.

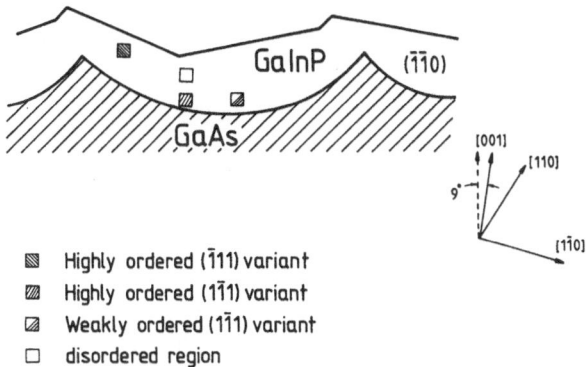

Highly ordered ($\bar{1}11$) variant
Highly ordered ($1\bar{1}1$) variant
Weakly ordered ($1\bar{1}1$) variant
disordered region

Fig. 4.18. Scheme of the InGaP/GaAs(011) cleavage face. Indicated by squares are regions which have been analysed by transmission electron microscopy [4.101]

Recently, such a Raman study was performed for 10 μm thick Ga$_{0.51}$In$_{0.49}$P layers (Fig. 4.18) which were grown by MOVPE on a grooved GaAs substrate [4.101]. The grooves running along the [011] direction of the GaAs substrate were prepared by wet chemical etching of a GaAs(100) wafer misoriented by 9° in [0$\bar{1}$1] direction. Thereafter the sample was cleaved and placed under the Raman microprobe to analyse the cross section on the (011) face.

The 676.4 nm (1.83 eV) line of a Kr$^+$-ion laser was used which is close to the E_0 resonance of the overlayer. Figure 4.19 shows spectra recorded at a position in the middle of a groove close to the interface and apart, respectively. From TEM analysis, the latter position is known to correspond to disordered, the former to ordered material. Both spectra consist of three phonon lines (pseudo-two-mode behaviour) corresponding to a TO-like mode at the low frequency side, a LO-like mode at the high frequency side, and a additional LO-like mode in between. The spectra were recorded for crossed polarisations of incident and scattered light where the TO scattering in a zincblende crystal is symmetry-allowed on the (011) cleavage surface and the LO mode occurs due to higher order scattering mechanisms. The disordered phase can be identified from the occurrence of the huge TO mode in the Raman spectra which is suppressed to a large extent on the ordered domain. The difference of the Raman spectra can be explained by two effects: Firstly, the symmetry properties of the layer change with ordering. The disordered phase is of zincblende type symmetry whereas the ordered is of C$_{3v}$ symmetry [4.99]. Consequently, the Raman selection rules are different and, in fact, the TO mode is symmetry-forbidden in the ordered case. For the samples investigated here the shift of the fundamental energy gap towards lower energies upon ordering was confirmed by cathodoluminescence results. For the disordered regions they indicate a gap energy of 1.92 eV, which means transparence to the exciting laser line of the Raman experiments, whereas the gap of the ordered phase is 1.83 eV, thus the laser light is absorbed [4.101]. Conse-

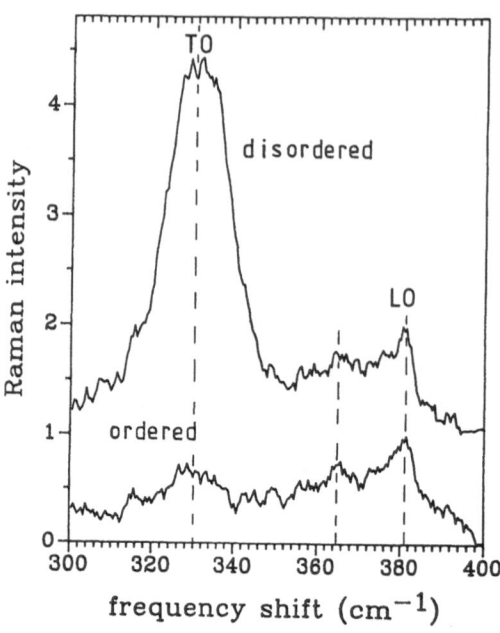

Fig. 4.19. Raman spectra of adjacent disordered and ordered $Ga_{0.51}In_{0.49}P$ domains on GaAs taken in the middle of a groove near the surface (top) and near the interface (bottom), respectively [4.101]

quently, for the ordered regions the scattering volume for Raman spectroscopy is reduced, which in addition gives rise to the lower scattering intensity in the spectra.

4.3.5 Detection of Reacted Phases

Since the optical phonons are a quite unique fingerprint of each material, the identification of the reacted phase is in many cases a straightforward procedure.

The chemical analysis of the interface has been applied e.g. for Pt on Si [4.103], where PtSi was identified after annealing and also for Ti on Si [4.104]. At the interface ZnSe/GaAs(100) Krost et al. detected, depending on the growth conditions, the presence of Ga_2Se_3 [4.105]. In analogy, In_2S_3 was observed at the CdS/InP(110) interface [4.64]. The corresponding Raman spectra exhibiting a broad vibrational band due to amorphous In_2S_3 was shown in Fig. 4.13.

Zahn et al. analysed MBE-grown CdTe/InSb(100) [4.106]. A crucial problem of this lattice-matched system is the abruptness and the stability of the heterostructure interface. PES analysis already indicated a nonabrupt interface [4.107], rich in In and Te, possibly containing In_2Te_3. The results of the

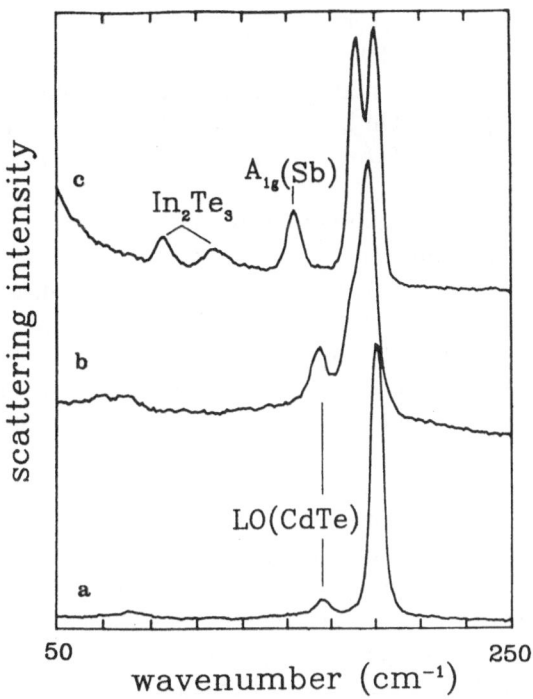

Fig. 4.20. Raman spectra of CdTe/InSb(100) heterostructures, grown by MBE at substrate temperatures of $75\,°C$ (**a**), $200\,°C$ (**b**), $300\,°C$ (**c**) using a CdTe single source. The CdTe layer thickness is approximately $30\,nm$ [4.106]

Raman analysis are shown in Fig. 4.20 for CdTe/InSb heterostructures with different growth conditions. All three spectra were taken from $300\,\text{Å}$ thick CdTe layers using a CdTe single source, but grown at different substrate temperatures. For $75\,°C$ (a), the spectrum displays besides the strong InSb LO phonon peak at $192\,\text{cm}^{-1}$ a feature at $168\,\text{cm}^{-1}$ attributed to the CdTe LO phonon. The broad lineshape of the latter is characteristic for polycrystalline CdTe formed at substrate temperatures below $150\,°C$. There is no indication for an interface reaction for growth at $75\,°C$. The spectrum (b) obtained at $200\,°C$ substrate temperature shows a more intense and somewhat narrower CdTe LO phonon peak revealing an improved crystalline qualitity of the layer. The InSb phonon, however, is broadened and shifted towards lower frequencies. This is interpreted as interfacial strain due to the formation of a thin In_2Te_3 interlayer, which has a large lattice mismatch of 4 percent to InSb. The interface reaction becomes more pronounced at higher substrate temperatures as shown in spectrum (c) for $300\,°C$. Here, the formation of the In_2Te_3 layer is clearly visible from the appearance of the according vibrational features. As a consequence of the interface reaction, Sb is liberated which is also detected in the Raman spectrum.

Studies of the chemical composition at surfaces were also performed during oxidation and annealing processes of III-V semiconductors. For GaAs the formation of crystalline and amorphous As was observed after annealing anodically oxidised substrates [4.108]. On InSb, Sb crystallites were seen after oxidation [4.109], while on oxidised InAs surfaces As crystallites were observed [4.110].

These examples are illustrative for the quite direct Raman spectroscopic identification of different chemical compounds which contain common elements. They can be specified unambiguously, since the phonon frequencies in the spectra are directly correlated to the crystal lattices. In contrast, photoemission spectroscopy, yielding electron binding energies, gives primarily an atomic identification. Different chemical compounds can only be identified from specific chemical shifts in the binding energy, which can be close to the spectral resolution.

However, it should be mentioned that for photoemission spectroscopy the detection limit is distinctly lower than for Raman scattering. The Raman detection limit strongly depends on the material, since the scattering efficiency shows wide variations. Besides, many materials (e.g. elementary In) cannot be identified by Raman spectroscopy, since they have no Raman active phonons. Furthermore, photoemission spectroscopy is strongly favourable for amorphous materials, which show only very broad Raman structures.

4.3.6 Monitoring of Growth

The development of optical multichannel detector systems with low noise and high sensitivity comparable to the photomultiplier standard has opened a new range of applications of Raman spectroscopy dealing with time-resolved spectroscopic analysis. Recently it has been demonstrated that a Raman spectrometer as shown in Fig. 4.8 in Sect. 4.2 can in fact be used for real-time growth monitoring of epitaxial semiconductor layers [4.111, 4.20]. Only few applications of this very new technique considering the growth of ZnSe layers on GaAs and of InSb on Sb have been reported so far. The online growth monitoring combines several advantages over conventional Raman analysis: The growth of the overlayer can be monitored without growth interruption which otherwise may influence the growth mode and thus modify the layer properties to be analysed. Furthermore, no movements or adjustments of sample or optical setup are required between the registration of the subsequent Raman spectra. This strongly improves the reliability of the absolute values of the scattering intensity. Finally, the time required to perform growth and analysis of a sample is reduced by far.

However, the online monitoring Raman technique requires a signal of sufficient intensity which allows to choose a small sampling time for one spectrum compared to the growth-induced changes of the sample to be analysed. In some cases the illuminating laser light may be a crucial quantity modifying the growth of the overlayer by photoinduced processes if a high light intensity is chosen.

Figure 4.21 shows Raman spectra taken during deposition of ZnSe from a single Knudsen source on GaAs(110) substrates kept at room temperature [4.111]. The clean substrate was prepared by cleavage in UHV. Subsequently, ZnSe was deposited. The first spectrum of the plot corresponds to the clean GaAs substrate. The polarisation of incident and scattered light were chosen along the [001] directions where both first order TO and LO scattering are symmetry forbidden. Moreover, since the excitation line of 2.71 eV is resonant to the ZnSe E_0-gap but off resonance to the GaAs optical gaps, the spectrum is featureless. After starting the ZnSe evaporation intensive scattering from the ZnSe LO and 2LO modes arise due to the strong Fröhlich interaction as well known from bulk ZnSe. The overall increase of these two features with ongoing deposition is of course due to the increasing overlayer thickness. The large background at the low energy side of the spectra occurring after some hours of deposition has been ascribed to near-bandedge luminescence of the ZnSe layer. Besides, the development of the spectra clearly shows intensity modulations with increasing overlayer thickness. These modulations are interpreted in terms of Fabry Perot interference effects enhancing or reducing the scattering intensity depending on layer thickness, light wavelength and refractive index of the overlayer. The appearance of the interference effects shows that the ZnSe overlayer grows in a smooth, two-dimensional layer. By

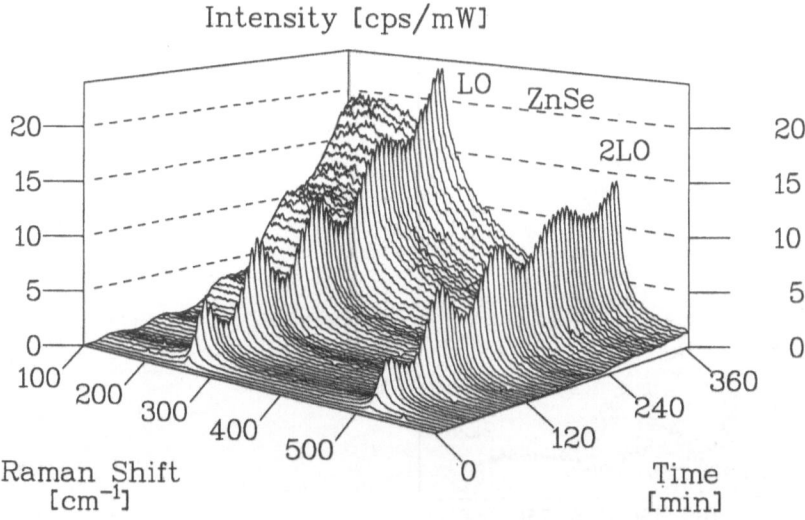

Fig. 4.21. Raman spectra taken during growth of ZnSe on GaAs(110) by MBE in UHV. The growth was performed at a substrate temperature of 300 K. The intensity oscillations are ascribed to Fabry-Perot interferences in the overlayer with increasing thickness [4.111]

comparing the interference effects calculated from the refractive index of bulk ZnSe to the intensity modulation the deposition rate has been estimated to 12Å/s for this experiment.

As mentioned above, the Raman efficiency is of great importance for the growth monitoring experiments. Here, a sufficient efficiency was achieved by utilizing the resonance enhancement at the E_0 gap of ZnSe. In a more general case of growth studies, of course, the substrate temperature is an important parameter which is not restricted to room temperature. Instead, elevated temperatures will be used and often the temperature must be adjusted to a value suitable to a specific epitaxy. A modification of the substrate temperature, however, will severely influence the Raman scattering efficiency because the optical gaps will shift to lower energy and thus the resonance condition is no longer valid. Consequently, the choice of excitation energy must depend on the material to study as well as on the substrate temperature. Figure 4.22 shows the Raman scattering intensity of the ZnSe LO phonon mode recorded as a function of temperature for several excitation lines of an Ar ion laser. The pronounced changes of the LO peak height underline the importance of the resonance effect. Especially at higher temperatures where the overall

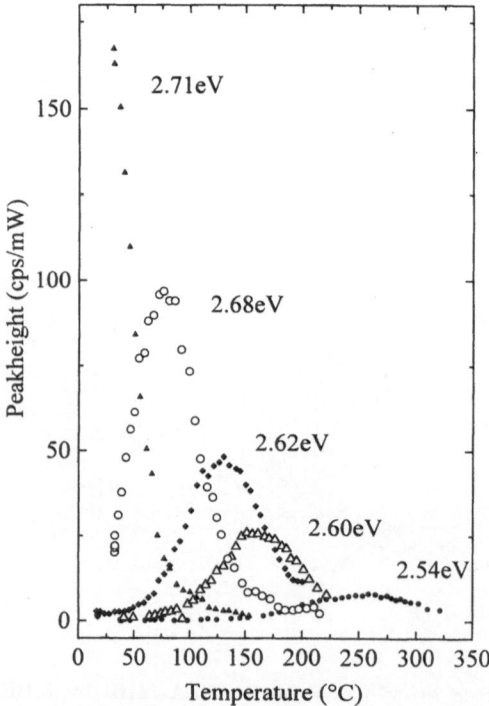

Fig. 4.22. Peak height of the ZnSe LO phonon mode in Raman spectra as a function of temperature recorded using different excitation lines of an Ar ion laser [4.111]

scattering intensity is reduced the resonance should be fulfilled in order to enable growth monitoring.

Another application of the growth monitoring has been reported for the growth of group III elements (In, Ga, Al) on group V substrates (Sb, Bi) [4.20, 4.112]. In contrast to the experiments mentioned above here a III-V-semiconductor overlayer is formed by chemical reaction of the adsorbate with the substrate. Figure 4.23 shows a set of spectra recorded during In deposition onto a Sb(111) substrate. The substrate was prepared in UHV by cleaving a Sb single crystal before deposition. A total amount of 450 ML In was deposited during a time interval of 7.5 hours.

Due to the polarisation of incident and scattered light being perpendicular to each other the A_{1g} phonon mode of Sb is forbidden whereas the E_g mode is allowed according to first order symmetry selection rules. Consequently, in the first spectrum of the set corresponding to the clean surface a strong peak at $113\,\mathrm{cm}^{-1}$ and a weak contribution at $153\,\mathrm{cm}^{-1}$ is attributed to the E_g and A_{1g} phonon modes, respectively. For subsequent In deposition an increase of the A_{1g} scattering intensity appears due to a relaxation of the symmetry selection rules. Such effects can be attributed to a depolarisation of the light caused by roughening of the surface which is a first indication of the surface modification by chemical reaction. At the same time even more drastic changes occur in the frequency range of the InSb phonon modes around $185\,\mathrm{cm}^{-1}$. Initially, a broad spectral band develops which splits into two distinct peaks at $180\,\mathrm{cm}^{-1}$ and $191\,\mathrm{cm}^{-1}$ as growth progresses.

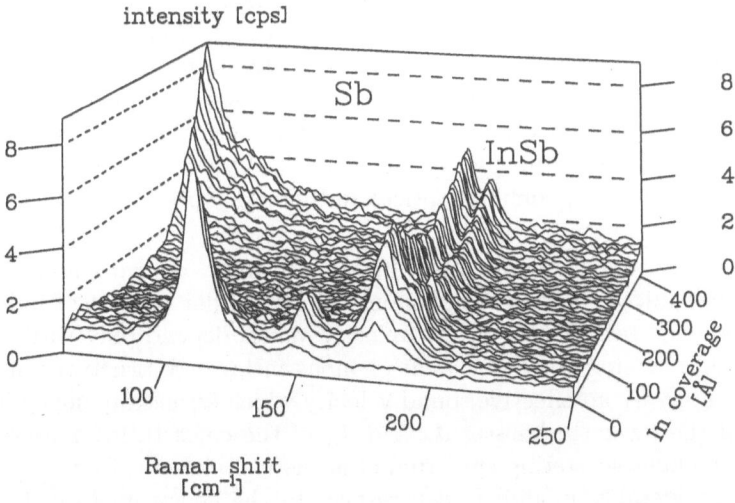

Fig. 4.23. Raman spectra taken during deposition of In onto cleaved Sb(111) in UHV at a substrate temperature of 300 K. New peaks developing at $180\,\mathrm{cm}^{-1}$ and $191\,\mathrm{cm}^{-1}$ are attributed to the TO and LO phonon modes of InSb formed by chemical reaction. The intensity oscillations are ascribed to Fabry-Perot interferences in the overlayer with increasing thickness [4.20]

These peaks are attributed to the TO and LO phonon modes of InSb. Their occurrence obviously demonstrates the formation of InSb due to interdiffusion, already at room temperature. The clear splitting of these peaks indicates a fairly good crystalline structure of the InSb overlayer even for this rather low growth temperature. Further information on the growth morphology can be derived from the appearance of the scattering intensity modulation with increasing layer thickness. The intensity modulation is attributed to Fabry-Perot interferences in the overlayer. Excellent agreement with the experimental data is obtained from intensity calculations, based on a two-dimensional growth model. Thus, the InSb growth is essentially two-dimensional and island formation is excluded. The strong damping of the interference structure is due to the light absorption in the InSb layer. The layer thickness which was deduced from the interference period corresponds very well to the amount of deposited In. Similar experiments for Ga deposition on Sb(111) show that at 300 K only inferior-quality GaSb is obtained and that the growth is not two-dimensional. Increased temperatures are required to improve the GaSb quality. For Al deposition no AlSb formation was observed at 300 K. From a systematic temperature-dependent study the minimum temperature for the detection of interdiffusion-induced compound formation was determined to 373 K for AlSb, 293 K for GaSb, and 268 K for InSb.

4.3.7 Low-Dimensional Effects

In addition to the features described above, in the Raman spectrum of lattice vibrations from superlattices (SL) and multi-quantum wells (MQW) a variety of new phenomena occurs which can be used to analyse different aspects of these structures such as the layer thicknesses and interface sharpness. Since these phenomena already have been reviewed extensively [4.6, 4.113, 4.114, 4.115, 4.5, 4.116, 4.117], we will give only a short overview, covering basic aspects of folded acoustical, confined optical and interface phonons.

4.3.7.1 Folded Acoustical Phonons. In SL and MQW of two materials A and B the dynamics of acoustical phonons is very similar to that of bulk samples of A and B, because the sound velocity values depend only weakly on the material. Therefore the acoustical phonons in layered structures are described accurately by an effective sound velocity which essentially depends on the ratio of the layer thicknesses d_A and d_B of the constituting materials. However, the light scattering spectrum of acoustical phonons from these structures is considerably modified with respect to the spectrum from the bulk because the periodicity length is enlarged from the lattice constant a to $D = d_A + d_B$, which implies a reduced Brillouin zone length and thus a folding of the phonon dispersion curves in the reduced zone scheme, as is shown schematically in Fig. 4.24. Consequently, in the spectral region of the acoustical-phonon dispersion curves a series of Folded Acoustical Phonon (FAP) peaks can occur in the Raman spectrum in addition to the acoustical peaks from the bulk at very low frequency, which are commonly observed by Brillouin spectroscopy. The positions of the FAP peaks are very sensitive to the periodicity length $D = d_A + d_B$ of the layer structure, since this determines the length of the new Brillouin zone and thus the quantitative shape of the folded dispersion curve. Therefore FAP modes are used as a sensitive probe for the periodicity length without being affected noticeably by the detailed structure of the individual layers. In contrast to the frequencies of the optical phonons, those of the FAP modes strongly depend on the phonon wavevector, as can be seen from Fig. 4.24. Therefore, due to the q-conservation in the light scattering process, the experimentally observed frequencies also depend on the wavelength of the incident phonon. The q-transfer in light scattering is indicated by the dash-dotted semi-vertical line in Fig. 4.24. Consequently, for the quantitative evaluation of the layer periodicity from the experimentally observed FAP frequencies, the applied laser line has to be taken into account. An example of FAP modes is shown in Fig. 4.25. Here, Raman spectra from a sample containing a stack of 10 InGaAs/InP bilayers shows three pairs of peaks at frequencies corresponding to the circles in Fig. 4.24. From the peak frequencies a periodicity length D of 8.6 nm was deduced [4.119]. Moreover, the extremely narrow halfwidths confirm the absence of a thickness gradient in the stack.

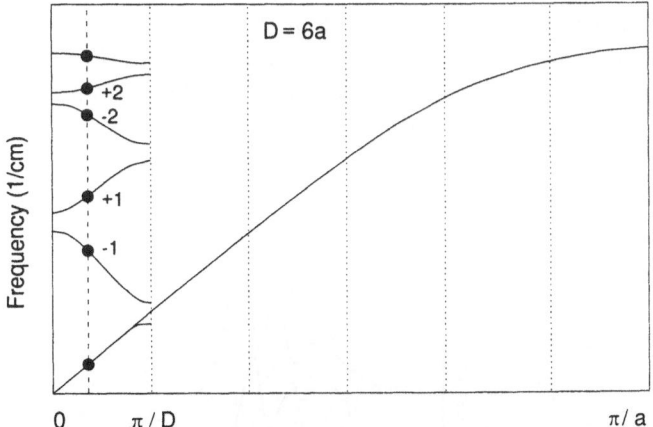

Fig. 4.24. Folding of the acoustical phonon dispersion curve due to the reduction of the Brillouin zone for a stack of layers with stack periodicity D. In this example the periodicity D was chosen as 6 times the lattice constant a. This leads to reduction of the Brillouin zone from π/a to $\pi/D = 1/6 \times \pi/a$. The dotted line is the edge of the new Brillouin zone. The dashed vertical line symbolises the q-transfer in Raman scattering. The circles denote the frequency and wavevector values of the observed folded phonon modes.

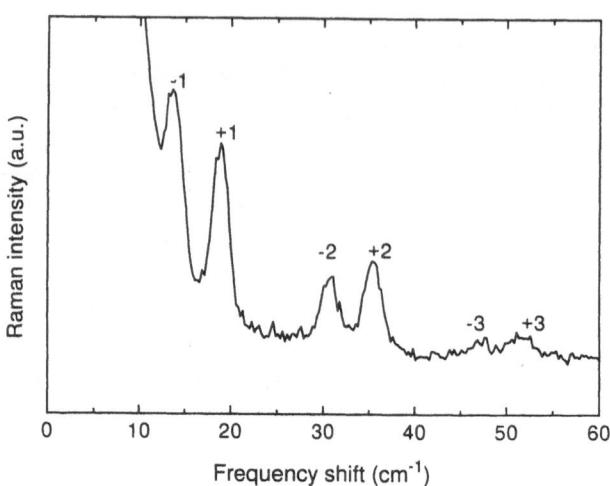

Fig. 4.25. Raman spectrum of folded acoustical phonon modes in a stack of 10 InGaAs/InP bilayers. The peak frequencies correspond to a periodicity length of 8.6 nm [4.119]

4.3.7.2 Confined Optical Phonons. Since optical phonons induce a distortion of the atomic basis, their eigenfrequencies strongly depend on the material. This implies that for many combinations of materials A, B in SL the optical phonon frequency regions of both materials do not overlap. In this case the optical phonons cannot propagate across the interfaces because at their eigenfrequencies no eigenmode can exist in the neighbouring layer. Therefore the resulting optical vibration modes are confined in material A and in B, respectively. For these modes the constraint of a vanishing vibrational amplitude in the neighbouring layer imposes wavelengths which are deter-

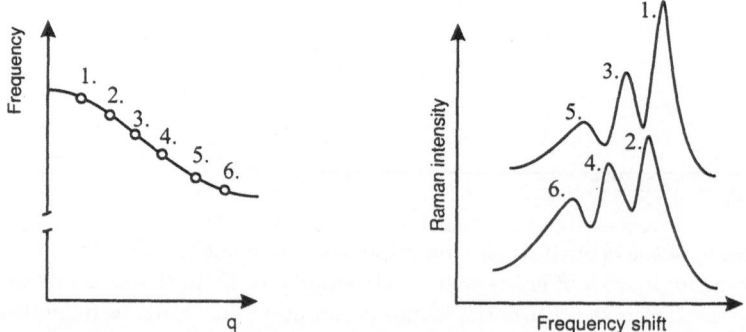

Fig. 4.26. Optical phonon dispersion curve Ω with allowed q-values 1...6. and corresponding Raman spectra. The odd-index modes appear in off-diagonal polarisation configuration, while the even-index modes require the diagonal configuration

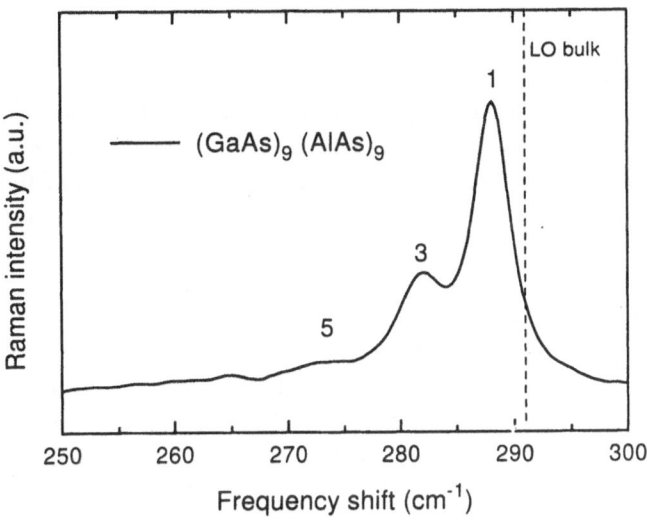

Fig. 4.27. Confined GaAs modes of a GaAs/AlAs superlattice (9 ML/9 ML, 50×). For offdiagonal configuration the odd-order confined Ga-As modes are observed. The even-order modes would appears diagonally. The dashed vertical line indicates the bulk LO frequency (from [4.120])

mined by the layer thicknesses d_A and d_B, resulting in series of harmonics with wavelenghts λ such that the layer thicknesses are approximately an integral multiple N of $\lambda/2$. These phonon wavelengths correspond to wavevectors $q = N \cdot \pi / d_A$ and $q = N \cdot \pi / d_B$, in the layers of material A and B, respectively. Therefore, in the Raman spectrum series of peaks may be observed. Their frequencies are determined by the optical phonon dispersion curves. These imply a decrease of the eigenfrequencies for increasing wavevector q, which means a peak-frequency decrease with increasing N. This is shown schematically in the left part of Fig. 4.26, where circles on the phonon dispersion curve symbolize the allowed q-values. As a result of the symmetry-derived selection rules, the modes with even N values appear in diagonal configuration and those with odd N values offdiagonally when the incident light is polarised along a main crystal axis (e.g. [001]). For this reason in the right part of Fig. 4.26 two Raman spectra are drawn, one for the odd-index modes, the other for those with even index. Figure 4.27 shows the confined GaAs modes in the Raman spectra of a GaAs/AlAs superlattice (9 ML/9 ML, 50×). Since the spectrum was taken in off-diagonal polarisation configuration, only the odd-index peaks 1, 3, 5 appear. The dashed line, which indicates the bulk LO frequency shows that already the first confined mode is distinctly shifted with respect to the bulk value.

The confined optical phonon modes have been applied as a sensitive probe for layer thicknesses [4.89], and interface sharpness and broadening effects [4.118]. Their potential for the characterisation of III-V and Si-Ge superlattices is emphasised in the review article by Fasolino and Molinari [4.117].

Since for strained layer SL they are shifted due to the strain in the individual layers, they were also used for strain analysis [4.121, 4.122]. Furthermore, for short-period Si/Ge SL the annealing-induced diffusion across the interface was analysed in great detail through the optical phonon intensities and frequency shifts [4.96], revealing the transition from the Si/Ge SL to the binary SiGe compound.

4.3.7.3 Interface Phonons. The modulation of the dielectric function in multi quantum wells and superlattices leads to the appearance of phonon modes, propagating in the plane of the constituent layers, which can be understood as a generalisation of surface phonons. They are referred to as interface phonons. Their eigenfrequencies, which are derived within the framework of the effective dielectric function of the layer stack [4.5], are located in those frequency ranges, where the dielectric functions of the materials A and B have opposite sign. Therefore, if the reststrahlen bands of both materials do not overlap, the interface modes appear between the TO and LO frequencies of the individual materials A and B. An overlap of the reststrahlen bands may well occur for material combinations in which ternary or even quaternary compounds are involved. The interface modes show a characteristic dependence on the thickness ratio d_A/d_B, since this ratio affects the effective dielectric function. Besides, their eigenfrequencies show a clear q-dependence already in the q-range which is accessible to light scattering experiments [4.125]. Therefore, a variation of the laser line induces a shift of the interface-phonon peaks in the Raman spectrum, which allows their unambiguous identification.

Excitation of modes which propagate in the plane of the heterostructure layers can be performed in accordance with wavevector conservation when the light is focused on a cleaved side face in a micro-Raman experiment (see Fig. 4.6). Interface phonons were observed experimentally e.g. for GaAs/AlGaAs [4.123] and GaAs/AlAs [4.124]. Also the theoretically predicted frequency shift for laser-line variation was confirmed by experimental results [4.126].

4.4 Analysis of Electronic Properties

Raman analysis can be a versatile alternative to electrical measurements to characterise the electronic properties of heterostructures, since it offers several important advantages: (i) no electrical contacts are required, (ii) single layers can be analysed separately in stacks with varying doping, (iii) lateral resolution can be achieved.

Electronic characterisation by Raman scattering from phonons can be performed by utilizing the resonance behaviour of the Raman cross section discussed in Sect. 4.1.1. Furthermore, Raman scattering from electronic single particle excitations (SPE) and collective excitation modes (plasmons) provide insight into electronic properties.

The resonance of the Raman efficiency of phonon scattering can be used to determine critical points of the electronic band structure of the solid. This was established investigating bulk semiconductors [4.55–4.57] and has been more recently extended for characterisation of epitaxial layers. In Sect. 4.4.1 some examples of resonant Raman scattering are described.

Raman scattering from SPE can be due to free carriers or due to impurity-bound electrons. The latter type of Raman scattering has been shown to be applicable for the characterisation of the concentration and incorporation of residual impurities in nominally undoped materials [4.190, 4.128, 4.132, 4.133]. In Sect. 4.4.2 the analysis of electronic excitations of shallow acceptors and donors in semiconductors is discussed.

The second group is based on collective excitations which are sensitive to electronic properties. In polar compound semiconductors the longitudinal phonon (LO) modes are such excitations. As mentioned in Sect. 4.1.2 the LO modes are accompanied with a macroscopic electric field which gives rise to the so called Fröhlich scattering mechanism [4.134, 4.135, 4.3]. Fröhlich scattering can also be utilised to determine the impurity concentration in highly doped semiconductors since the LO phonon modes interact via their electric field with charged impurities [4.190, 4.136].

Furthermore, the electric field of the LO phonon induces a coupling to the collective plasmon excitation of free carriers. In the presence of free carriers two new eigenmodes of the solid arise which are called coupled plasmon-LO-phonon-modes (PLP). Raman scattering at PLP modes (Sect. 4.4.3) sensitively depends on the carrier concentration which determines the eigenfrequency of these modes [4.4].

Low-dimensional structures can be analysed as well by Raman scattering from electronic transitions between subbands and from collective electronic excitations [4.137] which is briefly reviewed in Sect. 4.4.2.

4.4.1 Electronic Band Structure

Besides analysing the lattice dynamical properties the Resonance Raman scattering from phonon modes can also be utilized to gain information about the electronic band structure of a solid. The Raman scattering process can be regarded as consisting of three steps, the excitation of an electron-hole pair, scattering of the electron-hole pair into a second electron-hole pair state by electron phonon coupling, and generation of a scattered photon via recombination of the electron-hole pair.

Therefore, as explained in Sect. 4.1.1 the electronic band structure in the spectral regime of the incident and scattered photon is of crucial importance for the Raman cross section. In most cases where the energy shift between incoming and scattered photons is small compared to the width of optical gaps of a semiconductor the scattering efficiency shows broad maxima for photon energies close to critical points of the electronic band structure.

For a quantitative understanding of the Raman scattering efficiency, the generalised susceptibility (also termed Raman susceptibility) must be considered. According to Eq. (4.8) in Sect. 4.1.3 the Raman efficiency depends on the derivative of the susceptibility with respect to the photon energy E and the deformation potential of the optical gap. Both quantities can be determined by optical methods such as spectroscopic ellipsometry, for instance, and the Raman efficiency calculated.

On the other hand, the Raman efficiency can be measured by Resonance Raman experiments for various excitation photon energies around the optical gap. Before the Raman susceptibility can be evaluated further factors influencing the measured scattering intensity such as the spectral throughput of the setup, the light reflection at the sample surface, the scattering volume determined by the light absorption and the phonon occupation factor (Bose distribution function) have to be known.

Resonant Raman experiments have been performed for several bulk semiconductors, such as GaAs [4.55], InP [4.56] and Si [4.57]. Resonance experiments require a number of different excitation lines to cover the spectral range of the optical gap to be investigated. In the visible spectral range a combination of both an Ar and Kr ion laser and, additionally, cw-dye lasers have been used for these studies. Figure 4.28 shows the scattering efficiency as a function of photon energy for Si. The measurements were performed for the first order symmetry-allowed TO phonon scattering mediated by the deformation potential scattering described in Sect. 4.1.2. According to the susceptibility

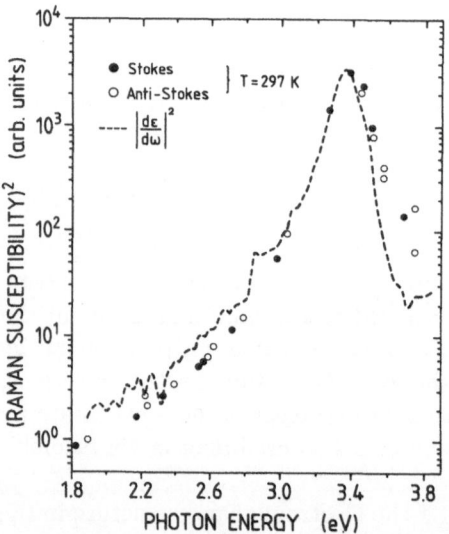

Fig. 4.28. Squared Raman susceptibility of Si at room temperature for both Stokes and anti-Stokes data. The straight line represents the squared derivative of the dielectric function determined by spectroscopic ellipsometry [4.57]

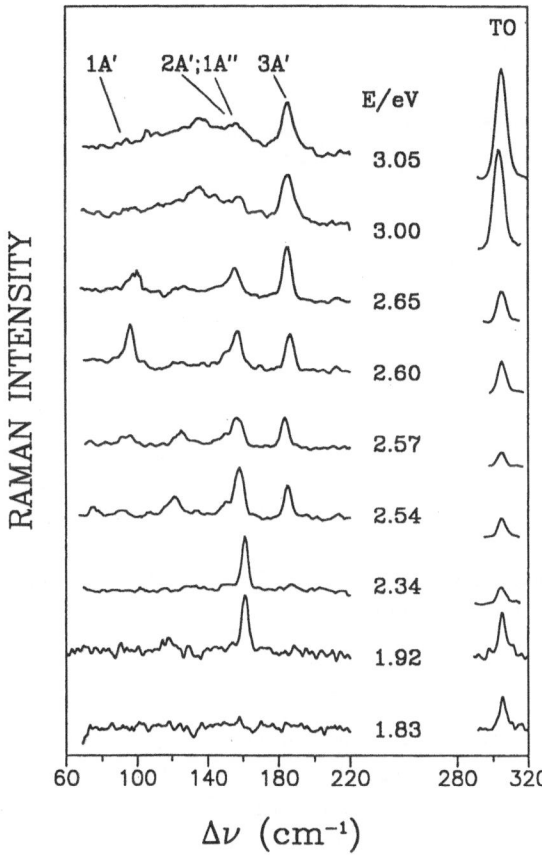

Fig. 4.29. Raman spectra of 1 ML Sb on InP(110) recorded for different excitation photon energies as indicated in the figure. Besides the monolayer vibrational modes the InP TO phonon mode occurs at $304\,\mathrm{cm^{-1}}$ [4.58]

model the Raman susceptibility of the first order scattering should be given by the derivative of the susceptibility (or dielectric function) with respect to energy. In fact the results show a good agreement to the square of the Raman susceptibility derived from the Si dielectric function determined in separate experiments by spectroscopic ellipsometry. Similarly good agreement between susceptibility theory and Resonance Raman experiments was also found for InP and GaAs, where additionally in the energy regime below the fundamental gap excitonic effects yield a significant contribution.

The results underline the importance of the optical properties manifested in the dielectric function of a solid for the Raman analysis. On the other hand, the energy dependence of the Raman efficiency can be utilized to characterise electronic properties of the sample. More recently, this approach was used to test the resonances of the Sb monolayers on III-V-(110) surfaces. The Raman monitoring of the surface vibrational modes is described in Sect. 4.3.1. With varying laser photon energy strong changes in intensity of the monolayer vibrational lines in the Raman spectra are found. Figure 4.29 shows a pile of spectra obtained for various lines of an Ar and a Kr laser [4.58].

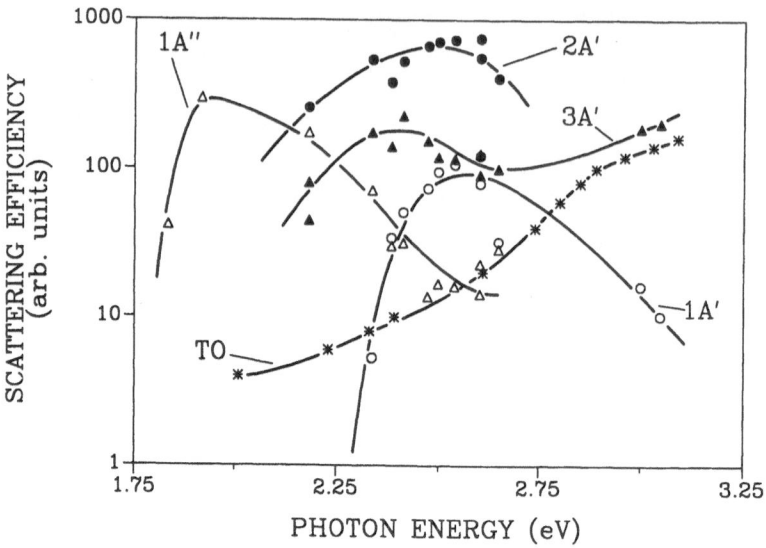

Fig. 4.30. Raman scattering efficiency as a function of the incident laser photon energy for the different monolayer vibrational modes. The solid lines are guidelines to the eye [4.58]

The intensity maxima of the monolayer vibrations are specific for each vibrational mode and are clearly different to the resonance of the InP substrate [4.56]. The Raman scattering efficiency is shown in Fig. 4.30.

The pronounced maxima of the Raman efficiency must be attributed to critical points in the two-dimensional electronic surface band structure. The surface bands in the accessible energy range are derived from the Sb-P and Sb-In bonds [4.39]. Regarding the deformation-potential scattering mechanism a coupling between the surface vibrations and the surface electronic bands is evident. The monolayer vibrations, of course, mainly modulate the bonds within the monolayer, i.e. Sb-Sb, Sb-P, Sb-In, which build up the electronic surface band structure. The Sb-Sb-derived bands have critical points at larger energies than the visible range investigated and thus do not significantly contribute to the resonances. The fact that the monolayer modes give rise to different resonance maxima shows that the electron phonon interaction described by the deformation potential in Eq. (4.8) depends sensitively on the normal coordinates of the specific monolayer vibration.

The resonance experiments can thus be used to determine different critical points of the surface bands. In contrast to the bulk semiconductors discussed above the dielectric function of the monolayer can not be determined straight forward. An optical method with sufficient sensitivity is of course the Reflectance Anisotropy Spectroscopy which, however, monitors only differences of the dielectric function for directions within the surface plane. This technique was recently applied to study the monolayer surface transitions [4.59].

The RAS experiments show in fact strong evidence for Sb-induced surface transitions between 2 eV and 2.5 eV where the Raman resonances have been observed.

4.4.2 Impurities

For nondestructive probing of nominally undoped and of p-doped GaAs the Electronic Raman Scattering (ERS) is a versatile technique [4.127–4.130]. In ERS electronic single particle excitations at acceptor states in a semiconducting material are investigated. In contrast to investigations of the local vibrational modes of impurity atoms [4.128] the ERS technique is not limited to elements with lower mass numbers than the semiconductor matrix atoms.

At a sufficiently low temperature, electronic transitions at the shallow acceptors can be observed in ERS if the acceptor ground state is empty (electrically neutral). In thermal equilibrium the shallow acceptor states in Semi-Insulating (SI) GaAs are filled due to the compensation by the EL2 levels. By photoexcitation, however, a metastable depopulation of the acceptor ground states can be achieved. For that purpose sub-bandgap illumination can be used as the exciting source in the Raman experiments [4.128].

Figure 4.31 shows ERS-spectra of different SI GaAs samples from bulk crystals grown by the Liquid Encapsulated Czrochalski (LEC) technique [4.128]. In the spectra features due to C and Zn acceptor incorporation are

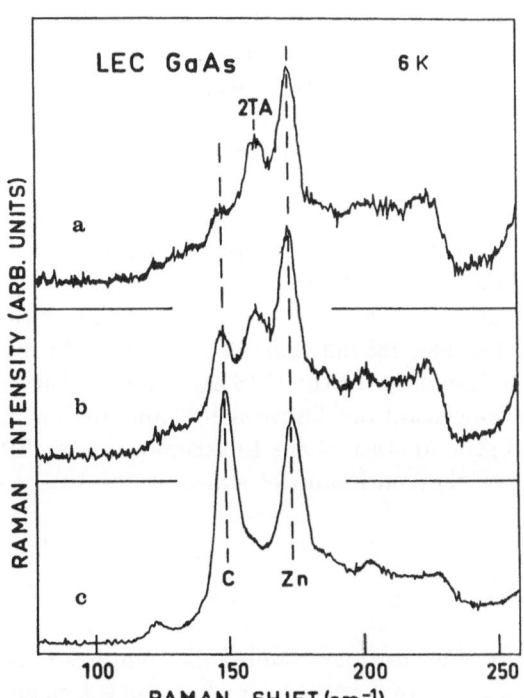

Fig. 4.31. Raman spectra of three different undoped GaAs samples. The spectra were excited at 1064.4 nm [4.128]. By IR absorption the C_{As} concentration was determined to $4 \cdot 10^{14} \, cm^{-3}$ (a), $1.2 \cdot 10^{15} \, cm^{-3}$ (b), and $6 \cdot 10^{15} \, cm^{-3}$ (c), the Zn concentration is in the range of $1 \ldots 3 \cdot 10^{15} \, cm^{-3}$

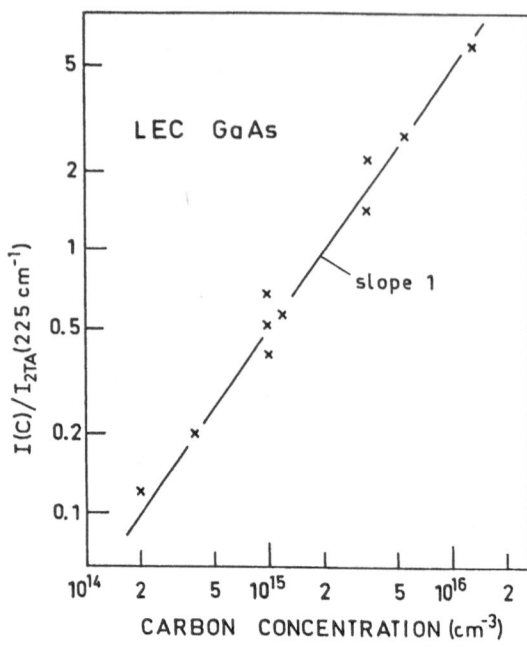

Fig. 4.32. Normalised ERS intensity for the C_{As} acceptor plotted versus C_{As} concentration determined by local vibrational mode IR absorption [4.128]

observed besides the 2TA phonon mode. The ERS scattering intensity is expected to obey a simple linear relationship to the C and Zn concentration. The scattering intensity of the 2TA mode can be used as a reference signal since it is independent of the impurity concentration. By calibrating the ERS at a sample of well-known impurity concentration a quantitative technique is established. Figure 4.32 shows the normalised ERS intensity of C_{As} for samples whose concentration had been determined by IR absorption. It proves that the postulated linear relation is in fact valid for a carbon concentration range of $10^{14}\,\mathrm{cm}^{-3}$ to $10^{16}\,\mathrm{cm}^{-3}$.

Other examples for ERS characterisation have been reported by Olego et al. [4.129, 4.130]. They investigated In (n) and Li (p) doped ZnSe layers grown by MBE on GaAs(100) substrates. For the Raman investigation of the Li doped p-type ZnSe layers, for example, the samples were mounted in a He closed cycle cryostat at 12 K [4.130]. The 488 nm (2.54 eV) line of an Ar ion laser supplied subbandgap illumination used for the ERS experiments (ZnSe: $E_0(12\,K) = 2.82\,\mathrm{eV}$). For doping concentrations between 10^{16} and $10^{17}\,\mathrm{cm}^{-3}$ electronic transitions from the 1S ground state of the Li acceptors to 2S, 2P and 3S excited states as well as to the continuum of states at the valence band maximum were identified.

4.4.3 Free Carriers

Among the Raman active elementary excitations coupled Plasmon-LO-Phonon modes (PLP) play a key role. They are discussed here because the eigenfrequency of the PLP modes sensitively depends on the concentration of the free carriers in a semiconductor. Thus Raman scattering from PLP modes can be used for a quantitative determination of the free carrier concentration in doped samples.

The PLP modes arise from a coupling between the LO phonon and the collective plasmon excitation of the free carriers through the macroscopic electric fields of both excitations [4.138]. The original LO phonon and plasmon eigenmodes are replaced by two new eigenmodes. In the following they are referenced as Ω^- (PLP mode with lower eigenfrequency) and Ω^+ (PLP mode with higher eigenfrequency). Both frequencies may strongly depend on the doping level [4.139]. Being longitudinal modes, Ω^- and Ω^+ are found at those frequencies, for which the dielectric function $\epsilon(\omega, q)$ is zero.

The frequencies of the PLP modes can be derived from the dielectric function $\epsilon(\omega, q)$, which for doped polar materials contains contributions of valence electrons (ϵ_∞), phonons ($\chi_{ph}(\omega)$), and free carriers ($\chi_{FC}(\omega, q)$). When plasmon damping effects are neglected, one obtains [4.140]:

$$\epsilon(\omega, q) = \epsilon_\infty + \frac{\omega_{TO}^2 \cdot (\epsilon_s - \epsilon_\infty)}{\omega_{TO}^2 - \omega^2 - i\omega\Gamma} - \frac{\omega_p^2}{\omega^2 - 3/5 \cdot v_F^2 \cdot q^2} \tag{4.14}$$

Here Γ is the phonon damping constant, q the wavevector, v_F the Fermi velocity, and ω_p the free carrier plasma frequency, which essentially depends on the free carrier concentration n and effective mass m^*:

$$\omega_p = \sqrt{\frac{n \cdot e^2}{\epsilon_0 \cdot \epsilon_\infty \cdot m^*}} \tag{4.15}$$

The description of the free-carrier susceptibility $\chi_{FC}(\omega, q)$ which is used in (4.14) results from an extended Drude model. The wavevector dependent term reflects the additional restoring force in plasmon oscillations for finite q-values due to carrier concentration gradients. It is referred to as spatial dispersion and described here by the hydrodynamical theory [4.141].

The observation of the PLP modes in Raman scattering may result from different scattering mechanisms based upon the phonon as well as the plasmon properties. The main scattering mechanisms are the Deformation Potential (DP) scattering due to the lattice deformation, the Electro-Optic (EO) scattering due to the longitudinal electric field of the PLP mode and additionally the plasmon-related Charge Density Fluctuations (CDF). The latter mechanism gives rise to diagonal components in the Raman tensor, while the tensor elements from the former two are off-diagonal in the main crystal axis system. The resonance behaviour of the DP and EO mechanism is analogous to the LO phonon case. The CDF scattering mechanism shows in III-V semiconductors no resonance at the E_1 and $E_1 + \Delta_1$ gap, since the carriers are located at the Γ-point of the Brillouin zone. Therefore their fluctuations only affect the dielectric susceptibility near the E_0 and $E_0 + \Delta_0$ gap [4.142].

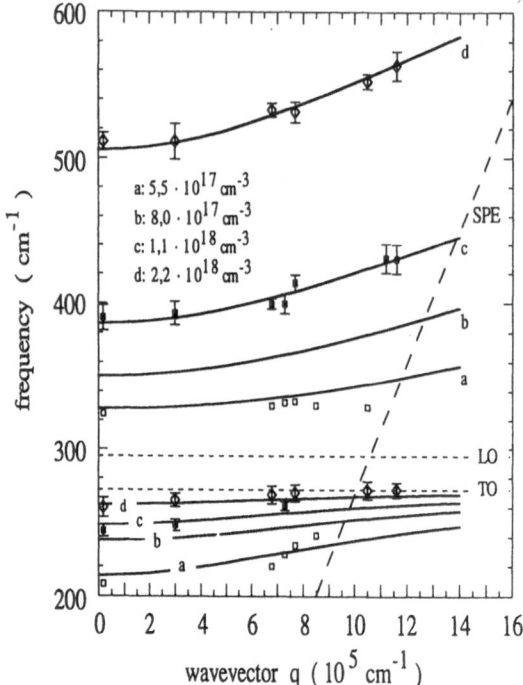

Fig. 4.33. Eigenfrequencies of the coupled plasmon-LO-phonon modes Ω^- and Ω^+ of n-GaAs in dependence of the wavevector q. Points: Results from Raman scattering with different Ar- and Kr-laser lines; the temperature of the samples was kept at 90 K; full lines: calculation according to [4.144]

In Raman experiments excitations with a finite, nonzero wavevector are involved, which is determined by the momentum conservation law (see Sect. 4.1.1). In contrast to the weak dispersion of the optical phonon modes, significant frequency shifts can occur for PLP modes in the Raman spectra [4.143, 4.4]. Therefore, the description of the relevant eigenfrequencies $\Omega^\pm$ requires the consideration of the dispersion of the PLP modes. Different approaches have been tested to describe the dispersion of the PLP modes which is observed in the Raman spectra simply by using different excitation lines in the visible range supplied by an Ar- and Kr-laser. The hydrodynamical theory was shown to reproduce the PLP dispersion within the experimental accuracy limit if the basic Eq. (4.14) is extended such that the effect of an energy-dependent effective mass due to the nonparabolicity of the conduction band is taken into account [4.144, 4.191].

Figure 4.33 shows the dispersion of the PLP modes from differently doped GaAs samples in comparison to a calculation based on this model [4.143, 4.144]. The Raman spectra were taken from n-doped GaAs homoepitaxial layers for a range of doping concentrations between $1 \cdot 10^{17}$ and $2 \cdot 10^{18}$ cm^{-3} cooled to liquid nitrogen temperature. In this case both Ω^- and Ω^+ modes were observed. The dispersion of the PLP modes is large (small) for eigenfrequencies well separated from (close to) the LO or TO phonon frequencies.

This example also demonstrates the capability of Raman scattering from PLP modes to determine carrier concentrations. Since the PLP modes close to the phonon frequency depend only weakly on the carrier concentration, the determination should be performed using the well-separated mode, i.e. the Ω^--mode up to 10^{18} cm^{-3} and the Ω^+-mode above. Especially in the concentration regime above $2 \cdot 10^{17}$ cm^{-3} an accuracy of better than 5 percent can be achieved, whereas for carrier concentrations below $1 \cdot 10^{17}$ cm^{-3} the PLP modes can hardly be observed in the Raman spectra any more. The latter is due to the fact that for these concentrations the scattering intensity of the Ω^- mode is very weak, while the Ω^+ mode nearly coincides with the LO phonon.

Recently, it has been shown for GaAs and InP samples that the observation of PLP modes and the determination of the carrier concentration is possible when the Raman experiments are performed at room temperature [4.145,4.146,4.19]. Due to the higher damping at room temperature, however, the scattering efficiency is larger when the experiments are performed at low temperature. This limits the accuracy of the room temperature experiments especially for low doping concentrations, i.e. below approximately $5 \cdot 10^{17}$ cm^{-3}.

The observation of coupled phonon plasmon modes has also been reported for p- and n-doped II-VI semiconductors. For Li doped p-type ZnSe and In doped n-type ZnSe layers grown by MBE on GaAs(100) substrates, for example, the carrier concentration as a function of temperature was studied by Raman scattering from the PLP modes [4.129, 4.131]. For dopant concentrations between approximately 10^{16} and 10^{17} cm^{-3} the frequency shift of the Ω^+-mode which is found close to the frequency of the unscreened LO phonon can be utilized to evaluate the increasing carrier concentration for the temperature rising from 12 K to 300 K.

The modeling of the momentum dependence can be avoided, if the Raman spectra are recorded with a laser excitation line in the infrared spectral range where the momentum transfer is sufficiently small to consider the approximation for $q = 0$ [4.147]. Figure 4.34 shows the doping dependence of the PLP frequencies for n-doped GaAs. The dots represent the TO phonon mode which is not influenced by the carrier concentration, the open circles denote the Ω^- and Ω^+ and the full line is a numerical fit derived from Eq. (4.14). For $n \leq 2 \cdot 10^{17}$ cm^{-3} the L$^+$ mode frequency nearly coincides with the LO phonon and the Ω^- mode with the plasmon frequency. Beyond $2 \cdot 10^{18}$ cm^{-3} the Ω^+ mode approaches the plasmon. The Ω^- mode coincides with the TO phonon frequency, since it corresponds to a screened LO phonon, i.e. a LO vibra-

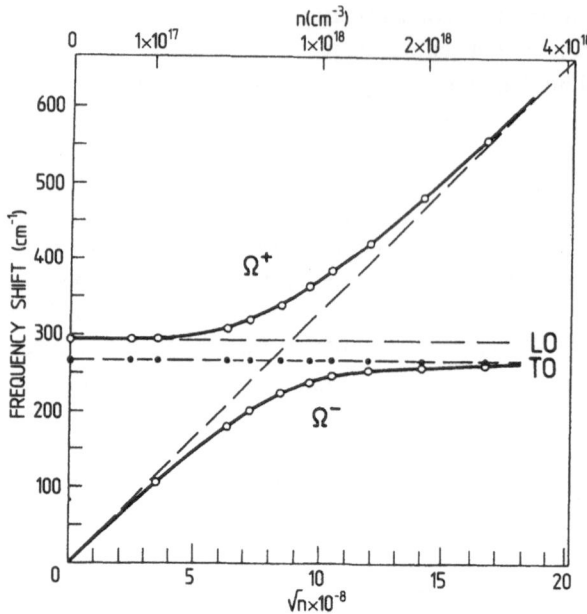

Fig. 4.34. Eigenfrequencies of the coupled plasmon- LO-phonon modes Ω^- and Ω^+ of n-GaAs for $q = 0$ as a function of the free carrier concentration n. Circles: Raman scattering results for $\lambda_{\text{laser}} = 1.06\,\mu\text{m}$; full lines: calculation with (4.14) [4.147]

tion, whose macroscopic electric field is fully compensated by the free carrier oscillation.

Figure 4.34 can be understood as a plot of the starting points at $q = 0$ of the dispersion curves shown in Fig. 4.33. Under standard experimental conditions where the exciting laser line is in the visible range, the $q = 0$ approximation in Fig. 4.34 can be used as a rough guide to the carrier concentration. Furthermore, it demonstrates which of the PLP modes is the suitable one for the determination of the carrier concentration, i.e. most strongly varying.

4.4.4 Low Dimensional Effects

Inelastic light scattering from free carrier excitations may be used to characterise the electronic properties of low-dimensional semiconductor structures. A detailed description has been given in an extensive review by A. Pinczuk and G. Abstreiter [4.137]. Therefore, only the basic ideas shall be outlined here to provide an overview on the information which can be obtained by light scattering.

Single or multiple potential wells are usually produced by growing heterostructures or by modulation of the doping concentration during growth

[4.149]. Due to the potential modulation along the growth direction the free carrier motion becomes restricted. The confinement along z-direction has the consequence that electronic subbands occur while the motion in x- and y-direction remains essentially unchanged ("two-dimensional carriers"). For parabolic bands, the energy of an electron is then given by:

$$E = E_n + \frac{\hbar^2 \cdot k_x^2}{2m^*} + \frac{\hbar^2 \cdot k_y^2}{2m^*}. \tag{4.16}$$

A further restriction of the "2D"-carrier motion may be introduced e.g. by lateral patterning processes by which quantum wires ("1D"-system) or quantum dots ("0D"-system) can be fabricated confining the carriers in x or x and y direction, respectively.

The electronic subband energies E_n can be obtained in first order by a self consistent solution of the Schrödinger equation using an effective mass approximation and the Poisson equation [4.150,4.149]. The spectrum of subband energies depends crucially on the shape of the confining potential. Therefore, it characterises the potential well structure.

The inelastic light scattering is a suitable tool to determine the spectrum of subband energies if intersubband transitions are responsible for the scattering. Such transitions can either be mediated by electronic single-particle excitations (SPE) between the subbands, similar as discussed in Sect. 4.4.2 for impurities, or by collective Spin-Density (SDE) or Charge-Density Excitations (CDE) [4.137,4.148,4.151]. Figure 4.35 shows Raman spectra obtained from a 2-D electron gas in an AlGaAs/GaAs single quantum well. The intersubband excitations observed for parallel polarisations of incident and scattered light are activated by charge-density fluctuations, whereas for orthogonal polarisations spin-density fluctuations are the relevant mechanism. Besides, Raman scattering from single particle intersubband-excitations (SPE) contributes in either polarisation configuration. The SPE give rise to a spectral band centered at $E_{01} + q_p v_F$ where E_{01} denotes the subband spacing of the zeroth to first subband at the Brillouin zone center, q_p denotes the momentum transfer parallel to the quantum well and v_F the Fermi velocity. The spectra shown in Fig. 4.35 correspond to backscattering geometry where the momentum transfer is perpendicular to the quantum well. Consequently, q_p equals to zero and the center of the SPE line corresponds to the subband spacing at the center of the Brillouin zone. The collective CDF or SDF excitations are shifted in energy in the order of 1–2 meV with respect to E_{01} which is caused by exchange Coulomb interactions [4.148]. The assignment of the SDE, CDE and SPE contributions has been verified experimentally by exploiting the energy dispersion of these lines with respect to q_p. This can be done by tilting the sample into off-backscattering geometry leading to a small but non-zero q_p momentum transfer [4.148].

Quite recently, inelastic light scattering experiments were performed on "2D" GaAs/GaAlAs quantum wells in high magnetic fields and for very low

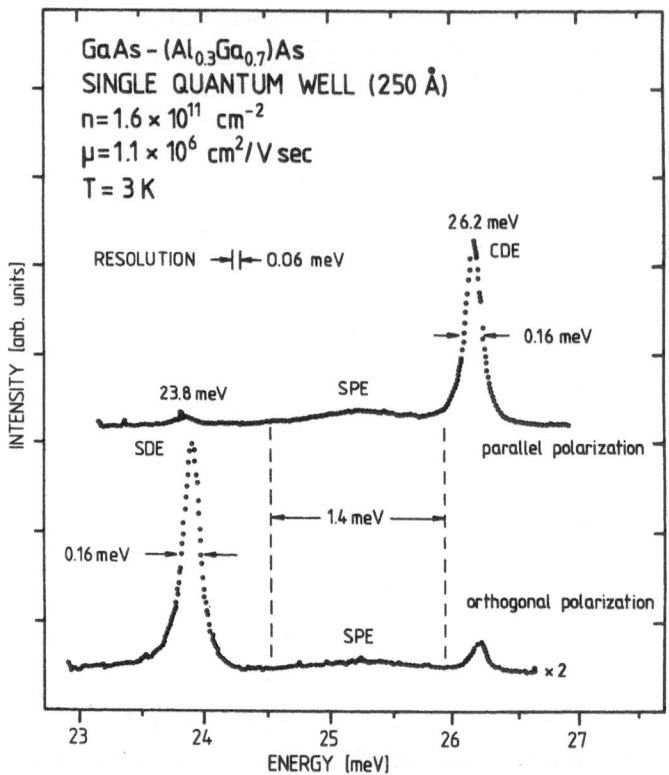

Fig. 4.35. Inelastic light scattering spectra of intersubband excitations of the high-mobility electron gas in a GaAs single quantum well. The peaks of Spin-Density (SDE), Charge-Density (CDE), and Single-Particle Excitations (SPE) are shown for parallel and orthogonal polarisation of the incident and scattered light, respectively [4.148]

temperatures. The magnetic field can be utilized to break the translation invariance in the direction parallel to the quantum well, thus confining the electrons into Landau levels. In the magnetic field range where the quantum Hall effect occurs inter-Landau-level excitations were observed by Raman scattering. In the fractional quantum Hall regime, modes were observed possibly due to neutral excitations of the fractionally charged quasiparticles, the so-called "magnetorotons" [4.153]. Probably these excitations have a vanishing dipole moment which prevents their observation by for example IR-spectroscopy techniques. However, since Raman scattering is not dependent on the direct photon-phonon interaction, the dipole moment is fortunately no limitation to the inelastic light scattering experiments.

Besides the intersubband transitions, also coupled plasmon-phonon modes can be observed from low dimensional structures [4.137]. Similarly as discussed in Sect. 4.4.3 for 3D samples, they allow to determine the concentration of the free carriers, which can be used for characterisation of semiconductor heterojunctions [4.152]. In heterojunctions, 2D-confinement of free carriers can occur through thin film growth but also through electric fields present at semiconductor interfaces (see Sect. 4.5).

4.5 Band Bending at Interfaces

An important aspect considering semiconductor interfaces is the formation of an electrostatic potential barrier, the band bending. The band bending effects were firstly investigated for metal-semiconductor interfaces, the so-called Schottky contacts [4.154–4.158]. However, such potential barriers are a much more general property of a semiconductor. They may exist at semiconductor-metal, -insulator, -semiconductor interfaces as well as at semiconductor surfaces covered with gas adsorbate or even under clean UHV conditions.

The origin of the band bending potential is a charge transfer across the interface. If, for instance, a heterostructure consists of an overlayer with an electronic work function Φ_O larger than Φ_{SC}^n of the n-doped semiconductor substrate, electrons flow into the overlayer until the electrostatic barrier generated by the charge transfer, e.g. band bending potential V_S, equalizes the work function difference. In a more sophisticated approach, the band bending built up is not simply due to work function differences but severely influenced by electronic interface states capturing electronic charge. For a quantitative description the band bending potential is mostly replaced by the Schottky barrier defined as the energy difference between the conduction band (or valence band) edge and the Fermi level at the interface for n-type (or p-type) material, respectively [4.156]. By the Schottky barrier definition the dependence on the substrate doping concentration is avoided yielding a material specific value. The detailed mechanisms of the Schottky barrier formation are not yet fully understood, in spite of an enormous amount of experimental as well as theoretical work [4.156–4.158].

A simple approach describing the band bending variation as a function of depth is the Schottky model. The barrier originates from a negative (positive) charge which is accumulated at the surface or interface of a n-type (p-type) semiconductor. Accordingly, a surface or interface layer of finite thickness depleted from free carriers is generated, the depletion layer [4.159,4.156]. This layer carries a positive (negative) charge due to the ionised donor (acceptor) states and, therefore, it is also called space charge layer. The Schottky model assumes a uniform charge distribution in the space charge layer, neglecting the fact that the charge density at the end of the space charge layer is smeared out instead of forming a step function. Solving Poisson's equation within this model yields a parabolic increase of the band bending potential $V_S(z)$ (given in eV) towards the surface:

$$V_S(z) = \frac{e^2 \cdot N}{2 \cdot \epsilon_s \cdot \epsilon_0} \cdot (d_r - z)^2 \tag{4.17}$$

The width d_r of the space charge layer depends both on the doping concentration N and the band bending potential V_S according to the relation:

$$d_r = \sqrt{\frac{2 \cdot \epsilon_s \cdot \epsilon_0}{e^2 \cdot N}} \cdot V_S(z = 0) \qquad (4.18)$$

For a semiconductor of a given doping concentration, the width d_r of the space charge layer is thus a measure of the band bending potential V_S.

Usually the determination of a band bending or Schottky barrier is assumed to be a domain for Photoemission Spectroscopy (PES). The PES technique is directly sensitive to the band bending potential V_S at the surface since the electrostatic potential appears as an offset on the binding energies of electronic states at the semiconductor surface. Thus, the band bending determination is limited by the spectral resolution and the finite escape depth of the photo-electrons. The latter leads to a broadening of the photoemission lines due to the depth variation of the band bending potential, $V_S(z)$. Especially the use of synchrotrons as light sources supplying narrow and intense excitation lines with variable photon energies of typically 20 eV up to beyond 200 eV is favourable for this purpose (kinetic electron energy = 10–200 eV). The synchrotron use allows for the high energy resolution required and, furthermore, for performing the experiment with extremely high surface sensitivity. The small escape depth of electrons limits the PES analysis to the very first stage of the interface formation, up to overlayer thicknesses of only a few ML [4.24, 4.160]. Therefore, the surface sensitivity of PES is a necessary condition to allow the accurate measurement of the potential barrier, but simultaneously, investigating an adsorbate-semiconductor interface, the electron escape depth limits the information depth to few atomic adsorbate layers. Even nowadays, when miniaturising is highly developed, such interfaces are far apart from device contacts.

Due to the large penetration depth which can be obtained by photons, Raman scattering overcomes the experimental limitations of the electron spectroscopy methods. In spite of the large penetration depth of photons, information about band bending at interfaces can be obtained from Raman scattering experiments monitoring light scattering from excitations modified by the surface or interface condition. Two mechanisms can be utilised to extract the band bending from Raman scattering experiments, either via the coupled Plasmon-LO-Phonon modes (PLP) or through field-induced Fröhlich scattering from LO phonon modes (EFIRS) [4.3, 4.4, 4.161]. As described below, the PLP method determines the width of the depletion layer d_r, whereas EFIRS is sensitive to the surface electric field E_S.

4.5.1 Band Bending Determination by Plasmon-LO-Phonon Modes

The PLP modes can be used for the interface analysis in different ways. According to Sect. 4.4.3 the eigenfrequency is very sensitive to the doping level and is therefore an accurate measure for the carrier concentration. The scattering intensity, on the other hand, can be used to evaluate the band bending potential at the interface.

In the presence of a band bending, the Raman spectra show contributions of the unscreened LO phonon mode from the surface depletion layer and of the PLP modes from the semiconductor bulk, respectively. The relative intensities of LO and PLP modes are a measure for the depletion layer width and thus for the surface band bending potential $V_S(z = 0)$ [4.162, 4.3]. With increasing band bending the intensity of the PLP modes is attenuated and the intensity of the LO phonon is enhanced. The intensity ratio is given by [4.163]:

$$\frac{I(\mathrm{LO})}{I(\mathrm{PLP})} = \frac{I_0(\mathrm{LO})}{I_0(\mathrm{PLP})} \cdot \frac{1 - e^{-(2 \cdot (d_r - \Delta)/D)}}{e^{-(2 \cdot d_r/D)}} \tag{4.19}$$

Here $I_0(\mathrm{LO})$ is the LO scattering intensity for $d_r \gg D$, and $I_0(\mathrm{PLP})$ is the PLP intensity for zero band bending. D is the attenuation length of the incident laser light and d_r is the width of the depletion layer, while the correction Δ accounts for the smearing out of the electron concentration at the end of the depletion layer. According to (4.19) the band bending at a semiconductor interface can be determined if a calibration is performed.

In order to demonstrate that this method can be used for a quantitative determination of the band bending at interfaces semitransparent Ni electrodes have been deposited on GaAs(100) which allowed the application of an externally controlled bias voltage [4.4]. Figure 4.36 shows Raman spectra of the GaAs LO phonon and the PLP modes for different voltages. The spectra reveal strong variations of LO phonon and PLP scattering intensities in dependence of the bias voltage. As a result, this method can be used to analyse the band bending at GaAs(100) surfaces.

For example, the LO phonon/PLP intensity ratio was used to investigate the band bending at clean GaAs(100) surfaces [4.146] which is nonzero, in contrast to GaAs(110) surfaces. For clean GaAs(100) surfaces prepared by MBE growth, several different reconstructions of the surface are known to exist dependent on the growth parameters. As an alternative preparation, thermal desorption (As decapping) of a protective arsenic layer from GaAs substrates with a doping concentration of $n = 1 \cdot 10^{18} \, \mathrm{cm}^{-3}$ was used in the Raman study to prepare (100) surfaces [4.146]. Again different reconstructions have been observed dependent on the As content of the surface, which in turn can be controlled by the desorption temperature. The corresponding band bending values determined by the PLP method are found in the range of $0.71 \pm 0.06 \, \mathrm{eV}$ for all the reconstructions studied, which is again in agreement with SXPS data [4.164]. An oxidised sample was used to calibrate the Raman

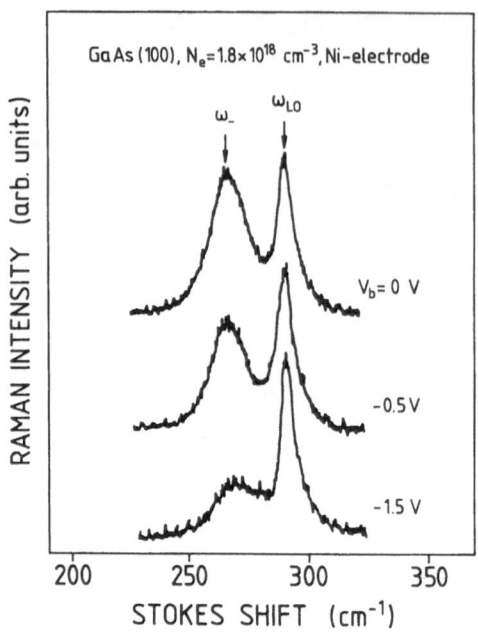

Fig. 4.36. Plasmon LO phonon spectra of an n-GaAs(100) sample with a semitransparent Ni electrode. The band bending potential is varied by the externally applied bias voltage [4.4]

spectra by comparison with SXPS data which report a band bending of $V_S = 0.78\,\text{eV}$ after oxidation [4.165].

Farrow et al. [4.163] used the Raman technique to monitor passivation effects of GaAs(100) surfaces etched by $Na_2S \cdot 9\,H_2O$. A reduction of the band bending from $0.78\,\text{eV}$ before passivation to $0.45\,\text{eV}$, or $0.51\,\text{eV}$ after passivation is reported for samples with doping concentrations of $n = 1.14$ or $1.46 \cdot 10^{18}\,\text{cm}^{-3}$, respectively.

Another application was reported by Olego [4.166] who studied the band bending at the lattice-matched II-VI/III-V combination ZnSe on GaAs(100) ($\Delta a/a = 0.27\%$ at $300\,\text{K}$) [4.166, 4.167]. The Raman experiments were performed *ex-situ*, with ZnSe layer thicknesses between $0.05\,\mu\text{m}$ and $5\,\mu\text{m}$ [4.168, 4.169]. The critical thickness for the system ZnSe/GaAs is $0.19\,\mu\text{m}$. The investigation of the interface below relatively thick ZnSe overlayers is possible because of the transparency of ZnSe for the applied laser line of $488\,\text{nm}$ ($h\nu = 2.54\,\text{eV}$, band gap of ZnSe $E_0 = 2.7\,\text{eV}$), whereas the penetration depth in the GaAs substrate is only $30\,\text{nm}$. The studies on n-doped substrates ($n = 1.3 \cdot 10^{18}\,\text{cm}^{-3}$) show for thicknesses beyond the critical thickness d_c ($0.19\,\mu\text{m}$) an increase of the band bending V_S, which is attributed to the generation of defects due to the relaxation of the overlayer [4.168].

Furthermore, Olego reports a change of the electrical surface fields with time because of instabilities at the interface due to atomic exchange on a time scale of several days [4.169]. Investigations for different ZnSe layer thicknesses show that this time-dependence of V_S is reduced with increasing thickness, which was interpreted as an oxidation of the ZnSe layer stimulated by the laser light.

4.5.2 Band Bending Determination by Electric-Field Induced Raman Scattering

Besides the scattering by PLP modes, Raman scattering from LO phonon modes via the Fröhlich mechanism can also be invoked for band bending investigations. At (110)-surfaces of zincblende-type semiconductors (III-V- and II-VI-semiconductors) the first order LO scattering can not be observed due to the symmetry selection rules discussed in Sect. 4.1.4. In spite of the symmetry selection rules, however, commonly a non-vanishing LO phonon intensity is observed due to symmetry-forbidden Fröhlich scattering. The Fröhlich interaction becomes active through additional external disturbances which may be regarded as reducing the crystal symmetry. For instance, Fröhlich scattering can be induced by the finite wavevector of the LO phonon or by an additional static electric field [4.3]. The latter mechanism is called Electric Field Induced Raman Scattering (EFIRS), which is the base for the analysis of electric fields in surface regions, caused for instance by a band bending potential V_S.

Of crucial importance for the observation of EFIRS is the resonance behaviour, i.e. the dependence of the efficiency on the incident laser frequency. As already mentioned in Sect. 4.1.1, the resonance of EFIRS is more pronounced than that of symmetry-allowed scattering processes. Its Raman tensor components are proportional to the third derivative of the dielectric susceptibility with respect to the frequency [4.173]. Therefore, excitation frequencies very close to an electronic gap are required.

The EFIRS mainly arises from a tilting of the energy bands due to the additional electrostatic band bending potential at the surface. The tilting leads to a modification of the electronic wavefunctions from Bloch to Airy functions thus modifying the optical properties of the semiconductor. This modification is called Franz-Keldysh effect (FKE) [4.170, 4.171]. Since the electric field acts in one crystal direction, the scattering contribution can be attributed to the induced symmetry lowering. For moderate field strengths the efficiency is proportional to the square of the static electric field [4.172].

The EFIRS and more generally the Fröhlich scattering can be distinguished experimentally from the deformation potential scattering by an appropriate choice of the polarisation directions of the incident and scattered light, since the two mechanisms have no common Raman tensors elements. While for deformation potential scattering only the offdiagonal tensor components in the main crystal axis system are nonvanishing (see Sect. 4.1.4) the Fröhlich scattering on the contrary gives rise to diagonal components [4.2, 4.3, 4.190].

For a quantitative analysis the field-independent contributions of symmetry-forbidden LO phonon scattering must also be taken into account. For the LO phonon scattering the resulting scattering intensity is the sum of both [4.174]:

$$I_{\text{LO}} = C_0 + C_1 \cdot (\mathbf{E}_S)^2 \tag{4.20}$$

Due to the depth dependence of the surface electric field and the light attenuation in the sample, the total EFIRS intensity is obtained by a weighted integration over the depletion layer. With a light absorption coefficient α it follows [4.161, 4.45, 4.174]:

$$I_{LO} = C_0 + C_1 \cdot \int_0^{d_r} (\mathbf{E}_S(z))^2 \cdot \exp(-2 \cdot \alpha \cdot z)\, dz \qquad (4.21)$$

When a semiconductor interface is investigated, the quantity of interest is the potential barrier V_S rather than the electric field $\mathbf{E}_S$. This can be achieved using the Schottky model to describe the electric field and the resulting potential barrier at the semiconductor surface. Comparing (4.20) with (4.17) it is found that the LO scattering intensity is directly proportional to the band bending potential V_S. For lightly doped substrates where the thickness of the space charge layer is large in comparison to the penetration depth of the light, the simple proportionality can in fact be used for the evaluation of the band bending potential V_S instead of the integration over the space charge layer [4.4]. The offset C_0 and the proportionality constant C_1 describing the scattering efficiency for field-induced Fröhlich scattering depend on many parameters, such as doping, temperature, and laser frequency, and are not known a priori. Therefore, a calibration with another method, e.g. Photoemission Spectroscopy (PES), must be performed at first to obtain the values C_0 and C_1, before the EFIRS technique may be applied to determine absolute values of V_S. In contrast to the method based on the PLP/LO phonon intensity ratio, EFIRS requires two calibration points of different band bending values V_S to determine both constants C_0 and C_1.

Practically, EFIRS is most convenient to be applied to (110)-surfaces of III-V-semiconductors for two reasons: Firstly, the clean (110) surfaces do not exhibit any band bending (except for GaP), which allows the determination of the constant C_0 in a simple and reliable manner [4.175, 4.176]. The calibration constant C_1 can then be obtained by comparing results from EFIRS and PES for a situation where a large band bending at the semiconductor interface is present. In general, the calibration can be performed using two calibration points with nonzero band bending. However, a comparable accuracy would require a large difference in V_S between both calibration points, which is hardly possible to obtain in practice. Secondly, on (110)-surfaces the first-order Raman scattering from the TO phonon can be observed simultaneously with the EFIRS-signal from the LO phonon [4.2, 4.3, 4.161]. The TO phonon scattering is not affected by the electric field and can be used as an intensity reference. By the normalisation the determination of absolute scattering intensities can be avoided, which is an important aspect of the reliability of the EFIRS technique. Furthermore, the normalisation is a reliable technique to compensate the effect of light attenuation in the overlayer with increasing coverage.

In spite of the drawbacks of the external calibration, EFIRS has some important advantages. It can detect changes in band bending with a higher accuracy than PES. Besides, the main advantage of EFIRS is the applicability for higher coverages. The maximum overlayer thickness is determined by the light absorption which leads to a typical information depth of about 100 ML in the case of metallic overlayers [4.177, 4.178, 4.4, 4.62].

A large number of band bending investigations with different techniques have been performed for the Sb/GaAs(110) interface. For coverages up to a few monolayers such studies were extensively performed by photoemission spectroscopy [4.179–4.181] and by scanning tunneling spectroscopy [4.41]. They show that surfaces with a 1 ML coverage exhibit a band bending which, however, can be considerably reduced by annealing at 600 K, inducing an improvement of the order in the monolayer [4.182].

Raman scattering is able to extend these band bending studies to higher coverages. Especially for Sb on GaAs(110) it is interesting to apply EFIRS, since Raman studies of the overlayer growth have shown different structural modifications dependent on overlayer thickness and substrate temperature (Sect. 4.3). For deposition at room temperature, for example, the amorphous phase growing on top of the epitaxial first monolayer was shown to

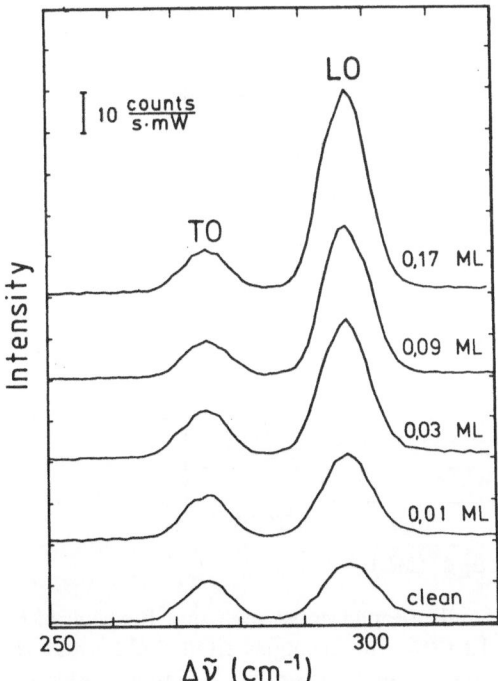

Fig. 4.37. Raman spectra of the GaAs phonons in the polarisation configuration $110(1\bar{1}0,1\bar{1}0)\bar{1}\bar{1}0$. The spectra were taken at 90 K using the $\lambda = 406.7$ nm line of a Kr^+ laser [4.44]

undergo a phase transition to crystalline Sb for coverages beyond approximately 15 ML. In contrast to PES, the EFIRS technique is well suited to monitor the band bending in the crucial coverage region where the Sb overlayer crystallises.

Corresponding EFIRS spectra for a p-doped GaAs sample ($p = 4 \cdot 10^{17}\,cm^{-3}$) are shown in Fig. 4.37 [4.44]. For these experiments (and for analogous experiments on n-doped substrates, $n = 1 \cdot 10^{17}\,cm^{-1}$) the samples were cooled to 90 K during the measurements to meet resonance condition with the 406.7 nm line of a Kr^+ laser where as for Sb deposition they were warmed up to room temperature. While the LO scattering intensity at the clean surface can be attributed to E-independent Fröhlich scattering, the strong increase with Sb coverage is due to the Schottky barrier being built up upon Sb deposition. In Fig. 4.38 the resulting Fermi level positions for n- and p-doped GaAs substrates are shown. The EFIRS data show a pronounced dip in the band bending on n-GaAs which is correlated with the crystallisation of the overlayer. For comparison, data from SXPS and I-V measurements are included, which confirm the EFIRS results in the low- and high coverage regime, respectively.

The remarkable variation in band bending induced by the crystallisation occurs for Sb coverages where neither photoemission nor I-V/C-V ex-

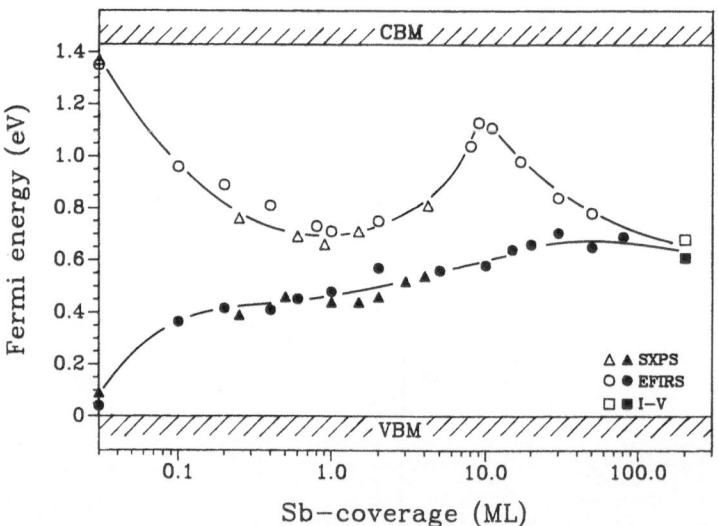

Fig. 4.38. Surface Fermi level for n- and p-doped substrates as a function of Sb coverage. The dots are taken from EFIRS-, the triangles from SXPS- and the squares from I-V-data. An excellent agreement between the different methods is demonstrated, as well as their limitation to characteristic coverage regimes. The pronounced dip at around 10 ML for n-GaAs shown in the EFIRS-data coincides with the crystallisation of the overlayer [4.44]

periments allow an accurate determination. The reason for this dip in band bending could be either structural changes at the interface or electronic properties of the overlayer being modified by the crystallisation. Recently it was found that in fact the electronic properties of the overlayer are changed from semiconducting to semimetallic through crystallisation [4.183–4.185]. Therefore, the modification of the band bending in the substrate was attributed to the effect of the overlayer metallisation.

An example for EFIRS applied to semiconductor heterojunctions is Ge on GaAs(110). The growth characterisation was already treated in Sect. 4.3.1. Both materials exhibit nearly perfect lattice matching (GaAs: $a_0 = 5.6533$ Å; Ge: $a_0 = 5.6577$ Å at 300 K) and have considerably different energy gaps of 1.45 eV and 0.67 eV, respectively. The valence band offset is known to be 0.33 eV; however, the exact potential variation across the interface for (110) orientation has been a point of discussion, since different photoemission experiments yielded contradictory results [4.60, 4.61].

Therefore an analysis of the band bending by *in-situ* EFIRS experiments for layer thicknesses up to 250 Å was performed [4.63]. In contrast to the study for Sb on GaAs(110) described above, these EFIRS experiments used the incident laser line of 413.1 nm which is in resonance to the E_1-gap of GaAs at room temperature. The band bending development determined via the LO intensity from EFIRS is shown in Fig. 4.39 for different growth temperatures. In the submonolayer region, Ge deposition leads to a strong increase of the band bending, which is reversed with further deposition on n- doped substrates for high growth temperatures. For a deposition temperature of 650 K flatband conditions are nearly restored at 100 ML. P-type substrates, on the contrary, exhibit no decrease of the band bending [4.63].

Of crucial importance for the explanation of these results is the Ge overlayer structure. It was monitored simultaneously by Raman scattering as described in Sect. 4.3.1. For 100 K deposition temperature amorphous growth, for 300 K polycrystalline, and above 670 K high quality epitaxial growth is achieved. The persisting band bending for growth temperatures below 600 K can therefore be attributed to defects at the interface due to the irregular Ge structure. In contrast, the flatband condition on n-doped GaAs for high coverages is indicative for the perfect epitaxial Ge growth at high temperatures which avoids the generation of defect states at the interface. Therefore, band bending at the interface can only originate from charged states at the Ge surface, and vanishes with sufficient overlayer thickness.

The persistence of the band bending on p-doped GaAs even for crystalline Ge is explained in terms of the band lineup between the substrate and the Ge film, which is assumed to be n-type [4.63]. The difference between the E_F^S values for p- and n-type GaAs is nearly equal to the Ge band gap.

The $Si_x Ge_{1-x}$-alloys on GaAs allow to extend the study of the GaAs/Ge system to strained heterojunctions. The large lattice mismatch between GaAs and Si of 4% allows to tune the mismatch and accordingly the critical thick-

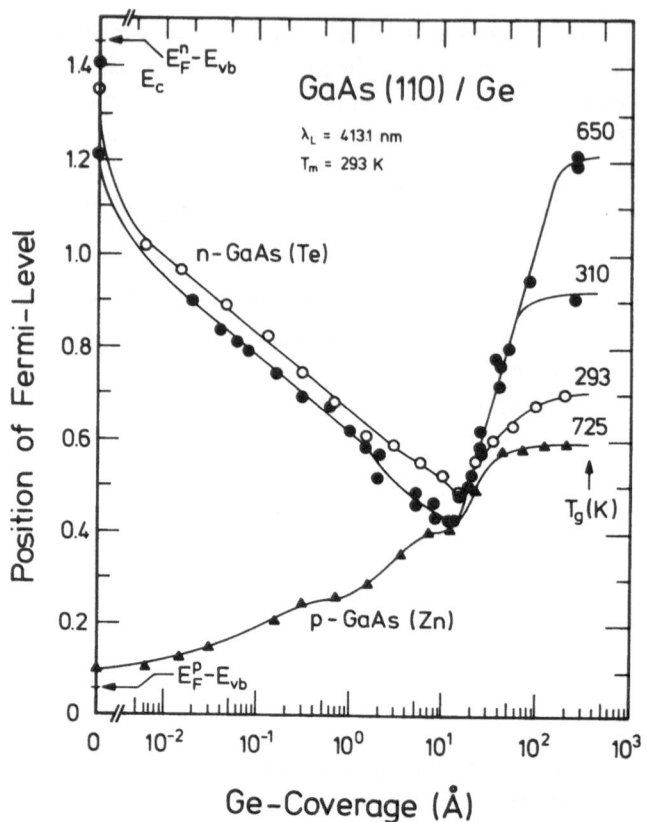

Fig. 4.39. Surface Fermi level E_F^S of n and p-doped GaAs for Ge coverage up to 20 nm for different deposition temperatures, derived from EFIRS. For 650 K deposition near flatband is restored for n-GaAs [4.63]

ness of the overlayer to any value above 8 Å by increasing the Si content x [4.186]. In this case the EFIRS technique can be used as an excellent monitor for the onset of the overlayer relaxation, because the band bending is extremely sensitive to the density of defects at the interface which are generated by the relaxation.

Indeed the EFIRS investigations on n-GaAs/$Si_x Ge_{1-x}$ show a pronounced, sharp increase of the band bending correlated with the onset of the relaxation of the strained $Si_x Ge_{1-x}$ layer [4.63]. The EFIRS-derived values for the critical thickness h_c are plotted vs Si content x in Fig. 4.40. The relaxation process turns out to extend over a wide thickness range, e.g. from 200 Å ($x = 0.8$) up to 2000 Å ($x = 0.4$). Below $x = 0.4$ the light penetration depth in the overlayer is too low to observe the relaxation of the layer, while beyond $x = 0.8$ no epitaxial growth is achieved. The abrupt increase of the band bending shows that the defects created by the relaxation are responsible for the band

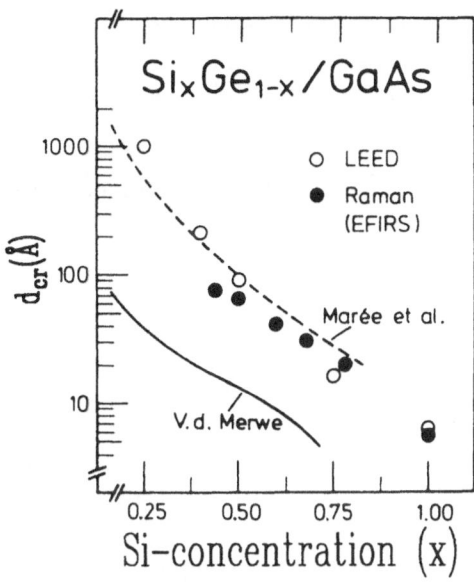

Fig. 4.40. Critical thickness h_c of Si_xGe_{1-x} on GaAs, derived from the EFIRS signal due to the increase of interface defects [4.63]

bending being built up. Therefore, EFIRS gives a quite sharp criterion for the determination of h_c.

More EFIRS investigations were performed for metal-semiconductor interfaces, for instance Au and Ag on GaAs(110) by Schäffler [4.161], and for gas-covered semiconductor surfaces such as O_2 on GaAs(110) [4.178], Cl_2 and O_2 on InAs(110) [4.188] and H on GaAs(110) [4.189]. A detailed description of these EFIRS applications can be found in a recent review article [4.192].

4.6 Summary

This overview has shown that Raman scattering is a very useful, nondestructive technique for the analysis of a wide variety of relevant heterostructure properties, offering in-depth and lateral resolution. Different interaction mechanisms were treated between electromagnetic radiation and elementary excitations such as phonons, electronic excitations and coupled plasmon phonon modes.

Various examples were discussed for Raman scattering from phonons applied to study crystalline properties of semiconductor layers such as structure, order, orientation, composition, and chemical reactivity. Besides for structural analysis, Raman spectroscopy was also shown to be suitable for the characterisation of electronic properties, giving information about the electronic band structure, the concentration of free carriers, the incorporation of dopants, and the electronic band bending at semiconductor surfaces and interfaces.

The unique advantage of Raman scattering is thus the broad range of obtainable information, which is important for applications in semiconductor heterostructure fabrication. Other important demands to a method to be applied for heterostructure characterisation are its sensitivity to extremely thin films, its applicability to various environments realised in semiconductor growth and processing equipment, and its resolution limit in a laterally resolved analysis. In all of these aspects Raman scattering offers an reasonable solution:

— The sensitivity of Raman scattering for recording vibrational modes was demonstrated to reach for some material systems even monolayer or sub-monolayer coverages.

— As an optical technique in the visible spectral range it can be adapted to a usual growth equipment, no matter whether extremely clean vacuum conditions, reactive gas environments, or simply atmospheric ambients are required.

— By light diffraction the lateral resolution of Raman spectroscopy is limited to the micrometer regime. Although this is poor compared to electron-based techniques such as e.g. Transmission Electron Microscopy, the lateral resolution of Raman scattering is still excellent in comparison to other spectroscopic techniques in the infrared or X-ray regime. In respect to lateral resolution the scanning nearfield optical microscopy might provide an essential improvement in near future.

Especially the development of reliable laser light sources in combination with fast and highly sensitive optical multichannel detector systems, allowing the simultaneous registration of a considerable spectral interval, has increased the potential of Raman scattering. Nowadays, even the structural and electronic analysis of single monolayer growth as well as real-time *in-situ* growth studies enabling online monitoring of the interface chemistry and the crystalline quality of growing layers have been realized.

5. Far-Infrared Spectroscopy

Bernd Harbecke, Bernhard Heinz, Volkmar Offermann, Wolfgang Theiß

Infrared spectroscopy (spectral range 10 to $10000\,\mathrm{cm}^{-1}$) has played an important role for the characterisation of epitaxial layers since the early days of epitaxial growth. Infrared transmission and reflection spectra yield information on electric dipole allowed elementary excitations. This technique complements inelastic light scattering techniques which were presented in the previous chapter. The information which is extracted from infrared spectra concerns

- the concentration, effective mass and mobility of free carriers
- the concentration and identification of impurities and defects
- the oscillator strength, frequency and damping of optical phonons
- the thicknesses of multi-layer-structures .

In addition this information may be obtained with a spatial resolution both laterally as well as along the growth direction down to the μm range under favourable conditions.

Of course infrared spectroscopy (IR) is nondestructive like all the other techniques discussed in this book. In the spectral range of interest prism or grating spectrometers have been used in the past. However, in the last twenty years with the advent of powerful desktop computers Fourier transform spectroscopy has become more and more important, particularly in the Far Infrared (FIR). This technique has even gained importance for routine inspection of epitaxial layers on large wafers in semiconductor industry. Recent developments like IR microscopy improve the achievable spatial resolution and open new possibilities for lateral investigations. The development of low loss infrared fibres promises even a further extension of the range of applications of Fourier transform spectroscopy.

Parallel to the improvement of experimental methods the required software for a fast extraction of relevant parameters from measured spectra has been developed. These developments are quite important, since IR spectra of multilayer structures are so complex that for their analysis in general a numerical simulation is required.

This Chapter is organized as follows. In the Sect. 5.1 an extensive theoretical description of the formalism based on Maxwell's equations and the classical frequency dependent dielectric functions for the interpretation of IR spectra of multilayer structures is presented. The Berreman effect and surface polaritons are treated as well. This is followed by a brief introduction in Fourier Transform Infrared Spectroscopy (FTIR). Only the principle ideas on FTIR are presented since excellent reviews exist about this topic [5.1, 5.2, 5.3, 5.4]. The most common FTIR spectrometers commercially available are discussed.

In the following Sections we concentrate on problems like the determination of layer thicknesses, carrier and impurity concentrations. Infrared spectroscopy on confined electron systems (two-dimensional carriers in heterostructures and quantum wells) are only briefly presented since several excellent reviews exist. Calculations on the optical properties of superlattices were reviewed by Appel and Hunderi [5.5], a complete review on experimental aspects has been recently published by Dumelow et. al. [5.6]. In the last Section of this Chapter porous silicon is discussed in some detail as an example for an inhomogeneous material system.

5.1 Theoretical Foundations

In this Chapter the theoretical background for the interpretation of the experimental data will be developed based on classical electrodynamics [5.7–5.10]: Maxwell's equations with the commonly adopted definitions of charges and currents, the linear response of matter to the fields, dispersion relations and their relevance for the characterisation of films. Maxwell's equations together with the constitutive equations are solved for the ansatz of a plane wave and the dispersion relation is obtained. A discussion of the energy flow follows subsequently and a short survey of the boundary conditions for electric and magnetic fields. To conclude the rather general part the formalism of multiple reflections of a layered structure is presented including a discussion of coherent and incoherent superposition of partial waves.

The rest of the chapter is devoted to a discussion of the Berreman effect and surface waves.

5.1.1 Maxwell's Equations

The propagation of electromagnetic waves in homogeneous media is described by Maxwell's equations

$$\operatorname{curl} \mathbf{E}(\mathbf{r}, t) = -\frac{\partial}{\partial t} \mathbf{B}(\mathbf{r}, t) \tag{5.1}$$

$$\varepsilon_0 \operatorname{div} \mathbf{E}(\mathbf{r}, t) = \varrho_{FC}(\mathbf{r}, t) - \operatorname{div} \mathbf{P}_{BC}(\mathbf{r}, t) \tag{5.2}$$

$$\frac{1}{\mu_0} \operatorname{curl} \mathbf{B}(\mathbf{r}, t) = \varepsilon_0 \frac{\partial}{\partial t} \mathbf{E}(\mathbf{r}, t) + \frac{\partial}{\partial t} \mathbf{P}_{BC}(\mathbf{r}, t) + \mathbf{j}_{FC}(\mathbf{r}, t) + \operatorname{curl} \mathbf{M}(\mathbf{r}, t) \tag{5.3}$$

$$\operatorname{div} \mathbf{B}(\mathbf{r}, t) = 0. \tag{5.4}$$

Here, $\mathbf{E}$ denotes the electric and $\mathbf{B}$ the magnetic field.

The conduction current density $\mathbf{j}_{FC}$ is due to the freely moving charge carriers with charge density ρ_{FC}. They obey the equation of continuity

$$\frac{\partial}{\partial t}\varrho_{FC}(\mathbf{r},t) + \operatorname{div}\mathbf{j}_{FC}(\mathbf{r},t) = 0. \tag{5.5}$$

This can be seen by taking the time derivative of (5.2) and the divergence of (5.3) and substituting both equations into each other.

The polarisation $\mathbf{P}_{BC}$ and the magnetisation $\mathbf{M}$ result from the bound electrons and nuclei. The polarisation current density $\partial\mathbf{P}_{BC}/\partial t$ occurs if the clouds of electrons and the nuclei move around their equilibrium positions and molecular dipole moments (or higher multipole moments) are formed. The corresponding charge density is given by $-\operatorname{div}\mathbf{P}_{BC}$. The equation of continuity is fulfilled by definition.

The orbital motions and the spins of the electrons and nuclei contribute to the magnetisation $\mathbf{M}$ of the matter. The magnetisation current density $\operatorname{curl}\mathbf{M}$ is not accompanied by an extra charge density.

Studying the propagation of electromagnetic waves, the distinction between $\mathbf{j}_{FC}$ and $\partial\mathbf{P}_{BC}/\partial t$ becomes meaningless for sufficiently high frequencies. Whether bound or not, the charges wiggle tightly about an equilibrium position. Thus we often put $\mathbf{j}_{FC}$ and $\partial\mathbf{P}_{BC}/\partial t$ together and consider it as a generalised current density which is the time derivative of a generalised polarisation density,

$$\mathbf{J}(\mathbf{r},t) \stackrel{\text{def}}{=} \mathbf{j}_{FC}(\mathbf{r},t) + \frac{\partial}{\partial t}\mathbf{P}_{BC}(\mathbf{r},t) \stackrel{\text{def}}{=} \frac{\partial}{\partial t}\mathbf{P}(\mathbf{r},t). \tag{5.6}$$

Consequently, ϱ_{FC} and $-\operatorname{div}\mathbf{P}_{BC}$ form a generalised charge density

$$\varrho(\mathbf{r},t) \stackrel{\text{def}}{=} \varrho_{FC}(\mathbf{r},t) - \operatorname{div}\mathbf{P}_{BC}(\mathbf{r},t) = -\operatorname{div}\mathbf{P}(\mathbf{r},t). \tag{5.7}$$

It is often convenient to define the auxiliary fields

$$\mathbf{D}(\mathbf{r},t) \stackrel{\text{def}}{=} \epsilon_o\mathbf{E}(\mathbf{r},t) + \mathbf{P}(\mathbf{r},t) \quad \text{and} \quad \mathbf{H}(\mathbf{r},t) \stackrel{\text{def}}{=} \frac{1}{\mu_o}\mathbf{B}(\mathbf{r},t) - \mathbf{M}(\mathbf{r},t). \tag{5.8}$$

Maxwell's equations then take the form

$$\operatorname{curl}\mathbf{E}(\mathbf{r},t) = -\frac{\partial}{\partial t}\mathbf{B}(\mathbf{r},t) \tag{5.9}$$

$$\operatorname{div}\mathbf{D}(\mathbf{r},t) = 0 \tag{5.10}$$

$$\operatorname{curl}\mathbf{H}(\mathbf{r},t) = \frac{\partial}{\partial t}\mathbf{D}(\mathbf{r},t) \tag{5.11}$$

$$\operatorname{div}\mathbf{B}(\mathbf{r},t) = 0. \tag{5.12}$$

An important property of Maxwell's equations is the fact that they contain the wave equations for the electric and magnetic fields. Taking the curl of (5.1) and the time derivative of (5.3) and equating leads to (in the equations which follow, the explicit dependence of the variables on $\mathbf{r}$ and t is omitted where the meaning is clear)

$$\text{curl } \text{curl } \mathbf{E} + \epsilon_0 \mu_0 \frac{\partial^2}{\partial t^2} \mathbf{E} = -\mu_0 \frac{\partial}{\partial t} \left(\mathbf{j}_{\text{FC}} + \frac{\partial}{\partial t} \mathbf{P}_{\text{BC}} + \text{curl } \mathbf{M} \right). \tag{5.13}$$

With the vector identity curl curl $\mathbf{E} = \text{grad } \text{div } \mathbf{E} - \Delta \mathbf{E}$ and the abbreviation $c_0^2 \stackrel{\text{def}}{=} 1/\epsilon_0 \mu_0$, the speed of light in vacuum, we get

$$\Delta \mathbf{E} - \frac{1}{c_0^2} \frac{\partial^2}{\partial t^2} \mathbf{E} = \text{grad } \text{div } \mathbf{E} + \mu_0 \frac{\partial}{\partial t} \left(\mathbf{j}_{\text{FC}} + \frac{\partial}{\partial t} \mathbf{P}_{\text{BC}} + \text{curl } \mathbf{M} \right). \tag{5.14}$$

With the help of (5.2) and (5.6) we can transform (5.14) into

$$\Delta \mathbf{E} - \frac{1}{c_0^2} \frac{\partial^2}{\partial t^2} \mathbf{E} = \frac{1}{\epsilon_0} \text{grad } \rho_{\text{FC}} + \mu_0 \frac{\partial}{\partial t} \left(\mathbf{J} + \text{curl } \mathbf{M} \right). \tag{5.15}$$

Equation (5.15) represents an inhomogeneous wave equation where the driving forces are determined by the charges and the currents.

Equivalent substitutions with the magnetic field $\mathbf{B}$ lead to

$$\Delta \mathbf{B} + \frac{1}{c_0^2} \frac{\partial^2}{\partial t^2} \mathbf{B} = -\mu_0 \text{curl}(\mathbf{J} + \text{curl } \mathbf{M}). \tag{5.16}$$

5.1.2 Constitutive Equations and Dispersion Relations

Maxwell's equations, together with suitably chosen boundary conditions, are sufficient to calculate the electromagnetic fields due to given currents and charges. In an optical experiment the situation is different. Here, a given external electromagnetic wave is incident on a sample, its response is observed and has to be interpreted.

Therefore, we complete the set of equations (5.1 to 5.4) by constitutive equations which describe the response of the matter to the applied electromagnetic fields. For most purposes it is sufficient to consider a linear response:

$$\mathbf{R}(t) = \int_{-\infty}^{\infty} dt' G(t - t') \mathbf{F}(t'). \tag{5.17}$$

$\mathbf{F}$ represents one of the fields $\mathbf{E}$, $\mathbf{B}$. The kernel $G(t - t')$ of the integral is called response function. It gives the response $\mathbf{R}(t)$ at time t to an idealised pulse of the field $\mathbf{F}(t) = \mathbf{F}_0 \delta(t')$ at time t'.

It is assumed that the response depends only on the difference between the times t and t'. This assumption requires that the macroscopic properties of the matter, such as pressure, temperature and other thermal quantities must be time invariant.

Fourier transforming the convolution integral (5.17) leads to a product of the corresponding Fourier components,

$$\mathbf{R}(\omega) = \chi(\omega) \mathbf{F}(\omega). \tag{5.18}$$

The Fourier transformation of the response function

$$\chi(\omega) = \int_{-\infty}^{\infty} d\tau\, G(\tau) \exp(i\omega\tau) \qquad (5.19)$$

is called susceptibility. We shall not introduce new symbols for the Fourier transformed fields $\mathbf{F}$ and responses $\mathbf{R}$. We always write the Fourier transformation $t \leftrightarrow \omega$ in the form

$$b(\omega) = \int_{-\infty}^{\infty} dt\, a(t) \exp(i\omega t) \quad \text{and} \quad a(t) = \frac{1}{2\pi} \int_{-\infty}^{\infty} d\omega\, b(\omega) \exp(-i\omega t). \quad (5.20)$$

A large number of phenomena can be described and explained with the help of three (or less) constitutive equations, namely

$$\mathbf{j}_{FC}(\omega) = \sigma(\omega)\mathbf{E}(\omega) \qquad (5.21)$$

$$\mathbf{P}(\omega) = \epsilon_0 \chi_e(\omega)\mathbf{E}(\omega) \qquad (5.22)$$

$$\mathbf{M}(\omega) = \frac{1}{\mu_0}\chi_m(\omega)\mathbf{B}(\omega). \qquad (5.23)$$

Here, σ is the conductivity of the free carriers, χ_e and χ_m are the electric and magnetic susceptibilities, respectively.

These susceptibilities are, in general, complex quantities. The requirement that the response $\mathbf{R}(t)$ must be real leads to $\chi(-\omega) = \chi^*(\omega)$, i.e. the real part of χ must be an even function, and the imaginary part an odd function of the frequency.

A further restriction of the susceptibilities is imposed by the requirement of causality of the response function (5.17), i.e. $G(t - t') = 0$ for $t' > t$. This is the foundation of the so-called dispersion relations (Kramers-Kronig relations) which play an important role in experimental and theoretical optics [5.11, 5.12].

Although we do not apply the dispersion relation for the interpretation of the measured spectra in this article, it is meaningful to present a short derivation of them. It was shown recently [5.13] that they can be used not only for the characterisation of the optical properties of bulk material, but also for a thin film on a substrate.

The result of the Fourier transformation of the even function $\text{Re}[\chi(\omega)]$ in frequency space to time space is again an even function $G_{even}(\tau)$; similarly, the odd function $i \cdot \text{Im}[\chi(\omega)]$ gives an odd function $G_{odd}(\tau)$. Thus we get $G(\tau) \equiv G_{even}(\tau) + G_{odd}(\tau)$.

This decomposition is entirely based on the property that $G(\tau)$ must be a real function of τ. The additional requirement that $G(\tau)$ is also a causal function can only be fulfilled, if the even and the odd part of $G(\tau)$ cancel for $\tau < 0$. Consequently, they must be identical for $\tau > 0$, i.e.

$$G_{even}(\tau) \equiv G_{odd}(\tau) \equiv \frac{1}{2}G(\tau) \qquad \text{for} \quad \tau > 0. \qquad (5.24)$$

This identity is the basis of the dispersion relation: for a generation of the response function $G(\tau)$ it is sufficient to know either the real part or the imaginary part of the susceptibility $\chi(\omega)$ for the complete frequency range. Consequently, the real and the imaginary part of the susceptibility are not independent of each other.

A simple representation of this relationship [5.14–5.16] reads

$$\chi(\omega) = \frac{1}{2\pi} \int_{-\infty}^{\infty} d\tau \exp(i\omega\tau) \cdot \text{sign}(\tau) \int_{-\infty}^{\infty} d\omega' \chi(\omega') \exp(-i\omega'\tau). \qquad (5.25)$$

This representation is also very well adapted to computer calculation with the help of the Fast Fourier Transform algorithm (FFT).

The usual representation is the Hilbert transformation

$$\chi(\omega) = \frac{i}{\pi} \int_{-\infty}^{\infty} d\omega' \frac{\chi(\omega')}{\omega - \omega'}. \qquad (5.26)$$

Although the Hilbert transformation consists of only a single integration, the evaluation with the help of a computer is much more tedious and time consuming than that of the two Fourier transformations.

5.1.3 Plane Waves in an Isotropic and Homogeneous Medium

Let the electric and magnetic fields be the plane waves,

$$\mathbf{F}(\mathbf{r}, t) = \text{Re} \left\{ \mathbf{F} \exp(i\mathbf{k} \cdot \mathbf{r} - i\omega t) \right\}. \qquad (5.27)$$

Then we can simplify (5.17) to

$$\mathbf{R}(\mathbf{r}, t) = \text{Re} \left\{ \chi(\omega)\mathbf{F} \exp(i\mathbf{k} \cdot \mathbf{r} - i\omega t) \right\}, \qquad (5.28)$$

with analogous representations for the three constitutive equations (5.21)–(5.23).

Later, we shall no longer mention the 'Re' explicitly. As long as the formulae only contain the fields and currents as linear forms, we are allowed to perform calculations with the complex fields and take the real part just at the end of the calculation.

For this ansatz the wave equation (5.14) takes the form

$$-(\mathbf{k} \cdot \mathbf{k})\mathbf{E} + \mathbf{k}(\mathbf{k} \cdot \mathbf{E}) + \frac{\omega^2}{c^2}\mathbf{E} = -i\omega\mu_0\sigma\mathbf{E} - \mu_0\epsilon_0\omega^2\chi_e\mathbf{E} + i\mathbf{k} \times \chi_m(-i\omega\mathbf{B}). \quad (5.29)$$

Up to now, we have not explicitly made use of the fact that the wave propagates in an isotropic medium, i.e. that the susceptibilities are scalar quantities. We can then simplify (5.29) further:

$$[-(\mathbf{k} \cdot \mathbf{k})\mathbf{E} + \mathbf{k}(\mathbf{k} \cdot \mathbf{E})] (1 - \chi_m) + \frac{\omega^2}{c^2}\left(1 + \frac{i\sigma}{\epsilon_0\omega} + \chi_e\right)\mathbf{E} = 0. \qquad (5.30)$$

Applying the Maxwell equation $\epsilon_0 \mathbf{k} \cdot \mathbf{E} = -(\mathrm{i}\,\rho_{\mathrm{FC}} + \mathbf{k} \cdot \mathbf{P})$, the continuity equation $-\omega\rho_{\mathrm{FC}} + \mathbf{k} \cdot \mathbf{j}_{\mathrm{FC}} = 0$ and the constitutive equations $\mathbf{j}_{\mathrm{FC}} = \sigma\mathbf{E}$ and $\mathbf{P} = \epsilon_0\chi_e\mathbf{E}$ we obtain

$$\left(1 + \frac{\mathrm{i}\,\sigma}{\epsilon_0\omega} + \chi_e\right)\mathbf{k} \cdot \mathbf{E} = 0. \tag{5.31}$$

Equation (5.31) tells us that $\mathbf{k} \cdot \mathbf{E} = 0$ as long as $1 + \dfrac{\mathrm{i}\,\sigma}{\epsilon_0\omega} + \chi_e \neq 0$. These waves are called 'transverse' plane waves.

As this quantity often occurs in the calculations we define a generalised susceptibility $\chi(\omega)$ and a generalised dielectric function $\varepsilon(\omega)$

$$\varepsilon(\omega) \stackrel{\mathrm{def}}{=} 1 + \chi(\omega) \stackrel{\mathrm{def}}{=} 1 + \frac{\mathrm{i}\,\sigma(\omega)}{\epsilon_0\omega} + \chi_e(\omega). \tag{5.32}$$

Furthermore, it is convenient to define a quantity 'magnetic permeability'

$$\mu(\omega) \stackrel{\mathrm{def}}{=} \frac{1}{1 - \chi_m(\omega)}. \tag{5.33}$$

In textbooks [5.7, 5.9] the electric and magnetic permeabilities are usually introduced as the constitutive equations $\mathbf{D} = \epsilon_0\varepsilon\mathbf{E}$ and $\mathbf{B} = \mu_0\mu\mathbf{H}$, which is in agreement with (5.8).

With these definitions, and the requirement $\varepsilon(\omega) \neq 0$, the wave equation (5.30) becomes

$$\left[\mathbf{k} \cdot \mathbf{k} - \frac{\omega^2}{c_0^2}\varepsilon(\omega)\mu(\omega)\right]\mathbf{E} = 0. \tag{5.34}$$

Thus, a transverse plane wave in an isotropic and homogeneous medium can only propagate if the periodicity in space, the wave vector $\mathbf{k}$, is related to the periodicity in time, the frequency ω, as

$$\mathbf{k} \cdot \mathbf{k} = \frac{\omega^2}{c_0^2}\varepsilon(\omega)\mu(\omega). \tag{5.35}$$

Equation (5.35) is called 'dispersion relation'.

If k is real, $\lambda = 2\pi/k$ is the wavelength and $c = c_0/\sqrt{\varepsilon\mu}$ is the phase velocity of the wave in the medium. This leads to the definition of the index of refraction

$$n(\omega) \stackrel{\mathrm{def}}{=} \sqrt{\varepsilon(\omega)\mu(\omega)} \tag{5.36}$$

where $n = c_0/c = \lambda_0/\lambda = k/k_0$. The symbols λ_0 and k_0 denote the wavelength and the wave vector of a wave with frequency ω in vacuum.

If the condition $\varepsilon(\omega) \neq 0$ does not hold, i.e. if $\varepsilon(\omega) = 0$, then (5.30) demands that $-(\mathbf{k} \cdot \mathbf{k})\mathbf{E} + \mathbf{k}(\mathbf{k} \cdot \mathbf{E}) \equiv \mathbf{k} \times (\mathbf{k} \times \mathbf{E}) = 0$. These plane waves with $\mathbf{k} \times \mathbf{E} = 0$ describe the longitudinal elementary excitations like plasmon-polaritons or longitudinal optical (LO) phonon-polaritons which cannot be excited directly in IR experiments under normal incidence conditions. Nevertheless for $\varepsilon(\omega) = 0$ a pronounced feature is in general observed ("plasma edge").

5.1.4 The Energy Balance

Starting from Maxwell's equations we now derive a continuity equation for the energy of the electromagnetic field in matter. We multiply curl $\mathbf{E}$ by $\mathbf{B}$ and curl $\mathbf{B}$ by $\mathbf{E}$, apply the vector identity $\mathrm{div}(\mathbf{a} \times \mathbf{b}) = \mathbf{b}\,\mathrm{curl}\,\mathbf{a} - \mathbf{a}\,\mathrm{curl}\,\mathbf{b}$, and get

$$\frac{\partial}{\partial t}\left(\frac{1}{2}\varepsilon_0 \mathbf{E} \cdot \mathbf{E} + \frac{1}{2}\mu_0 \mathbf{B} \cdot \mathbf{B}\right) + \mathrm{div}\left[\mathbf{E} \times \left(\frac{1}{\mu_0}\mathbf{B} - \mathbf{M}\right)\right]$$
$$= -\mathbf{E} \cdot \left(\mathbf{j}_{\mathrm{FC}} + \frac{\partial}{\partial t}\mathbf{P}_{\mathrm{BC}}\right) + \mathbf{M} \cdot \frac{\partial}{\partial t}\mathbf{B}. \tag{5.37}$$

Here, $\varepsilon_0 \mathbf{E}^2/2$ is the energy density of the electric and $\mathbf{B}^2/2\mu_0$ that of the magnetic field. The term in the divergence

$$S \overset{\mathrm{def}}{=} \mathbf{E} \times \left(\frac{1}{\mu_0}\mathbf{B} - \mathbf{M}\right) \equiv \mathbf{E} \times \mathbf{H} \tag{5.38}$$

is called Poynting vector. It describes the energy flow density and consists of two parts: a contribution from the electromagnetic field components $\mathbf{E} \times \mathbf{B}/\mu_0$ — which would also exist in vacuum —, and a contribution $-\mathbf{E} \times \mathbf{M}$ due to a coupling between the electric field and the magnetisation of matter.

For a complete analysis of the right hand side of (5.37), the response of the material to the electromagnetic fields (constitutive equations) has to be defined. It consists partly of an energy density, and partly of a sink term; in special cases, like spatial dispersion, the right hand side of (5.37) contributes to the energy flow.

We assume that all quantities of (5.37) vary harmonically in time according to

$$\mathbf{F}(\mathbf{r}, t) = \mathrm{Re}\left\{\mathbf{F}(\mathbf{r})\exp(-\mathrm{i}\omega t)\right\}. \tag{5.39}$$

On averaging (5.37) over one period in time, $T = 2\pi/\omega$, all terms which are total time derivatives, i.e. the energy densities, vanish. The nonvanishing terms are given by

$$\mathrm{div}\,\overline{\mathbf{E} \times \mathbf{H}}^T = \overline{-\mathbf{E} \cdot \left(\mathbf{j}_{\mathrm{FC}} + \frac{\partial}{\partial t}\mathbf{P}_{\mathrm{BC}}\right) + \mathbf{M} \cdot \frac{\partial}{\partial t}\mathbf{B}}^T \overset{\mathrm{def}}{=} -W_{\mathrm{diss}}, \tag{5.40}$$

where on the right side only the Ohmic losses W_{diss} remain.

If, in a next step, the properties of matter $\mathbf{j}_{FC}$, $\mathbf{P}_{BC}$ and $\mathbf{M}$ can be described (see (5.18,5.28)) by

$$\mathbf{j}_{FC}(\mathbf{r}, t) = \mathrm{Re}\left\{\sigma(\omega)\,\mathbf{E}(\mathbf{r})\exp(-\mathrm{i}\,\omega t)\right\} \tag{5.41}$$

$$\mathbf{P}_{BC}(\mathbf{r}, t) = \mathrm{Re}\left\{\varepsilon_0\chi_e(\omega)\,\mathbf{E}(\mathbf{r})\exp(-\mathrm{i}\,\omega t)\right\} \tag{5.42}$$

$$\mathbf{M}(\mathbf{r}, t) = \mathrm{Re}\left\{(\chi_m(\omega)/\mu_0)\,\mathbf{B}(\mathbf{r})\exp(-\mathrm{i}\,\omega t)\right\} \tag{5.43}$$

with complex (and scalar) susceptibilities σ, χ_e and χ_m, straightforward manipulations of (5.40) lead to

$$W_{\mathrm{diss}}(\mathbf{r}) = \frac{1}{2}\left(\sigma' + \omega\varepsilon_0\chi_e''\right)\mathbf{E}(\mathbf{r})\cdot\mathbf{E}^\star(\mathbf{r}) + \frac{\omega}{2\mu_0}\chi_m''\mathbf{B}(\mathbf{r})\cdot\mathbf{B}^\star(\mathbf{r}). \tag{5.44}$$

The last step consists of the calculation of the average Poynting vector $\overline{\mathbf{S}}^T = \overline{\mathbf{E}\times\mathbf{H}}^T$ for plane waves, see (5.28). We get

$$\overline{\mathbf{S}(\mathbf{r})}^T = \frac{1}{2\mu_0\omega}\,\mathrm{Re}\left\{(1 - \chi_m)\left[\mathbf{k}(\mathbf{E}(\mathbf{r})\cdot\mathbf{E}^\star(\mathbf{r})) - \mathbf{E}(\mathbf{r})(\mathbf{k}^\star\cdot\mathbf{E}(\mathbf{r}))\right]\right\}. \tag{5.45}$$

Equation (5.45) simplifies further for $\mathbf{k}^\star\cdot\mathbf{E} = 0$:

$$\begin{aligned}
\overline{\mathbf{S}(\mathbf{r})}^T &= \frac{1}{2\mu_0\omega}\,\mathrm{Re}\left\{(1 - \chi_m)\mathbf{k}\right\}\mathbf{E}(\mathbf{r})\cdot\mathbf{E}^\star(\mathbf{r}) \\
&= \frac{1}{2\mu_0\omega}\,\mathrm{Re}\left\{\mathbf{k}/\mu\right\}\mathbf{E}(\mathbf{r})\cdot\mathbf{E}^\star(\mathbf{r}) \\
&= \frac{1}{2\mu_0\omega}\,\mathrm{Re}\left\{\mathbf{k}/\mu\right\}\mathbf{E}\cdot\mathbf{E}^\star\exp\left(-2\,\mathrm{Im}(\mathbf{k})\cdot\mathbf{r}\right).
\end{aligned} \tag{5.46}$$

Equation (5.46) is used for the calculation of the reflectance and transmittance in s-polarisation (Sect. 5.1.6). In the case of p-polarisation calculations can be carried out more easily with the field $\mathbf{H}$ instead of $\mathbf{E}$. In this formulation the Poynting vector is given by

$$\overline{\mathbf{S}(\mathbf{r})}^T = \frac{1}{2\varepsilon_0\omega}\,\mathrm{Re}\left\{\frac{\mathbf{k}(\mathbf{H}(\mathbf{r})\mathbf{H}^\star(\mathbf{r})) - \mathbf{H}(\mathbf{r})(\mathbf{k}\mathbf{H}^\star(\mathbf{r}))}{1 + \mathrm{i}\,\sigma/\varepsilon_0\omega + \chi_e}\right\}. \tag{5.47}$$

With $\varepsilon \equiv 1 + \chi_e + \mathrm{i}\,\sigma/\varepsilon_0\omega$ and the assumption that $\mathbf{k}\cdot\mathbf{H}^\star = 0$ holds, (5.47) becomes

$$\begin{aligned}
\overline{\mathbf{S}(\mathbf{r})}^T &= \frac{1}{2\varepsilon_0\omega}\,\mathrm{Re}\left\{\mathbf{k}/\varepsilon\right\}\mathbf{H}(\mathbf{r})\cdot\mathbf{H}^\star(\mathbf{r}) \\
&= \frac{1}{2\varepsilon_0\omega}\,\mathrm{Re}\left\{\mathbf{k}/\varepsilon\right\}\mathbf{H}\cdot\mathbf{H}^\star\exp\left(-2\,\mathrm{Im}(\mathbf{k})\cdot\mathbf{r}\right).
\end{aligned} \tag{5.48}$$

To complete this chapter we want to add a word of caution. Equations (5.46) and (5.48) have been derived for a single plane wave. If there is a superposition of two (or more) waves, the resulting energy flow is in general not the sum of the Poynting vectors of the single waves. In absorbing media we get interference terms between the waves. Only in a nonabsorbing medium the separate Poynting vectors can be added and the total energy flow be obtained; the interference terms vanish in this special case.

5.1.5 Boundary Conditions

Up to now, we have considered plane waves in a single homogeneous medium. If more than one medium is present, we calculate the appropriate field distributions by the following procedure. First, we search for the solutions that are allowed by Maxwell's equations inside each material. Then we have to match the field amplitudes $\mathbf{E}$ and $\mathbf{B}$ at the boundaries. For the interpretation of the spectra presented in the following it is not necessary to assume special surface conductivities or susceptibilities.

The derivation of the appropriate boundary condition can be found in every standard textbook of electrodynamics (e.g. [5.9]). Therefore we shall only quote the results.

Maxwell's equation (5.4) leads to the continuity of the components of $\mathbf{B}$ normal to the surface,

$$\mathbf{B}_1 \cdot \mathbf{n} = \mathbf{B}_2 \cdot \mathbf{n} \qquad \text{or} \qquad B_{1n} = B_{2n}, \tag{5.49}$$

where $\mathbf{n}$ is a unit vector normal to the surface.

The continuity of the components of $\mathbf{E}$ tangential to the surface is required by Maxwell's equation (5.1):

$$E_{1t} = E_{2t}. \tag{5.50}$$

The boundary condition for E_n follows from (5.2):

$$\varepsilon_0(E_{2n} - E_{1n}) = \sigma_{\mathrm{FC}} - (P_{BC_{2n}} - P_{BC_{1n}}) \equiv P_{2n} - P_{1n}. \tag{5.51}$$

σ_{FC} is the surface charge density which is induced by the the free carriers ρ_{FC}.

Finally, Maxwell's equation (5.3) leads to

$$\frac{1}{\mu_0}(B_{2t} - B_{1t}) = M_{2t} - M_{1t}, \tag{5.52}$$

and with the field $\mathbf{H} = \mathbf{B}/\mu_0 - \mathbf{M}$

$$H_{1t} = H_{2t}. \tag{5.53}$$

5.1.6 Coherent and Incoherent Reflection and Transmission of Layered Structures

The dielectric properties of matter determine the reflection and transmission of layered structures. Figure 5.1 shows a stack of homogeneous and isotropic

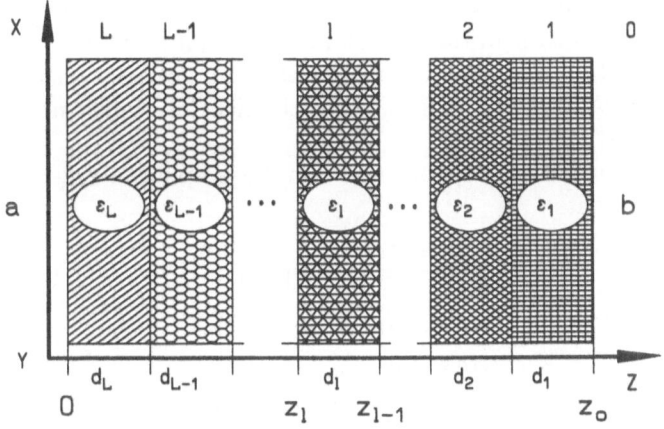

Fig. 5.1. Scheme of the stack of layers with thicknesses $d_1 \ldots d_L$ and dielectric functions $\varepsilon_1 \ldots \varepsilon_L$

layers with thicknesses d_l, dielectric functions $\varepsilon_l(\omega)$ and magnetic permeabilities $\mu_l(\omega)$. All layers are assumed to extend infinitely into x- and y-direction; the first medium (a) is transparent and has a real and constant refractive index $n_a \geq 1$.

A plane electromagnetic wave

$$\mathbf{E}_i(\mathbf{r}, t) = \mathbf{E}_i \exp\left(\mathrm{i}\,\mathbf{k} \cdot \mathbf{r} - \mathrm{i}\,\omega t\right) = \mathbf{E}_i \exp\left(\mathrm{i}\,k_x x\right) \exp\left(\mathrm{i}\,k_{az} z - \mathrm{i}\,\omega t\right) \quad (5.54)$$

is incident on the stack of layers at an angle α to the z-direction. Then there is a reflected wave E_r in the front medium a and a transmitted wave E_t at the back b,

$$\mathbf{E}_r(\mathbf{r}, t) = \mathbf{E}_r \exp\left(\mathrm{i}\,k_x x\right) \exp\left(-\mathrm{i}\,k_{az} z - \mathrm{i}\,\omega t\right) \quad (z < 0) \qquad (5.55)$$

$$\mathbf{E}_t(\mathbf{r}, t) = \mathbf{E}_t \exp\left(\mathrm{i}\,k_x x\right) \exp\left(+\mathrm{i}\,k_{bz} z - \mathrm{i}\,\omega t\right) \quad (z > z_0). \qquad (5.56)$$

In analogy to (5.54) and (5.55), the electric field in every single layer can be represented as a superposition of two waves. The tangential component of the wave vector

$$k_x = \frac{\omega}{c_0} n_a \sin \alpha \qquad (5.57)$$

is the same in all layers. This is a consequence of the mere existence of boundary conditions, whatever their detailed nature may be [5.9]. The normal component of the wave vector is different in all layers and is fixed by the dispersion relation (5.35),

$$k_z = \frac{\omega}{c_0} \sqrt{\varepsilon(\omega)\mu(\omega) - n_a^2 \sin^2 \alpha} \stackrel{\mathrm{def}}{=} \frac{\omega}{c_0} \widetilde{N}(\omega). \qquad (5.58)$$

The generalised index of refraction $\widetilde{N} = N + \mathrm{i}\,K$ defined above is a function of frequency which is different for each layer and which characterises its optical properties.

The tangential components of **E** and **H** are assumed to be continuous at the boundaries (see Sect. 5.1.5). In the case of s-polarisation (TE wave), the electric field is parallel to the interfaces; in the case of p-polarisation (TM wave) analoguous calculations are performed for the field **H**, which now is parallel to the interfaces.

It has been shown in [5.17, 5.18] that the complex-amplitude reflection and transmission coefficients

$$r_{ab} \overset{\text{def}}{=} \frac{E_r(z=0)}{E_i(z=0)}, \qquad t_{ab} \overset{\text{def}}{=} \frac{E_t(z=0)}{E_i(z=0)} \qquad (s\text{-pol.}) \tag{5.59}$$

$$r_{ab} \overset{\text{def}}{=} \frac{H_r(z=0)}{H_i(z=0)}, \qquad t_{ab} \overset{\text{def}}{=} \frac{H_t(z=0)}{H_i(z=0)} \qquad (p\text{-pol.}) \tag{5.60}$$

can be represented as a succession of matrix multiplications: the transformation of the fields across the interface between two layers, followed by the transformation from the back to the front of the layer

$$\left\{ \begin{matrix} 1 \\ r_{ab} \end{matrix} \right\} = \left\{ \begin{matrix} 1/\tau_{a,L} & \rho_{a,L}/\tau_{a,L} \\ \rho_{a,L}/\tau_{a,L} & 1/\tau_{a,L} \end{matrix} \right\} \cdot \left\{ \begin{matrix} \phi_L^{-1} & 0 \\ 0 & \phi_L \end{matrix} \right\} \cdots$$

$$\cdots \left\{ \begin{matrix} \phi_1^{-1} & 0 \\ 0 & \phi_1 \end{matrix} \right\} \cdot \left\{ \begin{matrix} 1/\tau_{1,b} & \rho_{1,b}/\tau_{1,b} \\ \rho_{1,b}/\tau_{1,b} & 1/\tau_{1,b} \end{matrix} \right\} \cdot \left\{ \begin{matrix} t_{ab} \\ 0 \end{matrix} \right\}. \tag{5.61}$$

Here,

$$\phi_l = \exp\left(i\,(\omega/c)\widetilde{N}_l d_l \right) \tag{5.62}$$

with the generalised index of refraction $\widetilde{N}_l$ given by (5.58) describes the phase and the amplitude propagation, which gives rise to oscillations in the optical spectra with periodicity

$$\Delta\nu = \frac{\Delta\omega}{2\pi c_0} = \frac{1}{2\,\mathrm{Re}(\widetilde{N})d}. \tag{5.63}$$

The transformation across the interfaces is expressed with the help of Fresnel's coefficients s-polarisation (TE):

$$\rho_{l+1,l} = \frac{\widetilde{N}_{l+1}/\mu_{l+1} - \widetilde{N}_l/\mu_l}{\widetilde{N}_{l+1}/\mu_{l+1} + \widetilde{N}_l/\mu_l} = -\rho_{l,l+1} \tag{5.64}$$

$$\tau_{l+1,l} = \frac{2\widetilde{N}_{l+1}/\mu_{l+1}}{\widetilde{N}_{l+1}/\mu_{l+1} + \widetilde{N}_l/\mu_l} = 1 + \rho_{l+1,l} \tag{5.65}$$

p-polarisation (TM):

$$\rho_{l+1,l} = \frac{\widetilde{N}_{l+1}/\varepsilon_{l+1} - \widetilde{N}_l/\varepsilon_l}{\widetilde{N}_{l+1}/\varepsilon_{l+1} + \widetilde{N}_l/\varepsilon_l} = -\rho_{l,l+1} \tag{5.66}$$

$$\tau_{l+1,l} = \frac{2\widetilde{N}_{l+1}/\varepsilon_{l+1}}{\widetilde{N}_{l+1}/\varepsilon_{l+1} + \widetilde{N}_l/\varepsilon_l} = 1 + \rho_{l+1,l}. \tag{5.67}$$

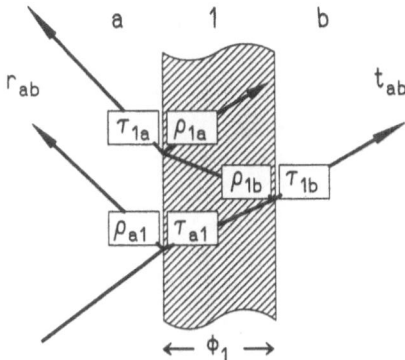

Fig. 5.2. Multiple reflections within a single layer (radiation incident from the left, for identification of symbols see text)

For the special case of a plane-parallel faced plate (see Fig. 5.2), the evaluation of (5.61) gives

$$r_{ab} = \rho_{a1} + \frac{\tau_{a1}\tau_{1a}\rho_{1b}\phi_1^2}{1 - \rho_{1b}\rho_{1a}\phi_1^2} \tag{5.68}$$

$$t_{ab} = \frac{\tau_{a1}\tau_{1b}\phi_1}{1 - \rho_{1b}\rho_{1a}\phi_1^2} . \tag{5.69}$$

The representation of the denominator as geometrical series

$$\frac{1}{1 - \rho_{1b}\rho_{1a}\phi_1^2} = 1 + \rho_{1b}\rho_{1a}\phi_1^2 + (\rho_{1b}\rho_{1a}\phi_1^2)^2 + \dots \tag{5.70}$$

can be visualised as the sum of multiple reflected lightwaves within the slab.

The representation of r and t with the formalism of multiple reflections is not restricted to a single layer (cf. Fig. 5.3). If, in a next step, a third interface is introduced, we can write

$$r_{ab} = \rho_{a2} + \frac{\tau_{a2}\tau_{2a}r_{2b}\phi_2^2}{1 - r_{2b}\rho_{2a}\phi_2^2} \tag{5.71}$$

$$t_{ab} = \frac{\tau_{a2}t_{2b}\phi_2}{1 - r_{2b}\rho_{2a}\phi_2^2}, \tag{5.72}$$

where r_{2b} and t_{2b} are the complex amplitude coefficients (5.68) and (5.69) of layer 1, where layer 2 plays the role of the front medium a.

In analogy, further films can be introduced successively until finally the reflection and transmission coefficients of the complete system are obtained.

In many cases, one of the stack's layers (e.g. the substrate) has a thickness d_s which is much larger than the wavelength of light. This leads to very narrow interferences in the spectra which usually are not resolved experimen-

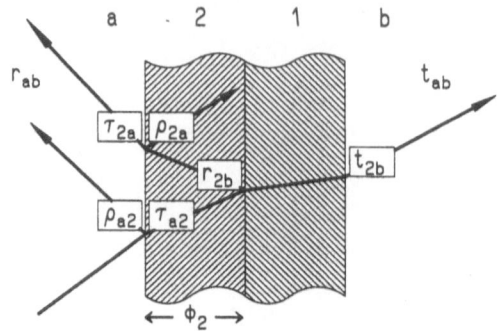

Fig. 5.3. Multiple reflections within a layer system (front a/layer 2/layer 1/back b)

tally. Thus the simulated spectra have to be convoluted with the function describing the resolution of the spectrometer.

For that case a very convenient calculation can be derived with the help of the formalism of multiple reflections [5.17]. It was proven that the averaged spectra are directly obtained by adding not the amplitudes but the intensities of the partial waves within the thick film, i.e. their phases are not taken into account (cf. Fig. 5.4). In this sense this thick film is called 'incoherent'. With $K = 2\pi\,\omega/c_0 \cdot \mathrm{Im}(\widetilde{N})$ we write

$$R_{\mathrm{incoh}} = |r_{as}|^2 + \frac{|t_{as}t_{sa}r_{sb}|^2 \exp\{-2K_s d_s\}}{1 - |r_{sa}r_{sb}|^2 \exp\{-2K_s d_s\}} \tag{5.73}$$

$$T_{\mathrm{incoh}} = Q \cdot \frac{|t_{as}t_{sb}|^2 \exp\{-K_s d_s\}}{1 - |r_{sa}r_{sb}|^2 \exp\{-2K_s d_s\}}. \tag{5.74}$$

r_{as}, r_{sb}, t_{as}, t_{sa} and t_{sb} are the complex amplitude reflection and transmission of the coherent stacks in front of and behind the incoherent layer, which are

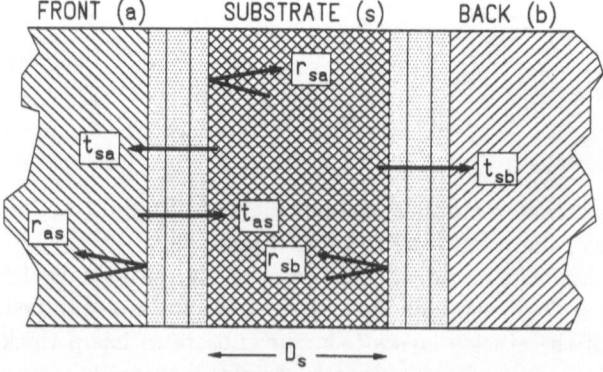

Fig. 5.4. Calculation of the reflectance and transmittance of a stack of layers with one thick ('incoherent') layer

calculated by the $\rho - \tau$ formalism discussed above. The factor Q in (5.74) is different from unity if the medium b in the back is different from the transparent medium "a" in front of the stack. For s- and p-polarisation (see (5.46) and (5.48)) Q is given by

$$Q = \text{Re}\left\{\frac{\widetilde{N_b}}{\mu_b}\right\} / \frac{n_a \cos\alpha}{\mu_a} \quad (s\text{-pol.}) \tag{5.75}$$

$$Q = \text{Re}\left\{\frac{\widetilde{N_b}}{\varepsilon_b}\right\} / \frac{n_a \cos\alpha}{\varepsilon_a} \quad (p\text{-pol.}). \tag{5.76}$$

For the special situation of a free, incoherent film f (with vacuum on both sides, $(a = b \stackrel{\text{def}}{=} v)$) (5.74) can be simplified using Fresnel's coefficients (5.64) and (5.65) or (5.66) and (5.67), $t_{as} = \tau_{vf} = 1 + \rho_{vf}$, $t_{sb} = \tau_{fv} = 1 - \rho_{vf}$, $r_{sa} = r_{sb} = \rho_{fv} = -\rho_{vf}$ and $Q = 1$ leading to

$$\begin{aligned} T_{f,\text{incoh}} &= \frac{|1 - \rho_{vf}^2|^2 \exp(-K_f D_f)}{1 - |\rho_{vf}^2|^2 \exp(-2K_f D_f)} \\ &= \frac{(1 - R)^2 + 4\,\text{Im}(\rho)}{1 - R^2 A^2} \cdot A, \end{aligned} \tag{5.77}$$

where $A \stackrel{\text{def}}{=} \exp(-K_f D_f)$ and $R \stackrel{\text{def}}{=} \rho_{vf}\rho_{vf}^\star$.

Equation (5.77) is exact. In cases where there is little damping and the term $4\,\text{Im}\,\rho$ in the nominator of (5.77) can be neglected, we get

$$T_{f,\text{incoh}} \approx \frac{(1 - R)^2}{1 - R^2 A^2} \cdot A = (1 - R)^2 A(1 + R^2 A^2 + R^4 A^4 + \ldots). \tag{5.78}$$

Often, it is sufficient to take into consideration the first term only in the geometrical sum, i.e.

$$T_{f,\text{incoh}} \approx (1 - R)^2 \exp(-K_f \cdot D_f). \tag{5.79}$$

5.1.7 The Dielectric Function $\varepsilon(\omega)$

If the elementary excitations of the material do not interact with each other, the dielectric function $\varepsilon(\omega)$ is a linear superposition of susceptibility contributions due to lattice vibrations $\chi_{PM}(\omega)$, free carriers $\chi_{FC}(\omega)$, and bound valence electrons $\chi_{VE}(\omega)$. In the case of interacting elementary excitations sometimes a renormalisation is possible which leads to similar formulas. However, in this case, the interpretation in terms of microscopic parameters is nontrivial. In the infrared, the contribution of valence electrons is approximately real and frequency independent, hence we define $1 + \chi_{VE} = \varepsilon_\infty$ and write

$$\begin{aligned} \varepsilon(\omega) &= 1 + \chi_{VE} + \chi_{FC}(\omega) + \chi_{PM}(\omega)\ldots \\ &\stackrel{\text{def}}{=} \varepsilon_\infty + \chi_{FC}(\omega) + \chi_{PM}(\omega)\ldots \end{aligned} \tag{5.80}$$

Contributions due to free carriers and phonons can be described by simple models discussed in the following which are nevertheless sufficient for an interpretation of most semiconductor spectra.

5.1.7.1 The Susceptibility χ_{PM} of Lattice Vibrations.

Electromagnetic radiation can excite collective vibrations of lattice atoms if these are connected with a dynamic dipole moment. In a simple linear approximation their contribution to the dielectric function can be written as a sum of Lorentzian oscillators [5.10]

$$
\begin{aligned}
\chi_{\mathrm{PM}}(\omega) &= \sum_{i=1}^{n} \frac{\Omega_{pi}^2}{\Omega_{oi}^2 - \omega^2 - \mathrm{i}\,\Omega_{\tau i}\omega} \\
&= \sum_{i=1}^{n} \frac{(\Omega_{oi}^2 - \omega^2) \cdot \Omega_{pi}^2}{(\Omega_{oi}^2 - \omega^2)^2 + \Omega_{\tau i}^2 \omega^2} + \mathrm{i} \sum_{i=1}^{n} \frac{\Omega_{pi}^2 \cdot \Omega_{\tau i} \cdot \omega}{(\Omega_{oi}^2 - \omega^2)^2 + \Omega_{\tau i}^2 \omega^2}.
\end{aligned} \tag{5.81}
$$

The resonance frequency Ω_0 of each oscillator is equal to the frequency of a transverse optical (TO) phonon polariton and depends on the coupling force between individual atoms. Ω_p determines the oscillator strength, and Ω_τ is a phenomenological parameter describing the system's damping. In general the TO-frequencies are found in the far and mid infrared. For most semiconductors an excellent description of the dielectric behaviour is possible within this simple model.

5.1.7.2 The Susceptibility $\chi_{\mathrm{FC}}(\omega)$ of Free Carriers.

The susceptibility $\chi_{\mathrm{FC}}(\omega)$ of free carriers is represented by the Drude-Lorentz model [5.19]. Although rather simple, this model is very useful for the description of semiconductor spectra.

$$
\chi_{\mathrm{FC}}(\omega) = \frac{\omega_p^2}{-\omega^2 - \mathrm{i}\,\omega_\tau\omega} = \frac{-\omega_p^2}{\omega^2 + \omega_\tau^2} + \mathrm{i}\,\frac{\omega_p^2 \omega_\tau}{\omega(\omega^2 + \omega_\tau^2)}. \tag{5.82}
$$

The plasma frequency ω_p and scattering frequency ω_τ of the free carriers are correlated with the parameters carrier concentration n, effective mass $m^\star$, damping τ and the mobility μ via

$$
\omega_p = \sqrt{\frac{ne^2}{\varepsilon_0 m^\star}} \qquad \omega_\tau = \frac{e}{m^\star \mu}. \tag{5.83}
$$

Similar classical dielectric functions have been used for the description of excitonic resonances in multiple-quantum-well structures [5.20].

5.1.8 The Berreman Effect

Reflectance and transmittance spectra of thin films recorded at oblique incidence in p-polarisation exhibit a special structure which is called Berreman effect [5.21–5.23]. It is due to a strong absorption for frequencies in the vicinity of the maximum of the energy loss function $\mathrm{Im}(-1/\varepsilon_f)$ of the film. The absorption process can be understood as follows:

In p-polarisation the electric field has both a normal component E_n and a tangential component E_t relative to the surface of the sample (cf. Fig. 5.5). We express the electric field $\mathbf{E}$ in terms of the field $\mathbf{H}$ which has merely a tangential component H_t and is continuous at the boundaries. At the interface between film (f) and substrate (s), the components of the electric field are

$$E_n^f \;=\; -\sin\alpha \cdot Z_0 H_t/\varepsilon_f \qquad \text{in the film,}$$
$$E_n^s \;=\; -\sin\alpha \cdot Z_0 H_t/\varepsilon_s \qquad \text{in the substrate,}$$

and

$$E_t^s \;=\; E_t^f = \sqrt{\varepsilon_s - \sin^2\alpha} \cdot Z_0 H_t/\varepsilon_s$$

with $Z_0 = \sqrt{\mu_0/\epsilon_0}$.

Inserting E_n^f, E_n^s and E_t^s into (5.44) finally yields the energy dissipation near the film-substrate interface for p-polarisation

$$W_{\mathrm{diss}}^p = \frac{\varepsilon_0\omega}{2}\left[\frac{\varepsilon_s - \sin^2\alpha}{|\varepsilon_s|^2}\cdot \mathrm{Im}(\varepsilon_f) + \sin^2\alpha \cdot \mathrm{Im}(-1/\varepsilon_f)\right] Z_0^2\,|H_t|^2. \qquad (5.84)$$

This formula will now be discussed for two cases of special interest.

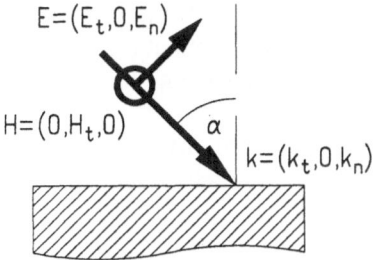

Fig. 5.5. Electric and magnetic field vector in p-polarisation

5.1.8.1 The Free Standing Film ($\varepsilon_s = 1$). In the case of a free standing film, (5.84) can be simplified to

$$W_{\mathrm{diss}}^p = \frac{\varepsilon_0\omega}{2}\left[\cos^2\alpha \cdot \mathrm{Im}(\varepsilon_f) + \sin^2\alpha \cdot \mathrm{Im}(-1/\varepsilon_f)\right] Z_0^2 \cdot |H_t|^2.$$

As expected, we observe absorption structures at those frequencies where the imaginary part of the film's dielectric function reaches its maximum (see Fig. 5.6). However, in p-polarisation additional structures are observed at frequencies where the energy loss function $\mathrm{Im}(-1/\varepsilon_f)$ has a maximum, i.e. at the zeroes of the film's dielectric function.

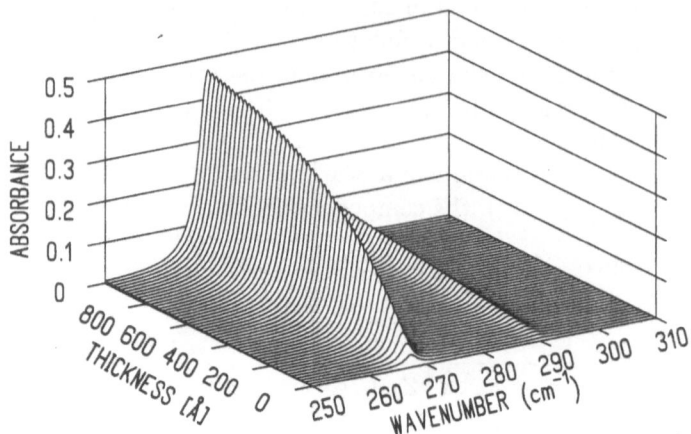

Fig. 5.6. Simulated energy dissipation in the system vacuum/GaAs/vacuum for various film thicknesses. The angle of incidence is $\alpha = 70°$. ($\Omega_0 = 268\,\mathrm{cm}^{-1}$, $\Omega_p = 375\,\mathrm{cm}^{-1}$, $\Omega_\tau = 2.5\,\mathrm{cm}^{-1}$). The spectrum exhibits the usual bulk TO phonon absorption at $268\,\mathrm{cm}^{-1}$ and a Berreman mode at about $292\,\mathrm{cm}^{-1}$

5.1.8.2 Metal Substrate ($|\varepsilon_s| \gg 1$). In the case of a metal substrate with a high conductivity the energy dissipation near the metal-film interface is approximately given by the second term of (5.84). The contribution of the first term is strongly suppressed because the tangential component of the electric field is shorted by the metal (see Fig. 5.7).

It has been shown [5.24] that for thin, weakly absorbing films on an ideal metal substrate (cf. Fig. 5.7) the reflectance is approximately given by

$$R = 1 - 8\frac{\omega}{c_0}\frac{\sin^2\alpha \cdot N_a^3}{\cos\alpha \cdot n_f^3}\kappa_f d_f,$$

where N_a is the generalised index of refraction of the front medium from which the radiation is incident and κ_f is the imaginary part of the refractive index of the film. Obviously, for measurements at oblique incidence with a metal substrate, the sensitivity is enhanced by a factor

$$V = 4\frac{\sin^2\alpha \cdot N_a^3}{\cos\alpha \cdot n_f^3}.$$

This explains why Infrared Reflection Absorption Spectroscopy (IRRAS) experiments are such a powerful tool for the investigation of very thin films [5.25].

An extraordinary high sensitivity is obtained if the radiation is incident from a medium with a high refractive index. However, in order to measure at large angles of incidence, a special reflection element — e.g. a prism — is necessary. The sample has to be brought into direct contact with this ele-

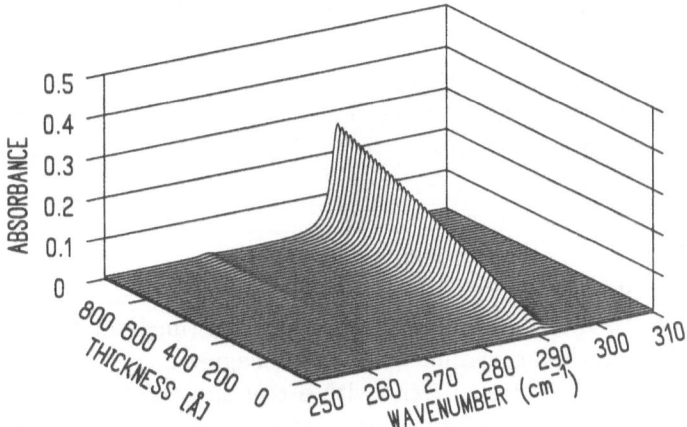

Fig. 5.7. Energy dissipation in the system vacuum/GaAs/metal at $\alpha = 70°$ angle of incidence. The absorption due to the Berreman mode is strongly enhanced by the metal

ment. Hence in experiments of this type no surface polaritons (cf. Sect. 5.1.9) are excited, the above mentioned absorption near the LO frequencies is only enhanced. In spectral regions with negligible energy loss function there is total reflection at the prism-sample interface. Therefore the experiments are referred to as Attenuated Total Reflection (ATR) [5.26].

Figure 5.8 shows one possible realisation. Both sides of the substrate are wedged by an angle of 10° which allows measurements at internal angles of

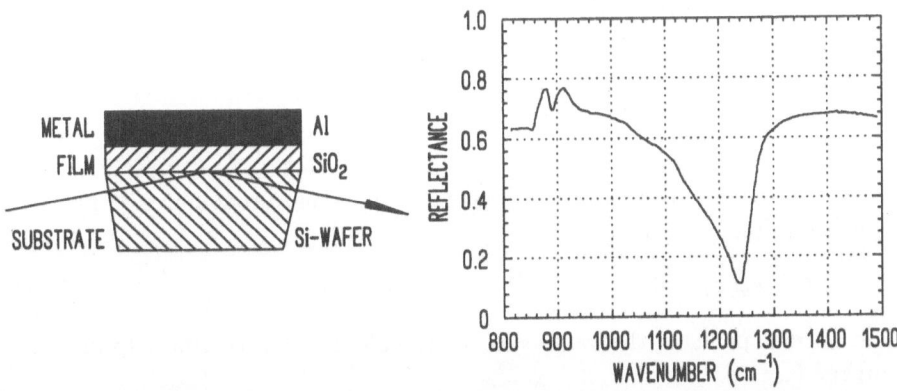

Fig. 5.8. Schematic diagram showing Brendel's internal reflection element (l.h.s.). The angle of incidence is 80°. Extremely thin SiO_2 films can be analysed because the radiation is incident from a material with high refractive index and the metal backing causes a further enhancement of the fields within the oxide film. In the spectrum on the right hand side a 1.3 nm thin film causes 60% reflectance breakdown [5.24]

incidence of about 80°. This substrate is covered by a thin oxide and a metal film. It has been possible to investigate with this experiment a 1.3 nm thin SiO$_2$ film on a silicon wafer [5.24]. Although the film is extremely thin, the Si-O vibrations of the layer generate a considerable reflectance breakdown. Enhancements of sensitivity up to $V = 600$ are possible.

5.1.9 Surface Waves

Let us consider the situation displayed in Fig. 5.2, especially the reflection and transmission of the waves at the interface between the film (medium 1) and the back medium b. All media are assumed to be nonmagnetic ($\mu = 1$) and, as usual, the front medium a is transparent with a real and constant index of refraction $n_a > 1$. For these conditions, we examine the case where the denominator of reflection and transmission coefficients of the partial waves (5.66) and (5.67) vanishes, i.e. $\widetilde{N}_f/\varepsilon_f = -\widetilde{N}_b/\varepsilon_b$, or with the definition (5.58) of $\widetilde{N}$,

$$\sqrt{\varepsilon_f(\omega) - n_a^2 \sin^2 \alpha}/\varepsilon_f(\omega) = -\sqrt{\varepsilon_b(\omega) - n_a^2 \sin^2 \alpha}/\varepsilon_b(\omega). \tag{5.85}$$

For nonmagnetic media, surface waves can only occur in p-polarisation.

Vanishing denominator of ρ and τ means that a reflected and transmitted wave exists without being excited by an incoming wave. Thus, (5.85) describes an eigenmode of the system medium 1/medium b (both semiinfinite to the $-z$ and $+z$ direction, respectively).

To discuss these modes in detail, we first state that ε_f and ε_b must be opposite in sign. Then we solve (5.85) for $n_a \sin \alpha$ and notice that $(\omega/c_0)n_a \sin \alpha = k_x$ is the component of the wave vector tangential to the surfaces,

$$k_x = \frac{\omega}{c_0} \sqrt{\frac{\varepsilon_f(\omega) \cdot \varepsilon_b(\omega)}{\varepsilon_f(\omega) + \varepsilon_b(\omega)}}. \tag{5.86}$$

As k_x has to be real, the expression under the square root must be positive, which requires that

$$\varepsilon_f(\omega) + \varepsilon_b(\omega) < 0. \tag{5.87}$$

The normal parts of the wave vector are calculated with the help of (5.35) and read

$$k_{f,z} = \frac{\omega}{c_0} \sqrt{\frac{\varepsilon_f(\omega)^2}{\varepsilon_f(\omega) + \varepsilon_b(\omega)}} \tag{5.88}$$

$$k_{b,z} = \frac{\omega}{c_0} \sqrt{\frac{\varepsilon_b(\omega)^2}{\varepsilon_f(\omega) + \varepsilon_b(\omega)}}. \tag{5.89}$$

Together with (5.87) this means that $k_{f,z}$ and $k_{b,z}$ are purely imaginary and the corresponding waves are exponentially decaying from the interface.

Therefore, the eigenmodes defined by the dispersion relations (5.86), (5.88), (5.89) are called surface waves. The assumption that both, ε_f and ε_b are real and one of them is negative can only be approximately fulfilled. Especially the dielectric function which has a negative real part has a possibly small imaginary part. In fact, it is the imaginary part of ε which is responsible for the breakdown of the reflectance in the experiment.

After these general considerations we turn to the experimentally most relevant geometries, which are used to excite surface waves (cf. Fig. 5.9). The so-called Kretschmann geometry [5.27] consists of vacuum as back medium, i.e. $\varepsilon_b = 1$; the film is the medium which supports the surface waves (the 'active' medium), i.e. $\varepsilon_f < -1$. Mostly, this medium is grown as a thin film on a transparent prism or hemisphere (medium a) on which the wave is incident. The necessity of the front medium will become clear below.

In the so-called Otto geometry [5.28], the active medium is the bulk material of the back b and the 'film' is vacuum. Again, medium a must be a transparent prism or hemisphere with $n_a > 1$. The reason is as follows. Equation (5.87) reads for this special case where the film is vacuum

$$\varepsilon_b(\omega) < -1. \tag{5.90}$$

(For the Kretschmann geometry there is the analogous relation for the active medium $\varepsilon_f < -1$.) Thus, the magnitude of the tangential part of the wave

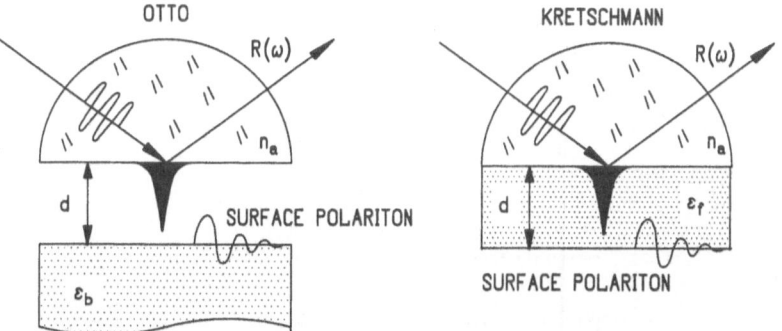

Fig. 5.9. Principle of ATR experiments: surface polaritons at the sample-vacuum interface are excited by an evanescent wave either in the vacuum ('Otto geometry' [5.28]) or in the sample ('Kretschmann geometry' [5.27]). This evanescent wave is obtained by total reflection of the incident radiation at the interface between vacuum (sample, resp.) and a medium with high refractive index n_a. If surface polaritons are excited, energy is dissipated in the sample and the observed reflectance $R(\omega)$ is less than unity

vector k_x, which is defined by the incident wave, see (5.85) and (5.86),

$$k_x = \frac{\omega}{c_0} \sqrt{\frac{\varepsilon_b(\omega)}{1 + \varepsilon_b(\omega)}} = \frac{\omega}{c_0} n_a \sin \alpha \qquad (5.91)$$

must be larger than the magnitude of the wave vector in vacuum $k_v = \omega/c_0$. Consequently, the right-hand part of (5.91) can only be fulfilled, as a necessary condition, if $n_a \sin \alpha > 1$. This is the condition of total reflection at the interface between medium a and vacuum. Therefore this excitation technique is called excitation by total reflection. A more general discussion can be found elsewhere [5.29].

Once $k_x = (\omega/c_0)n_a \cdot \sin \alpha$ is fixed by the prism and the angle of incidence of the incoming wave, the normal part of the wave vector in vacuum is $k_{v,z} = (\omega/c_0)\sqrt{1 - n_a^2 \sin^2 \alpha}$. Together with (5.88), $k_{f,z} = (\omega/c_0)\sqrt{1/(1 + \varepsilon_b)}$, this gives the relation

$$\varepsilon_b(\omega) = -\frac{n_a^2 \sin^2 \alpha}{n_a^2 \sin^2 \alpha - 1}, \qquad (5.92)$$

which can be read as an implicit relation for the frequency where surface waves are excited.

The quantity measured in such an experiment is the system's reflectance $R(\omega)$. At frequencies far from the resonance 100% reflectance is observed, but when surface polaritons are excited a large amount of energy is dissipated in the sample. This leads to a sharp reflectance breakdown. Hence experiments of this type are called ATR (Attenuated Total Reflection). Figure 5.10 shows a typical ATR spectrum of GaAs.

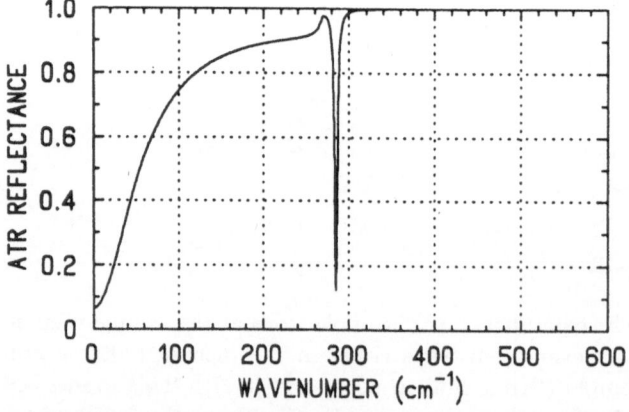

Fig. 5.10. Simulated ATR spectrum of GaAs with a Ge-prism at 60° angle of incidence in p-polarisation (Otto-geometry). At 285 cm^{-1} surface polaritons are excited and the total reflection is attenuated. Far from the resonance we observe 100% reflectance

It should be mentioned that in ATR experiments the surface polaritons are strongly influenced by the prism. There exists an optimum coupling gap thickness d, which is of the order of a few tenths of the wavelength. Due to the influence of the prism, resonance frequencies can be shifted away from the value given by (5.92) by a considerable amount.

5.1.10 Interpretation of Measured Spectra

The relevant material properties of a sample are the layer thicknesses d_i and their dielectric functions $\varepsilon_i(\omega)$. It has been shown that for layered structures the connection between measured spectra and material properties is quite complicated, and a direct inversion is generally not possible. Hence simulations have to be performed starting with reasonable models (cf. Sect. 5.1.7) for the dielectric functions of the layers and estimates of their thicknesses. Model spectra are calculated using the $\rho - \tau$ formalism derived in Sect. 5.1.6 and are compared to the measurements. Finally the parameters for the individual layers are varied until there is optimum agreement between measurements and simulations.

5.2 Fourier Transform Spectroscopy

The aim of optical spectroscopy is the measurement of the transmission or reflection of light of a sample as a function of frequency. In the visible and ultraviolet spectral range usually grating monochromators are used to disperse the radiation from a broad-band light source.

In the infrared region a different kind of spectroscopy has turned out to be advantageous and is used nowadays almost exclusively. This is the Fourier Transform InfraRed Spectroscopy (FTIR) which is explained here briefly using the schematic diagram Fig. 5.11. Thorough descriptions are given in Refs. [5.1–5.3].

5.2.1 Principle

In a Fourier transform spectrometer the radiation from a broad-band light source is divided into two parts by a beam splitter. After reflection from a fixed and a movable mirror, respectively, the two partial waves are combined by the same beam splitter and focussed on a detector. There they interfere and yield the total intensity. For different positions of the movable mirror the two partial waves get different phase shifts with respect to each other. Therefore, on the detector the radiation field is superimposed with a time-delayed copy of itself. Hence what is basically measured when the detector signal is recorded while the mirror moves is the *autocorrelation function of*

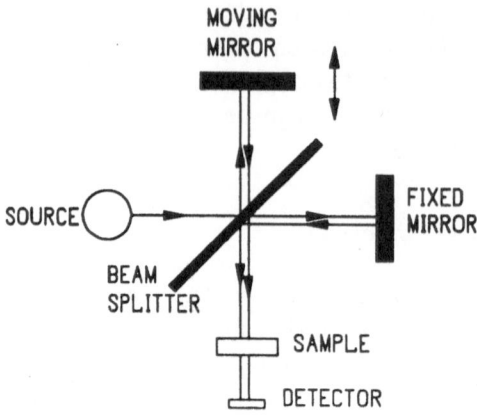

Fig. 5.11. Scheme of a Fourier transform spectrometer. Light waves emitted from the light source are divided by the beam splitter. One fraction is reflected by the fixed mirror, the other by the movable mirror. The partial waves interfere on the detector, where the total intensity is measured. The sample is usually placed between the interferometer and the detector

the radiation field (which is called the interferogram in FTIR spectroscopy). The Fourier transform of this autocorrelation function is the desired power spectrum in the frequency domain (Wiener-Khintchine theorem [5.30, 5.31]).

Figure 5.12 shows a typical interferogram and the corresponding intensity spectrum. The maximum of the interferogram corresponds to equal distances of the two mirrors to the beam splitter which causes constructive interference

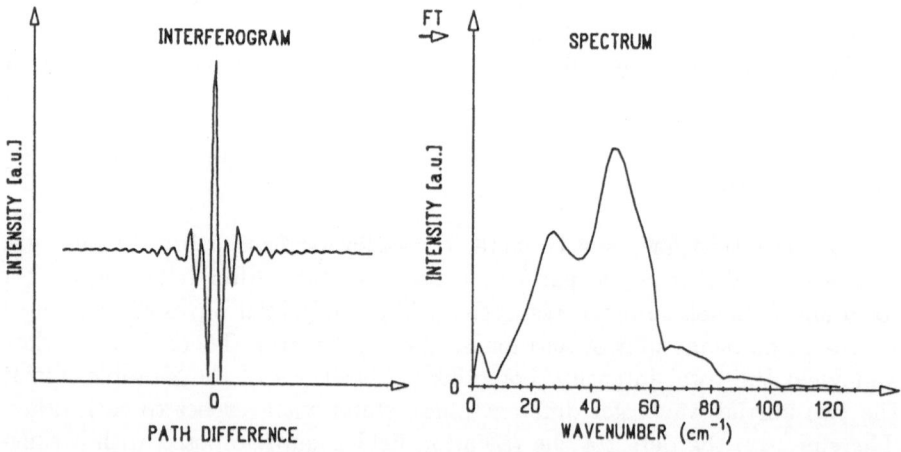

Fig. 5.12. A typical interferogram obtained by a FTIR spectrometer and the corresponding spectrum

of the two partial waves for all wavelengths. For this reason the peak is called 'White Light Position' (WLP). Outside this central structure the phase difference between the partial waves reflected at the two mirrors depends on the wavelength. For a broadband source their superposition to the interferogram contains less and less structures the further away one gets from the WLP. From the Fourier transformation it follows that a narrow WLP correlates to a broad frequency distribution whereas a broad interferogram correlates with a narrow, "monochromatic" frequency distribution. The spectral resolution in FTIR is determined by the inverse of the total path of the movable mirror. No mechanical slits are required in contrast to conventional grating instruments.

FTIR spectroscopy makes use of the following two advantages in comparison to grating instruments:

- In Fourier transform spectroscopy light from all spectral intervals contributes to the detector signal. This is called the *multiplex advantage* which is not available in slit based grating instruments. However, grating instruments equipped with an Optical Multichannel Analyser (OMA) which are commonly used in the visible and the near infrared may exhibit this multiplex advantage, too.

- Since no slit is necessary in FTIR large apertures can be realized resulting in a large throughput. This is the so-called *throughput advantage*.

These features reduce the time needed to record a spectrum considerably. Latest developments in Fourier transform spectroscopy extend the range of applicability from the IR to the visible and even to the UV, where mechanical instabilities of the moving mirror make a correct superposition of the partial waves with wavelengths smaller than $1\,\mu$m very difficult, and a dynamic alignment of the moving mirror is generally necessary.

5.2.2 Instrumentation

A large number of different FTIR spectrometer designs have been published in the literature. Three commercially available spectrometers are described more detailed in the following.

A compact and simple spectrometer used for routine spectroscopy in semiconductor and chemical industry is shown in Fig. 5.13.

The infrared radiation from a SiC lamp enters the Michelson interferometer which is equipped with a Ge/KBr beamsplitter. It is focussed onto the sample (transmission or reflection as shown) and then reaches the pyroelectric DTGS detector. The spectral range extends from $400\ldots4800$ cm^{-1} with a resolution better than 2 cm^{-1}.

The interferogram from the He-Ne laser beam with periodicity $\lambda/2$ (where λ is the laser wavelength) is used to record the position of the moving mirror by counting peaks from the laser interferogram using the WLP of an auxiliary broad band source as starting position.

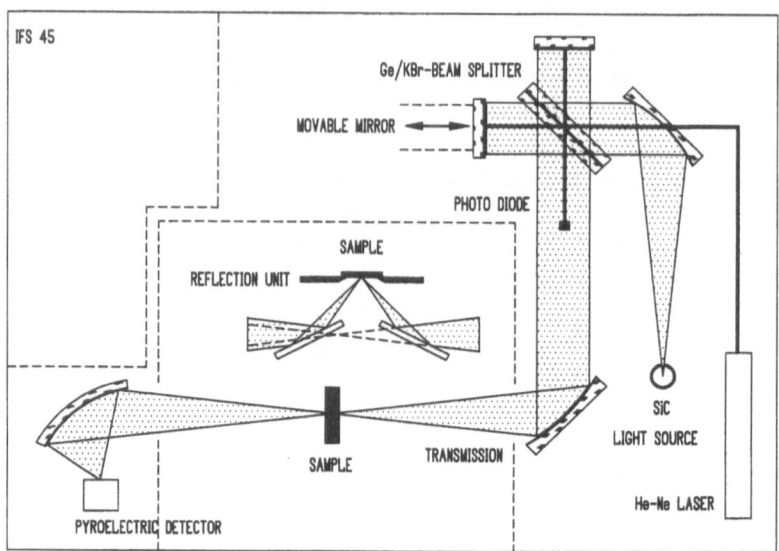

Fig. 5.13. A typical Michelson interferometer (Bruker IFS 45). The beamsplitter is positioned in the parallel beam. The interferogram from the He-Ne laser is used to measure precisely the position of the moving mirror

The optical setup of the vacuum FTIR spectrometer shown in Fig. 5.14 differs from the previous more conventional Michelson-type interferometer. The beam splitter is positioned in the focal plane of the two interfering beams and both mirrors of the interferometer are combined into one moving mirror. This design offers a factor two of increased resolution, because the optical path difference is four times the geometrical path. Moreover, at the beam splitter the intersecting beams are less inclined and consequently the output is less polarised in comparison to conventional interferometers. The standard frequency range from 10 cm^{-1} to 5000 cm^{-1} can be extended from 4 cm^{-1} to 40000 cm^{-1}. The resolution is better than 0.03 cm^{-1}.

In Fig. 5.15 another version of a Michelson interferometer is shown which offers a dynamical alignment of the moving mirror and thus an extended frequency range up to 50000 cm^{-1}.

Recently FTIR microscopes were developed with a diffraction limited lateral resolution of about 8 μm, by using special Cassegrain objectives (see

Fig. 5.14. A vacuum FTS interferometer (modified Bruker IFS 113). Here both in-▶ terferometer mirrors move. Thus the optical path in one arm is increased, whereas in the other it decreases. Hence a larger resolution is obtained with the same geometrical path. The position of the movable mirrors is measured with the auxiliary interferometer shown in the insert. The beamsplitter is positioned in the focus

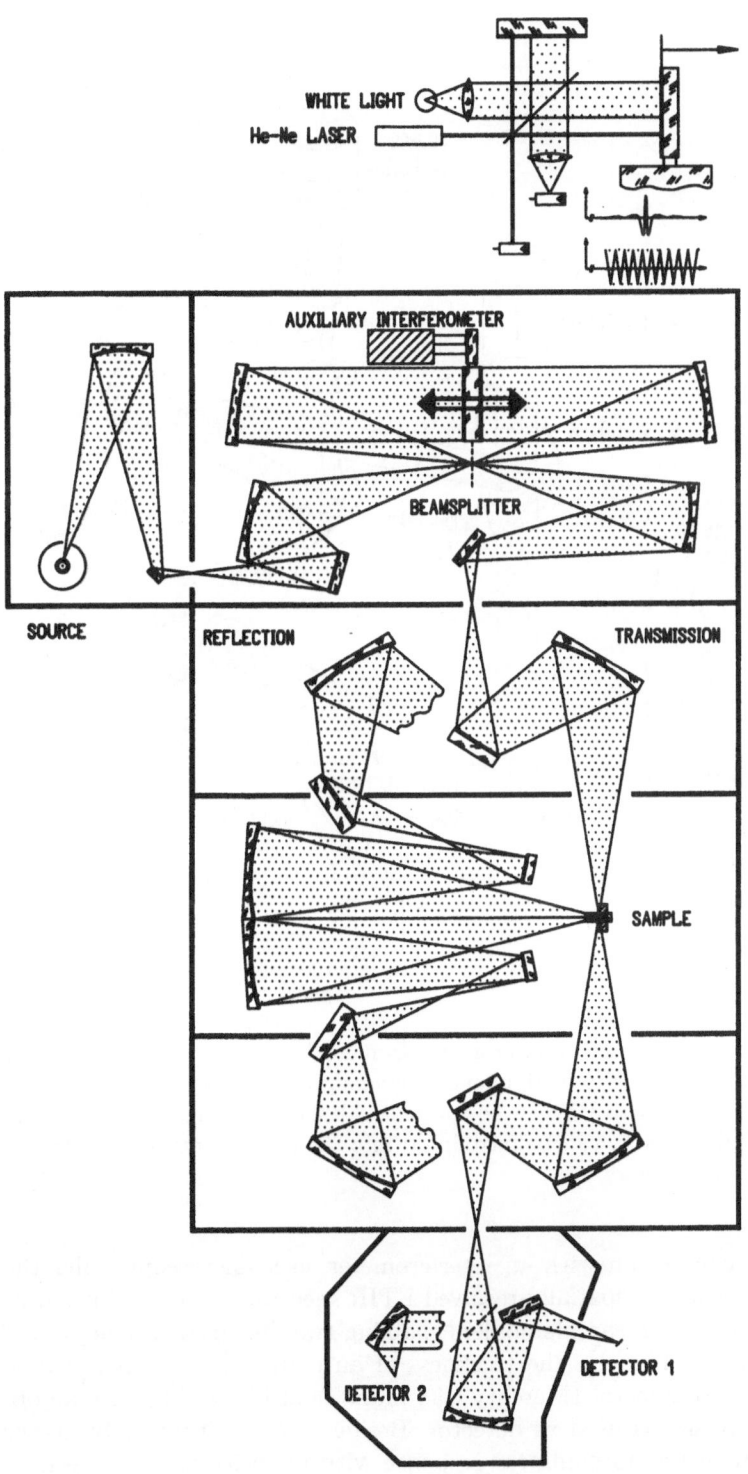

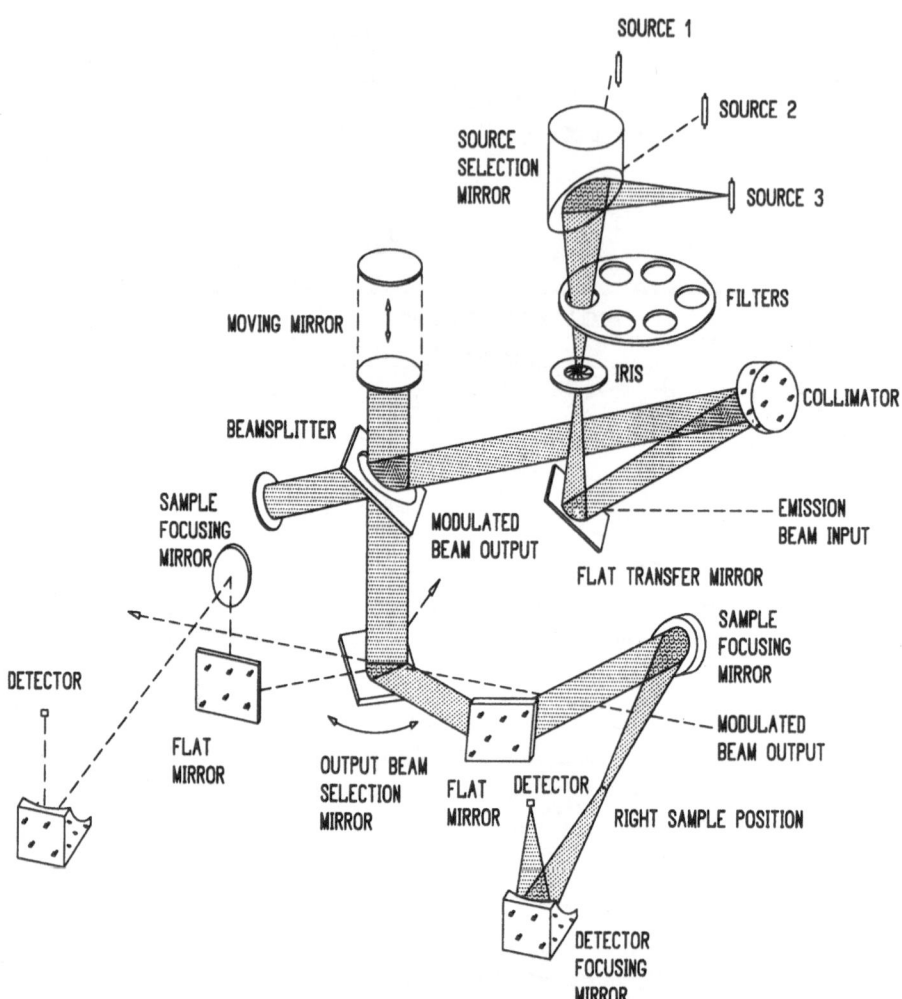

Fig. 5.15. The vacuum FTS interferometer Bomem DA3 with the beam splitter in the parallel beams. Due to a "dynamic" alignment of the moving mirror a wide spectral range from FIR to UV can be covered with 0.01 cm^{-1} resolution. Several light sources and detectors can be introduced into the optical path by electrically switched selection mirrors

Fig. 5.16). In connection with an interferometer such microscopes offer the possibility of recording spatially-resolved FTIR spectra. IR microscopes usually are combined with sample holders enabling two-dimensional mapping of physical quantities like e.g. the thickness of an epitaxial layer. A sensitive cryogenic infrared detector (Mercury Cadmium Telluride, MCT) with an optimised element size is used as detector. Besides the sensitivity advantages of the MCT detector its nonlinear response with intensity poses sometimes problems to the linearity requirements of Fourier transformation. This may

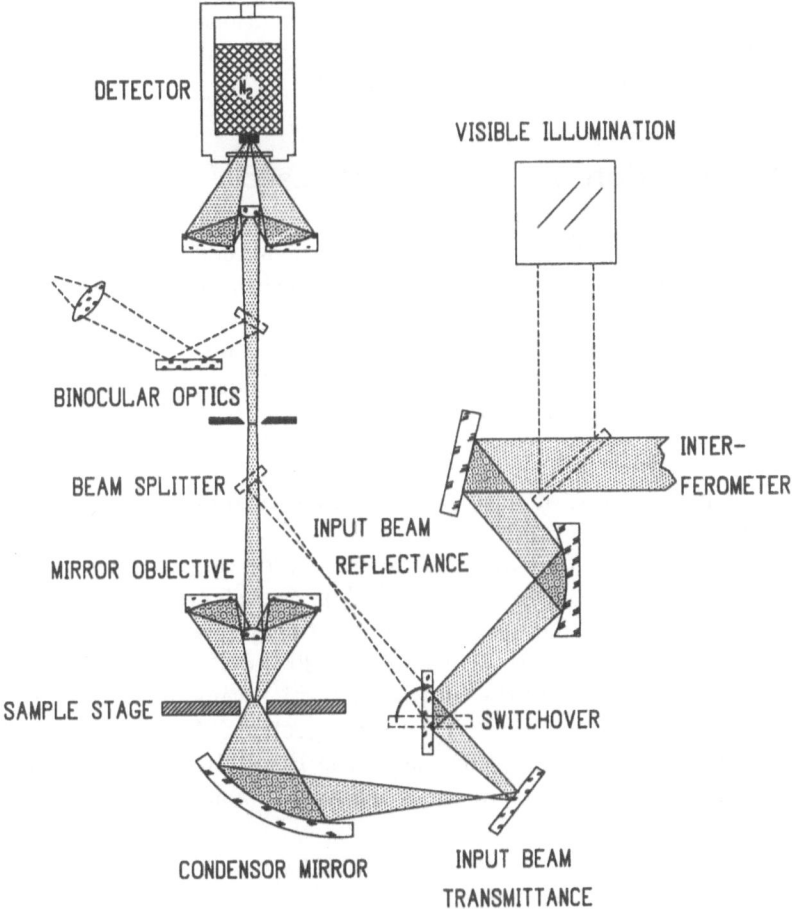

DETECTOR

VISIBLE ILLUMINATION

BINOCULAR OPTICS

BEAM SPLITTER

MIRROR OBJECTIVE

INPUT BEAM
REFLECTANCE

INTER-
FEROMETER

SAMPLE STAGE

SWITCHOVER

CONDENSOR MIRROR

INPUT BEAM
TRANSMITTANCE

Fig. 5.16. IR microscope for the spectral range $600\ldots5000$ cm^{-1}. The radiation coming from the interferometer is focussed onto the sample by a Cassegrain objective. Spot sizes down to $8\,\mu$m and a simultaneous observation with visible light are possible

cause reflection or transmission spectra with values larger than 100%. For the quantitive interpretation of the spectra it is necessary to know the nonlinear characteristic of the detector and to correct the original 'nonlinear' interferogram to a 'linear' one. The procedure must be done before the interferogram is Fourier transformed. Optical fibres also provide promising prospects for the future of FTIR spectroscopy. They allow easy access to samples which cannot be brought into the sample chamber of the spectrometer. Whereas for the visible, the near and mid infrared region flexible fibres have been already successfully introduced (e.g. silver halides for the MIR), the development in the far infrared still suffers from the lack of suitable materials with low absorption in a sufficiently wide wavenumber range. Current solutions using fibres with a maximum length of about 2 m have turned out to be rather unstable, but improvements seem feasible.

5.3 Determination of Layer Thicknesses

Since a phase factor of $2Nd$ appears in the equations on reflectivity and transmission (see e.g. 5.61) the thickness of epitaxial single or multilayers manifests itself in the frequency dependence of either reflectivity and/or transmission. The thickness d can be determined if the frequency dependence of the refractive index N is known. This allows various kinds of thickness determination from optical spectra presented in the following.

5.3.1 Simple Evaluation of Fabry-Perot Interferences

The most simple case for the thickness determination is given by a single epitaxial layer on a substrate. As an example Fig. 5.17 shows the reflectance spectrum below and above the fundamental energy gap of an epitaxial PbSe layer on a BaF$_2$ substrate in the range $700\ldots3000$ cm^{-1} at normal incidence together with a sketch of the important partial waves contributing to the Fabry-Perot interferences that dominate the spectrum. Above 2000 cm^{-1}, the onset of the interband absorption makes the PbSe film opaque and the interferences disappear. The periodicity of the well-pronounced interferences

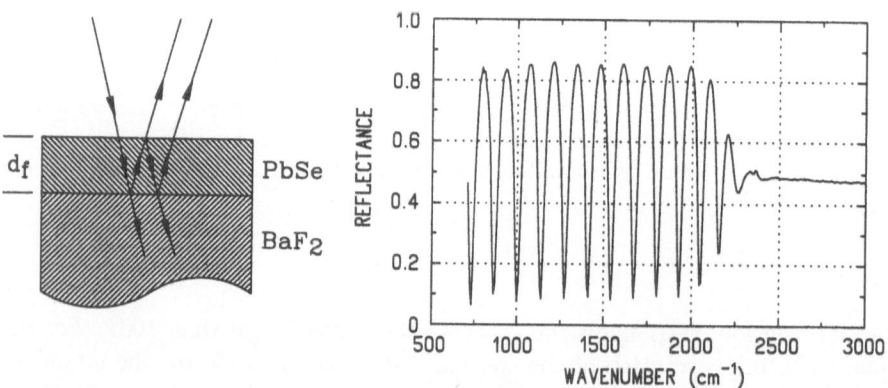

Fig. 5.17. A PbSe layer on top of a BaF$_2$ substrate. A sketch of the sample with important partial waves (left) and the measured reflectance spectrum (right) are shown. In the spectral range up to 2000 cm^{-1} interferences within the PbSe epi-layer are observed. They are used for the thickness determination ($d = 7.7\,\mu$m). Above 2000 cm^{-1} the interference fringes disappear due to the interband absorption within the epi-layer

below 2000 cm^{-1} is given by (cf. 5.63) $\Delta\nu = 1/2Nd$ where d is the geometrical thickness and N the real part of the generalised refractive index. For the example in Fig. 5.17 one obtains for the thickness d using

$$d = \frac{1}{2N\Delta\nu} \tag{5.93}$$

a value of 7.7 μm where a constant value of $N = n = 4.8$ was used as refractive index.

5.3.2 Thickness Determination by Fourier Transforms

Instead of evaluating the distances of extrema in $R(\omega)$ or $T(\omega)$ using (5.63), a Fourier transformation of the spectra fulfills an equivalent purpose and yields directly the optical thickness of an epi-layer [5.32]. As an example we show in Fig. 5.18 a Fourier transform of the spectrum in Fig. 5.17.

Clearly the path difference $2nd_f$ (and multiples of it) can be identified which is due to partial waves traveling from top of the layer to the bottom and back. If the refractive index n is known, one can determine the geometrical thickness d from the peak position in the Fourier transform of the spectrum.

The advantages of this method become apparent if the IR spectra of multilayer samples have to be analysed. In this case a visual inspection of the reflectance or transmittance spectrum will not be successful since in this case several periodic Fabry-Perot interferencesare superimposed. This is exemplified in a model calculation given in Fig. 5.19 where reflectance spectra of two- and three-layer systems consisting of GaAs epi-layers with different doping levels and their corresponding Fourier transforms are shown. Quite convinc-

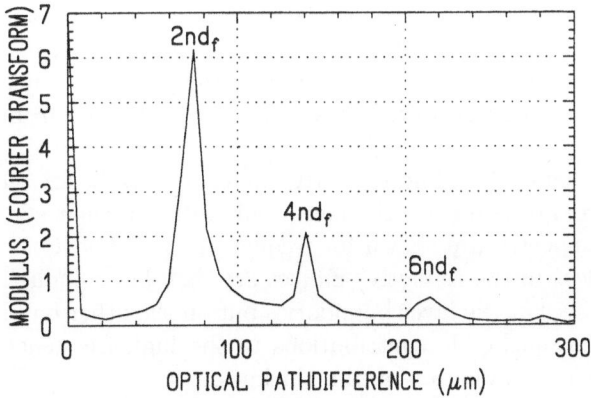

Fig. 5.18. The absolute modulus of the Fourier transform of the spectrum shown in Fig. 5.17. Due to multiple internal reflections peaks appear at multiple values of $2nd$

	d[μm]	ω_p[cm^{-1}]	ω_τ[cm^{-1}]
layer 1	3.5	700	50
layer 2	3.0	1200	50
substrate	∞	2000	50

	d[μm]	ω_p[cm^{-1}]	ω_τ[cm^{-1}]
layer 1	4.0	700	50
layer 2	3.5	1200	50
layer 3	3.0	1700	50
substrate	∞	2000	50

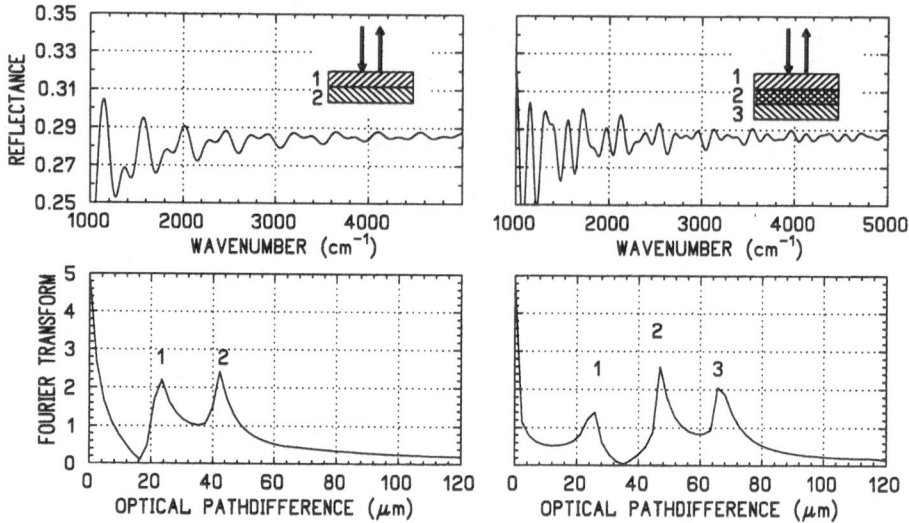

Fig. 5.19. Simulated reflectance spectra of GaAs multilayer systems and their Fourier transforms. As explained in the text the positions of the peaks marked are used to determine the various film thicknesses. Peaks in the Fourier transforms are found at path differences $2nd_1$, $2nd_1 + 2nd_2$

ingly the Fourier transforms reveal the signatures of the multilayers which are hidden in the frequency spectra. For the two-layer system shown on the left the peaks marked as '1' and '2' can be assigned to $2nd_1$ and $2nd_1 + 2nd_2$. In the three-layer case the additional third peak is due to the path difference $2nd_1 + 2nd_2 + 2nd_3$. Thus the unknown thicknesses can be determined from the peak positions.

For simplicity we have used the same refractive index $n = \sqrt{\varepsilon_\infty}$ for all layers. This is a good approximation since the layers only differ in their carrier concentration and the spectra are shown for frequencies well above the plasma edges. The differences in the refractive indices are then large enough to cause reflected partial waves at the layer boundaries but on the other hand are sufficiently small to give negligible contributions to the high frequency dielectric constant used for the evaluation of film thicknesses.

5.3.3 Direct Interferogram Analysis

In the previous Section the epi-layer thickness was obtained from the Fourier transform of the spectrum which itself is a Fourier transform of the interferogram. From the properties of Fourier transformation it follows that structures at path differences equivalent to the optical thicknesses should occur already in the original interferograms. The measured interferogram, however, is modified by the characteristics of the light source, the beamsplitter and the detector. Therefore it is different from the above mentioned Fourier transform which has been obtained from a normalised reflectance spectrum.

Nevertheless well pronounced interferences can be identified directly in the interferogram which enable a very fast direct analysis of the measurement instead of performing two Fourier transforms. Fast thickness determination is important e.g. in online process control and in two-dimensional scans of a surface usually performed with IR microscopes. For these applications several thousands of interferograms are taken which cannot be analysed using time consuming algorithms.

A typical interferogram of a sample with one epitaxial layer on a substrate is shown in Fig. 5.20. Apart from the above mentioned 'center burst' (or WLP: White Light Position) there are replicas of it to the right and left ('side bursts' or 'satellites').

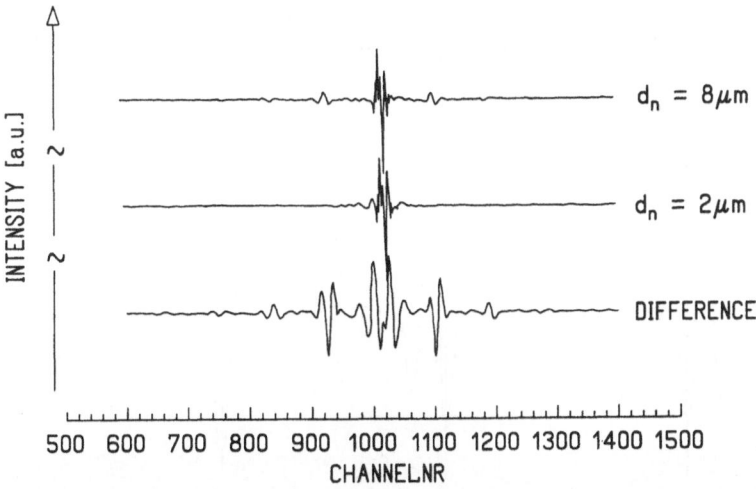

Fig. 5.20. Typical interferograms of a sample with an GaAs epitaxial layer (n) on a substrate (n^+). The distance from the 'side burst' to the 'center burst' is related to the optical thickness of the epitaxial layer. For a 8 μm thick layer (top) the satellites are clearly separated from the center burst. For a 2 μm layer (middle) they are located closer to the center burst and are not distinguishable in the interferogram. The difference (bottom) of both interferograms clearly reveals all side bursts [5.33]

These can be explained as follows: the interferometer divides the radiation from the light source into two partial waves. Therefore outside the white light position the two contributions experience a time delay with respect to each other. Now the sample itself splits up both incoming waves at the layer boundaries and — with respect to the directly reflected wave at the boundary between vacuum and epi-layer — time-delayed replicas appear with a retardation equivalent to the time necessary to pass through the epitaxial layer to the substrate and back (of course also multiples of this delay occur). If for a certain mirror position the 'interferometer delay' matches a 'sample delay' the second partial wave from the interferometer superimposes with a replica of the first one constructively. This results in a replica of the center burst at the corresponding path difference. For each delay in the sample there is a corresponding satellite structure in the interferogram. Obviously, the distance between white light position and the first satellite structure in the interferogram equals twice the optical thickness nd of the epitaxial layer.

In many cases an exact definition of the distance between center burst and side burst is not possible because the side burst shows a different appearance than the central structure of the interferogram. Figure 5.21 shows a comparison of satellite structures originating from 6 μm thick epitaxial silicon layers on three differently doped silicon substrates. It is clear from this figure that there is no distinct position from where to measure the distance to the center WLP.

The main reason for the deformation of the side bursts is the additional phase shift of the partial wave reflected at the boundary between epitaxial layer and substrate. This phase shift Φ_{12} is given by

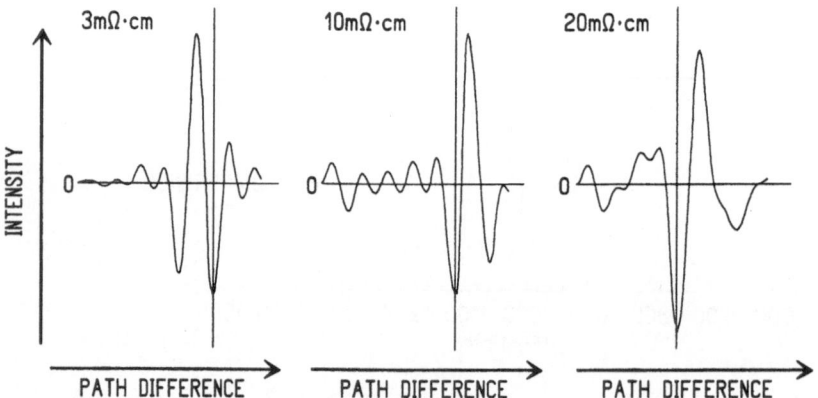

Fig. 5.21. Sidebursts of 6 μm undoped Si epitaxial layers deposited on three differently doped Si substrates. Obviously the satellites are distorted by the phase change occurring when then the partial waves are reflected at the interface between epi-layer and substrate [5.33]1990)

$$\tan \Phi_{12} = \frac{2n_1\kappa_2}{\kappa_2^2 + n_2^2 - n_1^2} \tag{5.94}$$

where the refractive index of the moderately doped epitaxial layer is a real quantity n_1 and the corresponding quantity of the highly doped substrate has a real part n_2 and also an imaginary part κ_2. The latter depends strongly on carrier concentration and frequency, and so does the phase shift.

In order to obtain a correct interpretation this additional phase shift must be known. The geometrical thickness can be obtained in this case from the frequency spectrum using the following equation (5.93) :

$$d = \frac{m}{2N(\nu_1 - \nu_2)} \cdot \left(1 - \frac{\Phi_{12}(\nu_1) - \Phi_{12}(\nu_2)}{2\pi m}\right) \tag{5.95}$$

where ν_1 and ν_2 denote the frequency positions of two interference maxima which differ by m interference orders.

5.3.4 Full Numerical Simulation of Reflectance Spectra

In the general case of a reflectance spectrum, however, a direct evaluation is no longer possible and the full formalism developed in Sect. 5.1 must be taken into account. Model calculations of reflectance spectra have to be performed and compared with the experiments. The unknown film thickness is then obtained by a variation of the parameters and selection of the best fit. Figure 5.22 shows a typical situation for an undoped epitaxial Si layer on a

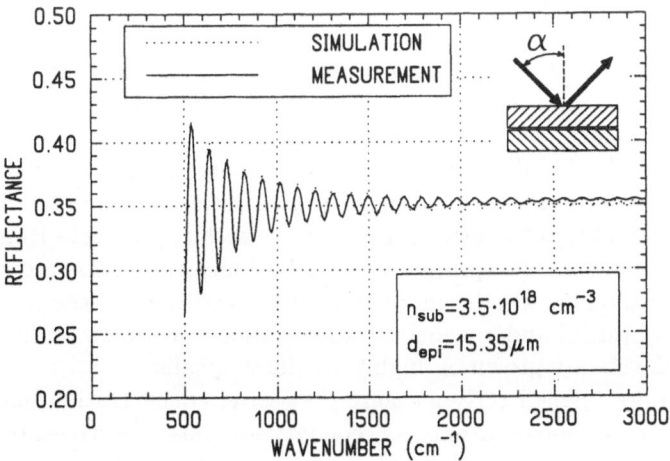

Fig. 5.22. Calculated ($\cdots$) and measured (——) reflectance of an undoped epitaxial silicon layer on a silicon substrate. With the parameters given in Tab. 5.1 an excellent agreement between simulation and experiment is obtained, such that the simulation can be hardly distinguished from the experiment

Table 5.1. Model parameters for the calculated reflection spectrum of the epitaxial silicon wafer (cf. Fig. 5.22)

	d [μm]	ω_p[cm^{-1}]	ω_τ[cm^{-1}]	
substrate	500	1100	350	incoherent
epitaxial layer	15.35	—	—	coherent

highly doped Si substrate. The measured reflectance spectrum is described in the model calculation by using a high frequency dielectric constant $\varepsilon_\infty = 11.7$ for both epi-layer and substrate and for the latter an additional Drude term representing free carrier contribution (cf. (5.82)). The complete set of parameters for the fits is given in Table 5.1.

The assumption made in the numerical analysis is of course that all properties change abruptly at the interfaces. Due to diffusion this is quite often not the case for the carrier concentrations. In this case a gradient of the carrier concentration must be taken into account. This will be described in Sect. 5.4.

The disadvantage of this evaluation method is that it requires a large amount of computational work. Therefore it is not very fast and hence not applicable for an online analysis. In order to speed up the procedure the problem can be solved by using a library of precalculated spectra for utilizing real time comparisons. Since there are very efficient library search algorithms available this evaluation method is quite appropriate to obtain layer thicknesses.

5.4 Determination of Carrier Concentrations

The carrier concentration and its temperature dependence is one of the basic parameters for characterising the electronic properties of epitaxial semiconductor layers. Conventionally, carrier concentrations are measured by the Hall effect either in the standard Hall or in van der Pauw geometry. For reliable measurements it is often necessary to use special procedures for fabricating ohmic contacts and in addition to pattern the epilayers with appropriate Hall masks. Consequently a part of the epilayer used for Hall measurements is lost for other analysing techniques. Furthermore only an average value of the carrier concentration is obtained and in general Hall effect measurements cannot be applied if the epilayers are grown on highly conducting substrates.

A method which does not directly measure the carrier concentration but rather the concentration of dopant atoms is Secondary Ion Mass Spectroscopy (SIMS). This technique even allows a determination of the dopant distribution as a function of depth and complete multilayer systems can be analysed. For profiling sputtering techniques have to be used and thus a part of the sample is lost for further studies. In addition, for a quantitative analysis a calibration procedure is necessary, too. A further commonly used technique is capacitance-voltage profiling (C–V), which also requires contacts.

This Section describes how IR reflectance measurements can be used to determine carrier concentrations in epilayer systems. The evaluation is most simple in case of thick layers where the reflectance signal is just due to the front surface. First this case is treated and subsequently partially transparent films and multilayers are discussed.

5.4.1 Semi-Infinite Samples

A thick layer where no radiation is reflected from the epilayer-substrate interface can be treated as semi-infinite. For high carrier concentrations semiconductor substrates with thicknesses in the range of some hundred microns are optically equivalent to a semi-infinite medium. In that case the reflectance of a semiconductor does not depend on geometry and is determined by the concentration of free carriers (plasma edge). If the epilayer material has infrared active optical phonons, these lattice vibrations contribute to the reflectivity, too, and are commonly described by plasmon-phonon-modes. Figure 5.23 shows the dielectric functions calculated according to the equations in Sect. 5.1.7 together with the corresponding half-space reflectance spectra.

First we discuss the left part of Fig. 5.23, where the dielectric function is the sum of the free carrier contribution and the dielectric background ε_∞ (here $\varepsilon_\infty = 10.95$, the value for GaAs).

For a negative real part of the dielectric function electromagnetic waves cannot propagate in the medium but are reflected almost completely. Hence the reflectance is close to unity at low frequencies, because only a small portion of the intensity is absorbed. $\text{Re}(\varepsilon)$ changes sign approximately at the frequency $\omega_p^\star$ which determines the position of the so-called plasma edge in the reflectivity spectrum

$$\omega_p^{\star 2} = \frac{ne^2}{m^\star \varepsilon_0 \varepsilon_\infty} \quad . \tag{5.96}$$

Usually, $\text{Re}(\varepsilon)$ passes zero with a relatively steep slope and the value 1 is passed at a frequency very close to $\omega_p^\star$. In most cases $\text{Im}(\varepsilon)$ is small in this region, and consequently the material is optically equivalent to vacuum and the reflectance vanishes. This drop of the reflectance from a value close to unity when $\text{Re}(\varepsilon) < 0$ to almost zero for $\text{Re}(\varepsilon) = 1$ is called the plasma edge, since it appears at the frequency $\omega_p^\star$ which is the plasma resonance frequency. From the position and form of the plasma edge the spectroscopic parameters ω_p and ω_τ and hence $n/m^\star$ and τ can be determined. In order to obtain reliable values model calculations have to be performed using a parameter fit. Since in the simple Drude model only the two parameters ω_p and ω_τ enter, a fast and completely automatic fitting procedure is possible. However, for

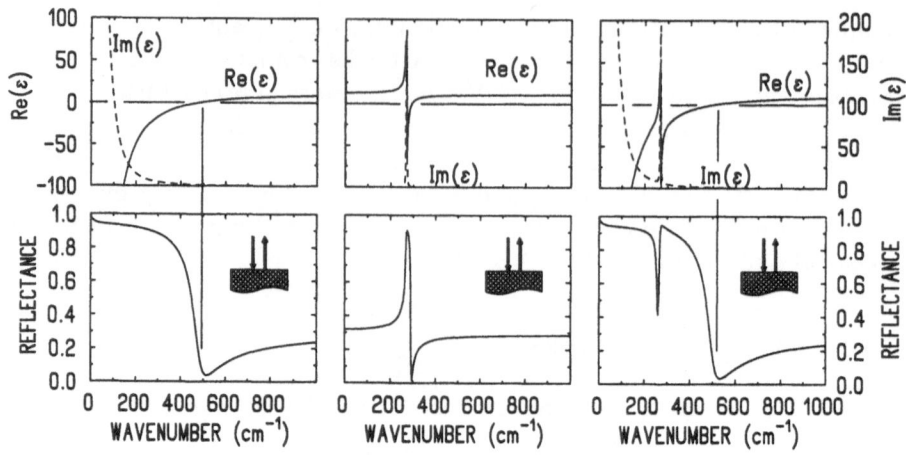

Fig. 5.23. Top: Calculated dielectric function considering free carriers (left) ($\omega_p = 1600$ cm^{-1}, $\omega_\tau = 50$ cm^{-1}), phonons (middle) ($\Omega_{TO} = 268$ cm^{-1}, $\Omega_P = 375$ cm^{-1}, $\Omega_\tau = 2.5$ cm^{-1}) and coupled plasmon-phonon modes (right) ($\varepsilon_\infty = 10.95$ in all cases, GaAs parameters). Bottom: the corresponding calculated reflectance spectra for semi-infinite sample at normal incidence of light. Spectral ranges with Re(ε) < 0 correspond to high reflectance

very high carrier concentrations (e.g. in GaAs beyond $5 \cdot 10^{18}$ cm^{-3} [5.34]) band nonparabolicity becomes important and such a simple two parameter analysis fails.

In the central part of Fig. 5.23 the frequency dependence of the dielectric function is shown which is typical for an infrared active phonon (see Sect. 5.1.7). There is a spectral range where the real part of the dielectric function is negative. As explained above, for these frequencies the electromagnetic radiation is almost totally reflected and this region is called the reststrahlen band.

Finally on the right side in Fig. 5.23 we show the sum of the contributions of a strong infrared active phonon and free carriers with parameters typical for GaAs. A strong phonon contribution can cause positive values of Re(ε) near the TO resonance frequency even if it is below the plasma edge, leading to a strong dip at the TO mode frequency in the high reflectance spectral range.

5.4.2 Multilayers

The situation becomes more complicated if the sample does not consist of a semi-infinite material but of a stack of layers with different carrier concentrations. The spectra show more complicated structures which require the use of model calculations in order to evaluate the carrier concentrations. These structures may be generated either by different carrier concentrations in the substrate and the epitaxial layers of the same material (homoepitaxy) or additionally by different lattice contributions in heteroepitaxial systems. An important basic condition for the characterisation of a certain layer is of course the requirement that the radiation reaches this layer, i.e. the layers in front of the one of interest must be partially transparent. Consequently deeply buried layers may escape such an analysis. Moreover it turns out to be difficult to characterise layers with low carrier concentrations embedded in layers with high carrier concentrations, because the plasma reflectivity edge of the layer with the highest concentration covers to a certain extent all others. However, using oblique incidence of p-polarised electromagnetic radiation, the longitudinal modes of the different carrier plasmas may be excited and appear at different spectral positions.

A particularly challenging problem is the determination of carrier concentrations in an epilayer system deposited on a substrate of the same chemical composition, i.e n-GaAs/n$^+$GaAs/n$^+$GaAs. In such a case IR spectra obtained under normal incidence conditions are not particularly helpful. As an example, we show a system of one moderately and one highly Si-doped GaAs layer on a highly Si-doped GaAs substrate. We show in the following that by using oblique incidence information on the carrier concentrations in the epilayers can be obtained. Fig. 5.24a shows the reflectance spectra for 70° incidence for s- and p-polarisation. While for s-polarised light the spectral shape turns out to be similar to that at normal incidence (see Fig. 5.24b), in p-polarised light additional features appear.

Before the IR analysis was done it was known for this particular specimen that the carrier concentration decreases from the substrate to the top layer. Therefore the light can pass the two top layers and reach the substrate whose plasma edge ω_p^* is around 600 cm^{-1}. Using (5.96) this indicates a carrier concentration of $2 \cdot 10^{18}$ cm^{-3}, where $0.069\,m_0$ has been used for the electron effective mass in GaAs (this value for m^* does not take into account nonparabolicity nor the proper averaging for the plasma mass [5.34]).

The two dips in reflectance marked with B1 and B2 (cf. Fig. 5.24a are due to enhanced energy dissipation due to the Berreman effect (cf. Sect. 5.1.8). They occur at the zeroes of the dielectric function of the weakly doped top layer. These zeroes depend on carrier concentration in the epilayer and therefore can be used more directly for evaluation. Figure 5.25 shows the dependence of the position of the absorption lines on the carrier concentration resulting from a model calculation which can be used as a calibration standard.

The peak positions of the two Berreman modes B1 and B2 can serve as calibration curves for carrier concentration determination (cf. Fig. 5.26). They shift differently with concentration; the low frequency line (the zero below the TO-phonon frequency at 268 cm^{-1}) varies strongly for low concentrations whereas the high frequency absorption line varies much less with increasing

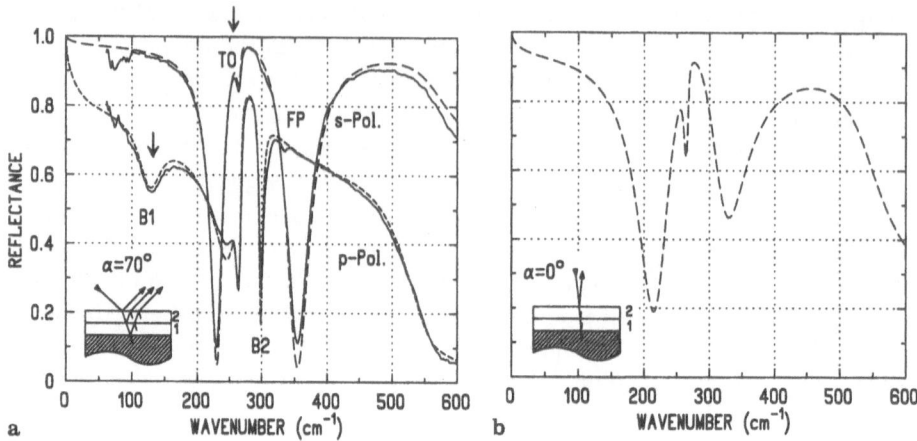

a   b

Fig. 5.24a. Reflectance spectra of a GaAs multilayer system ($d_1=d_2=2.1\mu$m) measured at oblique incidence of light (—). Spectra in s- and p-polarisation both exhibit Fabry-Perot interferences (FP) within the film and small structures at the GaAs TO frequency. In addition, in p-polarisation Berreman modes (B1, B2) are found. Their positions depend on the carrier concentration in the films. All spectra are compared to simulations (- -) (cf. Table 5.2), **b** simulated reflectance spectrum of the same sample at normal incidence of light. No Berreman modes are excited. Hence a determination of carrier concentrations from the spectrum would be more difficult

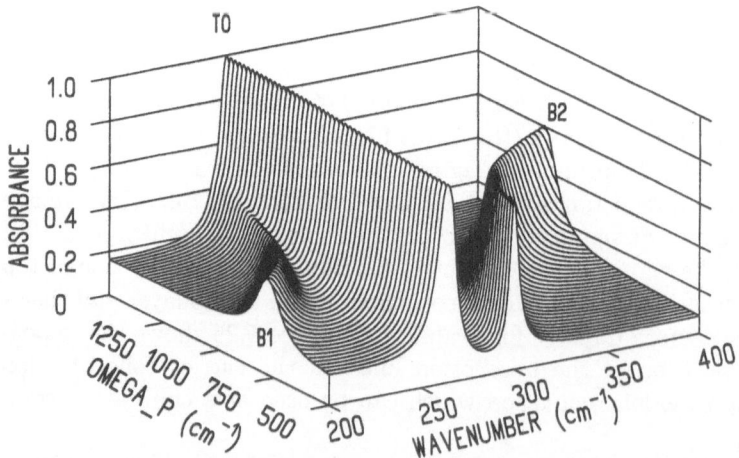

Fig. 5.25. Absorption in a thin doped GaAs film for various plasma frequencies. Besides the bulk TO-phonon absorption at 268 cm^{-1} two Berreman modes, whose positions depend on the doping (expressed in terms of ω_p), are observed

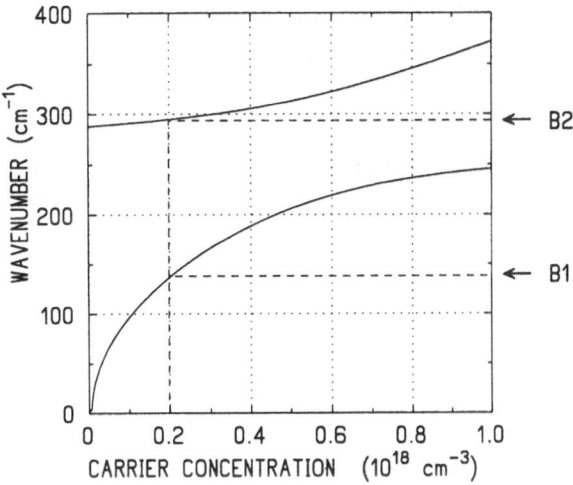

Fig. 5.26. Calibration curve for the determination of carrier concentration in a GaAs film: the positions of absorption lines B1 and B2 appearing in p-polarised spectra (cf. Fig. 5.24) can be used to obtain the carrier concentration

carrier concentration up to about $0.5 \cdot 10^{18}$ cm^{-3}. For concentrations higher than $0.5 \cdot 10^{18}$ cm^{-3} the situation is reversed.

In the spectra shown in Fig. 5.24 the absorption lines appear in p-polarisation around 125 cm^{-1} and 290 cm^{-1}, and thus both indicate a carrier concentration of about $0.2 \cdot 10^{18}$ cm^{-3} (see Fig. 5.26). This concentration can be assigned to one of the films. The carrier concentration of the other layer is close to the one on the substrate and the corresponding absorption lines coincide with the phonon absorption line and the plasma edge of the substrate, respectively. Hence the carrier concentration of the second epilayer cannot be determined.

However, one cannot decide which one of the two layers has the carrier concentration determined as described above. Only model calculations, which will be discussed below, can clarify these ambiguities.

Figure 5.24 shows the experimental and theoretical spectra of the GaAs multilayer system. The model parameters are given in Tab. 5.2.

In addition to the carrier concentration and damping parameters, the film thicknesses have been adjusted, too. For this purpose the spectral region from 600 cm^{-1} to 2000 cm^{-1} (not shown in Fig. 5.24) was also taken into account. In this region well pronounced interference fringes can be used to yield precise values of film thicknesses.

The parameters obtained by the fit procedure are compared to results obtained on the same sample with Secondary Ion Mass Spectrometry (SIMS), where a depth profile of the concentration of ^{28}Si dopant atoms is measured. Figure 5.27 shows the depth profile of the electron concentration used in

Table 5.2. Model parameters of the GaAs multilayer system shown in Fig. 5.24

layer		substrate	film1	film2
thickness	d [μm]	1000	2.1	2.1
	ε_∞	10.95	10.95	10.95
	Ω_{TO} [cm^{-1}]	268	268	268
phonon oscillator	Ω_P [cm^{-1}]	375	375	375
	Ω_τ [cm^{-1}]	2.5	2.5	2.5
free electrons	ω_P [cm^{-1}]	1720	1650	470
	ω_τ [cm^{-1}]	75	65	40

the IR model calculation together with the Si concentration from SIMS. For weakly doped GaAs each Si atom should contribute one electron to the electron gas — hence SIMS and IR spectroscopy yield the same results in the two films. The difference in the substrate is very likely due to the fact that at high doping rates some Si-atoms take different lattice sites in the GaAs host lattice and are consequently not ionised. Therefore the results of the two methods for the Si concentration and the carrier concentration in the substrate differ somewhat at higher concentrations as expected.

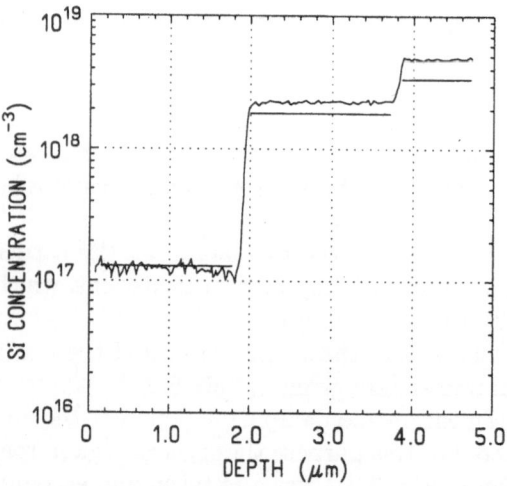

Fig. 5.27. Depth profile of the ^{28}Si concentration obtained by SIMS analysis (solid line) and the electron concentration profile used for the IR spectra simulation (dashed lines)

5.4.3 Carrier Concentration Profiles

Up to now we have treated systems with a sharp boundary between two adjacent layers, e.g. with an abrupt change of the carrier concentration between the layers. In many cases this approximation is not justified and IR spectra of such samples cannot be explained with the present simple layer model assuming abrupt interfaces. An example is given in Fig. 5.28 where the reflectance of a weakly doped silicon epilayer on a highly doped Si substrate is shown.

The best fit shown in Fig. 5.28 uses constant carrier concentration in the layers which change abruptly at the substrate interface. The interference structure in the experimental data decays more rapidly with increasing frequency than that in the simulation.

The assumption that the simple concept of a step-like carrier concentration profile used in the theory is not adequate for a description of the real system can be investigated experimentally: if a sample is annealed for several hours, considerable diffusion of the doping atoms takes place and the originally quite sharp concentration step is smeared out. Figure 5.29 compares the reflectance spectra taken before and after annealing: a strong diffusion correlates with a rapid decay of the amplitude of the interferences. Therefore in the theoretical description of the optical properties of such samples the concentration gradient inside the 'diffusion zone' between substrate and epitaxial layer must be included. This actually means that a spatially varying dielectric function has to be used in order to represent the gradient in the carrier concentration.

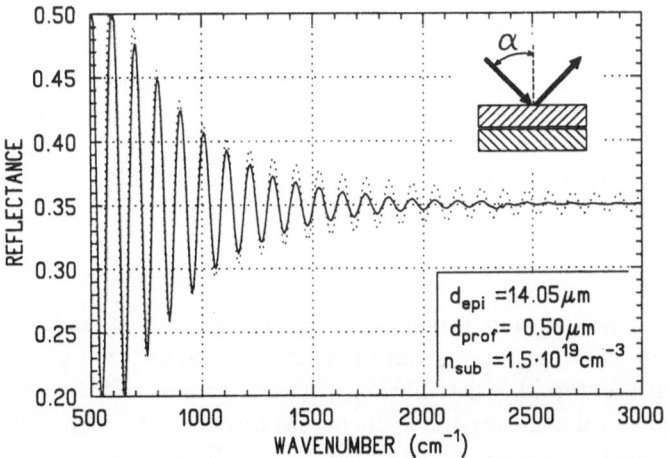

Fig. 5.28. Experimental reflectance spectrum (solid line) of a Si epi-layer on Si substrate sample ($\alpha = 30°$, p-polarisation) and the best fit (dotted line) assuming an abrupt interface between substrate and film. Obviously the model calculation cannot explain the measured interference amplitudes

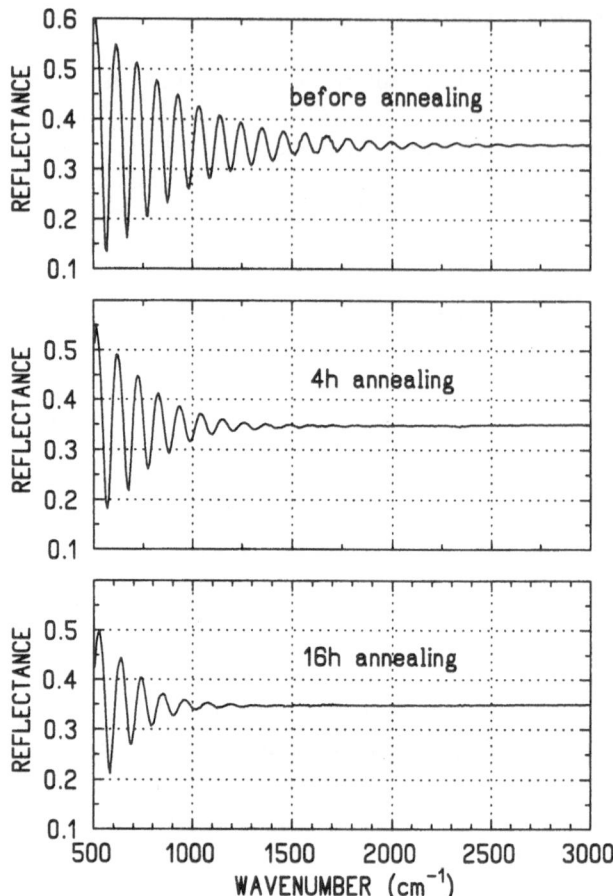

Fig. 5.29. Measured reflectance of the epitaxial silicon layer ($d_{\mathrm{epi}} = 13.6\,\mu$m) on a Si-wafer shown in Fig. 5.28 for various annealing times (annealing temperature 1000 °C). With increasing annealing time the interferences decay faster because the diffusion zone becomes broader (0 h: $d_{\mathrm{diff}} = 0.3\,\mu$m, 4 h: $d_{\mathrm{diff}} = 0.9\,\mu$m, 16 h: $d_{\mathrm{diff}} = 1.4\,\mu$m, for the definition of d_{diff} cf. Fig. 5.30)

The form of the concentration profile is of course not known and has to be determined. The most intuitive ansatz is probably the solution of the diffusion problem with a step-like distribution in the beginning. This seems to be the proper theoretical counterpart of the actual physical situation. The solution is the well-known error function, shown in Fig. 5.30. The width of the diffusion zone as it is used in the following is indicated by d_{prof} in this figure.

Now this assumption on the carrier concentration profile must be embedded in a model calculation that actually can be compared to experiments to decide whether the ansatz was correct or not.

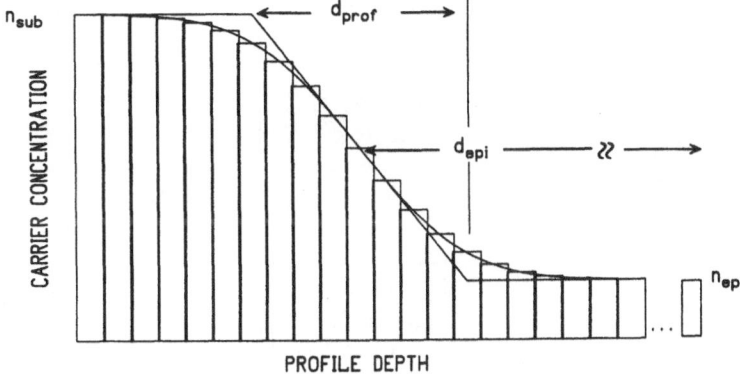

Fig. 5.30. Carrier concentration profile given by the error function (full line) and the definition of the diffusion zone width d_{prof}. The continuous distribution can be sampled as a stack of thin layers with constant concentration as sketched in the histogram

To calculate the optical properties the solution of the wave equation for the given spatially varying dielectricfunction has to be found. In general this cannot be done analytically and even with moderate computational effort numerical solutions are difficult to find.

An easy way to solve this problem is the following: the continuously varying carrier concentration profile is sampled as a stack of very thin layers with constant dielectric function within each layer. Then the formalism to calculate reflectance and transmittance of layer stacks (cf. Sect. 5.1.6) can be used further. To prevent artefacts from sampling, the layer thicknesses used must be much smaller than the smallest light wavelength inside the medium $\lambda_{\mathrm{min}}/\sqrt{\varepsilon_\infty}$. In addition the following check should be done: if by increasing the number of sampling layers (with a simultaneous decrease in film thicknesses) the calculated spectra do not change any more, the sampling is sufficient and there is no need to use smaller intervals.

Using this method, a very good agreement between theory and experiment can be achieved for the sample described in the beginning of this Section (see Fig. 5.28), as is demonstrated in Fig. 5.31. The model parameters used are also given in Fig. 5.31. The thickness of the sampling layers was chosen to be $0.1\,\mu\mathrm{m}$.

5.4.3.1 A Fast Evaluation Scheme for Diffusion Profiles. For a fast online analysis an algorithm has been developed that automatically fits the sample parameters to IR reflectance spectra under the following conditions: the form of the diffusion profile is given by an error function and the carrier concentration in the epitaxial layer is smaller than $1 \cdot 10^{17}\,\mathrm{cm}^{-3}$, i.e. it is negligible.

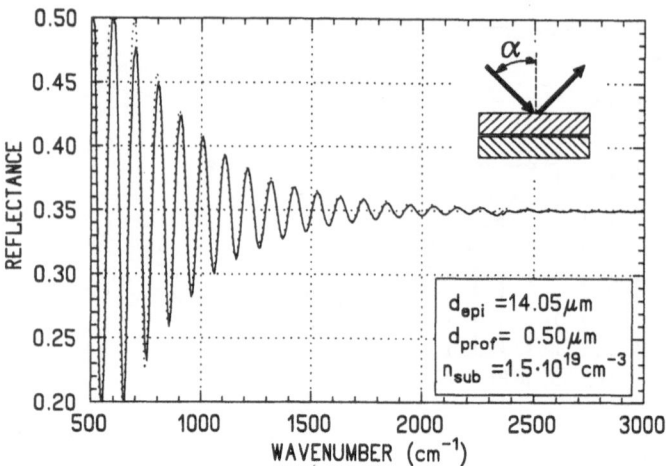

Fig. 5.31. Experimental reflectance spectrum (Si epilayer on Si, as Fig. 5.28) but with model calculations taking into account a diffusion zone with a distribution of carriers according to an error function: the agreement between the theory (dotted) and the experiment (solid line) is now excellent and hardly distinguishable from the experimental data (the parameters used for the substrate are $\omega_p = 1900$ cm^{-1} and $\omega_\tau = 400$ cm^{-1})

To understand the idea of the evaluation procedure, the influence of the relevant parameters on the reflectance has to be studied separately. We start with a system of a $10\,\mu$m thick undoped epitaxial layer in front of a highly doped substrate ($n_{sub} = 2 \cdot 10^{19}$ cm^{-3}). The diffusion profile is assumed to be step-like in this starting configuration and the corresponding reflectance spectrum is shown in Fig. 5.32a.

Now the individual parameters are varied: first we reduce the carrier concentration in the substrate while all other parameters are kept constant. The resulting spectrum is given in Fig. 5.32b.

A reduction of the epi-layer thickness by a factor 2 with respect to the starting parameters leads to spectrum (c) in Fig. 5.32, whereas the introduction of a diffusion zone of width $1\,\mu$m gives the situation shown in (d) in the same figure.

From an inspection of these model calculations one can establish three rules which will be essential for the automatic algorithm:

— The form of the envelope-function is independent of the thickness d_{epi} of the epitaxial layer.
— The amplitude of the interferences for low wavenumbers is determined by the carrier concentration n_{sub} in the substrate exclusively.
— The amplitude of the interferences for high wavenumbers depends strongly on the width d_{diff} of the diffusion profile.

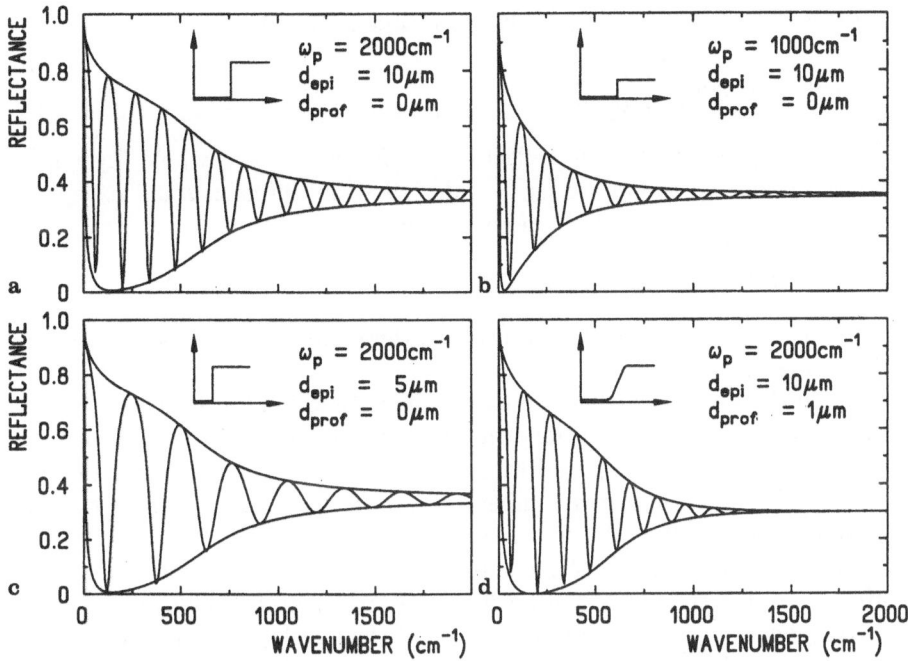

Fig. 5.32. Calculated reflectance of a silicon sample consisting of an undoped epitaxial layer on a doped substrate with a diffusion zone in between. Spectra (**b–d**) have been calculated by varying one of the parameters in (**a**). A decrease of the substrate's carrier concentration (**b**) influences mainly the amplitude of the interferences for low frequencies. Their period is determined by the thickness of the epilayer (**c**). The decay of the interferences' amplitude depends on the diffusion profile thickness (**d**)

The automatic determination of the parameters n_{sub}, d_{epi} and d_{diff} is performed in two steps.

First the parameters n_{sub} and d_{diff} are determined from the form of the envelope-function. As was shown above, the thickness of the epitaxial layer does not enter in this step, but will be calculated afterwards.

In order to calculate the envelope-function for a given pair of n_{sub} and d_{diff} it is useful to look at its meaning: it encloses all possible values for the reflectance which can be obtained by varying the thickness of the top layer, i.e. going from constructive to destructive interference of the waves reflected at the interface between vacuum and top layer and from the boundary between the top layer and diffusion zone respectively.

To calculate the upper part of the envelope-function — which is used here exclusively — one must choose an artificial thickness d^*_{top} ensuring constructive interference. A valid choice for an arbitrary wavenumber can be calculated in the following straightforward way: first the phase change at the interface between epi-layer and diffusion zone is calculated:

$$\Phi(\nu) = \arctan\left(\frac{\mathrm{Im}\ r_{\mathrm{res}}}{\mathrm{Re}\ r_{\mathrm{res}}}\right)$$

where r_{res} is the amplitude reflection coefficient for the whole layer stack (substrate and all layers are used to describe the diffusion zone) except the top layer. With this result the nominal thickness

$$d_{\mathrm{top}}^{\star} = \Phi(\nu)/4\pi\nu N$$

will lead to constructive interference and therefore to the desired upper part of the envelope-function.

With the ability to calculate the envelope-function for arbitrary frequencies one can proceed in the following way: from the experimental spectrum one chooses two frequency positions I1 and I2 with constructive interference, where I1 is chosen for a wavenumber as low as possible and I2 for a wavenumber as high as possible. A typical example is shown in Fig. 5.33.

For the determination of the substrate carrier concentration n_{sub} and the diffusion profile thickness d_{diff} an iterative procedure is employed. Since for low frequencies n_{sub} determines the form of the envelope-function this is first varied to match the reflectance at I1 where some reasonable start value for d_{diff} is used. Then I2 is used to fit d_{diff}. Then in turn one must reexamine n_{sub} and after this d_{diff} and so on, until the values converge.

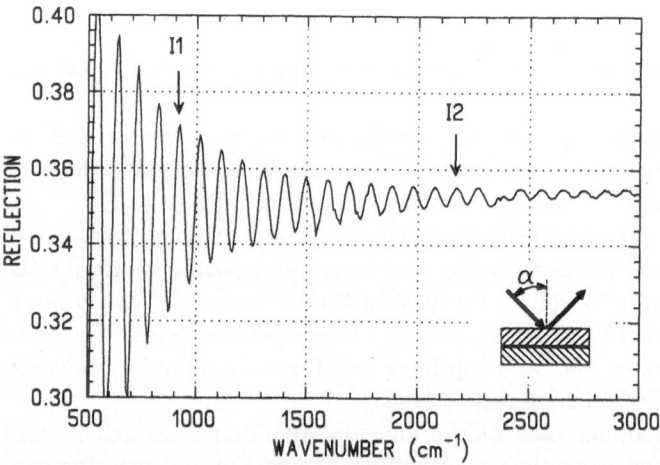

Fig. 5.33. Reflectance of an epitaxial silicon layer on a highly doped silicon substrate. The positions of the chosen points I1 and I2 (explanation in the text) are marked

After the determination of n_{sub} and d_{diff} the thickness of the top layer is easily calculated. One simply counts the number of interferences between I1 and I2 and then applies (5.95) where the phases can be calculated from n_{sub} and d_{diff}. As can be seen from Fig. 5.30, the thickness of the epitaxial layer is given by the sum of the top layer thickness and all thicknesses of the layers used to sample half of the diffusion profile.

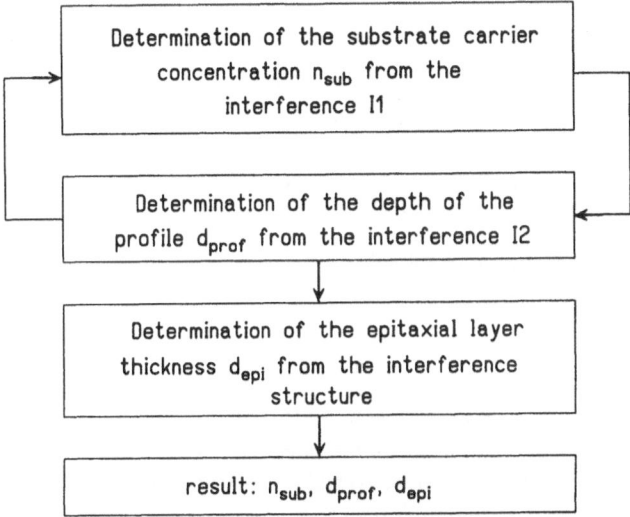

Fig. 5.34. Flow chart diagram describing the procedure to obtain n_{sub}, d_{prof} and d_{epi} from just two points in the IR spectrum

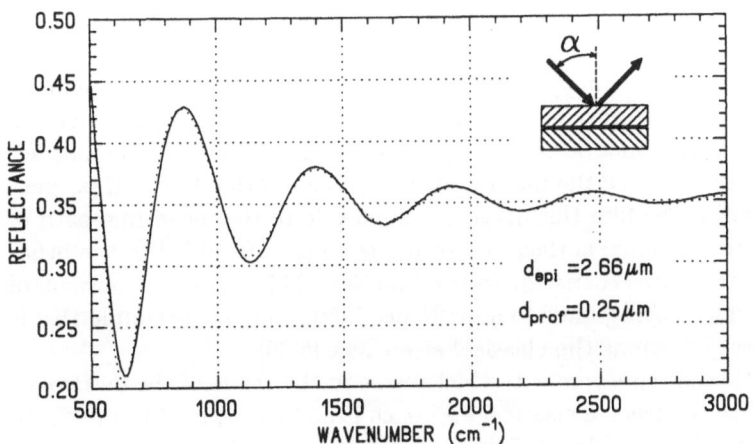

Fig. 5.35. Comparison of a model spectrum (dotted) based on the results of the fast algorithm described in this Section and the corresponding measurement (solid)

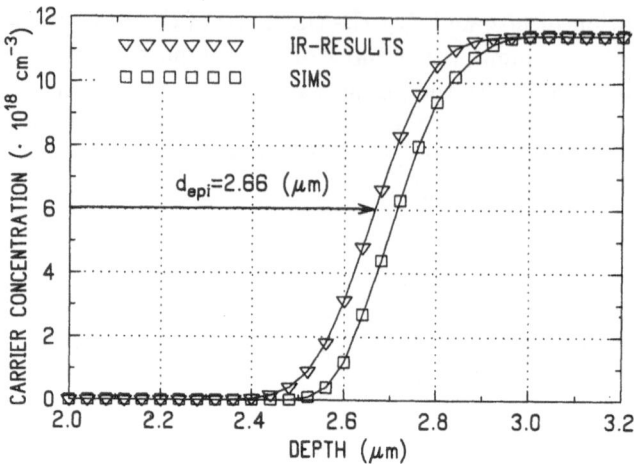

Fig. 5.36. Carrier concentration profile obtained by the IR analysis in comparison to SIMS data

The procedure just described is shown schematically in Fig. 5.34. It results in very reliable parameters which can be checked by calculating the whole spectrum and by comparing the results to the measurement. Figure 5.35 shows a typical fit. The carrier concentration profile is checked by a comparison with SIMS results (see Fig. 5.36). The form of both profiles agrees very well, however, the shift of about $0.05\,\mu$m between the two curves is not yet explained.

5.5 Confined Electron Systems

If the layer thickness of thin epitaxial layers is comparable to either the mean free path or to the de Broglie wavelength the classical description with a Drude-like dielectric function is no longer appropriate. The layer boundaries start to strongly influence the motion of the carriers in the direction perpendicular to them. If the film thickness is comparable to the mean-free path of the carriers, an additional surface-scattering term can be added to the other contributions in the free carrier dielectric function. This causes a reduction of the mobility which corresponds to an enhanced damping in the simple Drude model, an effect known as the classical size effect [5.35].

For even thinner epilayers with thicknesses in the range of the de Broglie wavelength of the carriers, a classical approach is no longer possible and quantum effects must be considered [5.36, 5.38, 5.39, 5.40]. With epitaxial growth techniques such as Molecular-Beam Epitaxy (MBE) or Metal-Organic chemical Vapour Phase Epitaxy (MOVPE) epilayer thicknesses in the range of a few atomic monolayers can be achieved with high accuracy. Indeed as a consequence of these achievements studies on heterostructures, quantum wells and superlattices have become a major part of todays semiconductor physics. Stimulated by the discovery of the quantum Hall effect, numerous studies of far infrared properties of carriers in low dimensional systems were published. Since quite recently an excellent review on this topic was published by Petrou and McCombe [5.36] we will restrict ourselves to a rather short description. In the following we sketch how the basic effects of carrier confinement manifest themselves in the IR spectra and which techniques are usually employed.

5.5.1 Properties of Confined Electrons

A strong confinement of electrons in one dimension (e.g. the z-direction) due to potential barriers leads to a quasi-two dimensional (2D) electron gas with a modified energy-momentum relationship as compared to the one of 3D electrons. In a 3D system for parabolic and isotropic bands the energy-momentum relationship is

$$E_{3D} = \frac{\hbar^2}{2m^*}(k_x^2 + k_y^2 + k_z^2) \tag{5.97}$$

leading to a density of states given by

$$\rho_{3D} = \frac{\sqrt{2}(m^*)^{3/2}}{\pi^2 \hbar^3} E^{1/2} . \tag{5.98}$$

A 2D electron gas shows electric subbands (index i) with discrete minimum energies E_i (due to the quantised motion in z-direction) and quasi-free-carrier-like terms for the k_x- and k_y-direction:

$$E_{2D} = E_i + \frac{\hbar^2}{2m^*}(k_x^2 + k_y^2) . \tag{5.99}$$

The density of states (including the twofold spin degeneracy)

$$\rho_{2D} = m^*/\pi\hbar^2 \tag{5.100}$$

is constant within each electric subband for a parabolic $E(k)$ dispersion relation. The separation between individual subbands depends on the shape and the width of the confining potential and is typically of the order of 10 to 100 meV. Therefore, FTIR spectroscopy is well suited for the investigation of the subband structure.

If a magnetic field B is applied in z-direction, a further quantisation occurs for the electron motion in the plane perpendicular to it, i.e., in x- and y-direction. In the 3D case one obtains for the density of states (neglecting damping)

$$\rho_{3D}(B) = \frac{\sqrt{2m^*}}{4\pi^2\hbar}\left(\frac{eB}{\hbar}\right)\left(\sum_N [E - (N + 1/2)\hbar\omega_c]\right)^{-1/2} , \tag{5.101}$$

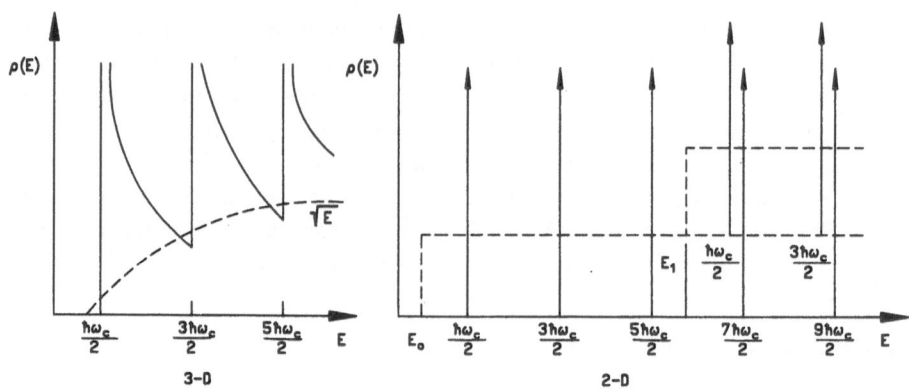

Fig. 5.37. Density of states for a three-dimensional (l.h.s.) and a two-dimensional system (r.h.s.) with (solid lines) and without (dashed lines) magnetic field [5.36]

with $\omega_c = eB/m^*$. In the 2D system the application of a magnetic field perpendicular to the plane in which the carriers are confined causes a completely quantised system:

$$\rho_{2D}(B) = \frac{m^* \omega_c}{\pi \hbar} \sum_N \delta(E - (N + 1/2)\hbar\omega_c) \,. \tag{5.102}$$

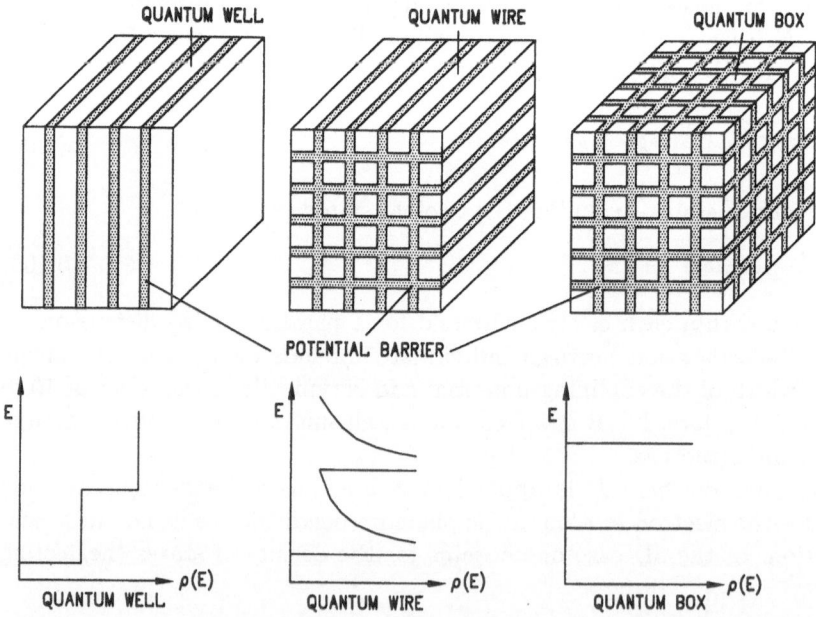

Fig. 5.38. Scheme of a quantum well stack, quantum wire and quantum box arrangement. Each has qualitatively different densities of states [5.36]

Figure 5.37 shows the corresponding graphs of the 3D and 2D densities of states.

Photon absorption due to transitions between different electronic states (intersubband transitions) are possible in isotropic parabolic systems if the electric field of the electromagnetic radiation contains a nonvanishing z-component, i.e. normal to the interface. The strength of the interaction with light is determined by the corresponding matrix elements. More details may be found e.g. in [5.36].

If the motion of the carriers is further reduced by potential barriers in either the x- or y-directions the electron gas will be confined to one dimension (1D: quantum wires), or even zero dimensions (0D: quantum dots, sometimes called 'artificial' atoms). Figure 5.38 shows schematically the corresponding densities of states (for zero magnetic field). In order to increase the interaction of electromagnetic radiation with the confined electrons as shown in Fig. 5.38, multiple low dimensional systems are studied. These can be arranged in a regular fashion (multiple quantum wells in 2D systems, periodic quantum wires, an array of quantum dots, etc). In such cases, the interaction of the confined electrons in the various layers has to be considered as well.

5.5.2 Spectroscopic Techniques

Since the total number of electrons which interact with electromagnetic radiation in a low-dimensional system is quite small, the observation of any changes of transmittance or even reflectance is in general difficult. Consequently differential measuring techniques (i.e. modulation techniques) have to be used.

In gated 2D systems the variation of the electron concentration by the applied gate voltage is a convenient means to achieve this goal, i.e., in order to correct for the characteristics of the measurement system the relative change of transmission $-\Delta T/T = (T(0) - T(n_{\mathrm{subs}}))/T(0)$ is determined, where n_{subs} is the carrier sheet density. This technique was originally employed with Si MOS samples (see Fig. 5.39). In such a case the sample itself with the 2D-system switched off is used as a reference [5.42].

Such a procedure has been used also with modulation doped GaAs/ Al-GaAs heterostructures with an added gate electrode on top. Depending on the type of FTIR spectrometer used (rapid scan or step scan) two procedures are employed for the modulation spectroscopy.

In a rapid scan spectrometer one takes one interferogram with the sample in "reference mode" switches to the "sample mode" and records the next interferogram, and repeats this switching co-adding reference and sample interferograms in different memory areas until the desired signal-to-noise ratio is achieved. This procedure works well if any signal fluctuations of the measurement system during the time needed to record one interferogram are smaller than the relevant signal in the spectrum.

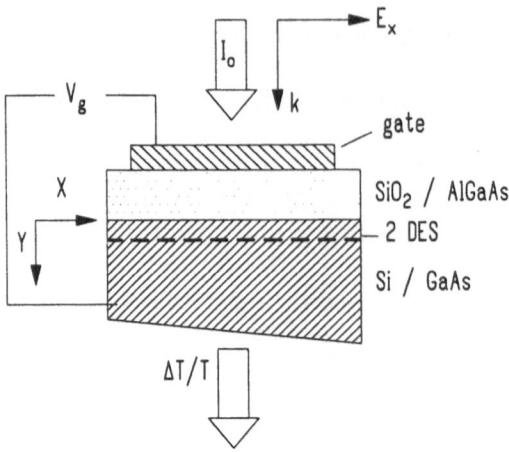

Fig. 5.39. Sketch of a MOS system or gated heterostructure which has been used in FIR spectroscopy. The infrared radiation is incident from the gate side. The charge density in the two-dimensional electron gas is modulated by an oscillating gate voltage. The gate electrode is a typically few hundred angstrom thick evaporated metal film which is transparent for the FIR radiation. The resulting transmittance change is measured. The sample is wedged on the substrate side in order to suppress interference phenomena [5.41]

The second option is to record only the differences between the reference and sample interferogram. This can be done in a step scan spectrometer. Here the movable mirror is stopped at the desired position to record an interferogram point for quite a long time during which the modulation is performed many times and the modulated signal is filtered from the noise by a lock-in amplifier.

These techniques can be used for the study of intraband phenomena or for interband transitions involving electronic states in the valence and conduction band. For the study of intersubband transitions within either the conduction or the valence band, in parabolic semiconductors a component of the electromagnetic field parallel to the z-direction is required. For the excitation of these intersubband transitions either an oblique incidence is necessary, which is achieved usually in a wave guide geometry [5.36] or a grating coupler technique developed by Heitmann et al. [5.43]. The latter technique is applied either to couple the incident light wave to nonradiative modes (surface plasmons, cf. Sect. 5.1.9) which would otherwise not be detectable by optical experiments, or to excite intersubband transitions. Experimentally, this is realised by arranging periodic conducting stripes in a layer parallel to the 2D system separated by an insulating material such as SiO_2 or deposited directly to form Schottky barriers. The schematic view in Fig. 5.40 shows the local electromagnetic fields induced by the grating with normal incidence of the infrared light wave. Due to the lateral periodicity of the system surface

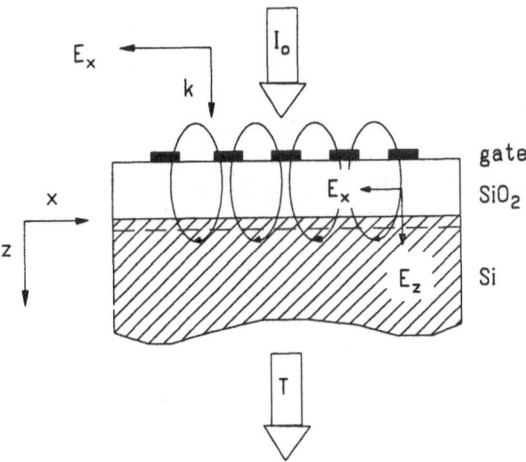

Fig. 5.40. MOS system with a grating coupler [5.41]. At the gate electrodes the electric field of the incident radiation is shorted. The resulting near field has a normal component E_z and is able to excite intersubband transitions. Surface plasmons can be excited due to the periodic structure, too

plasmons can be excited at certain frequencies according to their dispersion relation.

The incident electric field is shorted in the regions of the gate electrodes. The resulting near field has components perpendicular to the layers which are necessary for the excitation of transitions from one subband to another.

5.5.3 Results

In the past 10 years an enormous amount of far infrared spectroscopic investigations has been published on low-dimensional systems. Fourier transform spectroscopy was shown to be extremely useful and versatile since in comparison to FIR laser spectroscopy wider frequency regions are accessible and the lineshapes of electronic transitions can be studied at constant values of external parameters. In 2D systems at zero magnetic field intraband processes such as intersubband transitions, plasmon excitations and intra-impurity transitions were the main area of research. As reviewed by various authors [5.36, 5.41, 5.44, 5.45] such studies provide detailed information on the characteristic frequencies and the $\omega(k)$ relationship of 2D collective excitations (plasmons) and of the confinement potential through intersubband transitions. Such studies cannot only be performed in transmission but also in far infrared emission spectroscopy [5.44].

Furthermore, impurity level spectroscopy in 2D systems has been used to determine the impurity energy level separation like in the 3D case and to investigate the influence of the position of the dopants with respect to the confining potential in real space on these energies. The investigation of cyclotron resonance transitions in 2D systems has also been a major research topic in the past 10 years since it provides not only information on the effective masses but also on the carrier scattering time and thus on the dynamical conductivity in a quantizing magnetic field, for a review see [5.36] .

Since the discovery of the Quantum Hall effect (QHE) and the fractional quantum Hall effect particularly in GaAs/GaAlAs structures the cyclotron resonance position and shape has been investigated carefully as a function of the Landau level filling factor. Oscillations of the cyclotron linewidth at integer and fractional filling factors were reported which are still the subject of ongoing investigations. Magnetic fields applied in the plane of the layers have also been used to probe the confining potential and for the study of short period superlattices (Duffield et al. [5.37]). Intersubband transition energies can be determined indirectly as well by studying cyclotron resonance over a wide range of magnetic fields which are slightly tilted with respect to the surface normal. In such a case a coupling of the cyclotron motion with the intersubband oscillator occurs and anti-level crossing was observed for magnetic fields for which the cyclotron energy equals the intersubband transition energy [5.45]. Recently, tilted field geometries have been successfully used to study the effect of collective phenomena like the depolarization shift [5.46, 5.47], which is due to a perturbation of the effective confining potential caused by the FIR radiation which induces a perturbation of the local electron density $\rho(z)$ and final state correction due to an exciton-like interaction between the excited electron in the higher and the remaining hole in the lower subband [5.48]. Both effects shift the observed optical intersubband transition energies with respect to the single particle separation.

Out of the many reported investigations we show some few results. Figure 5.41 shows transmission spectra of a Si-MOS-system (Metal-Oxide-Semiconductor) where in the range from $50 \ldots 150$ cm^{-1} 2D plasmon resonances show up and for higher frequencies intersubband transitions are observed as indicated. The shift of the resonance positions with varying 2D carrier density is in reasonably good agreement with theoretical calculations.

Quite recently several experimental techniques have been used to fabricate periodic arrays of one-dimensional carrier systems. In the following we show results obtained by FIR spectroscopy on deep mesa etched quantum wires fabricated from high mobility GaAs/GaAlAs single and multiple quantum well structures. The actual MQW structure consists of a sequence of GaAlAs, AlAs and GaAs layers where the active GaAs modulation doped well has a width of 5 nm and a total period of about 200 nm which results in electronically decoupled wells. With the deep mesa etching technique wires of a geometrical width of about 550 nm and a periodicity of 1100 nm were produced homogeneously over areas of about 3 mm · 3 mm. From the investigation of dc transport studies as well as from luminescence studies it turned out that the active wire width was just 320 nm with a lateral depletion of

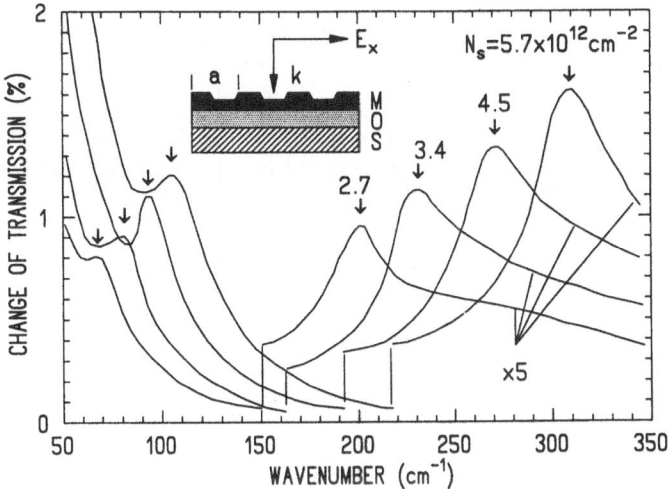

Fig. 5.41. Relative change of transmission $(\Delta T/T)$ versus wavenumber in 2D electron accumulation layers of a Si (111) MOS structure at different 2D carrier densities N_S as indicated. In the low frequency part 2D plasmons are observed $(q = 1.25 \cdot 10^5 \text{ cm}^{-1}, \mathbf{q} \parallel [110])$, at higher wavenumbers intersubband transitions $(E_0 \to E_1)$. For the (111) Si MOS structure excitation with normal incidence radiation is possible (E_x parallel to the surface) since the conduction band structure of Si (six ellipsoids of revolution oriented along the $\langle 100 \rangle$ directions) makes intersubband excitations possible through a finite angle between E_x and the $\langle 100 \rangle$ directions

about 100 nm on each side of the wire. Since for these sample parameters (assuming a harmonic 1D confining potential of parabolic shape) the subband energies are about 1.5 meV, FIR spectroscopy is well suited for studying such transitions.

The FIR measurements were performed on these wire structures in transmission with magnetic field applied normally to the surface of the laterally nanostructured samples. As an example Fig. 5.42 shows transmittance spectra $T(B)$ of a double-layered quantum wire system normalized to $T(B_0)$ where B_0 was chosen in such a way that this reference spectrum did not show any frequency dependence in the frequency region of interest. For $B=0$ two resonances at frequencies ω_{ri} are observed for radiation polarized perpendicular to the wires. They shift with increasing field to larger frequencies and apparently follow the relation $\omega_{ri}^2(B) = \omega_{ri}^2(B = 0) + \omega_c^2$, where ω_c denotes the cyclotron resonance frequency. The dependence of the resonance positions ω_{ri} on the square of the magnetic field is summarised in Fig. 5.43 for a one-layered, a two-layered and a five-layered system, where in the two and the five layered structure each of the two or five observed resonances follows the same quadratic dependence on magnetic field.

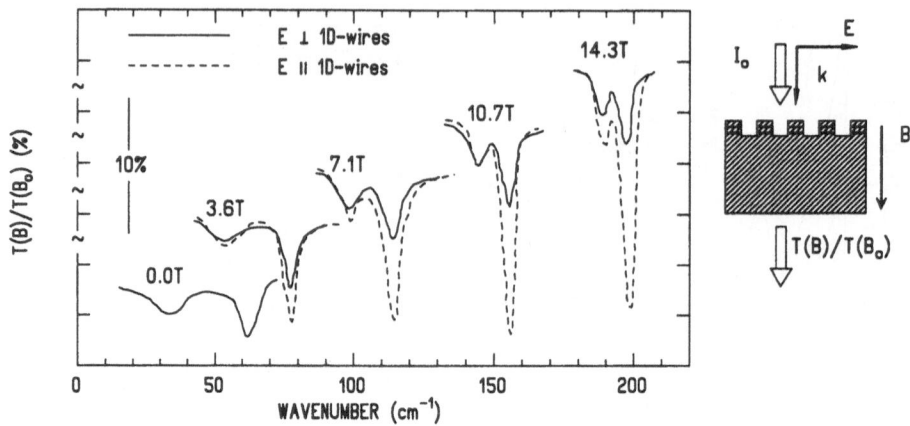

Fig. 5.42. 1D quantum wire system [5.42].Normalised transmission spectra on a periodic array of double layered quantum wires (i.e. in each of the wires two QWs are embedded) fabricated from ultrafine mesa etching of a double layered MBE grown GaAs/GaAlAs structure, to which a magnetic field is applied perpendicular to the surface. Infrared radiation incident onto the sample excites coupled intersubband-cyclotron resonance transitions. Full lines and dashed lines denote polarization of the incident FIR radiation with electric field vector perpendicular and parallel to the wires

Even at $B = 0$ the resonance positions ω_{ri} cannot be simply interpreted as giving directly the 1D intersubband transition energies. From Shubnikov de Haas measurements on the same samples it turns out that the intersubband separation is 1 meV and 1.5 meV for the one and two layered systems, respectively. However, the FIR resonance energies are 4 meV (one layered) and 8 and 4 meV (two layered system), respectively. Thus in this case there seems to be a particularly strong depolarization shift in the intersubband resonance as observed by FIR spectroscopy. Heitmann has argued that this strong collective contribution in the periodic 1D wires can be understood in terms of a local plasmon resonance, which also accounts for the splitting of the branches in the n-layered systems.

For the two layered 2D system, the plasmon dispersion was calculated and it was shown that for widely separated sheets of carriers ($q \cdot d \gg 1$, where d denotes the separation of the sheets, i.e. the MQW period) the plasmon branches fall onto each other whereas for small distances ($q \cdot d \approx 1$) this degeneracy is removed and the coupling leads to a splitting of the plasmon dispersion. A physical interpretation is the following: the upper branch in the two layered system corresponds to the longitudinal oscillation of both layers which are in phase, the frequency of the lower branch is determined by the coupling strength of the two layers. However, in the 1D system the observed FIR resonances are even higher than those calculated from the coupled 2D plasmon model. This additional shift is accounted for by assuming that the

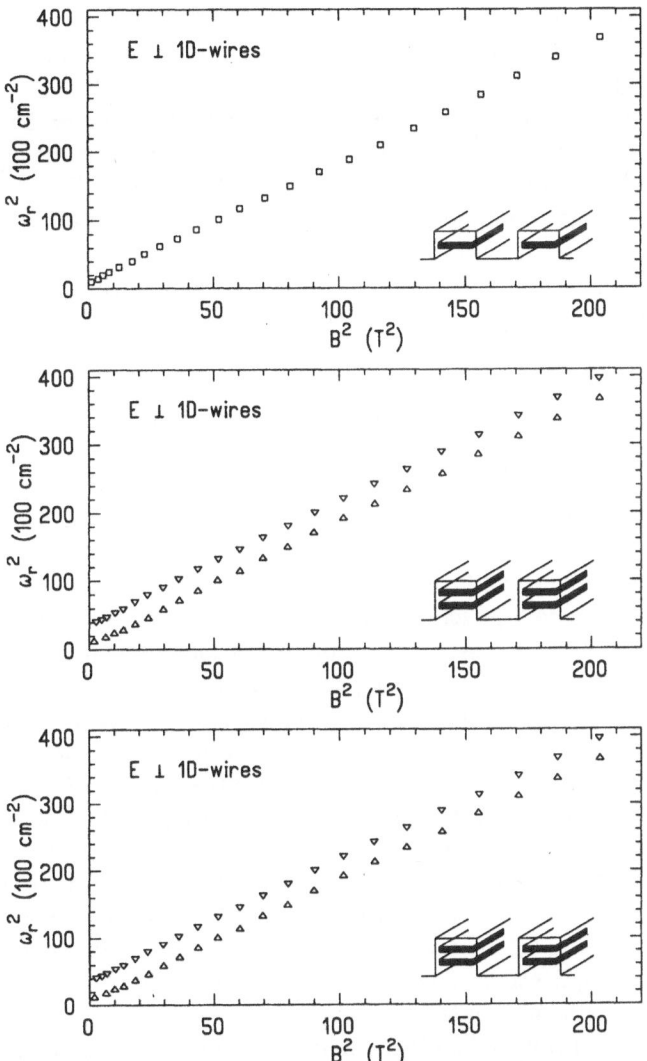

Fig. 5.43. Experimental FIR resonance positions for three periodic quantum wire samples obtained from transmission spectra like those in Fig. 5.42. For the one-layered wire system (top) one resonance is found for each value of the magnetic field. For multi-layered systems these modes split (middle, bottom). The number of resonances is equal to the number of layers (one, two or five quantum wells) within each of the wires [5.42]

upward shift of frequencies is caused by a localisation, i.e. by the fact that the two-dimensional electron gas is confined in the plane by the finite width of the wires. This lateral quantization imposes a restriction on the plasmon wavevector which is quantized in values of $q = \pi/w_e$ (w_e: effective wire width). This quantisation together with the assumption that Coulomb interaction with the neighbouring electron stripes can be neglected indeed accounts, at least qualitatively, for the observed frequency shift. More refined explanations are currently being worked out.

In 2D MQW and superlattice systems intersubband photon emission has been studied recently with the ultimate goal to achieve population inversion and hence an electrically pumped intersubband laser [5.49]. Recently Faist et al. [5.51] have indeed realized an intersubbandlaser at a wavelength of 4.5 μm. Furthermore intersubband absorption measurements have been carried out in superlattices, i.e. in systems where the wavefunctions of neighbouring quantum wells overlap considerably and minibands with a finite energy-momentum dispersion along the growth direction are formed. Recently Fourier spectroscopy has been used to study optical transitions between the two lowest minibands in strongly coupled n-type GaAs/GaAlAs superlattices. The inter-miniband absorption spectrum contains information about the dispersion along the growth axis and about the widths of the minibands. Indeed two absorption maxima related to the critical points of the mini-Brillouin zone were clearly observed [5.50].

Recently an extensive review on far infrared spectroscopy of phonons and plasmons in semiconductor superlattices has been published [5.6]. Appel and Hunderi [5.5] have extensivly discussed furthermore the local and nonlocal dielectric function of superlattices. Apart from the interest in the basic physical properties of intersubband transitions in low dimensional systems, their obvious application as an infrared photon detector has been recently investigated extensively. Levine [5.52] has studied carefully various schemes of MQW structures which can be used for IR detection and has shown that high sensitivity low dark current GaAs/GaAlAs quantum well infrared detectors for the mid infrared region can be fabricated. Recently it has been demonstrated that other III-V material systems but also Si/SiGe MQW structures [5.53–5.55] may represent promising alternatives for FIR and MIR detection which might replace for certain applications detectors based on interband transitions in narrow gap semiconductor materials.

Optical interband absorption and emission between confined carrier states in the conduction and valence band is of importance for the fabrication of single and multiple quantum well lasers [5.56]. For QW structures of narrow gap materials such transitions are located in the mid infrared region and thus can be conveniently investigated by Fourier spectroscopy. Recently, the step-like changes of the absorption constant as a function of frequency which reflect the 2D joint density of states were observed at the onset of various interband transitions between confined states. Furthermore it was shown that interference fringes observed in the transmission spectra of MQW structures can only be properly described if the cusps in the frequency dependence of the refractive index, which are related to the above mentioned absorption steps via the Kramers-Kronig relation, are taken into account [5.57, 5.58].

5.6 Determination of Impurity Concentrations

The evaluation of epi-layers requires knowledge on their impurity content. Two properties are of interest in this respect: the determination of the chemical nature and the concentration of impurities. Both informations are necessary for the epitaxial layers as well as for the substrates on which these are deposited. Using IR spectroscopy impurities are detected either through vibronic (i.e. local) excitations or through electric dipole allowed transitions of electrons between ground and excited states. The frequency specifies the transition and thus the chemical nature and the absorbance is proportional to the impurity concentration.

Among the semiconductor substrates used for the deposition of epitaxial layers, silicon and GaAs are the most widely used ones and consequently in the following mainly these two materials will be considered.

Impurities in a crystal lattice can occupy interstitial or substitutional sites. If the contaminant atom is displaced from its equilibrium position, a dynamic electric dipole moment occurs. Hence, vibrations of the impurity are excited by electromagnetic radiation of appropriate frequency. The resonance frequency Ω_0 is characteristic for the impurity atom involved. The positions of the absorption lines serve as fingerprints for the specific impurities at certain lattice sites. Their peak height correlates with the concentration. Most of the eigenfrequencies are found in the mid IR, where data can be taken quite conveniently.

5.6.1 Experimental

Since impurity concentrations are often relatively small, their contribution to the dielectric function of the material is also comparatively small. Hence (5.58...5.69) of Sect. 5.1 can be simplified, especially in the case of materials without IR-active phonons like Si or Ge. In fact the real part $\mathrm{Re}(\varepsilon)$ of the dielectric function is only slightly changed by the vibronic or electronic excitations of the impurities (i.e. $\mathrm{Re}(\varepsilon) \approx \varepsilon_\infty$), and we often find $\mathrm{Re}(\varepsilon) \gg \mathrm{Im}(\varepsilon)$ if the concentration of the impurities is not too high. Expansion of $\tilde{n} = \sqrt{\varepsilon' + i\,\varepsilon''}$ yields the approximate optical constants

$$n \approx \sqrt{\varepsilon_\infty} \tag{5.103}$$
$$\kappa \approx \mathrm{Im}(\varepsilon)/(2\sqrt{\varepsilon_\infty})\,, \tag{5.104}$$

and one also finds $\kappa \ll n$.

Within these limits the absorption index can now be decomposed into contributions by free carriers, phonon and multiphonon processes and impurities. All dissipative processes contribute linearly to the dielectric function (see (5.80)):

$$\text{Im}(\varepsilon) = \text{Im}(\varepsilon_{\text{FC}}) + \text{Im}(\varepsilon_{\text{Ph}}) + \text{Im}(\varepsilon_{\text{Imp}}) \,, \tag{5.105}$$

(ε_{FC}: free carriers, ε_{Ph}: phonons, ε_{Imp}: impurities).

From (5.104) it follows that one can then decompose κ or the corresponding absorption coefficient $K = 4\pi\nu\kappa$:

$$K = K_{\text{FC}} + K_{\text{Ph}} + K_{\text{Imp}} \,. \tag{5.106}$$

For not too high impurity concentrations $\text{Im}(\varepsilon_{\text{Imp}})$ is directly proportional to the impurity concentration N_{Imp}, and thus

$$N_{\text{Imp}} = f \cdot K_{\text{Imp}} \,. \tag{5.107}$$

f is a factor which has to be determined by calibration. A measurement of the absorption coefficient K_{Imp} at the characteristic eigenfrequencies of the impurities consequently yields their concentrations.

Usually, the absorption coefficient of the sample is obtained by measuring the transmittance of the wafer at normal incidence and applying (5.79). As $\kappa \ll n$ the sample's reflectance is not influenced by the impurities (i.e. R does not depend on κ). Therefore K_{Imp} can be determined by comparing the transmission spectrum of the sample T_s to that T_r of a reference wafer with the same thickness and with a very low concentration of unintentional impurities. This is realized for silicon by using material grown by the float-zone technique. Application of (5.79) yields

$$\ln T_s - \ln T_r = -(K_s - K_r) \cdot d \,. \tag{5.108}$$

Because both samples have the same phonon absorption we finally find

$$K_{\text{Imp}} = -(\ln T_s - \ln T_r)/d - K_{\text{FC}} \,. \tag{5.109}$$

Since free carriers cause only broad structures in the optical spectra, K_{Imp} can be extracted by this procedure using a local baseline in the frequency range of interest. From K_{Imp} finally N_{Imp} is obtained.

If more heavily doped semiconductors ($n > 2 \cdot 10^{16}\,\text{cm}^{-3}$) have to be investigated, the transmittance of the wafer is low, which makes experiments more difficult. In addition sophisticated baseline techniques are necessary [5.60].

The standard procedures for the determination of impurity concentrations require plane parallel wafers, which are polished on both sides. In that case, however, (5.79) is only an approximation, since it does not take into account multiple reflections within the wafer. If interference fringes are not resolved, the proper transmittance formula is given by (5.77). The absorption coefficient of the sample is then given by (again under the assumption that R does not depend on the absorption index κ)

$$K = -\frac{1}{d} \ln \left(\frac{\sqrt{(1-R)^4 + 4R^2T^2} - (1-R)^2}{2R^2T} \right) . \tag{5.110}$$

The contribution K_{Imp} of the impurities (cf. (5.106)) is obtained by comparing the data with a reference sample similarly as described above.

An even more complicated situation arises in the case of single-side polished wafers, because there is certain scattering loss at the rough wafer backside. An empirical formula for the transmittance was given in [5.33].

5.6.2 Impurities in Substrates

Si substrates are grown either by Czochralski or vertical zone melting techniques. In Czochralski grown material some crucible atoms are dissolved by the molten silicon, and the resulting unintentional impurities are incorporated in the crystal. Most important impurities are carbon and oxygen. Since carbon is in the same group of the periodic table as silicon, it can occupy silicon sites in the lattice. Oxygen is usually found at interstitial lattice sites after crystal growth, where it forms Si_2O "quasi-molecules". However, during epitaxial growth the substrate temperature can rise to quite high values and at temperatures of $800\ldots1200°$ C the concentration of interstitial oxygen (O_i) is larger than its solubility. Therefore part of the oxygen precipitates to SiO_2 complexes. This is accompanied by the formation of microdefects in the silicon lattice which strengthen the wafer and which act as gettering centers for impurities incorporated. Thus a certain amount of oxygen in the silicon is desirable. Because of the importance of silicon for semiconductor technology standard procedures were developed which guarantee comparable results on the carbon and oxygen impurity content of Si wafers.

In the following details of the concentration determination for carbon, interstitial oxygen and oxygen precipitates will be described. For the determination of the carbon concentration, in the interesting frequency region two-phonon absorption is quite strong, and multiple reflections have not to be considered. However, multiple reflections have to be taken into account for the determination of the oxygen concentration, since the total absorption coefficient at the frequency characteristic for oxygen impurities is quite small (cf. Fig. 5.46). Figure 5.44 shows a typical absorption spectrum of a Si wafer containing carbon and oxygen impurities illustrating the experimental situation.

5.6.2.1 Substitutional Carbon in Silicon.

Due to C-Si vibrations the strongest absorption band is located at 607 cm^{-1} [5.61]. As mentioned, it is overlapped by a strong two-phonon mode of the silicon at 610 cm^{-1} (see Fig. 5.45).

For the determination of the carbon content of the wafer the proper subtraction of multiphonon contributions is necessary, and as discussed above the sample and reference have to have identical thickness and the same surface finish. The presently accepted value for the conversion factor f is after

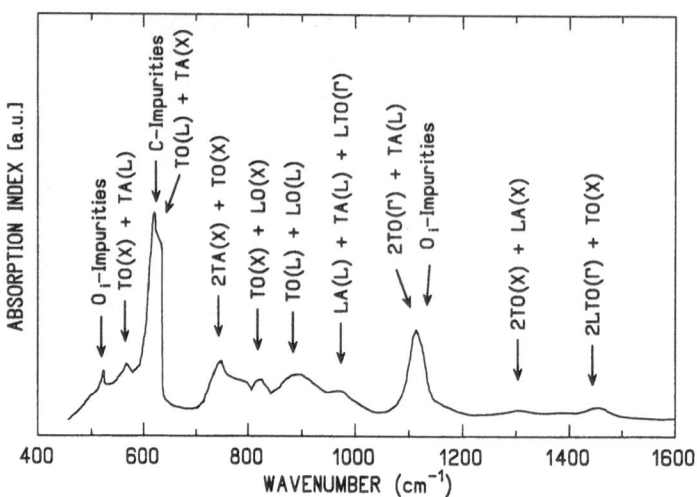

Fig. 5.44. Typical absorption spectrum of a silicon wafer in the mid IR [5.33]. Absorption peaks are due to multiphonon processes and impurities. (Assignment of multiphonon bands according to [5.59])

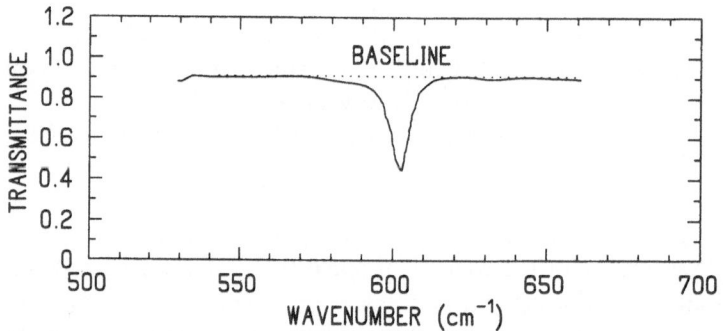

Fig. 5.45. Transmittance of a silicon sample with carbon impurities compared to that of a reference wafer with low carbon content [5.62]. The spectrum clearly reveals the absorption line at $607\,\mathrm{cm^{-1}}$ which is due to substitutional carbon in silicon

a round robin test [5.61]

$$f_C = 1.0 \cdot 10^{17}\,\mathrm{cm^{-2}}\,. \tag{5.111}$$

The lowest concentration measurable at room temperature is about $1 \cdot 10^{16}\,\mathrm{cm^{-3}}$. At low temperatures the sensitivity turns out to be much higher.

5.6.2.2 Interstitial Oxygen in Silicon. Most of the infrared absorption modes of oxygen in Si can be related to fundamental modes of the Si_2O quasi-

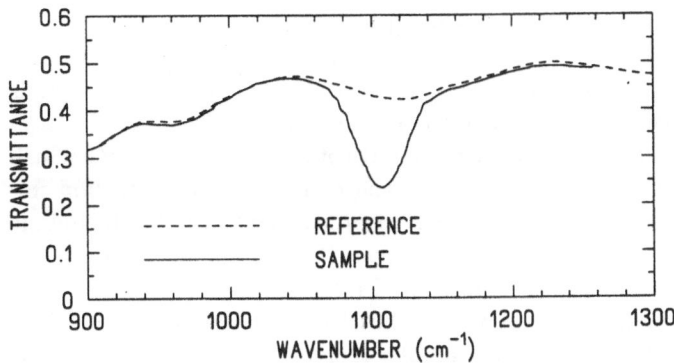

Fig. 5.46. Absorption peak of interstitial oxygen in silicon at 1107 cm^{-1}. Transmittance spectra of a Czochralski sample and a float-zone reference are compared [5.63]. The remaining absorption in the reference sample is due to the weak multiphonon absorption (2TO(Γ)+TA(L), cf. Fig. 5.44) in silicon. The Czochralski sample exhibits additional absorption due to interstitial oxygen

molecule [5.33]. Usually the strong mode at 1107 cm^{-1}, which is due to an antisymmetric stretching of the Si–O–Si, is used for concentration determination (see Fig. 5.46).

If corrections due to multiple reflections are taken into account, a reproducibility of better than 1% is achieved. Concentrations higher than $2.5 \cdot 10^{15}$ cm^{-3} can be detected. The conversion factor for interstitial oxygen is according to an international round robin test [5.63, 5.64]

$$f_{O_i} = 3.14 \cdot 10^{17}\,\text{cm}^{-2}. \qquad (5.112)$$

5.6.2.3 Oxygen Precipitates. Precipitated oxygen causes characteristic absorption lines at 1100 cm^{-1} and 1230 cm^{-1}. As the first one interferes with the O_i-absorption at 1107 cm^{-1}, the second line usually is used. However, at low temperatures because of narrowing effects and frequency shifts it is possible to distinguish between interstitial and precipitated oxygen [5.65]. This absorption is also utilized in IR microscopy where it was shown that oxygen precipitates are not distributed uniformly within the wafer but rather form clusters with an extension of the order of 100 μm [5.66].

5.6.3 Impurities in Thin Layers

The determination of the impurity species and their concentration in epitaxial layers is usually more difficult than in substrates because of their much thinner thickness. Conventional transmission experiments are not appropriate, since the absorption in the substrate strongly dominates the resulting spectra. There are two possibilities to overcome this problem.

For thick epitaxial layers ($d \approx 100\,\mu$m) the wafer can be cut into a slice and the transmission is measured with an infrared microscope perpendicular to the growth direction [5.67]. However, the disadvantage of this technique is that only comparatively thick epi-layers can be investigated.

The second technique involves attenuated total reflection (ATR) which was recently used for the measurement of the concentration of carbon impurities in thin GaAs-epi-layers ($d = 1.0\ldots1.5\mu$m) [5.68]. In this technique the internal reflectance of the interface between the sample and an optically denser medium (e.g. germanium) is measured (see Fig. 5.47).

The angle of incidence is chosen such that total reflection occurs at the interface between ATR-element and sample. Inside the sample, the electric field decays exponentially with distance from the interface, and thus most of the absorption occurs in the epi-layer [5.26]. (In this ATR experiment surface polaritons (cf. Sect. 5.1.9) cannot be excited since there is no coupling gap between the ATR-element and the sample.) Thus this method is particularly sensitive to impurities in the layer and not in the substrate. The sensitivity is further increased by a multiple reflection technique. In GaAs epi-layers carbon concentrations from $3 \cdot 10^{16}\,$cm^{-3} to $10^{19}\,$cm^{-3} were measured by using this method.

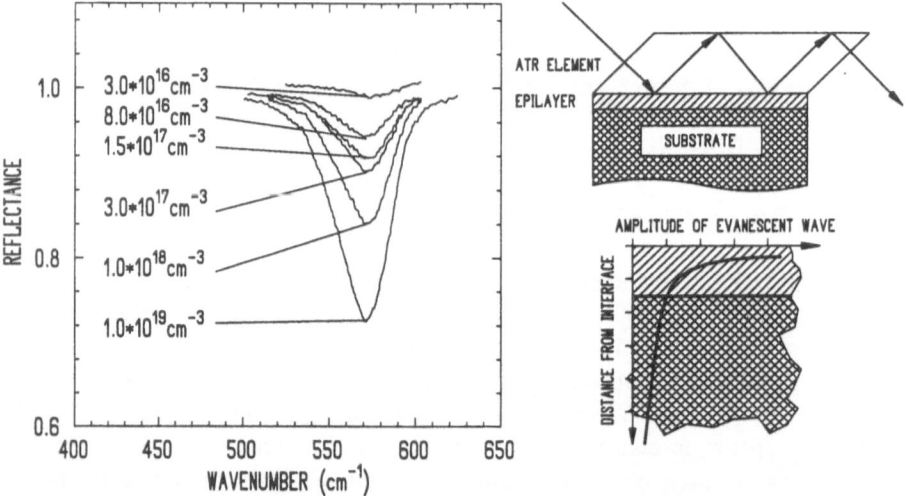

Fig. 5.47. Principle of the determination of impurities in thin layers by means of the ATR method. The reflectance of the interface between sample and ATR-element is measured. The fields decay exponentially inside the sample. Thus at sufficiently high angles of incidence only the film is investigated, because the electric field inside the substrate is already very small. The spectrum on the l.h.s. clearly reveals the reflectance dip at $573\,$cm^{-1} characteristic for the C-Ga vibration. Impurity concentrations between $3 \cdot 10^{16}\,$cm^{-3} and $10^{19}\,$cm^{-3} were investigated [5.68]

5.7 Shallow Donors and Acceptors

Shallow impurity states in semiconductors have been the subject of intensive investigations since the early days of semiconductor physics and numerous experimental and theoretical studies were devoted to this subject [5.69–5.71]. Since the binding energies range from less than 1 meV to about 50 meV the spectral range covered by Fourier transform spectrometers makes these instruments excellent tools for studying transitions between the ground and excited states. Furthermore, magnetic fields are quite often used for the study of the energy spectrum of shallow impurities. Such fields shrink the impurity wave functions and thus tend to enhance their energies.

Shallow impurities are characterized by impurity wave functions which extend over many lattice sites. In the simple hydrogen like model, the binding energies of the H-atom are scaled down by the effective mass of the electrons in the semiconductors and by the square of the dielectric constant which appears in the denominator of the expression for the hydrogen like binding energies. For the parameters of GaAs, a Bohr radius for donors of about 10 nm results, i.e. it is nearly 20 times larger than the GaAs lattice constant. Such considerations are used to justify the use of an effective mass Hamiltonian with a coulombic potential.

Since many excellent reviews exist on both experimental and theoretical investigations of shallow impurities [5.69–5.72] in this Section only a brief survey is presented.

Shallow impurity states have been experimentally investigated using absorption and photoconductivity spectroscopy in the infrared, Raman spectroscopy and luminescene. Apart from the influence of a magnetic field also the effect of uniaxial pressure, i.e. of elastic strain on impurity binding energies and their splitting has been studied, too. Of course for the interpretation of e.g. the different binding energies of group V donors in Si (P, As, Sb, and Bi) more refined models have to be used, which include the description of the central cell effects. Furthermore, in many-valley semiconductors like Si or Ge, the electron effective mass is not a scalar but a tensor. The conduction band minima (6, or 4, respectively) are located along the Δ-directions or at the L-points of the Brillouin-zone, respectively. A substitutional donor ground state should thus exhibit an either 6 or four fold degeneracy of the ground state. This degeneracy is lifted by the crystal potential. Considering that the symmetry of the impurity site is T_d, e.g. in the case of a donor in Si the $1s$ ground state multiplet splits into a singlet $1s(A_1)$, a doublet $1s(E)$ and a triplet $1s(T_2)$ component, where the notation A_1, E and T_2 refers to irreducible representations of T_d. For the calculation of the binding energies of holes in elemental group IV, III-V or II-VI semiconductors the complex valence band structure has to be taken into account.

Using Fourier transform spectrometers for spectroscopic investigations of impurity states requires either the measurement of transmission spectra or of photoconductivity spectra. In the latter method, photons are resonantly absorbed, raising the carriers from the ground states to excited states or to the conduction or valence band continuum. From the excited states the carriers can be further excited thermally into the conduction band, and this process changes the electrical conductivity of the sample. In absorption measurements donor concentrations down to $10^{11} \ldots 10^{12}$ cm^{-3} are detectable [5.73]. With the photoconductivity (or better the photothermal ionisation spectroscopy), where the sample is used as its own detector in the FTIR spectrometer, even lower impurity concentrations can be detected (less than 10^{11} cm^{-3}, for details see e.g. [5.74]).

5.7.1 Donors and Acceptors in Bulk Materials

A classical example for the investigation of donors and acceptors by means of Fourier transform spectroscopy are the spectra of group V donors in silicon. They are representative both for the method and for semiconductor materials. Figure 5.48 shows the absorption spectra of a phosphorus doped Si sample with concentration of $1.3 \cdot 10^{14}$ cm^{-3}. The measurements were made in a Fourier spectrometer with a resolution better than 0.06 cm^{-1} [5.75] at liquid helium temperature. In the frequency range $32 \ldots 45$ meV the spectra show a series of sharp peaks, which are due to the electric dipole transition from $1s(A_1)$ to the excited p states, i.e. the Lyman transition $1s(A_1) \rightarrow 2p$, $3p, \ldots$. The labeling of the peaks describes the final state of the electron. The

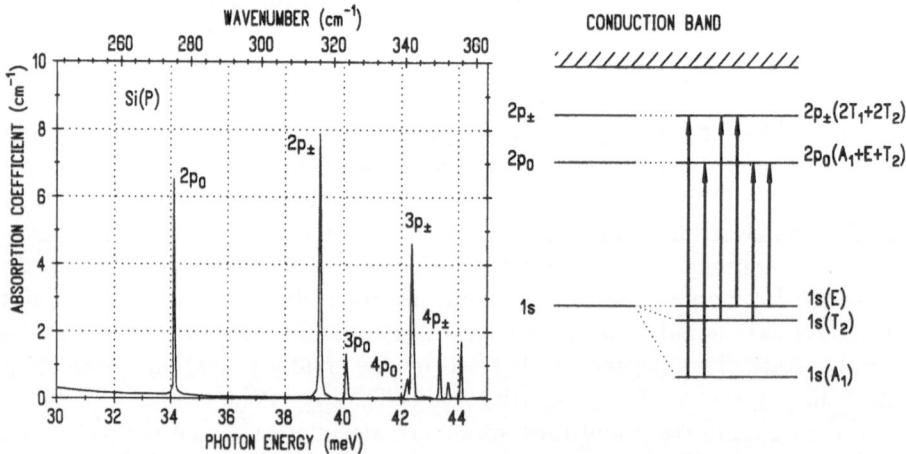

Fig. 5.48. l.h.s.: absorption coefficient vs. photon energy of is $n \approx 1.2 \cdot 10^{14}$ cm^{-3} [5.75],r.h.s.: Energy binding scheme of group V impurities in Si illustrating valley orbit splitting. It shows the transitions from the $1s$ states to the $2p$ states. In this spectral range only the transitions from the state $1s(A_1)$ can be observed [5.75]

splitting of the p state is a consequence of the fact that the effective mass is not a scalar but a tensor. It was observed that the relative intensities of the spikes of Si(P) and for comparison Si(As) samples and the spacing between them are identical within the experimental errors, however, an energy shift occurs. This shift in the absorption spectra arises from the chemical-species-dependent binding energy of the s state of the group V impurities in Si.

Many compound semiconductors (e.g. GaAs, InP, InSb, InAs, CdTe) have band minima at $k = 0$, i.e. at the center of the Brillouin zone, and their effective electron mass is isotropic. The absorption spectrum is then particularly simple because it resembles a strictly scaled down version of the Lyman spectrum of a H-atom (apart from central cell effects). In narrow gap semiconductors like InSb with extremely small effective masses very large Bohr radii result. The wave functions of the donors overlap then already for quite moderate donor concentrations and the ionisation energy decreases. The presence of a magnetic field (magneto-impurities [5.72]) splits the impurity levels resulting in quite complicated patterns of magnetooptical transitions. Theoretically the magneto-impurity problem does not have exact closed solutions even for the case of a parabolic and isotropic energy band and several approaches have been suggested in the literature which are reviewed by Zawadzki [5.72].

5.7.2 Donors and Acceptors in Quantum Wells

Like in bulk materials, information on unintentional and on intentional impurities in modulation doped heterostructures and quantum well structures is both of scientific and of technological interest. Due to the change of the band edges as a function of the spatial coordinate on both sides of a semiconductor interface, in such systems not only the control of the number and of the chemical species of impurities is important but also the control of their location. Indeed, advances in the MBE and MOCVD growth techniques have made it possible to achieve even delta doping, i.e. to locate the dopants during a growth sequence in the desired atomic layer.

In this respect two main issues are of interest:

— modulation doping: shallow impurities are incorparated in the barrier layers of QW or heterostructures separated from the hetero-interface by an intentionally undoped spacer layer.
— confinement effects: if shallow impurities are incorporated in the well layers then confinement effects occur whenever either the width of the confining potential or the distance of the impurity ion from the potential barrier are comparable to the extent of the impurity wavefunction.

Most of the experimental and theoretical work on shallow impurities has been performed on GaAs/GaAlAs structures. Far infrared magnetospectroscopy was almost exclusively used for the experimental investigations.

For the calculation of the energy states of hydrogen-like donors the effective mass approximation is used and the confining potential appears in the Hamiltonian. For the solution of the Schrödinger equation for this problem, a number of variational calculations [5.76–5.80] have been carried out for GaAs/GaAlAs QWs. The main qualitative results are:

– For an impurity which is centered in a well, the electron binding and transition energies increase with decreasing well width, because due to the additional confining potential the electron is on the average closer to the donor ion than in the 3D case.
– For constant well width, the binding energy decreases as the impurity ion is incorporated closer to a well edge. It decreases even further if the impurity is incorporated into the barrier, because the electron, which remains in the well in all cases, has a large average distance from the positive ion.

For the calculation of acceptor states, like in bulk material, a simple effective mass approximation is not possible but one has to start from the Luttinger Hamiltonian to which the Coulomb term and additionally the confining potential has to be added. Masselink et al. [5.82, 5.83] have solved the corresponding Schrödinger equation variationally. The main results are the following: in GaAs/GaAlAs structures the acceptor ground state splits (which becomes observable for GaAs well widths less than 10 nm) and the acceptor binding energy increases for the center doped case. Due to the fact, that already in bulk GaAs the acceptor ground state wave function is much less extended than the corresponding one of the donors, the confinement effect is less pronounced.

The theoretical results agree in general with the experiments which have been performed on well-center doped, top-edge and bottom-edge doped GaAs/GaAlAs MQW's [5.81, 5.36].

Figure 5.49 shows transmission spectra of a n-doped GaAs/GaAlAs MQW sample with GaAs well widths of 21 nm. The center-doped sample has a sharp dip at 162 cm^{-1} which is due to the $1s$-$2p$ (m_l=1) transition. The edge doped samples show completely different spectra. The spectrum of the bottom-edge doped sample exhibits dips at 160 cm^{-1}, 143 cm^{-1} and a long tail down to 110 cm^{-1}. The top-edge doped sample shows a line near 110 cm^{-1} and an absorption tail up to 160 cm^{-1}. In fact these results are in agreement with the trends on the binding energies outlined above. The more subtle line shape features also apparent from Fig. 5.49 were attributed [5.81] by the redistribution of the Si-donors along the growth direction during growth, i.e. in the bottom edge doped sample the actual density of the Si dopants is smaller than intended and the peak of the impurity distribution is shifted towards the center of the well, which explains the high frequency tail in the spectrum. The amount of impurity redistribution was estimated by the authors to be of the order of 3 to 4 nm. For further details the reader is referred to [5.36, 5.72].

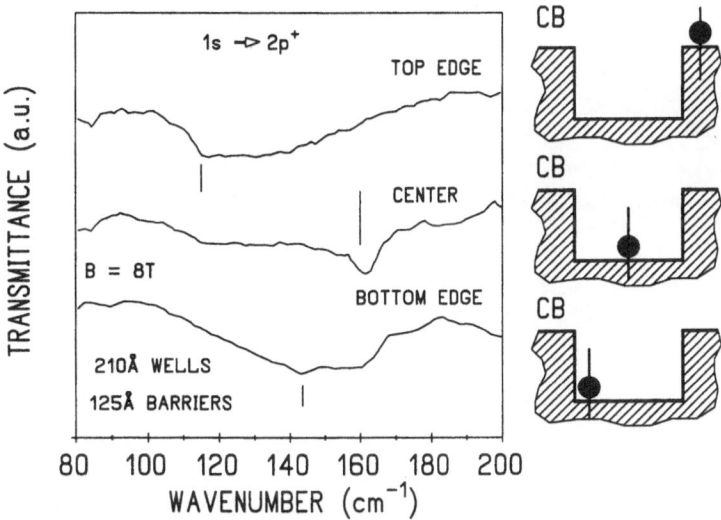

Fig. 5.49. Transmission spectra of Si-doped GaAs/Al$_{0.3}$Ga$_{0.7}$As multiple quantum wells with well width of 21 nm and 12.5 nm barrier width [5.81]. Different samples were doped at the center of the well, at the top and at the bottom edge of the barrier. Transmission spectra were recorded at 4.2 K in an 8 T magnetic field along the well axis. The dip for the center-doped sample at 162 cm^{-1} is attributed to the $1s \rightarrow 2p^+$ transition. The edge-doped samples exhibit broad absorption bands, and transitions at lower frequencies are observed

5.8 IR Characterisation of Porous Silicon Layers

Porous silicon, which is quite easily produced from standard silicon wafers by an electrochemical etching process using hydrofluoric acid, is a heterogeneous material which has received a lot of attention in recent years in respect to possible applications in semiconductor devices. The reason for this is the discovery of very strong photo- [5.84] and electroluminescence [5.85, 5.86] in the visible at room temperature. A realisation of optoelectronic devices based on silicon exclusively seems to be possible and hence this new kind of material is studied at present with great effort [5.87].

There is no generally accepted explanation for the occurrence of enhanced luminescence in porous silicon layers up to now but it is clear from the current discussion [5.88] that it must be connected to the extraordinary microscopic structure of the material (see Fig. 5.50). The samples showing the brightest luminescence are those with a high porosity (70... 80%) and remaining silicon structures with dimensions in the nanometer range, going along with a high specific surface area of several hundreds m^2 per g [5.90]. Similar nanostructures are known from porous glasses and aerogels [5.91].

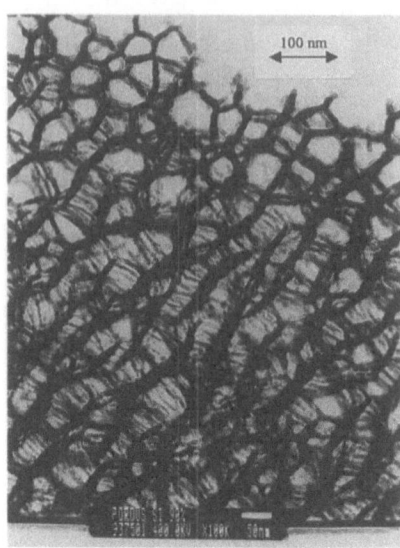

Fig. 5.50. TEM view of nanostructured porous silicon. The dark structures are the remaining silicon walls whereas the bright areas indicate the empty pores [5.89]

With this microstructure in mind some suggested "luminescence schemes" seem plausible: on one hand the occurrence of light emission in the visible at room temperature could be related to quantum size effects in the thin silicon walls which are considered to be quantum wires (or loosely connected quantum dots) with electronic states quite different from those of bulk silicon. Another assumption is the existence of special surface states enhancing the radiative recombination of electron-hole pairs. The formation of Si-O-H compounds (siloxenes) in the pores which show a luminescence behaviour quite similar to the one observed in porous silicon has been discussed earlier [5.92] but this model has lost a lot of attraction since bright luminescent porous silicon layers have been formed with no observable oxygen content [5.93].

Although a completely satisfying theory for the intense light emission has not been established up to now the development of techniques for the fabrication of silicon optoelectronic devices has already begun. Here of course one can profit from all the knowledge obtained in "normal silicon technology". It has been verified for example that the usual photolithography steps can be used to produce small porous structures (e.g. light emitting pixels for screen applications) and the necessary contacts for electroluminescent devices [5.94]. Recently it has been utilized that the current density during etching influences the porosity of the obtained porous layer: by a variation of the current density stacks of many thin successive layers with different porosities can be produced [5.95]. This can be used for an optimized design of a device. As an example we will show at the end of this section porosity superlattices with fascinating properties in the visible and infrared.

The search for a convincing luminescence model as well as the investigation of the necessary technological steps for porous silicon devices can both benefit from IR spectroscopy. As with homogeneous layers one can study in the infrared region the characteristics of the free carriers and detect atoms and molecules adsorbed on the huge surfaces in the pores. In addition one can determine layer thicknesses from interferences which are observable in many cases. Several authors have reported on IR transmittance measurements on porous silicon samples [5.96–5.98]. Transmittance measurements usually show absorption bands which can be interpreted easily if the thickness of the porous layer is large enough so that interferences are not observed in the experiment. A striking disadvantage is the need of a sample preparation: usually there is a non-transparent metallic contact layer covering the wafer backside (this is needed for the etching process) that has to be removed before the transmittance measurement. Alternatively the porous layer itself can be separated from the substrate if it is thick enough to be handled as a free-standing sample.

A more direct method is reflectance spectroscopy where samples can be taken as prepared [5.99,5.100]. An interpretation of the obtained spectra must then be done on the basis of model calculations with parameters which are varied until a good fit to the experimental data is achieved. Examples will be given for the standard reflection technique and ATR.

5.8.1 Effective Medium Theories

A complication arises with inhomogeneous materials since we cannot use susceptibilities the same way as for homogeneous materials. Obviously porous silicon is a mixed material consisting of the pores and a silicon skeleton with structures in the nanometer range. For light waves with wavelengths much larger than the dimensions of the inhomogeneities the heterogeneous material can be treated as a homogeneous one with a so-called effective dielectric function which depends on the dielectric functions of the constituents and the micro topology of the system. A detailed review on the use of effective dielectric functions in optical spectroscopy is given in [5.101].

The porous silicon is treated here as a two-phase composite: silicon microcrystals with dielectric function ε are embedded in an air matrix (the pores, dielectric function $\varepsilon_M = 1$). The calculation of the effective dielectric function ε_{eff} now depends on the assumption about the micro geometry: here effective medium theories differ from each other. The simplest choices are the well known Maxwell Garnett [5.102] and Bruggeman [5.103] formulas, which just use one single parameter to describe the topology, namely the silicon volume fraction f (obviously one has f=1-porosity):

$$\text{Maxwell Garnett:} \qquad \frac{\varepsilon_{\text{eff}} - \varepsilon_M}{\varepsilon_{\text{eff}} + 2\varepsilon_M} = f \cdot \frac{\varepsilon - \varepsilon_M}{\varepsilon + 2\varepsilon_M} \qquad (5.113)$$

$$\text{Bruggeman:} \qquad f \cdot \frac{\varepsilon - \varepsilon_{\text{eff}}}{\varepsilon + 2\varepsilon_{\text{eff}}} + (1 - f)\frac{\varepsilon_M - \varepsilon_{\text{eff}}}{\varepsilon_M + 2\varepsilon_{\text{eff}}} = 0. \qquad (5.114)$$

These concepts are often used, but in some cases more than just one parameter is needed to characterise the topology influence on the effective dielectric function. In the case of IR spectra of porous silicon layers the most important parameter of the topology is the grade of "connectivity" of the silicon component. Since there are silicon walls around the pores, the silicon must be more or less percolated. In the framework of the so-called Bergman representation [5.104–5.107] which is the most general ansatz for any effective medium theory the degree of percolation in the theory is a free parameter that can easily be adapted to experimental data. This is not possible for the Maxwell Garnett and the Bruggeman theory.

In the following model calculations, therefore, the Bergman representation is used. Bergman has shown that the effective dielectric function of any inhomogeneous medium can be written in the form

$$\varepsilon_{\text{eff}} = \varepsilon_M \left(1 - f \int_0^1 \frac{g(f,n)}{\frac{\varepsilon_M}{\varepsilon_M - \varepsilon} - n} \, dn \right). \tag{5.115}$$

The function $g(f,n)$ which is called spectral density describes the topology of the system: it is a positive-valued distribution function, which is normalised to unity and whose first moment is equal to $(1-f)/3$ for isotropically disordered systems. The spectral density only depends on the micro geometry and not on the material properties. Unfortunately $g(f,n)$ cannot be calculated exactly even if the topology is well known. On the other hand it was shown that in many cases very simple forms of the spectral density result in a very good agreement with experimental spectra [5.106, 5.107].

One feature has turned out to be very important, namely the above mentioned description of the percolation strength of the porous system. To show this it is useful to separate from the spectral density a term proportional to a δ-function at $n = 0$ and a finite function $g'(f,n)$:

$$g(f,n) = g_0(f)\delta_+(n) + g'(f,n), \tag{5.116}$$

The function $g_0(f)$ now describes the strength of the percolation, which can be demonstrated most easily considering a porous system of a conductor with electrical conductivity σ in a vacuum matrix: in this case the effective electrical conductivity σ_{eff} of the system according to the Bergman representation (which holds for electrical conductivities as well as for dielectric functions) is given by [5.101]

$$\sigma_{\text{eff}} = f g_0(f)\sigma. \tag{5.117}$$

Since $f g_0(f)$ determines the effective conductivity of the system which is certainly related to the "connectivity" of the solid component in the porous system this product is called the percolation strength. It is clear from the formula that a measurement of σ_{eff} gives the percolation strength if the conductivity σ of the compact material is known.

Systems with the same volume fraction, but with different micro geometries, obviously differ in their electrical conductivity and hence must be described with different values of $g_0(f)$. Simple mixing equations oversimplify the situation: the Maxwell Garnett formula (designed to describe systems of well separated spheres with low volume fraction) gives vanishing percolation for any volume fraction, the Bruggeman theory has zero-percolation for volume fractions below the percolation-threshold 1/3 and $g_0(f) = (3f - 1)/(2f)$ above [5.101]. To describe real systems a theory is needed with a flexible choice for $g_0(f)$ which can be done easily using the Bergman representation with a reasonable guess for the remaining function $g'(f, n)$. The meaning of the latter will not be discussed in this work since it is not very important in the discussion of porous silicon [5.106].

5.8.2 Examples

Using the Bergman representation a quantitative interpretation of IR spectra of porous systems is possible which is demonstrated in the following examples.

First we show the reflectance spectrum of a sample with a porosity of 36%, corresponding to a volume fraction of silicon of 0.64 (Fig. 5.51). The dominant feature of the spectrum is clearly the appearance of well-pronounced regular interference fringes indicating well defined boundaries between porous layer,

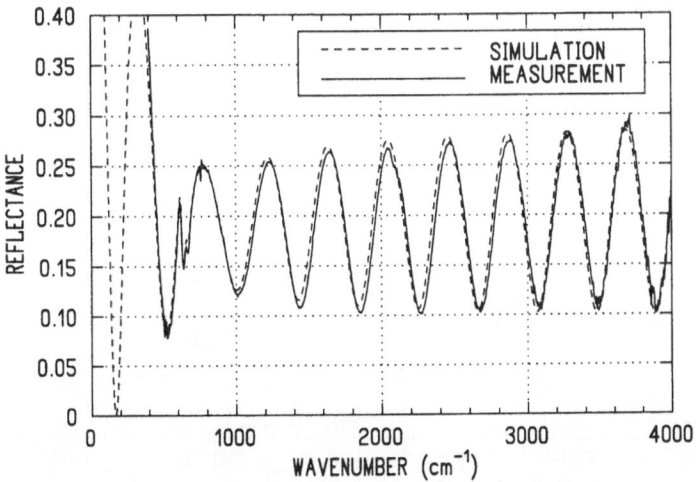

Fig. 5.51. Measured reflectance spectrum of a porous silicon layer (volume fraction of the solid component is 0.64) on a highly p-doped silicon substrate (solid line) and a simulated spectrum using the Bergman theory for effective dielectric functions with the spectral density shown in Fig. 5.53 (dashed). Besides the well-pronounced interferences there is a strong absorption structure around 650 cm^{-1} which is shown enlarged in Fig. 5.52

substrate and vacuum respectively. In the best theoretical fit which is shown as well in Fig. 5.51 a thickness of $4.85 \, \mu m$ was used.

In addition there is a quite strong structure around 650 cm^{-1} which looks like two absorption lines. These bands have been reported by several authors [5.96, 5.97, 5.92] and usually they are interpreted as Si-H vibrations. On the other hand, the well known Si-H vibrations around 2100 cm^{-1} are very weak indicating a low hydrogen concentration. Also peaks at 620 and 650 cm^{-1} are observed in Raman experiments [5.108] even after a heat treatment of samples to remove adsorbed hydrogen. As an alternative interpretation therefore the absorption bands can be considered as enhanced and slightly shifted multiphonon processes in silicon (see also Fig. 5.44). The enhancement is then due to surface effects (an enhancement of multiphonon absorption can also be found on rough silicon surfaces). It should be noted that in the porous microstructure the local fields can be very inhomogeneous and probably excitations occur that are not allowed in bulk material (like the efficient luminescence itself). A good example for such a mechanism is the grating coupler discussed earlier (see Fig. 5.40).

Independent of any interpretation, the absorption-like structures can be described as two additional oscillators in the dielectric function of the silicon component of the porous medium. This is shown in Fig. 5.52 where we have enlarged the relevant part of Fig. 5.51 and plotted in addition the used dielectric function (imaginary part) of silicon with two harmonic oscillators at 625 and 665 cm^{-1}. Figure 5.53 shows the spectral density.

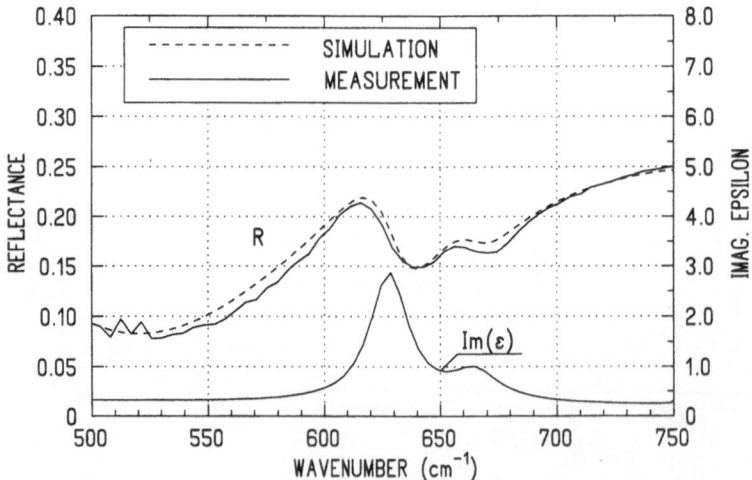

Fig. 5.52. Enlargement of Fig. 5.51. In addition the imaginary part of the dielectric function of the silicon constituent in the porous layer is shown. The two absorption bands are very probably due to multiphonon processes which are enhanced by the numerous surfaces in the pores (see text)

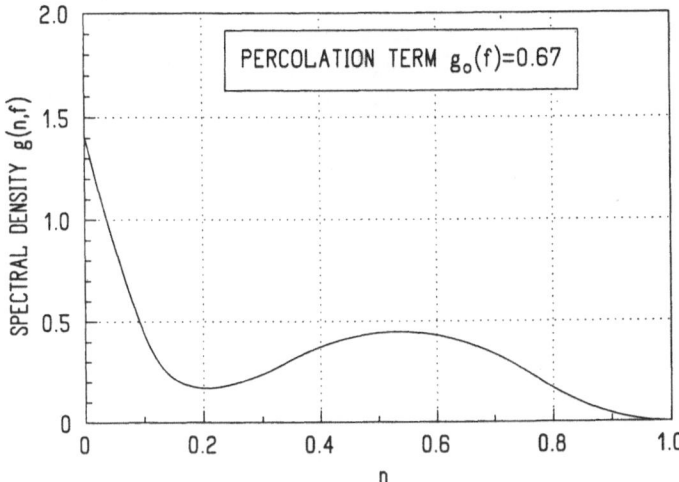

Fig. 5.53. Spectral density used to obtain the fit from Fig. 5.51. It is a smooth and simple function with a large percolation term indicating a strong silicon network

To demonstrate the significance of fitting parameters to experimental spectra Fig. 5.54 shows a comparison of three model calculations. In Fig. 5.54a we have used the same Drude parameters for the free carriers in the porous layer that have been fitted to the substrate. In (b) the carrier concentration was set to zero (i.e. the porous silicon is treated as being completely depleted) and the best fit in (c) has been achieved with large damping parameters for the free carriers, where the high damping can be explained as additional surface scattering. This is known as the classical size effect [5.35].

In samples with higher porosities no free carriers at all are needed for a good fit. Instead more absorption bands can be found (Si-H and Si-O vibrations). Figure 5.55 shows a reflectance spectrum of a high porosity sample (71%) which is typical for strongly luminescent samples. For the solid component of the porous silicon we have used the dielectric function given in Fig. 5.56 where the assignments of the bands are given too. We would like to stress again that the vibrations are quite strong. For this reason a straightforward evaluation of transmission experiments (which is working well in the chemical analysis of organic materials with weak oscillators [5.109]) is not an adequate method in the case of porous silicon layers. Instead the dielectric function is the suitable quantity for a discussion of frequency positions and oscillator strengths.

The intermixture of interferences (carrying layer thickness information) and absorption bands in reflectance spectra can be avoided using the above mentioned ATR technique with variable angle of incidence. As Fig. 5.57 shows the angle of incidence determines the penetration depth of the radiation and hence can be used to switch on and off the interferences. In a first step the strengths of the absorption bands in model calculations can be adjusted to

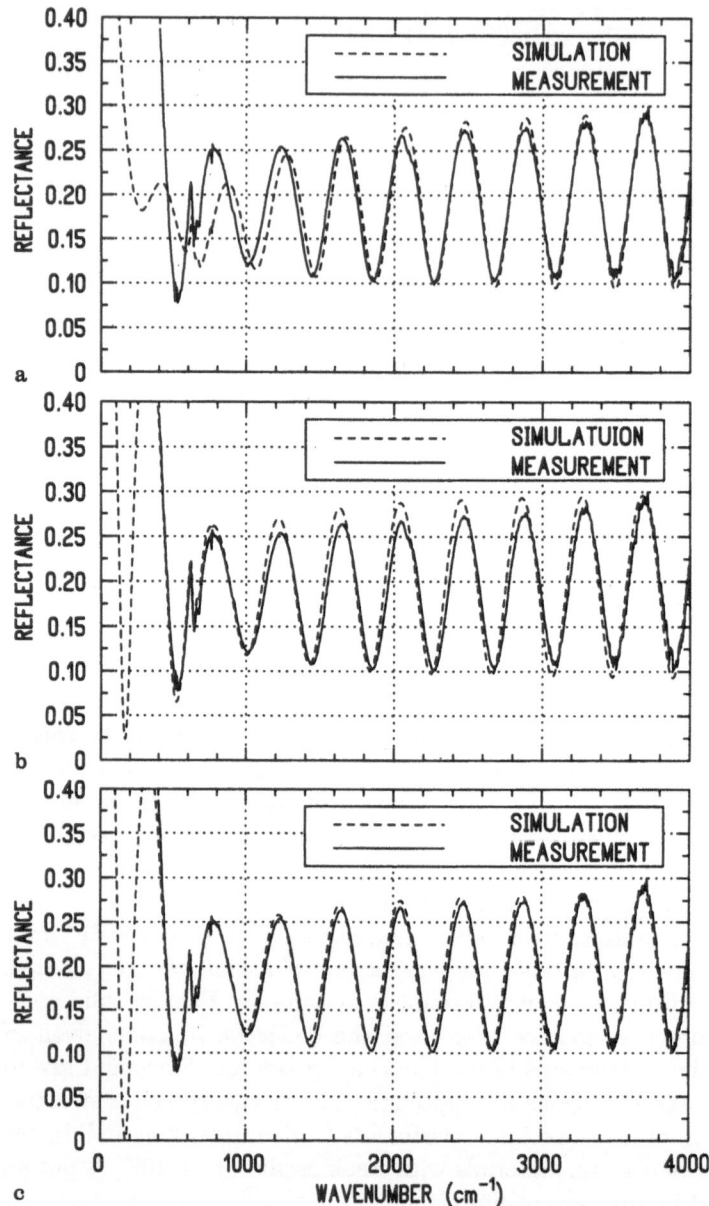

Fig. 5.54. Model calculations with different parameters for the free carriers in the film. **a** Unchanged substrate parameters, **b** no free carriers (porous silicon completely depleted), **c** carrier concentration as in substrate, but very high damping (classical size-effect). Clearly the last fit is best, indicating that there are still free carriers in the silicon skeleton (no complete depletion). Due to the numerous surface collisions, the motion of the carriers is strongly damped

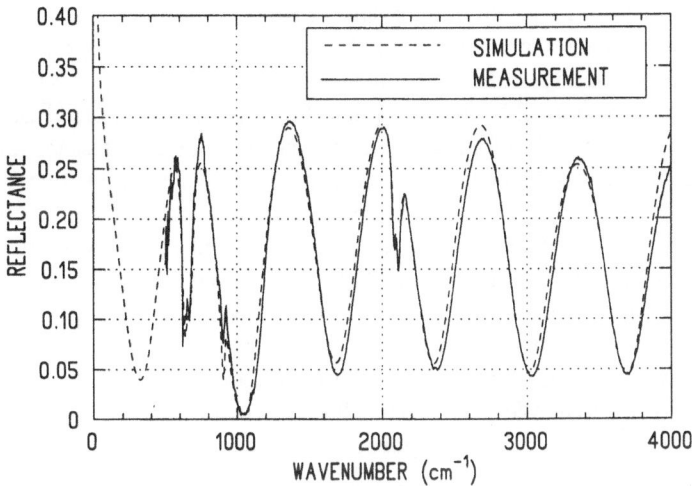

Fig. 5.55. Reflectance of a typical porous silicon sample where the volume fraction of the solid component is 0.29 (Solid line: measurement, dashed line: theory). Samples of this kind show quite strong (usually red) luminescence

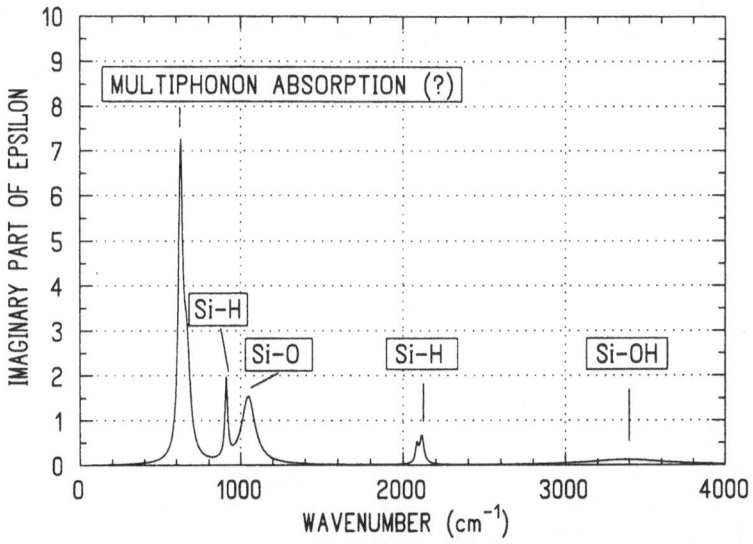

Fig. 5.56. Imaginary part of the dielectric function used to obtain the fit shown in Fig. 5.55: the microscopic vibrations are assigned to the bands

an ATR spectrum with oblique incidence of light, and afterwards the layer thicknesses can be fitted using the "normal incidence spectrum". Of course this procedure can work only for porous layers homogeneous in depth. For inhomogeneous cases recording ATR spectra with a variety of incidence angles can be used as a depth profiling techniqueas long as a careful data evaluation is provided. Besides the angle of incidence the refractive index of the ATR prism determines the penetration depth which is typically in the range from 0.1 to 5 μm. The spectra shown in Fig. 5.57 have been obtained using a KRS6 prism. Other widely used materials are undoped silicon, germanium or KRS5.

Now we turn to a more technological example and demonstrate the use of IR spectroscopy to monitor photolithography steps needed for contact formation on porous silicon layers [5.94]. As has been mentioned before, contacts on top of porous layers must be provided for electroluminescent samples. Under discussion are thin gold or indium-tin-oxide (ITO) layers, but also conducting polymers are a possible choice.

Here we present the basic steps for the deposition of a gold contact on top of a porous silicon layer. Figure 5.58 shows various stages of the process and the corresponding IR spectra. We start with a high porosity layer (Fig. 5.58a similar to the one shown in Fig. 5.55. The porous layer has a thickness of about 5 μm and is covered in the second step (Fig. 5.58b by a layer (thickness about 3 μm) of organic photoresist. The IR spectrum now shows many bands, due to vibrations in the photoresist. At the locations of the desired gold contact the photoresist is exposed to light, developed and removed. Figure 5.58c shows an incomplete removal of the photoresist (whose absorption bands still can be observed in the spectrum) and a strongly modified (i.e. damaged) porous layer after this processing step. Now gold is evaporated onto the system and shows its metallic reflectance in Fig. 5.58d. The remaining photoresist (which

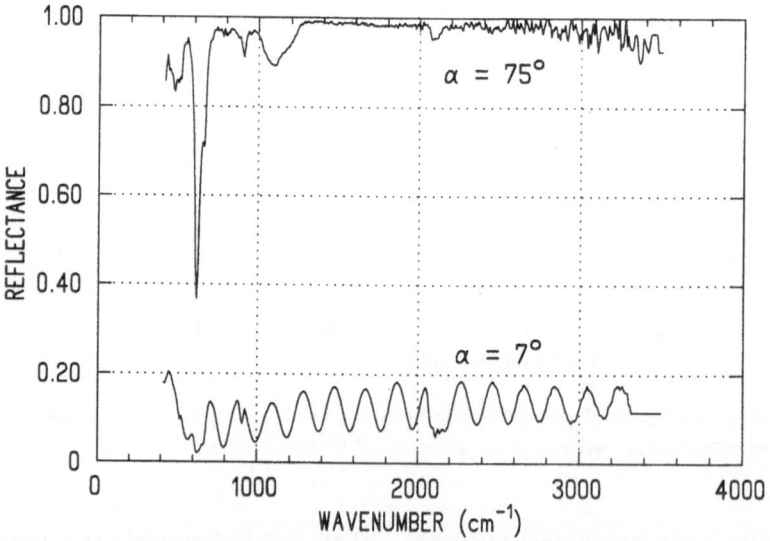

Fig. 5.57. ATR spectra of a 70% porosity sample (layer thickness: 17 μm): the top curve corresponds to an incidence angle of 75°, whereas the bottom curve has been recorded almost with normal incidence of light (7°). See text for details

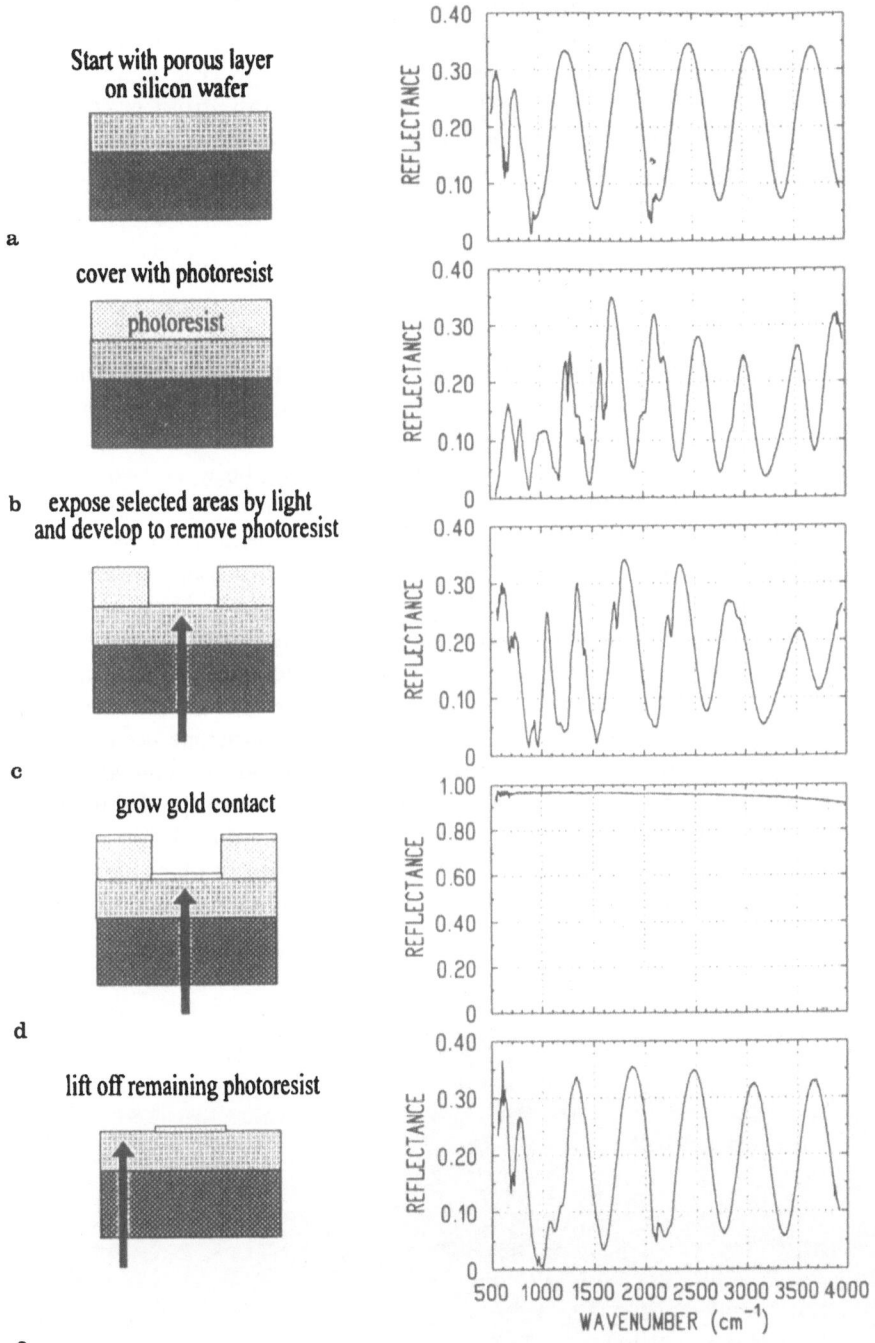

Fig. 5.58. Photolithography steps for contact formation on top of a porous silicon layer investigated by IR spectroscopy. The positions where the reflectance spectra have been taken are indicated by the arrows. See text for discussion

was not exposed to light) is finally removed by the usual lift off procedure and the initial porous layer reappears (Fig. 5.58e, this time only minor changes appear in the IR spectrum compared to the first stage (Fig. 5.58a. From this one can learn that the critical step in contact formation is the light exposure and photoresist developing. It should be noted that the porous layer is modified but usually does not loose its luminescence properties completely [5.94]. Nevertheless the observed modifications must be taken into account in a device design.

We close this section presenting a quite new development of porous silicon technology, namely the production of porosity superlattices [5.95]. These can be obtained very simply by a periodic variation of the current density during the etching process: since the achieved porosity depends on the current density it can be controlled by the latter. With a programmable current source almost any porosity profile can be produced. Here we present porosity superlattices which have been generated by a switching between two distinct current densities. This leads to a pair of porous silicon layers with different porosities that has been repeated up to 60 times. Surprisingly the layers that are produced first do not suffer significantly from the necessary transport of material through their pores during the etching of the layers produced last. Figure 5.59 shows a TEM cross section image of a porosity superlattice.

The optical properties of such porosity superlattices are remarkable: the reflected waves at the numerous interfaces do pile up by constructive interference to very sharp structures at certain frequency positions, as Fig. 5.60 shows. In the infrared even a reflectance equal to unity can be achieved which means that the superlattice reflects as good as the reference (gold mirror) at that particular frequency. For higher frequencies — in the visible and UV — the reflectance peaks decrease in height due to absorption processes. In the visible region porosity superlattices can be used obviously as optical fil-

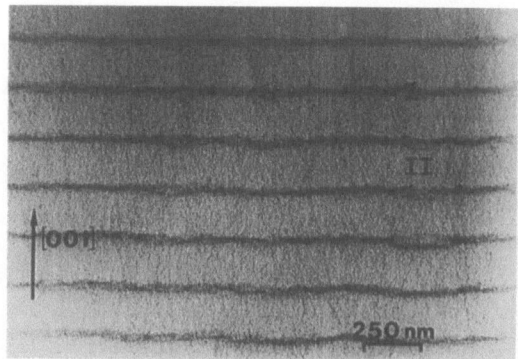

Fig. 5.59. TEM cross sectional view of a Si porosity superlattice. Clearly two layer types with different porosities (denoted as I and II) can be identified [5.95]

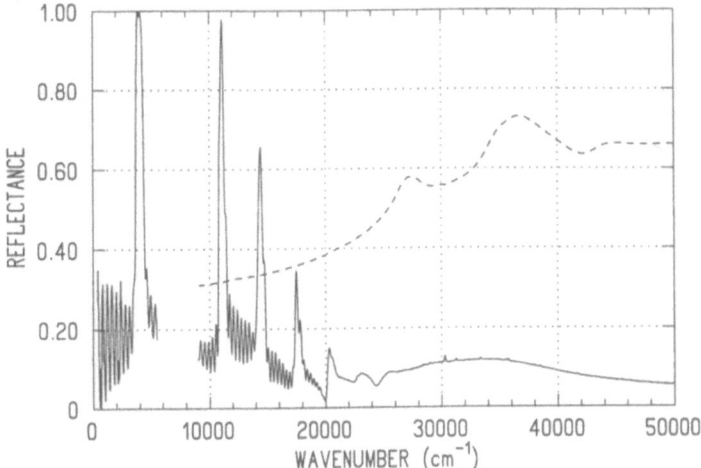

Fig. 5.60. Reflectance of a porosity superlattice from the infrared to the ultraviolet spectral region (solid line). For comparison also the reflectance of the initial silicon wafer is shown (dashed)

ters opening a wide range of possible applications with such kind of filters directly etched into a silicon chip. With this glimpse into the future of silicon optoelectronics we would like to close this section.

5.9 Summary

Infrared spectroscopy is a versatile tool for the nondestructive characterisation of semiconductor layers with respect to their thickness, carrier concentrations, impurities and lattice vibrations.

For the quantitative interpretation of the spectra of reflection and transmission, more or less sophisticated model calculations have to be performed in order to simulate the experimental data. The basis of these calculations are the solutions of Maxwell's equations together with constitutive equations which are derived from simple models (harmonic oscillator, Drude model).

For the instrumentation, almost exclusively Fourier transform spectroscopy is used as the experimental method of choice. The IR techniques and the evaluation of the spectra were considerably refined in recent years: under favourable conditions even the spatial variation of carrier concentration and of the chemical composition are accessible with this technique.

With advances in the growth of heterolayers twodimensional electronic systems became more and more important. Also in these, carrier concentration, their damping as well as impurities are accessible to IR spectroscopy.

Further lateral microstructuring which results in 1D or 0D structures produces inhomogeneous media. In this case for the evaluation of data — as was shown for porous silicon — the use of refined effective medium models based on the Bergman theory is required which is described, in detail too, in this review.

6. High Resolution X-Ray Diffraction

Alois Krost, Günther Bauer, Joachim Woitok

Conventional high resolution X-Ray diffraction has been developed into a powerful tool for the nondestructive *ex-situ* investigation of epitaxial layers, of heterostructures and superlattice systems:

The information which is obtained from diffraction patterns concerns the composition and uniformity of epitaxial layers, their thicknesses, the built-in strain and strain relaxation, and the crystalline perfection related to their dislocation density. Furthermore information on interfaces like interdiffusion and intermixingis obtained under certain circumstances as well. For the analysis of the diffraction patterns from epilayers, heterostructures and multilayers, the kinematical diffraction theory, although being still useful for a quick inspection of the data, in general can no longer be used for the quantitative description of the experiments. Instead dynamical theory is applied which takes into account extinction, multiple scattering, and the slight deviation of the refractive index from one.

The instrumentation has also been improved continuously and simple powder diffractometers using a focussing path for the X-Rays were replaced by double- and triple -axis spectrometers equipped with multiple crystal or channel cut monochromators and analyzers. Apart from investigations under normal Bragg conditions grazing angle incidence techniques both for the determination of layer thicknesses as well as for precise information on lattice constants of thin films have also been employed. X-Ray topography is used for imaging purposes of layers grown on large wafers.

There are a number of excellent reviews on the analysis of epitaxial layers by X-Ray diffraction: In the early reviews by Segmüller and Murakami [6.1] and Paine [6.2] the various diffractometer methods are described whereas in the most extensive one [6.3] the emphasis is on the mathematical description of the strain state of epitaxial layers, of inhomogeneous strain on periodic multilayers as well as on grazing incidence techniques. For the latter technique the latest developments are summarised by Segmüller [6.4] and Schiller et al. [6.5]. Tanner [6.6, 6.7] and Halliwell [6.8] highlight the Double-Crystal Diffractometer (DCD) for the analysis of thicknesses and strains of epitaxial layers whereas Fewster [6.9–6.11] describes the advantage of triple-axis spectrometers. Picraux et al. and Ryan et al. [6.13] gave a complete survey of the amount of information which can be extracted from high resolution X-Ray diffraction (HRXRD) including especially reciprocal lattice scans and the use of triple axis spectrometers whereas Wie [6.14] focus on the characterization of heterostructure interfaces and the analysis of diffuse scattering in the Bragg diffraction geometry. A summary of most recent developments can also be found in [6.15, 6.16].

In the first Section an overview on basic scattering geometries and advanced instrumentation which is nowadays commercially available is given. It follows a Section on kinematical and dynamical theory. The dependence of Bragg reflection intensities on thicknesses is described in Sect. 6.3 and in Sect. 6.4 strain phenomena and partial relaxation of strain are discussed. The application of HRXRD to single heterostructures is treated in Sect. 6.5 and the topic of Sect. 6.6 are multilayers including Ewald sphere constructions for the interpretation of DCD diffractograms, interdiffusion, lattice plane tilts, terracing and mosaic spread. The triple axis spectrometers have led to extensive studies of layers and multilayers through scans in the reciprocal lattice (Sect. 6.7). Further new developments like the analysis of periodic quantum wires or dots and the real time X-Ray diffraction for strain relaxation phenomena and growth processes are described in Sect. 6.8. The three final Sections are devoted to grazing incidence techniques including specular and anomalous reflection as well. Grazing incidence diffraction (GID) is a surface sensitive method which yields information on film properties parallel to surfaces or interfaces and achieves even monolayer sensitivity. New developments include also a contribution of high resolution X-Ray diffraction to topography for the analysis of layers on wafers. We refer to a recent review by Köhler [6.18] on this topic. Both the spectrometric tools as well as the crystalline quality of the materials investigated have reached fantastic limits already in the early 1970s: i.e. the full width at half maximum ($FWHM$) of a Si (422) asymmetric Bragg reflection corresponds to 0.16" [6.17]. In order to perform such a measurement an experimental resolution of 0.016" is required. Such an accuracy corresponds e.g. to an angular position of the earth on its orbit around the sun between two positions which are 3.7 km apart from each other. (The mean distance earth - sun is $1.5 \cdot 10^8$ km).

X-Ray standing wave technique which is especially interesting for surface and interface analysis, since the position of atoms which are either adsorbed or being constituents of thin epilayers can be determined with high accuracy is not presented here. A comprehensive review on this field is given by Malgrange and Ferret [6.19].

New instrumentation and the possibility to use high intensity synchrotron sources will lead in the near future to new *in-situ* applications as well as to transmission spectroscopy (Laue case) of HRXRD. However, the main purpose of this review is the demonstration of methods and results obtained on epitaxial systems with up to date but conventional laboratory instrumentation.

6.1 Principal Scattering Geometries

In this Section basic scattering geometries and advanced instrumentations are described such as double-crystal diffraction, triple-axis spectrometers etc.

6.1.1 $\omega - 2\Theta$-Scan and ω-Scan (Rocking-curve)

Any measurement of lattice spacing is in principle determined by Bragg's law [6.20]

$$2d_{hkl} \cdot \sin\Theta_B = n\lambda \tag{6.1}$$

where d_{hkl} is the spacing of lattice planes with Miller indices (hkl) and Θ_B is the corresponding Bragg angle. This equation follows from kinematic diffraction theory and neglects the fact that the refractive index of matter for X-Rays is less than 1 by a few parts in 10^{-6} and so the incident beam is refracted to an internal angle slightly smaller than the external one. In Fig. 6.1 the scattering geometry is shown. φ denotes the angle between the lattice plane (hkl) and the surface, $\mathbf{k}_i$ is the incident and $\mathbf{k}_s$ the scattered wavevector. In principle, the electric vector of the incident beam can be polarised perpendicularly (σ) or parallel (π) with respect to the incident plane.

The Bragg diffraction is called "symmetric" if $\varphi = 0$, i.e. the reflecting lattice planes are parallel to the surface. For $\varphi \neq 0$ the Bragg diffraction is defined as "asymmetric". The asymmetry factor is given by:

$$b = \frac{\gamma_i}{\gamma_s} = -\frac{\sin(\omega_+)}{\sin(\omega_-)} = -\frac{\sin(\Theta + \varphi)}{\sin(\Theta - \varphi)} \tag{6.2}$$

where γ_i and γ_s are the direction cosines of the incident (i) and scattered (s) wave with respect to the surface normal (o).

The corresponding Ewald sphere construction is shown in Fig. 6.2. $\mathbf{G}_{hkl}$ denotes the reciprocal lattice vector and ω is the angle between the incident

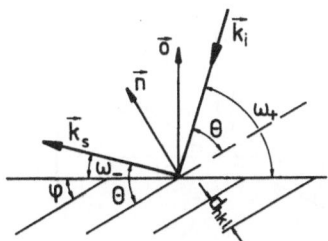

Fig. 6.1. Scattering geometry. $\mathbf{k}_i$: incident wavevector, $\mathbf{k}_s$: scattered wavevector, $\mathbf{o}$: surface normal, $\mathbf{n}$: normal on reflecting planes, Θ : Bragg-angle, φ : angle between surface and reflecting plane

wavevector $\mathbf{k}_i$ and the surface plane:

$$\mathbf{k}_s = \mathbf{k}_i + \mathbf{G}_{hkl} \qquad (6.3)$$

In Fig. 6.2 two possible scans for measuring the intensity of a Bragg reflection due to the reciprocal lattice point (hkl) are indicated:

(i) conventional powder diffractometers use a "$\Theta - 2\Theta$"-scan for measuring symmetric Bragg reflections ($\varphi = 0, \omega^+ = \omega^- = \omega = \Theta$). For such a scan, the detector is rotated twice as fast and in the same direction around the diffractometer axis as the sample. In reciprocal space, this conventional motion of sample and detector corresponds to a change of $\mathbf{k}_s$ in the following way: the tip of the vector $\mathbf{k}_s$ moves along the reciprocal lattice vector $\mathbf{G}_{hkl}$. During this motion the angle ω between the incident beam and the sample surface changes. For asymmetric (hkl) Bragg reflections ($\omega = \Theta \pm \varphi$), +: corresponds to the high incidence and − to the low incidence) the corresponding $\omega - 2\Theta$ scan direction runs also radial from the origin (000) of the reciprocal space along $\mathbf{G}_{hkl}$ (Fig. 6.2a).

(ii) In the ω-scan, the detector is fixed in position with wide open entrance slits and the sample is rotated, i.e. ω changes. In reciprocal space, this corresponds to a path as indicated in Fig. 6.2b by the bold arrow. The scan direction is transversal in reciprocal space. Thus the so-called rocking-curve is obtained.

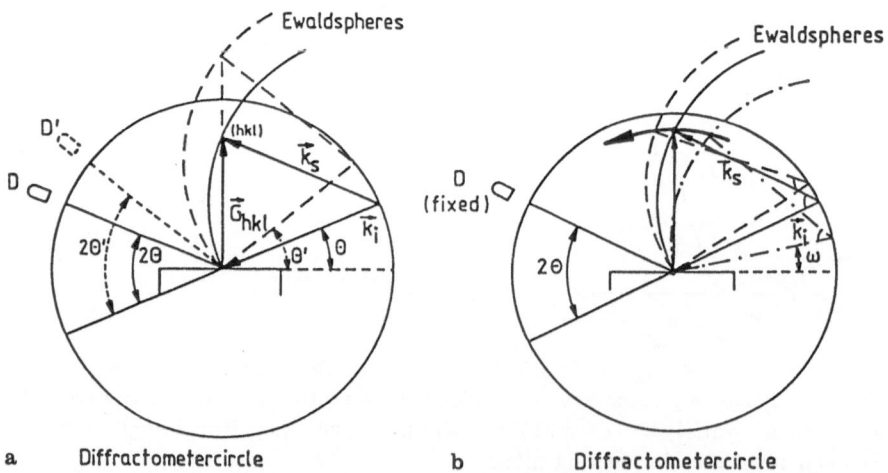

Fig. 6.2a. Ewald sphere construction (for symmetric reflections, i.e. reflecting planes are parallel to the surface) for the $\omega - 2\Theta$-scan geometry. Θ: angle between incident X-Rays and surface plane, $\mathbf{G}_{hkl}$: reciprocal lattice vector, (hkl): reciprocal lattice point, D: detector (on diffractometer circle in real space), **b** Ewald sphere construction for an ω-scan, the bold arrow indicates the movement in the reciprocal lattice

In Figs. 6.2a,b both the diffractometer circle (real space) as well as Ewald-spheres (reciprocal space) are shown. As can be seen, the two scan directions are perpendicular to each other. In contrast to a conventional focussing setup, e.g. Bragg-Brentano-, Seeman Bohlin-, Johanson-, etc. -configuration (e.g. [6.1,6.21]), the sample is illuminated by an intentionally perfect parallel beam.

The practical resolution of lattice constant determination of a perfect crystal is $\Delta d/d = 10^{-5}$. In epitaxial systems typically lattice constant variations between 10^{-2} and 10^{-4} have to be measured. The corresponding angular changes $\Delta\theta$ in the Bragg angle follow from the differentiation of Bragg's law:

$$\frac{\Delta d}{d} = \frac{\Delta\lambda}{\lambda} - \frac{\Delta\Theta}{\tan\Theta} \tag{6.4}$$

For high precision measurements of lattice constants and strains, the sample and X-Ray spectrometer should be temperature stabilised. In Table 6.1 we give some characteristic values which impose practical limitations on resolution:

Table 6.1. Characteristic examples and values for spectral resolution in X-Ray Diffraction (XRD) [6.22]

lattice perfection	$\Delta d/d \geq 10^{-7}$
thermal expansion coefficient	$\alpha = 10^{-6} K^{-1}$
X-Ray reflection (Fig. 6.10)	$\Delta\theta = 2'' \approx 10^{-5}$
line width of $CuK_{\alpha 1}$ radiation	$\Delta\lambda/\lambda = 3 \cdot 10^{-4}$
separation $CuK_{\alpha 1} - K_{\alpha 2}$	$\Delta\lambda/\lambda = 2.5 \cdot 10^{-3}$
single-crystal diffractometer	
with slit collimator	$\Delta\theta = 0.1° \approx 10^{-2}$

6.1.2 Double-Crystal Diffraction

In order to measure the X-Ray reflection of a single crystalline bulk sample as a function of the angle ω (the so-called "rocking-curve") the wavelength spread $\Delta\lambda$ has to be minimized. The method most often used is the Double-Crystal Diffraction (DCD) which was already described in [6.23, 6.24]. The first crystal is often a dislocation-free Ge or Si crystal sometimes cut for using an asymmetric Bragg diffraction for extremely high resolution (see below). The second crystal is a sample to be investigated in the so-called $(+m, -n)$ [6.25] scattering geometry (which is usually named double-crystal setting) (Fig. 6.3). The dispersion of a double-crystal spectrometer is defined by

$$\frac{\delta\omega}{\delta\lambda} = \frac{\delta\Theta_B^{II} - \delta\Theta_B^{I}}{\delta\lambda} \tag{6.5}$$

In the symmetric case $(+n, -n)$ the dispersion is zero. This is the mode which is used in the DCD.

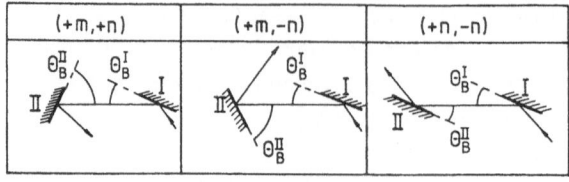

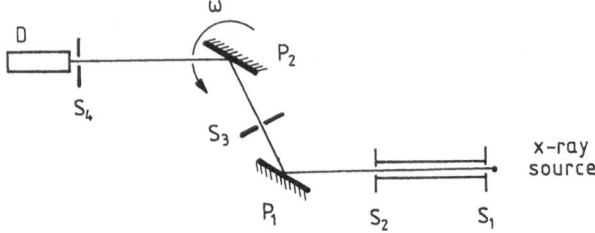

Fig. 6.3. Double-Crystal Spectrometer (DCS). Upper half: schematic presentation of X-Ray optics used in DCS (m, n represent modes corresponding to diffraction orders), with different dispersions; lower half: setup for DCS, $S_1 \ldots S_4$ denote slits, P_1 and P_2 monochromator crystal and sample, respectively

The divergent "polychromatic" radiation impinging on the first crystal is reflected according to the corresponding Bragg angles. The second crystal positioned in the equivalent ($-\Theta_B$) Bragg position reflects all wavelengths if it is oriented parallel to the first one (Fig. 6.3). Thus in the ($+n$, $-n$) geometry the X-Ray diffraction is nondispersive, i.e. X-Rays with different wavelengths are diffracted at the same ω setting if both the first crystal and the sample have the same lattice constant.

In the other configuration, the setup is dispersive, i.e. in the ($+m$, $+n$) position a reflection profile results which depends on the divergence of the primary radiation. It is evident that the proper choice of an asymmetric first reflection diminishes the *FWHM* of the reflection profile. Keeping the sample fixed, a beam with small divergence results which is used in triple-crystal and five-crystal (Bartels monochromator) arrangements.

The analysis of the diffraction condition of the double-crystal apparatus and others was performed by Bubakova [6.26] using DuMond's diagrams [6.25], which are in principle graphic presentations of Braggs law $\lambda(\Theta)$ taking into account the spectral broadening of the Bragg reflections. Successive reflections are easily represented. An in depth discussion can be found in DuMond's paper [6.25], including the operation principles of a four-crystal monochromator.

In DCD usually the ($+m$, $-n$) configuration is chosen since for heteroepitaxy the line width of the epilayers is much larger than those from the substrates so that extremely high resolution is unnecessary.

The ($+m$, $-n$) configuration, however, works only dispersionless if the primary monochromator and the sample to be investigated have the same lattice constant, which is a severe limitation of the applicability of this method. It should be mentioned that in order to achieve minimum *FWHM* line, a careful adjustment of the DCD is required [6.26–6.31].

Commercial instruments offer a stepping accuracy on the Θ-axis of one arcsec. This can be achieved by a tangent arm with a micrometer drive, which offers high accuracy over a limited angular range. The angular dispersion of the incident X-Ray beam is typically less than 10 arcsec due to the diffraction from the highly perfect primary crystal.

6.1.3 The 4+1 Crystal Diffractometer

DuMond [6.25] has proposed a four-crystal monochromator which has later been used by Beaumont and Hart [6.32] for wavelength selection of synchrotron radiation. Bartels [6.33] first realised a compact 4-crystal monochromator for high resolution X-Ray diffraction work. It uses 4 Ge crystals between the source and the sample which are cut and oriented in such a way that either (220) or (440) reflected intensities are transmitted (Fig. 6.4).

In the five-crystal diffractometer equipped with a Bartels monochromator, the first two crystals are in the $(+n, -n)$ setting and thus the whole spectral distribution passes as in the DCD. However, the third crystal is in a $(+n, +n)$ position with respect to the second one and thus only a small wavelength range can pass. The fourth crystal positioned with respect to the third one in a $(+n, -n)$ mode reflects the X-Rays in a direction which is parallel to the X-Rays emitted from the source. The successive diffractions produce an extremely monochromatic X-Ray beam ($\approx 5\%$ of the intrinsic width of the CuK$_{\alpha 1}$ line). The wavelength spread $\Delta\lambda/\lambda$ is $2.3 \cdot 10^{-5}$. In the Ge (440) setting the horizontal divergence is 5 arcsec whereas in the (220) setting which yields a factor of 30 higher intensity the intrinsic rocking-curve width is 12 arcsec.

The main advantage is the tunable parallel beam arrangement for a 2Θ scan ranging up to 160° with no loss in resolution in the (440) setting. Arbitrary sample materials can be investigated, independent of the monochromator ma-

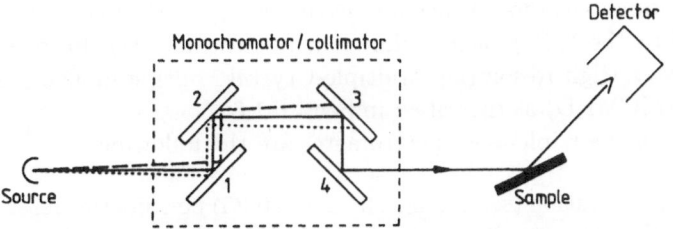

Fig. 6.4. $4 + 1$ crystal diffractometer [6.33]. The four Ge crystals use either (220) ($\Delta\theta = 12$") or (440) reflections ($\Delta\theta = 5$"). Dash-dotted line: $\lambda_0 \pm \Delta\lambda$

terial. One problem is the vertical divergence (perpendicular to the diffraction plane, i.e. the plane defined by incident and diffracted beams) since it reduces the resolution of the instrument and induces an experimental error. Thus a slit arrangement at the exit of the four-crystal spectrometer is necessary (Soller-slit). It is important to note that the X-Ray beam leaving the four-crystal monochromator in the (440)-setting is nearly completely σ-polarised. Rocking-curves of almost any lattice planes in any direction can be measured. With a suitable goniometer absolute lattice constant determinations are also possible by applying Bond's method [6.34]. Using a similar four-crystal monochromator equipped with 15° asymmetric cut Ge-crystals van der Sluis was able to enhance the intensity by a factor of 3.8 as compared to the original design [6.35].

An alternative beam conditioner is the so-called Channel-Cut Collimator (CCC) which utilizes multiple reflections from parallel crystal planes which are fabricated by cutting a channel into a single crystal. The multiple reflected X-Ray beam leaving the CCC is conditioned with respect to its wavelength and angular dispersion before it is finally monochromated [6.36]. For measurements of extremely narrow reflections the beam divergence may be too large and arrangements with asymmetric reflections should be used [6.37, 6.38].

6.1.4 Triple-Axis Spectrometer

Renninger [6.39, 6.40] suggested the use of a triple-axis spectrometer in order to measure the asymmetric form of the dynamical reflection profile of a single crystal. With a triple-axis spectrometer several arrangements are possible, which were analysed by Godwod [6.29] and Lefeld-Sosnowska [6.41].

For the investigation of semiconductor heterostructures, a spectrometer setting is used as shown in Fig. 6.5a. Recently, the triple-axis spectrometer system has been further upgraded by using a four-crystal spectrometer as a monochromator in front of the specimen and behind it a two-crystal analyser (Fig. 6.5b). Alternatively to the four-crystal monochromator in another realisation two channel-cut crystals can be adjusted using either eight symmetric (022) reflections from Si or two asymmetrically cut (022) channels with four reflections for high divergence or high interensity purposes (c). A channel-cut crystal is used as an analyser, too. In such an instrument up to thirteen reflections are performed by the X-Ray beam before reaching the detector. In so far this apparatus is also a High-Resolution Multiple-Crystal Multiple-Reflection Diffractometer (HRMCMRD) as described in Sect. 6.1.6.

The advantages of the triple-axis spectrometer are the following:

a) Improved angular resolution (see e.g. figs. 6.34 and 6.35) permits the observation of weak diffraction satellites. This feature also makes the triple-axis spectrometer quite useful for the analysis of extremely thin layers. E.g. using Ge(111) monochromators and analyser crystals Ryan et al. [6.42] have reported a wavevector resolution of $5 \cdot 10^{-4}$Å^{-1}.

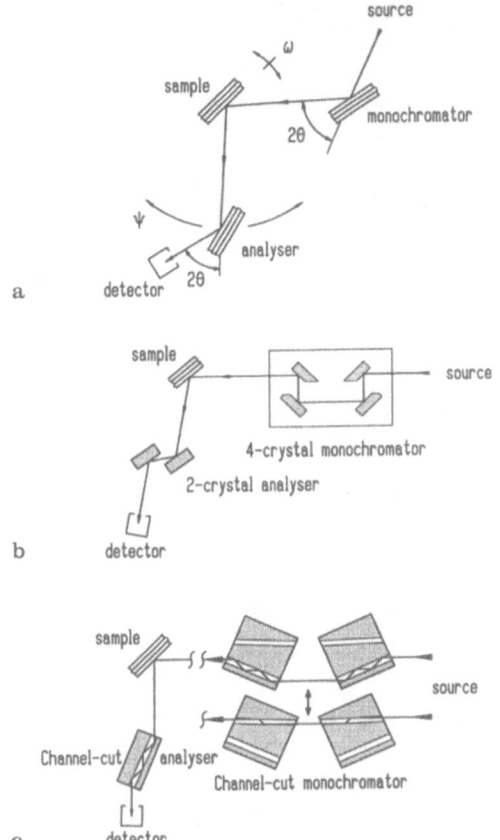

Fig. 6.5a. The triple-axis spectrometer is an extension of the DCS where an analyser crystal is inserted between the sample and the detector, **b** 4 + 1 + 2 spectrometer consisting of a 4-crystal Bartels monochromator for either (220) or (440) setting, the sample and a two-crystal analyser [6.11], **c** triple-axis diffractometer using either 8 symmetric (022) reflections from Si (5" divergence beam $\Delta\lambda/\lambda = 5.5 \cdot 10^{-5}$) or two asymmetrically cut Si (022) channels (4 reflections; 12" divergence) [6.36]

b) Mapping of the reciprocal lattice space: in order to measure intensity contour maps, keeping one of the Miller indices, e.g. l in the reciprocal lattice fixed, and varying h by $\pm\Delta h$, k by $\pm\Delta k$, the instrument just described is used in the following way: the third crystal (2Θ) is scanned for a sequence of different angular positions ω of the sample. A two-dimensional intensity map is thus obtained by measuring a number of $\omega - 2\Theta$ scans along the vector $q_\parallel[hkl]$ for a (hkl) reflection for different ω offsets (ω-scan direction: perpendicular to $q_\parallel[hkl]$) (Fig. 6.6). In Fig. 6.6 the reciprocal space maps thus obtained are shown. The conversion of a peak intensity position (ω,

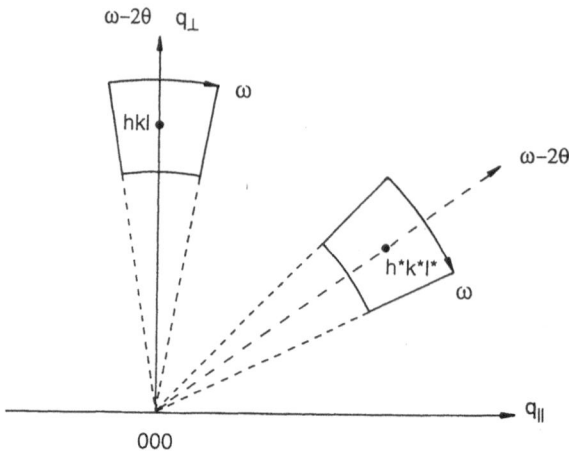

Fig. 6.6. Scans in reciprocal space for two different reciprocal lattice points hkl and $h^*k^*l^*$

2Θ) in reciprocal space coordinates is given by [6.43, 6.44]:

$$q_\perp = \frac{2}{\lambda} \sin \Theta \cos(\omega - \Theta) \tag{6.6}$$

$$q_\parallel = \frac{2}{\lambda} \sin \Theta \sin(\omega - \Theta) \tag{6.7}$$

For the investigation of epilayers, the substrate reflections are used as an internal standard because the absolute values of the angular scale of the diffractometer are unknown. The vector components $q_\perp$ and $q_\parallel$ refer to directions perpendicular and parallel to the growth plane. The region which is accessible in reciprocal space mappingdepends on the geometry (reflection or transmission), on the wavelength used as well as on the lattice constants of the epitaxial layer and the substrate. In Fig. 6.7 the region indicated by "C" is accessible for Bragg reflection measurements which is the most common case for the investigation of epitaxial layers. For this example, for the layer the lattice constant of GaAs and the CuK$_{\alpha 1}$ wavelength was used.

c) The instrument can be used to measure X-Ray reflectivity from the surfaces of semiconductors and amorphous materials and thus one can get information on surface roughness and film thickness.

d) The triple-axis diffractometer allows Bragg plane tilts and dilatations to be determined independently. Thus effects of wafer curvature and mosaicity can be separated [6.6]. A triple-axis spectrometer is most useful for the study of less perfect epitaxial layers and superlattices since, in this mode,

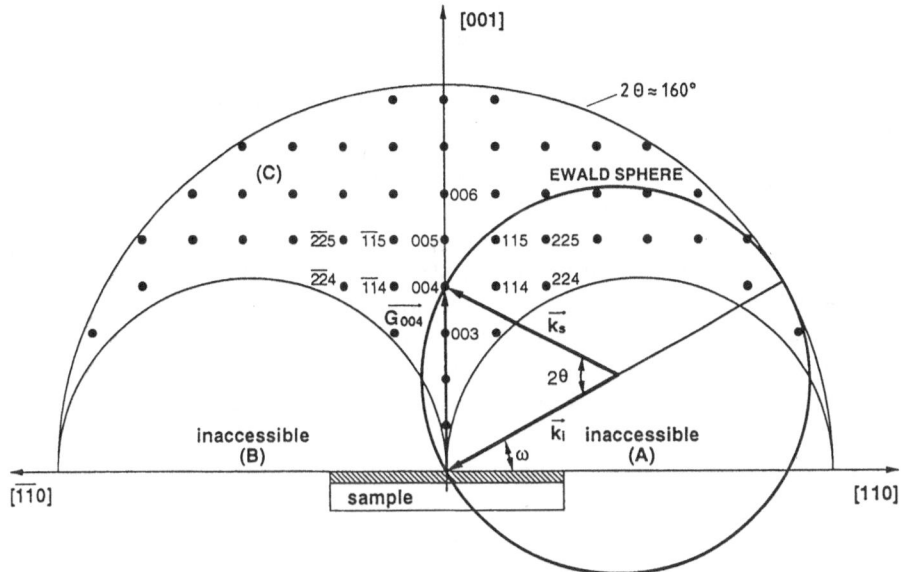

Fig. 6.7. Reciprocal space map showing accessible range for Bragg reflection measurements. The radius of the outer semicircle is limited by the maximum diffractometer angle ($2\theta \approx 160°$). The two inner regions defined by the two semicircles are not accessible in the Bragg case

not only scans as they are performed with double-crystal spectrometers but also reciprocal lattice grids are measured. A combination consisting of the $4+1+2$ or channel-cut arrangement (Fig. 6.5c) offers the advantage of a reduction of the background (due to fluorescence) and a signal-to-noise ratio of $10^5 \dots 10^6$ is obtainable [6.45].

e) Diffuse scattering e.g. originating from distorted interfaces can easily be separated from coherent Bragg scattering.

The disadvantage is the following: In comparison to a double-crystal spectrometer, the analyser crystals reduce the intensity and thus longer measurement times result unless the irradiance is increased by using a rotating anode source which provides an higher intensity. The adjustment procedure is quite difficult [6.29] and detailed information on the spectrometer theory is necessary. These difficulties can be avoided by the use of a double-crystal spectrometer equipped with a Position-Sensitive Detector (PSD) [6.46–6.48], placed on the 2Θ arm of that spectrometer (see Fig. 6.49).

Despite the fact that the sample is rotated in the conventional ω-direction, the PSD collects the scattered X-Rays simultaneously over a 2Θ range of 10°. However, it does not integrate over that range but resolves the scattered intensities with a resolution of 70 arcsec.

The main advantage is the fact that during a normal ω-scan using the PSD, information on the scattered intensity in 2Θ-direction is also obtained without any increase in measurement time. A disadvantage results from the reduced dynamic range of the PSD, which is smaller than that of a conventional X-Ray counting system by about two orders of magnitude. Another disadvantage is the relatively poor angular resolution, which causes artifacts in the reciprocal space maps.

Thompson et al. [6.48] have shown that a DCD equipped with a PSD directly yields a separation of mosaic structure from strain effects in rather imperfect layer systems (zone-melt recrystallised silicon sandwiched between SiO_2 layers on Si). The data can be collected within the same amount of time necessary for an ordinary rocking-curve analysis. Thus the combination of a DCD with a PSD offers an alternative to the reciprocal space mapping performed as outlined above.

Picraux et al. [6.12, 6.49] have recently demonstrated that such an instrument is particularly useful for the investigation of strained layers SL's. Although the resolution is much poorer, especially for the 2Θ-scan ($\Delta 2\Theta = 70$ arcsec), and thus poorer than the resolution in $\Delta \omega$ which is a few arcsec, intensity mapping of the reciprocal space is possible. This is done by ·transforming $\Delta \Theta$ and $\Delta \omega$ into $\mathbf{G}_\parallel$ and $\mathbf{G}_\perp$ using

$$|\mathbf{G}| = \left(G_\perp^2 + G_\parallel^2\right)^{\frac{1}{2}} = \frac{4\pi \sin(\Theta_B + \Delta\Theta)}{\lambda} \tag{6.8}$$

$$\frac{G_\parallel}{G_\perp} = \tan(\varphi - \Delta\Theta + \Delta\omega) \tag{6.9}$$

The advantage is a quicker determination of intensity contours at the expense of resolution.

Intensity contour maps provide interesting information on the states of strain of the epitaxial films, i.e. on asymmetries in the distribution of strains and on mosaicity. In real space the resolution is mainly determined by the dimensions of the source aperture. By using a DCD in combination with a microfocus X-ray tube and a narrow detector slit Itoh et al. [6.50] were able to analyze the mosaicity of GaN/(0001)-sapphire and GaAs/ZnSe films.

6.1.5 Renninger Scans

In a Renninger scan a crystal is rotated about the normal to a set of diffracting planes while diffraction from those planes is measured [6.51]. The principle of the technique is illustrated in Fig. 6.8. In such scans an intensity modulation due to multiple-beam interaction is observed which has been evaluated e.g. for accurate measurements of lattice constants and structure factors [6.52, 6.53]. Recently this technique has been applied to characterise epitaxial ZnSe/GaAs structures [6.54]. Morelhao et al. [6.55, 6.56] used a hybrid multiple diffraction in a Renninger scan to study the mosaic spread of a GaAs layer grown on

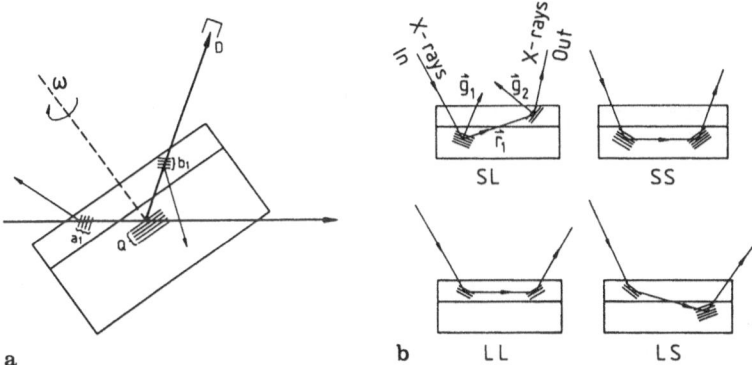

Fig. 6.8a. Renninger scan geometry for detection of modulation of diffracted intensity from substrate lattice planes (Q) due to diffraction within the epilayer a_1 and b_1 [6.54], **b** Geometry for hybrid multiple diffraction in Renninger scans of layer (L) on substrate (S) epitaxial systems [6.55]

a Si substrate. Hybrid diffractions occur when the beam first diffracted by a substrate or layer plane is rescattered by another substrate or layer plane towards the detector. These type of measurements are evaluated with the aid of Kossel diagrams (see e.g. [6.57]).

6.1.6 High-Resolution Multiple-Crystal Multiple-Reflection Diffractometer (HRMCMRD)

One of the latest developments in HRXRD instrumentation is the HRM-CMRD spectrometer (Fig. 6.5c), which is in principle a $4 + 1 + 3$ instrument using a total of 8 reflections [6.58, 6.9, 6.10].

The principal setup is shown in Fig. 6.9. It combines the possibilities of the four-crystal monochromators and a multiple-reflection analyser crystal to perform reciprocal lattice scans as well as X-Ray topographs in the same region of the sample.

Fewster has demonstrated that this diffractometer is particularly useful for distinguishing between the residual strain and the mosaic spread in imperfect crystals and avoids misinterpretation of rocking-curves obtained with DCD. As already discussed, the analyser crystal selects the angular range of the diffracted beam reaching the detector. For the analyser a symmetric (220) reflection from a perfect Ge crystal was used and the reduction of the tail intensities in the analyser reflectivity profiles for a twofold and threefold reflection were compared [6.25, 6.59, 6.58]. The benefits of this instrument are the large dynamic intensity range (1 count/s to $10^5 \ldots 10^6$ counts/s) and the size of the diffraction space probe (10") which is very useful for diffraction mapping in the reciprocal space. Further it avoids the so-called star pattern

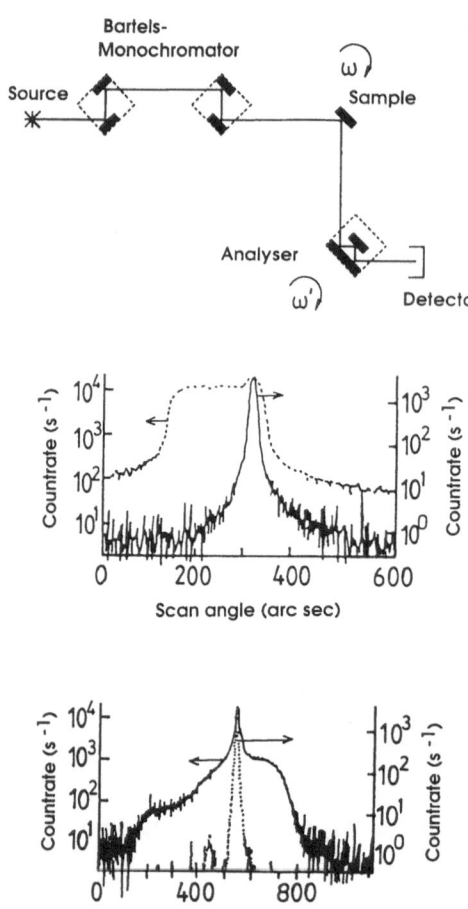

Fig. 6.9. Upper panel: High-Resolution Multiple-Crystal Multiple-Reflection Diffractometer (HRMCMRD) [6.59, 6.58]. Central panel: comparison of a (004) diffraction profile for a mosaic GaAs crystal as obtained with the 4 + 1 diffractometer (dashed line) and the HRMCMRD (solid line). Lower panel: similar comparison for a (004) diffraction from a ZnSe-layer on GaAs (4 + 1) diffractometer (solid line) and from an $\omega - 2\Theta$ scan obtained with the HRMCMRD (dotted line)

around the reciprocal lattice points found in three-crystal three-reflection spectrometers caused by the transfer function [6.60].

In Fig. 6.9 the (4 + 1) and HRMCMRD results are compared for two different material systems: a GaAs substrate, exhibiting mosaicity and a ZnSe epilayer on a GaAs substrate. Clearly, the application of the HRMCRD leads to a tremendous reduction of the linewidth, but, due to third analyzer, reflected intensity is lost in comparison to the (4 + 1) instrument, and thus the measuring time is increased.

6.2 Kinematical and Dynamical Theory

For the investigation of epitaxial layers deposited on comparatively thick substrates, the X-Ray diffraction is monitored in reflection geometry which is called "Bragg-case". The "Laue-case" denotes the measurement of diffracted intensities in transmission which will be not further considered here.

There are two theories describing the scattering of X-Rays (and electrons, neutrons) in crystals: the simpler kinematical and the dynamical theory. For thick perfect crystals one needs the latter one developed by Darwin [6.61], Ewald [6.62] and von Laue [6.63]. In this Darwin-Ewald-Laue theory for a perfect crystal, an exact solution of the wave equation within the crystal is attempted by expressing the wave field by Bloch functions with coefficients which are invariant with respect to the space coordinates. The wave field excited in the crystal can be expressed by a sum of two or more of these wave fields with slightly different values of k_i. As a result the incident and diffracted waves show an amplitude and phase modulation (Pendellösung) effect, first observed in the Laue-case by Kato and Lang [6.64] and in the Bragg-case by Batterman and Hildebrandt [6.65]. The theory was modified by Prins [6.66] for the case of absorption. A general description of the dynamical theory can be found e.g. in the books of Zachariasen [6.67] and Pinsker [6.57]. With recent advances in epitaxial crystal growth technology nearly perfect single crystalline films have become abundantly available and, therefore, the necessity for applying dynamical theory for such systems arose taking into account the finite thickness and interfacial boundaries.

If the interference conditions are fulfilled, a diffracted beam is produced which leads to a weakening of the incoming wave, a process which is called extinction. In this case, at each lattice plane, a part of the incident beam is reflected so that the incident beam arriving at the next plane has a smaller amplitude. Since with increasing depth from the crystal surface this process is repeated many times, the incident beam is finally reduced to negligible amplitude at a depth corresponding to the extinction length (6.22). The extinction length in the case of strong interference depends on the angles of incidence and of emergence of the X-Rays, the X-Ray wavelength, and the structure factor as well as the Debye-Waller factor (Sect. 6.3). The extinction length is typically a factor of ten smaller than the penetration depth due to normal absorption caused by the photoelectric effect. It is of the order of $1 \ldots 10\,\mu m$ for the materials of interest [6.33].

Therefore only a layer with a thickness smaller than the extinction length contributes to the diffracted intensity in the Bragg-case. The finite width of this layer is the origin of the width of a Bragg diffraction peak along the $\omega - 2\Theta$ scan direction (Darwin width) in dynamical theory.

In the kinematical theory, the crystal potential is treated as a small perturbation (first Born approximation) which is adequate in cases when the extinction length is large compared to the total thickness. It is also adequate for nonperfect (mosaic) crystals. In the kinematical theory the incident wave with wavevector $\mathbf{k}$ remains unattenuated whereas the reflected wave with wavevector $\mathbf{k} + \mathbf{G}$ ($\mathbf{G}$ denotes a reciprocal lattice vector) is assumed to be weak, but increases in intensity with increasing thickness (i.e. proportional to the square of the thickness). Its application is thus limited to cases for which the thickness is smaller than the extinction length. Moreover, for all calculations inside the crystals, the vacuum wavelength is used. This assumption does not hold e.g. for precision measurements of lattice constants, where the deviations of the refractive index from 1 (in the order of 10^{-6} to 10^{-5}) have to be taken into account. The possibility of multiple scattering, which becomes important in highly perfect crystals of large coherency distances, is neglected.

An alternative description applicable to crystals with small distortions was first given by Takagi [6.68, 6.69] and later by Taupin [6.70]. The Takagi-Taupin formalism is the one which is used nowadays almost exclusively for the calculation of dynamic scattering effects in high resolution X-Ray diffraction. The wave excited in the crystal is expressed by a single sum, but the coefficients are considered as a slowly varying function of depth position instead of being constant as would be in the case of kinematical theory. In the two-beam approximation the change of the amplitude of the incident wave D_i and scattered wave D_s with the depth z into the crystal are given by the Takagi-Taupin equations:

$$\frac{\mathrm{i}\lambda\gamma_s}{\pi}\frac{\partial D_s}{\partial z} = \Psi_i D_s + C\Psi_s D_i - \alpha_s D_s \tag{6.10}$$

$$\frac{\mathrm{i}\lambda\gamma_i}{\pi}\frac{\partial D_i}{\partial z} = \Psi_i D_i + C\Psi_s D_s \tag{6.11}$$

with the notations $\alpha_s(\omega) = -2\lambda(\Theta - \Theta_B)/(d_{hkl})$ and $\Psi_{i,s} = -\lambda r_e F_s/\pi V$, r_e being the electron radius, $F_{i,s}$ are the complex structure factors, and V is the volume of the unit cell. γ_i and γ_s are the direction cosines of the incident and scattered beams with respect to the internal surface normal. i, s denote a certain reflection (hkl) and consequently $F_{\bar{s}}$ is the structure factor for $(\bar{h}\,\bar{k}\,\bar{l})$. C denotes the polarisation factor ($C = 1$ for σ-polarisation and $C = |\cos 2\Theta_B|$ for π-polarisation). Combining these two equations for the Bragg-case a differential equation for the amplitude ratio D_s/D_i results:

$$-\mathrm{i}\frac{dX}{dT} = X^2 - 2\eta X + 1 \tag{6.12}$$

where X, η, T are complex quantities given by

$$X = \sqrt{\frac{F_{\bar{s}}}{F_s}}\sqrt{\left|\frac{\gamma_s}{\gamma_i}\right|}\frac{D_s}{D_i} \tag{6.13}$$

$$\eta = \frac{-b(\Theta - \Theta_B)\sin 2\Theta_B - \frac{1}{2}\Gamma F_0(1-b)}{\sqrt{|b|}C\Gamma\sqrt{F_s F_{\bar{s}}}} \tag{6.14}$$

$$T = \frac{\pi C\Gamma\sqrt{F_s F_{\bar{s}}}\, d}{\lambda\sqrt{|\gamma_i\gamma_s|}} \tag{6.15}$$

where $\Gamma = r_e\lambda^2/\pi V$, $b = \gamma_i/\gamma_s$. d is the crystal thickness, and F_0 the structure factor for (000). The departure from the Bragg angle Θ_B determines the deviation parameter η. The solution of the differential equation is given by

$$X_d = \eta + \sqrt{\eta^2 - 1}\,\frac{S_1 + S_2}{S_1 - S_2} \tag{6.16}$$

where

$$S_{1,2} = \left(X_0 - \eta \pm \sqrt{\eta^2 - 1}\right)\exp\left(\mp i T\sqrt{\eta^2 - 1}\right) \tag{6.17}$$

For layered structures the recursion X_d usually starts with the infinite thick substrate $(d \to \infty)$ [6.71]:

$$X_\infty = \eta - \text{sign}(\text{Re}(\eta))\sqrt{\eta^2 - 1} \tag{6.18}$$

The reflectivity R_s is finally given by

$$R_s = \left|\frac{\gamma_s}{\gamma_i}\right| \cdot \left|\frac{D_s}{D_i}\right|^2 = \left|\frac{F_s}{F_{\bar{s}}}\right| |X|^2 \tag{6.19}$$

The rocking-curve of the sample is determined by the reflectivity R_s as a function of the deviation parameter η.

Recently, on highly strained superlattices small discrepancies between the experimental and theoretical angular positions of higher-order satellite peaks have been observed using the conventional deviation parameter (6.14) in the Takagi-Taupin equations [6.72, 6.74, 6.73]. One reason was assumed to be the linear approximation

$$\sin\Theta - \sin\Theta_B \approx (\Theta - \Theta_B)\cos\Theta_B \tag{6.20}$$

for the conventional deviation parameter. Indeed, for symmetric reflections a higher-order approximation gives the correct Bragg positions for the superlattice peaks [6.73, 6.74]. For asymmetric reflections, however, only a new deviation parameter based on the solution of the dynamical equations for the amplitudes of the electric field in the crystal gives the correct result. Unlike the conventional solution, the new deviation parameter takes into account the scattering geometry before reducing the dispersion relation of 4th degree to second degree in the wavevector [6.75]. Thus a deviatiation parameter results which describes all scattering geometries correctly.

In order to illustrate the influence of the different orientations on the shape of the reflectivity as a function of ω in Fig. 6.10, we compare the reflectivity curve for a (333) Bragg diffraction with that of a (115) crystal diffraction. Since for both Bragg diffractions $h^2 + k^2 + l^2 = 27$ the only difference are their anisotropy factors b (6.2).

Especially, in the case of thin films Pendellösung fringes are caused by the interference of the wave fields. These fringes will be further modulated if two or more layers with different lattice parameters scatter the X-Rays. Then the wave fields originating from the single layers interfere as well and affect the Pendellösung fringes. Consequently, the interference phenomena in multilayers are extremely useful for the investigation of structural properties even of ultrathin layers (buried layers of monolayer thickness and heterointerfaces).

Recently, Chen and Bhattacharya [6.76] have shown that the Darwin-Prins formulation [6.77] of the dynamical X-Ray scattering is equivalent to the

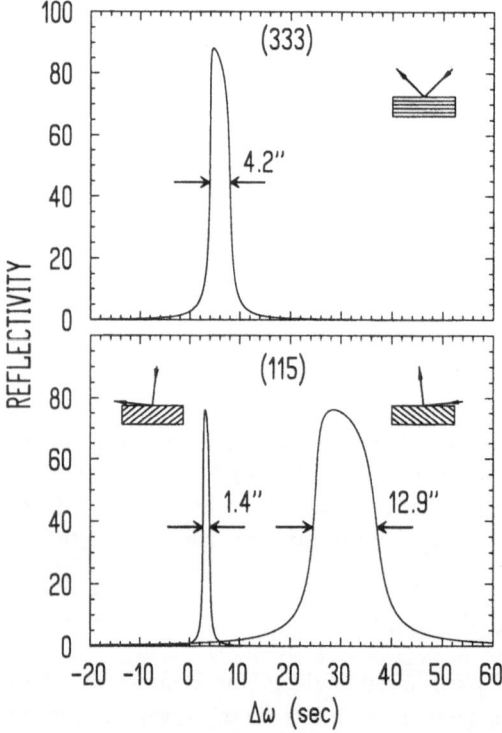

Fig. 6.10. Calculated reflectivity curves for σ-polarised CuK$_{\alpha 1}$ radiation of (111) oriented Ge for symmetric (333) and asymmetric (115) and ($\overline{1}\overline{1}5$) glancing incidence and glancing exit reflections (see Fig. 6.17). The latter ones demonstrate the influence of the asymmetry factor on the *FWHM*. σ-polarised radiation is the relevant one in most experiments. For π-polarised radiation the peak shape appears nearly symmetric around a central maximum [6.86]

Takagi-Taupin approach under the assumption that the crystal is a continuous medium. The computer programs nowadays available for fitting diffraction data of epitaxial films and an arbitrary number of epitaxial layers (even with different lattice parameters as a function of depth) are based on the Takagi-Taupin formalism. Sometimes further simplifying assumptions are made in so called "semi-kinematical theories". The original formulation is due to Petrashen [6.78, 6.79] for distorted crystals. Tapfer and Ploog [6.80, 6.81] have further developed this method for obtaining precise information on chemical composition, thickness, strain profile, and interface quality of heterostructures and multilayers. A first iteration of Taupin's equations for the amplitude ratio of the incident and diffracted waves $\mathbf{k}_i$ and $\mathbf{k}_s$ is used. This approach is valid if the thickness of the deformed layers is small compared with the extinction length. The main advantage of this procedure is the reduction of computation time for the simulation of diffraction curves. In the spirit of this "semi-kinematical" approach, the thin epitaxial film is treated kinematically whereas for the thick substrate a dynamical calculation is performed. Originally, for epitaxial layers, superlattices [6.82–6.84] and ion-implanted semiconductors [6.85] the kinematical model was quite successfully applied in analysing the measured rocking-curves. In [6.3], the kinematical method is described in detail and the limits of its applicability are discussed. Bartels et al. [6.71] have used the criterion that the observed peak reflectivities must be less than about 10% to 6% in order that the kinematical theory can be applied. Therefore, close to the $i = 0$ central peak of a superlattice or close to the substrate peak, an intensity analysis based on the kinematical approach may yield improper results. In recent years it was observed that there are small but important discrepancies in the Bragg peak position of a thin epilayer or the zeroth order superlattice peak in the results from the kinematical and the dynamical models [6.8, 6.87–6.89].

These discrepancies are very important for practical applications since from the angular distance of the Bragg peaks of the epilayer and that of the substrate lattice strain and changes in chemical composition are deduced. It was shown that in most cases a combination of kinematical theory for the layers and dynamical theory for the substrate can be used for the simulation of rocking-curves of the thin layer samples. However, an important point is the matching condition for the phases at the layer to substrate boundary. According to Wie and Kim [6.90] matching conditions for the amplitudes (dynamical amplitude for the substrate, kinematical amplitude for the epitaxial layer $\hateq$ Amplitude Boundary Conditions (ABC)) have to be used instead of the previously taken Intensity Boundary Conditions (IBC). Even with this recent improvement, it can be dangerous to use the previously accepted rule (e.g. [6.71, 6.3]) according to which for epitaxial films with less than 10% X-Ray reflectivity power, the full dynamical and the kinematical calculations yield identical results. Kim and Wie [6.90] have made a comparison of calculated X-Ray scattering results for AlGaAs/GaAs, GaInAs/InP heterostructures and a superlattice sample AlGaAs/GaAs using the semikinematical and a full dynamical calculation. Both the Bragg peak profile and the peak positions agree for both types of models using the amplitude boundary condition, if e.g. a layer sequence CAC of AlGaAs (C)/GaInAs (A)/AlGaAs (C) is deposited on a GaAs substrate. However, in the case where one of the layers is identical in chemical composition to the substrate (e.g. GaAs (B)/GaAlAs (A)/ GaAs substrate (B)), a semi-kinematical model is no longer adequate because the substrate peak (B) is calculated dynamically and a layer peak (also B, of the same chemical composition) is treated kinematically. Thus, in general, nowadays the dynamical theory should be applied.

Already in 1986, Macrander et al. [6.91] came to that conclusion and performed simulations of graded layers, which have no discontinuity in lattice constant at the interfaces. For such structures dynamical simulations are also required. The authors employed a computational procedure for the simulation of superlattice rocking-curves based on the Abeles' matrix method [6.92]. This method was originally developed for optics in the visible range (see Chapter on FIR spectroscopy).

In another method, used by Tapfer et al. [6.93], a recurrence formalism is used to calculate the diffraction pattern based on the work by Vardanyan et al. [6.94]. This is also a dynamical diffraction theory for (layered) crystals of arbitrary thicknesses which takes into account multiple reflection and the interference of the wave fields from the various layers and from the substrate, including the exact boundary conditions at the heterointerfaces and including lattice strain as well. (Also in the Takagi-Taupin formalism the effects of strain can be taken into account). Tapfer et al. [6.88, 6.93] have recently extended the theory by Varadanyan et al. [6.94] for very small glancing angles where the refraction effect cannot be neglected. In strongly mismatched heterostructure systems, the crystalline quality can be poor and even mosaic structured. In such a case (e.g. CdTe/GaAs, GaAs/Si) diffraction within one mosaic crystal is independent from the adjacent ones and the kinematical approach is well suited. A further approach due to Wie and Kim [6.95, 6.90] starts with the Takagi-Taupin equations, but solves the dynamical recursive formulae in the Bragg-case with the use of a matrix formalism which is particularly convenient for large period superlattices since then the algorithm is faster than the recursive formula approach.

Very recently the Darwin theory of dynamical diffraction was extended by Caticha for the symmetric Bragg case [6.96] to include the regions between Bragg peaks as well as situations of grazing and normal angles of incidence. In the modified theory the diffracting crystal is built up of N plates, the surface of which are normal to the z direction. The layers of thickness d are separated by infinitesimal gaps. The electric field in each gap is treated as a superposition of an incident and reflected plane wave, thus an involved many-beam dynamical diffraction calculation reduces to a two-beam calculation. It is shown that the theory reproduces the two-beam Laue dynamical theory in the vicinity of the Bragg peaks and the reflectivities between the Bragg peaks in agreement with the kinematical theory, which was exclusively used in this angular range so far.

6.3 Thickness Dependence of Bragg Reflections from Thin Films

Both intensity and Full Width of Half Maximum ($FWHM$) of Bragg reflections strongly depend on the thickness of the measured crystal. This dependence is caused by different effects: due to the photoeffect the penetrating X-Ray intensity is always weakened, independently of Bragg condition. This process is described by an exponential law. The absorption depth d_{abs} is defined as the thickness of a layer that reduces the intensity by a factor of e:

$$d_{\mathrm{abs}} = \left(\mu \left(\frac{1}{|\gamma_i|} + \frac{1}{|\gamma_s|} \right) \right)^{-1} \tag{6.21}$$

where μ is the linear absorption coefficient and $\gamma_{i,s}$ are the direction cosines as defined in (6.2). The absorption depth is of the order $< 100\,\mu$m.

As already discussed in the previous Chapter, in case interference conditions are fulfilled, the intensity is additionally weakened due to the production of diffracted beams. This process is called *primary* extinction and is described by the extinction length d_{ext}

$$d_{\mathrm{ext}} = \frac{V\sqrt{|\gamma_i \cdot \gamma_s|}}{r_e \lambda |F_s'|} \tag{6.22}$$

where F_s' contains only contributions of the real part of the structure factors. The extinction depth is of the order of $< 10\,\mu$m. Apart from the primary extinction there is *secondary* one, which is important in X-ray diffraction from polycrystalline materials or powders [6.97]. It results from the fact that a certain grain within the sample is illuminated by a smaller intensity, if a grain which differs in orientation is situated above oriented, scatters in Bragg condition. The reflectivity of InP epitaxial layers is shown in Fig. 6.11 as a function of layer thickness taking only extinction into account.

Thus, even without absorption, X-Rays penetrate only several thousand atomic layers into a single-crystalline film. This leads to the finite broadening of the reflected X-Ray intensity since just a finite number of scattering centres contributes:

$$FWHM = \frac{2.12 \cdot \lambda \cdot \gamma_s}{\pi d_{\mathrm{ext}} \sin 2\theta_B} \tag{6.23}$$

In Fig. 6.12 the reflectivity of thin InP crystals including absorption is shown.

The reflected X-Ray intensity is thus determined both by primary extinction as well as by absorption. The $FWHM$ of a rocking-curve is generally

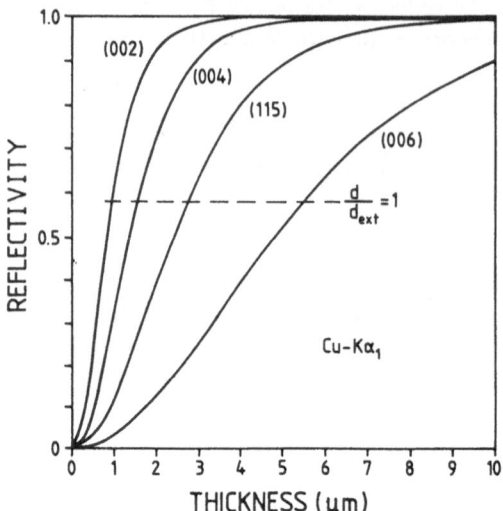

Fig. 6.11. Reflectivity versus layer thickness of an (001) oriented InP epitaxial layer for different Bragg reflections. Absorption assumed to be zero. d_{ext}: extinction depth [6.33]

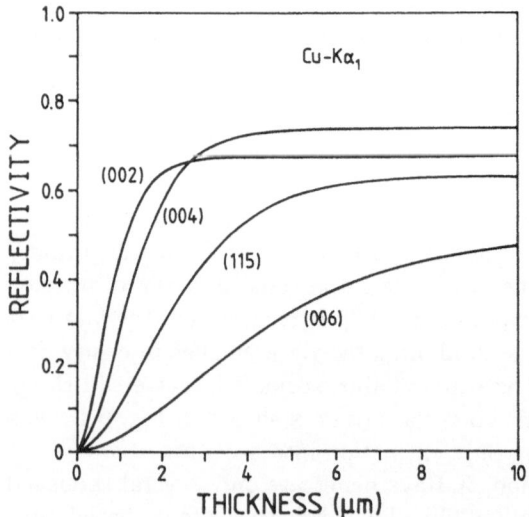

Fig. 6.12. As Fig. 6.11 but taking absorption and extinction into account [6.33]

taken as a measure for the crystal quality. For perfect crystals with thicknesses below the extinction length additional broadening occurs which increases with decreasing thickness. Therefore the proper interpretation of a half-width obtained from a thin film requires knowledge of the film thickness. It's worth mentioning that the *FWHM* of a certain Bragg peak is linked to the structure factor of the material. For a Bragg peak with high structure factor the number of contributing lattice planes is comparatively small and, consequently, the *FWHM* is large.

With an incident angle corresponding to a reflection position (Bragg angle), the extinction essentially determines the penetration depth of X-Rays in the perfect crystal. For practical applications it is quite useful to have the effects just described in mind. In Fig. 6.13 the *FWHM* versus film thickness is plotted for CuK$_{\alpha 1}$ radiation and a Bragg angle $\theta = 31.6°$ (InP, (004)). Consequently, a comparison of the width of X-Ray intensities, even from the same symmetric Bragg diffraction but from different films without specifying the thicknesses, can only be performed if the thickness dependence of the *FWHM* is taken into account.

In addition, broadening due to the experimental setup occurs. In an experimental setup the measured reflectivity $I(\omega)$ as a function of ω is the autocorrelation of $R(\theta)$:

$$I(\omega) \sim \int_{-\infty}^{\infty} R(\theta)R(\theta - \omega)\, d\theta \tag{6.24}$$

where the $(+n, -n)$ scattering geometry is assumed and the two crystals are exactly parallel. Thus $I(\omega)$ should be a symmetric function around ω_o. In a detector not the ideal intensity distribution $I(\omega)$ is recorded, but the intensity resulting from a convolution of $R(\theta)$ with a function depending on the special experimental setup $C(\theta, \alpha, \Phi, \Delta\lambda)$ where α denotes the horizontal divergence $(-\alpha_m, +\alpha_m$: few degrees) and Φ the vertical divergence $(-\Phi_m, +\Phi_m$ such that $\sin \Phi_m \approx \Phi_m)$ [6.26]. Due to those influences, $I(\omega)$ is in practice broader than $R(\theta)$.

Besides these broadening effects, reflectivity measurements of thin films exhibit Pendellösung fringes originating from the fact that the diffracted intensity oscillates periodically with thickness. The intensity modulation of the

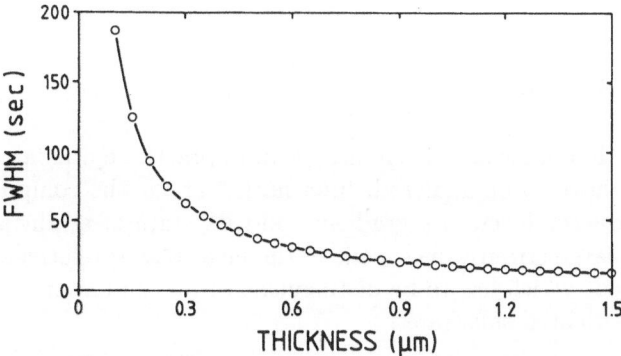

Fig. 6.13. *FWHM* of the (004) reflection of InP versus layer thickness for CuK$_{\alpha 1}$ radiation [6.98]

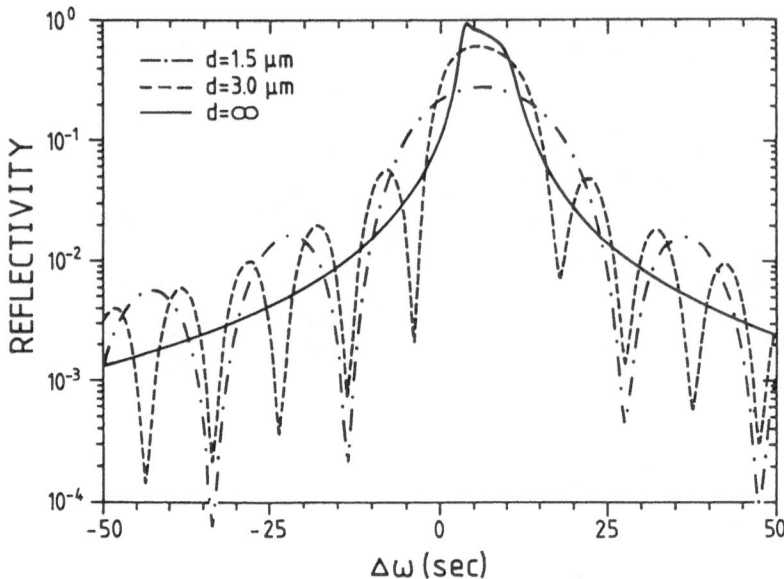

Fig. 6.14. Calculated rocking-curves for InP (004) Bragg reflection for three different thicknesses for CuK$_{\alpha 1}$ radiation [6.86]

reflected waves critically depends on the boundary planes of the crystalline film and its homogeneity. According to Bartels [6.33] for thin films with a thickness D small compared to the extinction thickness d_{ext} (neglecting absorption) the *FWHM* is:

$$FWHM = \frac{\lambda \cdot \gamma_s}{D \cdot \sin 2\theta} \tag{6.25}$$

The oscillation period of the reflected intensity is directly connected with the thickness as shown in Fig. 6.14 for a (004) diffraction of (001) oriented InP crystals.

6.4 Strain Phenomena

Until now only free-standing films have been discussed. In practice epitaxially grown heterostructures have to be analysed. Information about the composition of alloy semiconductor layers, its gradients and its state of strain is required. In the case of large strain values a tilt of the layer planes relatively to the substrate can occur which has to be distinguished from similar effects for layers grown on misoriented substrates.

6.4.1 Strains in Epitaxial Layers

For about 20 years strains in epitaxial layers have been studied by various X-Ray methods. The lattice constant of a thin film which grows coherently on a single-crystalline substrate is modified parallel to the growth direction. Due to strain the crystallographic properties of single crystal films become anisotropic and the measured X-Ray diffraction phenomena have to be analysed in order to deduce the lattice parameters and the reduction of symmetry due to strain (Fig. 6.15). It is assumed that the film is much thinner than the substrate and that therefore the substrate remains unstrained [6.99].

In most cases the growth direction is a [001] direction, sometimes also an [111] or an [110] direction is chosen. By X-Ray measurements information is obtained on the strains which are related to the stresses in a cartesian coordinate system by the following equation (for cubic symmetry) [6.100]

$$
\begin{pmatrix} \sigma_{xx} \\ \sigma_{yy} \\ \sigma_{zz} \\ \sigma_{yz} \\ \sigma_{xz} \\ \sigma_{xy} \end{pmatrix} = \begin{pmatrix} C_{11} & C_{12} & C_{12} & \cdot & \cdot & \cdot \\ C_{12} & C_{11} & C_{12} & \cdot & \cdot & \cdot \\ C_{12} & C_{12} & C_{11} & \cdot & \cdot & \cdot \\ \cdot & \cdot & \cdot & C_{44} & \cdot & \cdot \\ \cdot & \cdot & \cdot & \cdot & C_{44} & \cdot \\ \cdot & \cdot & \cdot & \cdot & \cdot & C_{44} \end{pmatrix} \begin{pmatrix} \epsilon_{xx} \\ \epsilon_{yy} \\ \epsilon_{zz} \\ \epsilon_{yz} \\ \epsilon_{xz} \\ \epsilon_{xy} \end{pmatrix} \tag{6.26}
$$

where the C_{ij} are the stiffness coefficients. For growth along z-direction with x, y in the plane of the film the stress tensor component σ_{zz} is always zero

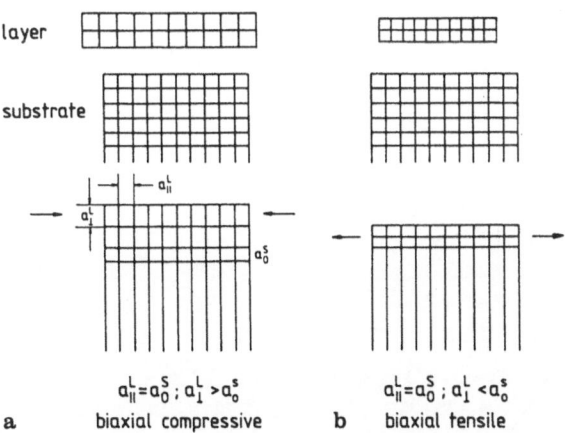

Fig. 6.15. Schematic illustration of the origin of biaxial compressive ($a_L > a_S$) (a) and biaxial tensile strain ($a_L < a_S$) (b). The in-plane lattice constants remain unchanged and the $a_\perp$ lattice constants are either increased (compressive in plane strain) or decreased (tensile in plane strain). Diagram corresponds to a tetragonal distortion

and thus

$$\sigma_{zz} = 0 = C_{12}\epsilon_{xx} + C_{12}\epsilon_{yy} + C_{11}\epsilon_{zz} \tag{6.27}$$

This yields a relation between $\epsilon_{zz}, \epsilon_{xx}$ and ϵ_{yy} provided that the elastic stiffnesses are known. Assuming a tetragonal distortion [6.101] the in-plane lattice spacings along the x [100] and y [010] direction are that of the substrate a_0^S, whereas the lattice spacing along the growth direction is modified (Fig. 6.16). In this case there exists a finite angle $\Delta\varphi$ between the [101] directions of the substrate and film. As can be seen from Fig. 6.16 for tetragonal distortion $\Delta\varphi_1 = \Delta\varphi_2$ results.

The statements given above can be expressed in the following way: If there are no forces in growth direction, then $\sigma_{zz} = 0$. For biaxial tension or compression $\sigma_{xx} = \sigma_{yy}$ and thus

$$\sigma_{zz} = C_{11}\epsilon_{zz} + C_{12}\epsilon_{xx} + C_{12}\epsilon_{yy} = 0 \tag{6.28}$$
$$\sigma_{xx} = C_{11}\epsilon_{xx} + C_{12}\epsilon_{yy} + C_{12}\epsilon_{zz} \tag{6.29}$$
$$\sigma_{yy} = C_{12}\epsilon_{xx} + C_{11}\epsilon_{yy} + C_{12}\epsilon_{zz} \tag{6.30}$$

and thus it follows $\epsilon_{xx} = \epsilon_{yy}$. Therefore in the strained case the lattice constants of the film in x and y direction are:

$$a_x = a_y = a_\| \tag{6.31}$$

For tetragonal distortion and pseudomorphic growth the face diagonals in the xy-plane [110] and [$\bar{1}$10] are of the same length. The components of the strain tensor are given by:

$$\epsilon_{xx} = \epsilon_{yy} = \frac{a_0^S - a_0^L}{a_0^L} \tag{6.32}$$

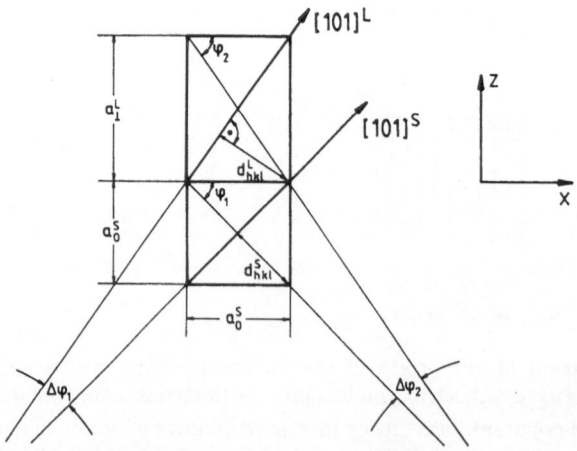

Fig. 6.16. Definition of angles in tetragonal distortion: for layer L on substrate S

and

$$\epsilon_{zz} = \frac{a_z^L - a_0^L}{a_0^L} \tag{6.33}$$

where a_z^L is equivalent to $a_\perp$. It is important to note that in the X-Ray investigation of thin epitaxial films on comparatively thick substrates usually no absolute measurements of the lattice parameters of both the film and the substrate are made but that the measurements are made relative to the substrate.

Therefore, apart from the conventional definition of the components of the strain tensor, following the work of Hornstra and Bartels [6.102] it has become convenient to define an X-Ray strain which compares lattice constants in the strained film with that of the unstrained substrate (parallel means within the surface plane, perpendicular means along the surface normal, i.e. along the growth direction):

$$\epsilon_{\parallel}^L = \frac{a_{\parallel}^L - a_0^S}{a_0^S} \tag{6.34}$$

$$\epsilon_{\perp}^L = \frac{a_{\perp}^L - a_0^S}{a_0^S} \tag{6.35}$$

Using (6.32) these X-Ray strains ϵ^L of the films are related to the actual strain in the epitaxial layer ϵ_L, which would be measured relative to an unstrained epitaxial film, by the following equation:

$$\epsilon_L = \left(\epsilon^L + 1 \right) \frac{a_0^S}{a_0^L} - 1 \tag{6.36}$$

where a_0^L is the lattice constant of the unstrained layer. $\epsilon_L^\perp$ and $\epsilon_L^\parallel$ are related to each other via the elastic constants of the epitaxial layer (for [001] oriented samples):

$$\epsilon_{\perp}^L = \left(1 + \frac{2C_{12}}{C_{11}} \right) \frac{a_0^L - a_0^S}{a_0^S} - \frac{2C_{12}}{C_{11}} \epsilon_{\parallel}^L \tag{6.37}$$

(note: the relation for the X-ray strains is different from that for the actual strain ϵ_L)

$$\epsilon_L^\perp = -\frac{2C_{12}}{C_{11}} \epsilon_L^\parallel \tag{6.38}$$

Using symmetric X-Ray diffraction, information on a_z^L is obtained. From (6.37) the unstrained lattice constant of the layer which is important for determining e.g. the composition of alloys is then given by the following relation:

$$a_0^L = \frac{C_{11}}{C_{11} + 2C_{12}} \left(a_z^L - a_0^S \right) + a_0^S \tag{6.39}$$

(Here the contribution resulting from $\epsilon_\parallel^L$ is zero). For growth direction along [111] and [110] similar relations are given by Segmüller and Murakami [6.99], Ortner [6.103] and Anastassakis [6.104].

Only in the case of pseudomorphic growth and a simple heterolayer sufficient information can be extracted from the symmetric X-Ray diffraction alone. Usually the growth of heterolayers changes from pseudomorphic to non-pseudomorphic beyond a certain critical thickness, and partial strain relaxation occurs. Therefore $a_\parallel^L \neq a_0^S$. In order to determine the state of strain, in addition to symmetric reflections, asymmetric ones are needed: the definition of angles for such an asymmetric situation is given in Fig. 6.17.

In the rocking-curve the angular difference $\Delta\omega$ between the Bragg diffractions (hkl) of the layer and the substrate is due to two components:

$$\Delta\theta_B = \theta_B^S - \theta_B^L \tag{6.40}$$

and

$$\Delta\varphi = \varphi^S - \varphi^L \tag{6.41}$$

The angular separation between the diffraction peaks is either $\Delta\theta_B + \Delta\varphi$ or $\Delta\theta_B - \Delta\varphi$ in reflection geometry A or B, respectively. Thus:

$$\Delta\omega_A = \Delta\theta_B + \Delta\varphi \tag{6.42}$$

$$\Delta\omega_B = \Delta\theta_B - \Delta\varphi \tag{6.43}$$

According to Bartels [6.33] the difference $\Delta\theta_B$ which corresponds to the differences in lattice spacing and $\Delta\varphi$ which determines the difference in lattice plane orientation is used to calculate $(\Delta a/a)_\perp$ and $(\Delta a/a)_\parallel$

$$\left(\frac{\Delta a}{a}\right)_\perp = \Delta\varphi \tan\varphi - \Delta\theta_B \cot\theta_B \tag{6.44}$$

$$\left(\frac{\Delta a}{a}\right)_\parallel = -\Delta\varphi \cot\varphi - \Delta\theta_B \cot\theta_B \tag{6.45}$$

where

$$\Delta\theta_B = \theta^S - \theta^L = \frac{1}{2}(\Delta\omega_A + \Delta\omega_B) \tag{6.46}$$

$$\Delta\varphi = \varphi^S - \varphi^L = \frac{1}{2}(\Delta\omega_A - \Delta\omega_B) \tag{6.47}$$

The Eqs. (6.44, 6.45) are approximations for small $\Delta\theta$ and $\Delta\varphi$. For large strains the errors in $(\Delta a/a)_\perp$ and $(\Delta a/a)_\parallel$ are of the order of several percent. In order to determine $\Delta\theta_B$ and $\Delta\varphi$ independently of each other, measurements in the geometries A and B have to be performed.

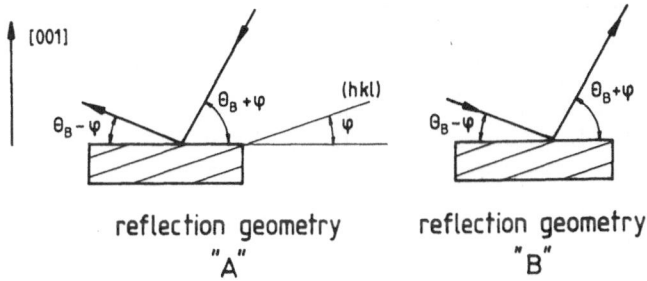

Fig. 6.17. Definition of diffraction geometry for asymmetric Bragg diffraction. φ: angle between plane (hkl) and sample surface, θ_B: Bragg angle. With the help of reflection geometries A ("high incidence") and B ("low incidence") strain, tilt and terracing measurements are performed

In order to confirm whether an epitaxial film on a (001) surface is tetragonally distorted, several asymmetric Bragg reflections have to be measured, e.g. those belonging to a [110] zone. Meyerheim [6.98] used a GaAs/Ga$_{0.4}$Al$_{0.6}$As structure and investigated the asymmetric (117), (115), (112), (335) and (444) diffraction and the symmetric (004) diffraction. From the latter the value of $(\Delta a/a)_\perp$ was determined using (6.44).

The differentiated Bragg equation

$$\left(\frac{\Delta d}{d}\right)_{llh} = \frac{\Delta\omega_A + \Delta\omega_B}{2\tan\theta_{llh}} \tag{6.48}$$

yields the value of the distortion normal to the lattice plane (llh). The following relations hold [6.102]:

$$\frac{\Delta\varphi}{(\Delta d/d)_\perp} = \cos\varphi\sin\varphi \tag{6.49}$$

$$\frac{(\Delta d/d)_{llh}}{(\Delta d/d)_\perp} = \cos^2\varphi \tag{6.50}$$

In Fig. 6.18 the measured values of these expressions are plotted vs. φ and compared with the trigonometric functions. The agreement between the measured and calculated values proves the validity of the models for the tetragonal distortion in this case.

There remains a further problem: namely whether or not under certain conditions a more general distortion, e.g. a monoclinic, triclinic or orthorhombic one, can be observed. In the case of anisotropic and inhomogeneous strain relaxation of In$_x$Ga$_{1-x}$As/GaAs structures Grundmann et al. [6.105] and Giannini et al. [6.106] have suggested such a distortion. Such a deformation would lead to non-equivalent angles $\Delta\varphi_1 \neq \Delta\varphi_2$ (Fig. 6.19).

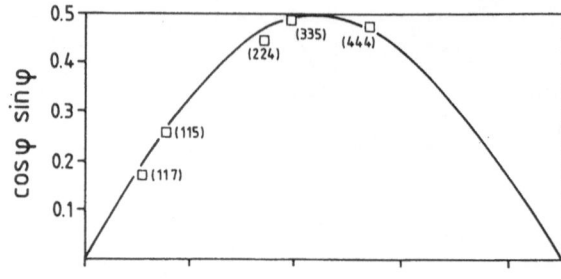

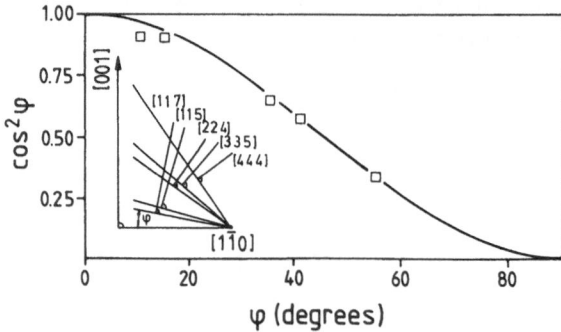

φ (degrees)

Fig. 6.18. Determination of elastic tetragonal distortion for a GaAs / Ga$_{0.4}$Al$_{0.6}$As heterostructure. Measurement quantities are $\Delta\varphi$ and $\Delta d/d_{hkl}$: upper half: $\cos\varphi\sin\varphi(=\Delta\varphi/(\Delta d/d))$ versus φ. lower half: $\cos^2\varphi(=((\Delta d/d)_{hkl})/(\Delta d/d))$ versus φ for 5 Bragg diffractions where $\Delta d/d$ denotes the misfit to the direction of the surface normal [6.98]

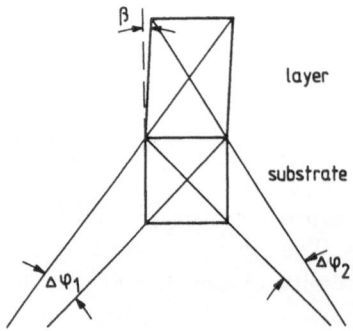

Fig. 6.19. Illustration of monoclinic distortion of cubic layer with respect to the substrate. In contrast to tetragonal distortion in Fig. 6.15 angles $\Delta\varphi_1 \neq \Delta\varphi_2$

For the Al$_{0.6}$Ga$_{0.4}$As/GaAs structures the equivalence $\Delta\varphi_1 = \Delta\varphi_2$ of which was checked by measurement of the (115) and ($\bar{1}\bar{1}5$) Bragg reflection as shown in Fig. 6.20 for the A and B geometries. Since the experimental result of Fig. 6.20 proves that:

$$(\Delta\omega_A - \Delta\omega_B)_1 - (\Delta\omega_A - \Delta\omega_B)_2 = \Delta\varphi_1 - \Delta\varphi_2 = 0 \qquad (6.51)$$

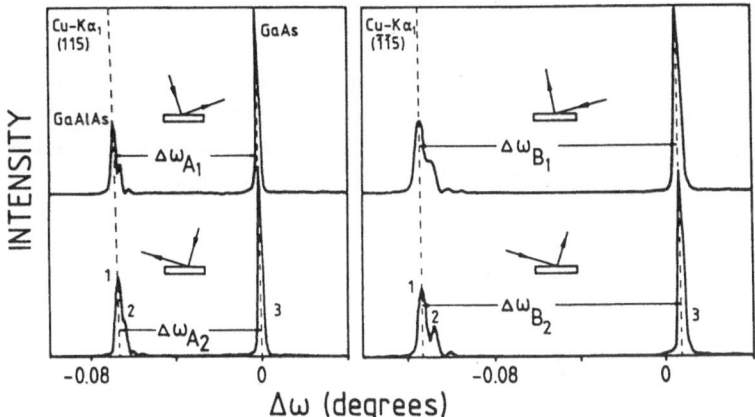

Fig. 6.20. Method for confirming the tetragonal distortion using (115) and ($\overline{1}\overline{1}5$) Bragg reflections in scattering geometries A and B (Fig. 6.17), [6.98]

the distortion is tetragonal in this case. The monoclinic angle β is given after Bartels and Nijman [6.109] by

$$\beta = \frac{\Delta\varphi_1 - \Delta\varphi_2}{\sin^2\varphi} - \frac{1}{2} \qquad (6.52)$$

Meyerheim [6.98] has suggested that, based on the symmetry of a (001) plane which is D_{4v}, no monoclinic distortion is possible. However, for planes which have no higher symmetry than C_2 or C_s [like a (211) plane] it can occur and therefore also for other vicinal surface planes.

In many cases the strains in epitaxial layers are of the order of 10^{-4} to 10^{-2}. However, high resolution X-Ray techniques are capable of determining even much smaller strain values, as low as 10^{-5}. Segmüller [6.4] has reported the observation of changes in the lattice strain in AlGaAs layers heavily doped with Sn grown on a GaAs substrate due to the emission of electrons from deep level (DX-) centres upon illumination. The angular distance between high and low conductivity state changed by $\Delta\theta = -0.0010° \pm 0.0003°$ for the (400) reflection using a Cu $K_{\alpha 1}$ radiation from a spot focus $1 * 1\,mm^2$. The measurements were performed with a triple-axis spectrometer at low temperatures (15 K) and the strain normal to the surface was deduced which corresponded to a change of lattice parameter a of $\Delta a/a = +(14 \pm 4) \cdot 10^{-6}$.

The lattice strain results from the population of the conduction band with electrons ejected from the DX centers (lattice expansion due to conduction band filling), with a concentration of about $1.5 \cdot 10^{18}\,cm^{-3}$ and partly from the strain contribution from the emptied DX centres.

This beautiful example shows the standard of high resolution X-Ray diffraction which has been achieved nowadays.

6.4.2 Partial Relaxation of Strain

With increasing layer thickness the elastic energy increases. Beyond a critical value misfit dislocations are formed and the strain is partly relieved [6.110–6.112]. In such a case, for [001] growth direction, the lattice parameters of the layer are still such that $a_x = a_y$ but different from the substrate.

We define a parameter δ

$$\delta = \frac{a_\parallel^L - a_0^S}{a_0^S} \tag{6.53}$$

as a quantity measuring the partial strain relief. For complete strain relaxation $a_\parallel^L = a_0^L$ holds, where a_0^L is the unstrained lattice parameter of the layer material.

In order to determine a_0^L in such a case [for (001) growth] we use

$$a_0^L = \frac{C_{11}}{C_{11} + 2C_{12}}[a_\perp^L - a_\parallel^L] + a_\parallel^L \tag{6.54}$$

which corresponds to (6.39). The measurement thus proceeds like in the pseudomorphic case using asymmetrical Bragg reflections in A and B geometry as described above.

Partial strain relief leads to a graded strain profile. If reciprocal lattice scans can be performed, then the determination of the lattice constants of epitaxials layers irrespective of their strain status is straightforward as shown in the following. We consider for simplicity the growth of an epilayer along the [001] direction on a (001) substrate and assume that a_L (bulk) is larger than a_S and that the layer growth is pseudomorphic. The in-plane lattice constant of the layer is identical to that of the substrate and a tetragonal distortion occurs (Fig. 6.21).

In the partially relaxed situation (Fig. 6.21 central part) the in-plane lattice constant of the layer is larger that of the substrate but there is still

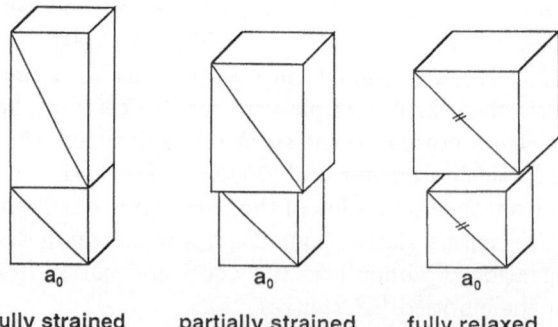

fully strained   partially strained   fully relaxed

Fig. 6.21. Scheme of different strain status for an epilayer on a substrate

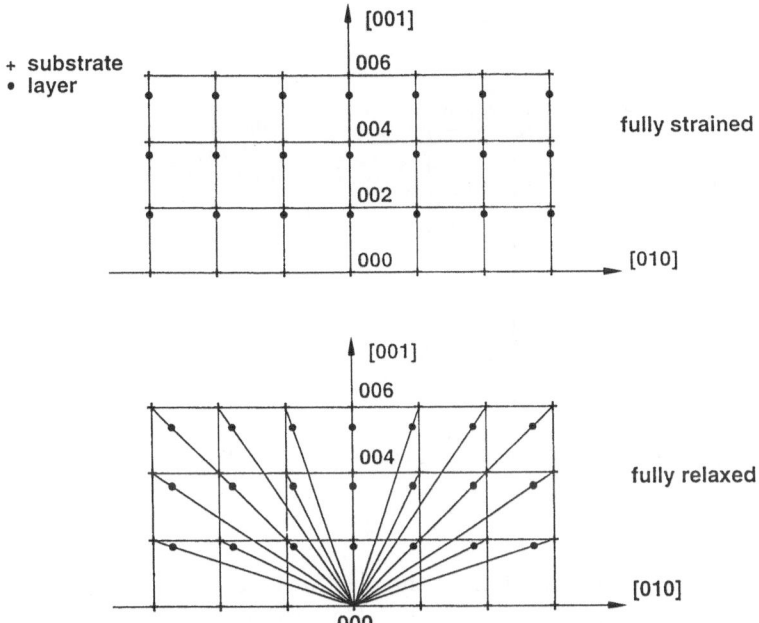

Fig. 6.22. Scheme of different strain status for an epilayer on a substrate in the reciprocal lattice

a tetragonal distortion whereas in the fully relaxed case the layer is cubic with its bulk lattice constant a_L different from a_S. In the reciprocal lattice in Fig. 6.22 the situation for pseudomorphic growth (fully strained) and for the case of full relaxation (lower part) are shown.

Since for pseudomorphic growth the in-plane lattice constants are identical, the tetragonal distortion of the layer becomes apparent from the relative positions of the layer reciprocal lattice point maxima with respect to those of the substrate. In this special cross section through the reciprocal lattice, the lattice constant of the layer along growth direction is larger than its equilibrium bulk value and larger than that of the substrate. For the fully relaxed case, the layer has regained its cubic structure and consequently any given direction within the substrate is parallel to the corresponding one within the layer as shown in the lower part of Fig. 6.22. In Fig. 6.23 we demonstrate how reciprocal space maps can be used to determine independently of each other the in-plane ($a_\parallel$) and perpendicular lattice constants ($a_\perp$) of an epitaxial layer without any knowledge of the elastic constants. As an example we use $Si_{1-x}Ge_x$ on (001) Si substrate.

For pseudomorphic layer growth the symmetrical (002) and asymmetrical (202) reciprocal lattice points of the SiGe layer are situated below, i.e. along the [001] direction, the corresponding ones of the Si substrate $\left(a_\parallel(Si) = a_\parallel(SiGe), a_\perp(SiGe) > a_\perp(Si)\right)$. If the relaxation process starts, the

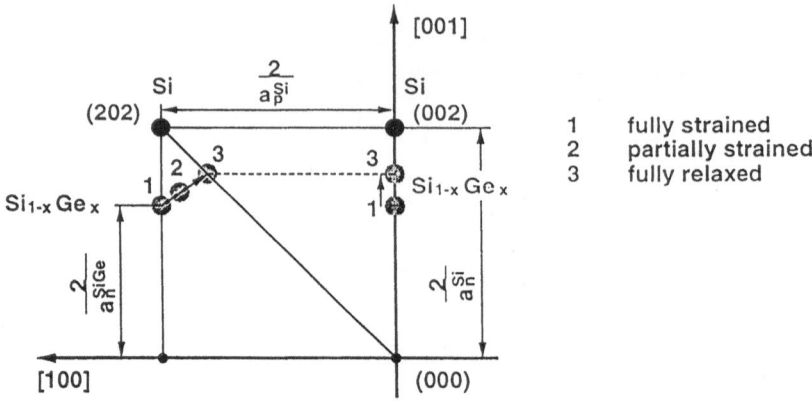

Fig. 6.23. Evaluation of lattice constants for $Si_{1-x}Ge_x$ on Si(001) without any knowledge of elastic constants

SiGe reciprocal lattice points move from the positions marked "1" to the positions "2". In the fully relaxed case, the $Si_{1-x}Ge_x$ REciprocal Lattice Points (RELP's) are at positions "3" (the [202] directions in the Si substrate and the SiGe epilayer are parallel to each other and therefore the (202) RELP of the $Si_{1-x}Ge_x$ layer lies on the line connecting the center (000) of the reciprocal lattice and the (202) RELP of the Si substrate). Finally, the reciprocal maps are ideal for detecting any crystallographic layer tilt. Using a map around a symmetrical reflection one immediately recognizes whether the layer (00n) RELP is situated on the line connecting the (000) origin with the substrate (00n) peak. If this is not the case (as shown in Fig. 6.24) the tilt angle is immediately apparent. Consequently, reciprocal space maps offer appreciable advantage for the identification of the strain status as well as of a possible tilt of epitaxial layers.

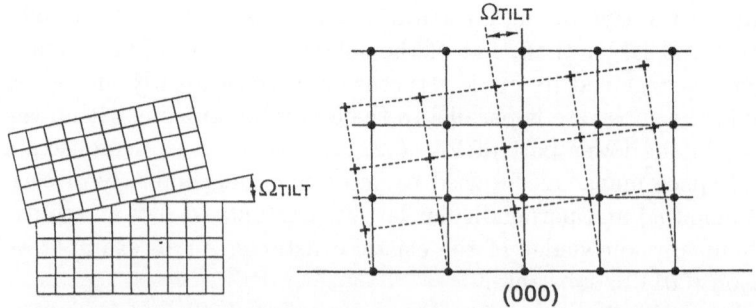

Fig. 6.24. Scheme for evaluation of tilt angle between epilayer and substrate by reciprocal lattice scans

6.5 Rocking-Curves from Heterostructures

As has been shown above in a rocking-curve from a single heterostructure, two peaks appear since the lattice constant of substrate and layer are normally different. Even if the film has the same crystalline perfection as the substrate, its diffraction peak is much broader due to the finite thickness effect discussed above. The angular distance $\Delta\omega$ between the two peaks yields direct information on chemical composition in the case of alloy semiconductors.

6.5.1 Single Heterostructures

In Fig. 6.25 a sequence of calculated diffraction patterns of AlGaAs/GaAs structures is shown for two Al contents (10% and 20%) and two different AlGaAs thicknesses ($d = 2\,\mu m$ and $d = 1\,\mu m$). For the thinner AlGaAs layer the finite thickness fringes are clearly visible, as well as the shift of the AlGaAs peak due to the different chemical composition. The interpretation of X-Ray scattering from more than one epitaxial layer has usually to be based on a careful analysis, because the amplitudes of the scattered waves interfere with each other and produce complicated patterns (Fig. 6.25 lower right corner).

However, it should be pointed out that for layers with thicknesses below about $1\,\mu m$, the X-Ray diffraction method for a simultaneous determination

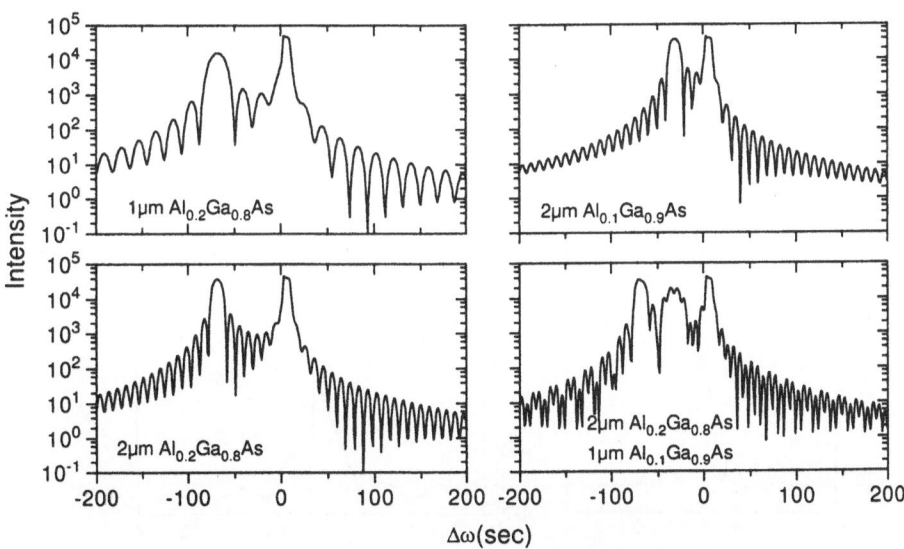

Fig. 6.25. Calculated diffraction pattern for AlGaAs layers of different compositions and thicknesses on a GaAs substrate. The lattice mismatch and finite thickness are reflected in a broadening of the main peaks and the appearance of interference fringes at characteristic positions

of alloy composition and mismatch based on the measurement of separation of diffraction peaks in the high resolution diffractograms has to be handled with care. Fewster and Curling [6.87] have reported the occurrence of a considerable peak shifting for identical mismatch between layer and substrate but different thicknesses. As the layer thickness is reduced, the layer peak shifts towards the substrate peak. Fewster and Curling [6.87] have shown that the observed peak shifts can be simulated within the solution of the Takagi-Taupin equations, but not within the kinematical approach.

This effect occurs if the layer grows coherently on the substrate, i.e. for comparatively low interfacial misfit dislocation density. As a numerical example, Fewster and Curling have shown that for a $0.2\,\mu m$ layer of $In_{0.524}Ga_{0.476}As$ on InP substrate a 10% error in mismatch would occur when derived from the peak position. Consequently, care should be taken if the composition of submicrometer layers has to be determined with an accuracy better than 10%. A reliable determination of strain and knowledge of relaxation or measurements of asymmetric Bragg reflections and alloy composition thus requires the simulation of diffraction profiles.

Subtle phenomena can be deduced from the Pendellösung fringes of heterostructures as shown by Tapfer et al. [6.93]. In Fig. 6.26 three different Bragg reflections a symmetrical (004), and two oblique ones (311) and (422)

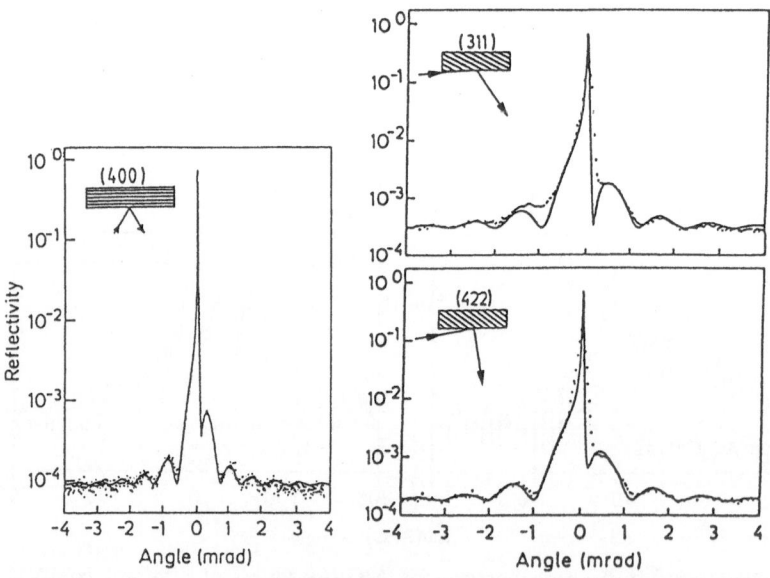

Fig. 6.26. Experimental (dotted line) and simulated (solid line) diffraction pattern of the symmetrical (400) and the asymmetrical (311) and (422) diffraction patterns of a 140 nm thick Si cap-layer deposited on 3 monolayers of Ge on Si(100) substrate. All patterns were simulated with the same parameter set [6.93]

are shown for a structure with a layer sequence: Si substrate, 3 monolayers Ge (= 0.437 nm), and on top a 140 nm thick Si layer. The Ge layers still grow pseudomorphic since its thickness is below the critical one for the formation of misfit dislocations. The Si substrate and the Si cap layer are separated by the strained Ge layers. The origin of the oscillations observed in Fig. 6.26 is the following: because the strained lattice constant of the 3 Ge monolayers in growth direction is different from the silicon one, the waves diffracted from the Si cap layer are phase shifted with respect to those diffracted from the Si substrate. The angular distance between two interference fringes is related to the cap layer thickness [6.93]. The observed spectra are compared with calculated ones based on the dynamical diffraction theory. For the same sample also asymmetric (311) and (422) diffraction patterns are shown in order to determine unambiguously the strain status. The symmetrical (004) reflection is sensitive to the lattice strain perpendicular to the layers while the two asymmetrical ones are influenced both by the lattice strain parallel and perpendicular to the layers. All three diffractograms were interpreted with the same parameter set and indeed, for the parallel strain $\epsilon_{xx} = 0$ was found, which shows that the distortion of the three Ge monolayers along the [001] direction is tetragonal. This example clearly demonstrates that the experimental data can only be understood on the basis of a rather elaborate model calculation. On the other hand it demonstrates the sensitivity of X-Ray diffraction for monolayer resolution.

6.5.2 Composition Gradients

Often the intensity oscillations accompanying the Bragg reflection of an epitaxial layer are asymmetrical. Such a behaviour always occurs if there are composition (or strain) gradients as assumed in Fig. 6.27. The calculated intensities are shown for a diffraction pattern from an one micron thick $In_xGa_{1-x}As$ layer deposited on InP with $x = 0.537$ corresponding to a lattice mismatch of $\Delta a/a$ of $+5 \cdot 10^{-4}$. Introducing a composition gradient the fringes become strongly asymmetric with an enhancement of the fringes on the left hand side of the peak. The In content was varied from 54.0% to 53.7% over the first $0.5\,\mu m$ and kept constant of the top $0.5\,\mu m$. An experimental example of a layer peak of 620 nm InGaAs on InP with a 1000 nm InP cap layer indicating both the composition gradient from the layer and the cap is shown in Fig. 6.28. A detailed discussion of these phenomena are given, e.g., in [6.113–6.116].

6.5.3 Characterisation of Epitaxial Layers Grown Tilted Relative to the Substrates

Often, epitaxial layers are grown on off-oriented substrates, the surface normal of which deviates from a low index crystallographic direction by as much as

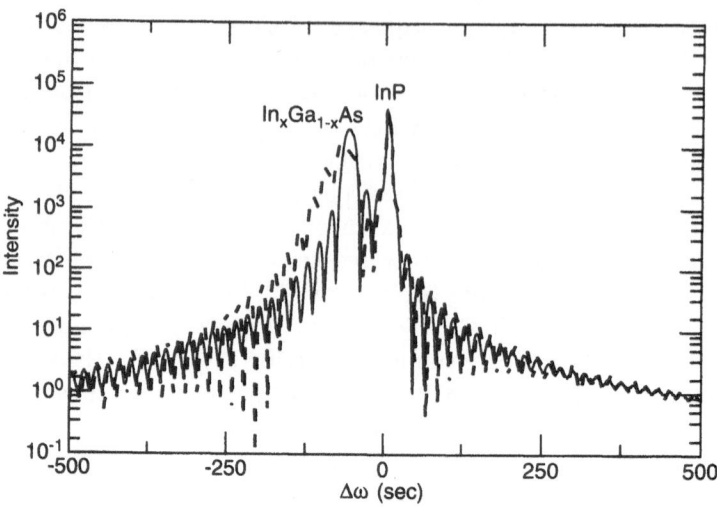

Fig. 6.27. Influence of composition gradient on reflectivity spectra. Full line: simulation of diffraction pattern of strained $In_{0.537}Ga_{0.463}As$ (1 μm) deposited on InP (CuK$_{\alpha 1}$, (004) diffraction). Finite thickness fringes are visible on both sides of the InGaAs peak. Dotted line: influence of a composition gradient within the InGaAs layer. Layer peak shape and fringe structure become asymmetric

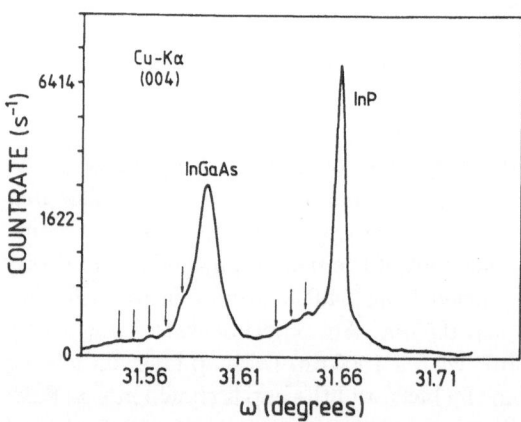

Fig. 6.28. Experimental diffraction pattern of a 620 nm thick InGaAs layer deposited on InP with a 1 μm thick InP cap layer. Finite thickness fringes yield $d = 620$ nm, asymmetric increase of reflection profile on the low angle side of the main diffraction peak indicates composition gradient [6.13]

a few degrees. As a consequence the asymmetric geometry has to be taken into account. On the other hand, epitaxial layers can also grow tilted with respect to the substrate. In both cases the experimental rocking-curves are similar and therefore difficult to distinguish. They can only be separated by use of more advanced techniques such as reciprocal space mapping. Further information on this topic is given in Sect. 6.7.

In Fig. 6.29 the diffraction pattern is shown of an $Al_xGa_{1-x}As$ layer with $x = 0.2$, grown on GaAs with a 2 degree misorientation in the nearest [110] direction. For diffraction conditions where the misorientation direction lies in the diffraction plane towards the incident X-Ray beam the full line results, whereas for the opposite orientation the dashed line would be observed. For a symmetrical Bragg reflection the asymmetry and/or tilt is easily verified by rotating the film around its surface normal: the diffraction feature exhibits a periodic modulation of peak positions. Nagai [6.118, 6.117] observed for the first time in InGaAs/GaAs layers that the epitaxial orientation was inclined relative to the substrate crystal.

He presented a model in which the tilt is a consequence of the surface steps and of the lattice mismatch. In this model the miscut direction of the substrate and the tilt direction of the epilayer are parallel to each other, the misorientation angle between the (001) planes and the substrate surface being α and the tilt angle between the substrate and epilayer planes 2β (Fig. 6.30). If the relative lattice mismatch is $\Delta a/a$, then $\tan 2\beta = \tan \alpha \cdot \Delta a/a$ (where for small α, $\cos \alpha = 1$ was taken [6.119]). The effect of the misorientation

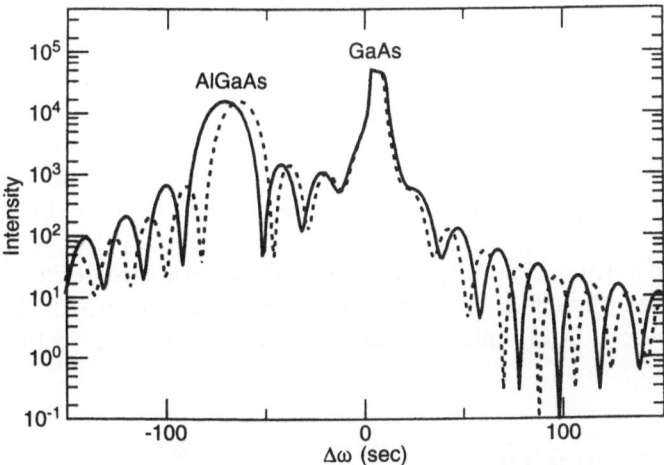

Fig. 6.29. Influence of misoriented GaAs substrate (miscut: 2°) on $Al_xGa_{1-x}As$ diffraction pattern: full line: miscut direction lies towards the incident beam and dashed line away from it. Analysis of the pattern yields two different x-values of 0.19 and 0.21 whereas the true value is 0.2

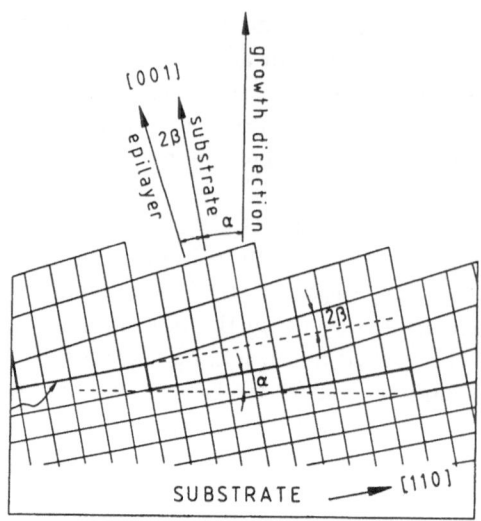

Fig. 6.30. Nagai's model for the growth of tilted layers on misoriented substrates; α is the miscut angle, 2β the tilt angle between substrate and epilayer [6.119]

angle and the step-edge orientation on the diffraction pattern of GaAs/AlAs superlattices has been studied by Auvray et al. [6.120] in order to evaluate the interface quality of the superlattice. They found that the [$\bar{1}$10] step-edge orientation yields the best interfaces.

The validity of Nagai's model was confirmed by Neumann et al. [6.121] for the growth of GaAsSb/GaAs superlattices on GaAs, and by Auvray et al. [6.119] for $Al_xGa_{1-x}As$/GaAs heterostructures and for AlAs/GaAs super-lattices. Pesek at al. [6.122, 6.123] have recently studied among other systems the epitaxial orientation of ZnSe with respect to a GaAs substrate and have found that the tilt directions of the substrate and of the epilayer are not parallel but that an azimuthal rotation exists between the directions of the relative tilt and the substrate miscut. They concluded that Nagai's model is only valid for small-misfit systems where the formation of misfit dislocations is yet excluded.

In Fig. 6.31 the determination of the tilt from X-Ray data is reproduced for ZnSe/GaAs as an example [6.122]. Similar effects are of importance in the anisotropic strain relaxation of epilayer quantum well systems, e.g. [6.105, 6.124].

6.6 Multilayer Structures

In this Section the application of HRXRD to multilayers including Ewald spheres constructions for the interpretation of DCD diffractograms and scans

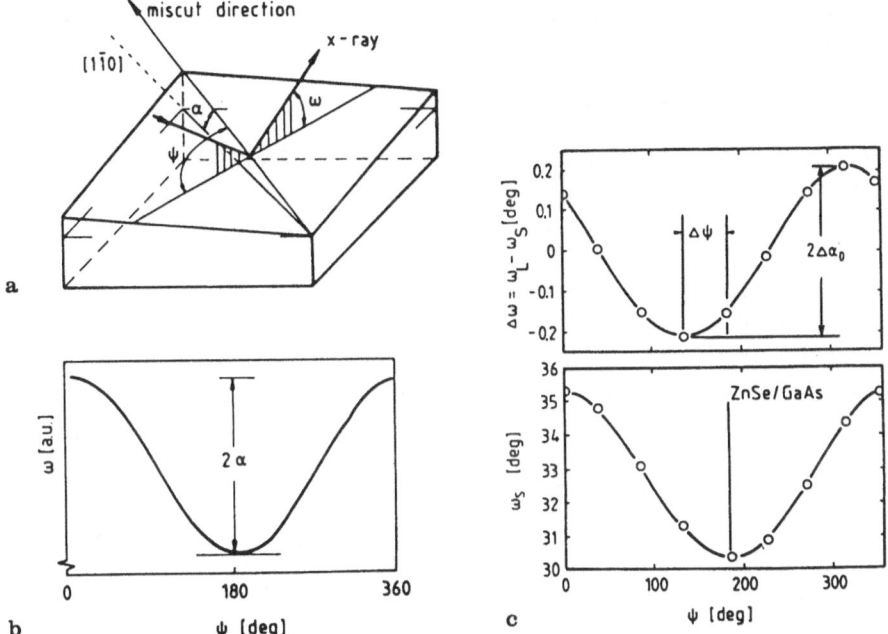

Fig. 6.31. Influence of tilted substrates on diffraction from epilayers. **a** Sketch of scattering geometry for (001) surface with $[1\bar{1}0]$ miscut direction; miscut angle: α; ψ is the angle between the projection of incident X-Ray beam on the sample surface and the miscut direction; ω is the angle between X-Rays and sample surface for the Bragg condition, **b** plot of ω versus ψ exhibits periodic variation of ω with period of 360° and a modulation corresponding to 2α for $\psi = 180°$, **c** Relative tilt $\omega = \omega_L - \omega_S$ from epilayer and substrate versus ψ (upper part) and tilt of substrate versus ψ (lower part) exhibit a phase difference $\Delta\psi = 50°$ [6.122]

in reciprocal space is given. Interdiffusion of superlattices (SL's), tilt, terracing, and mosaic spread are discussed.

6.6.1 Superlattices

Artificially structured multilayers have become an important class of new materials which offer within certain limits unique electronic, optical, magnetic and mechanical properties. Along growth direction, usually two layers of different chemical composition are alternatively deposited. The one-dimensional periodicity with period D is the origin of a one-dimensional periodic potential, which is superimposed on the three-dimensional crystal potential, the period of which is determined by the lattice constants of the materials. For all properties listed above the structural perfection of a multilayer system is decisive. The X-Ray diffraction pattern of such a periodic structure consists of a series of satellite peaks accompanying the main zero-order diffraction peak along the direction of chemical modulation. The period D is given by:

$$D = \frac{(L_i - L_j)\lambda}{2(\sin(\theta_i) - \sin(\theta_j))} \tag{6.55}$$

where L_i, L_j are diffraction order indices, $\sin(\theta_i)$ and $\sin(\theta_j)$ the corresponding Bragg angles of the satellites L_i, L_j. (Note: The angular distance between adjacent satellites peaks L_i, L_j need not be equidistant, an effect which becomes important for short period superlattices). Besides the position of the main superlattice peak and of the satellites, their intensities and *FWHMs* are experimentally accessible. These data provide in principle information on thickness variations, composition and composition variations throughout the multilayer structure, interface roughness and grading, interdiffusion, and strains within the layers. We start with a perfect superlattice structure of a finite total thickness and abrupt interfaces. We consider a periodic sequence of layers of materials A and B with layer thicknesses d_A and d_B, the period $D = d_A + d_B$. The lattice parameters of the two materials along growth direction are a_A and a_B. The structure consists of N periods as shown in Fig. 6.32. A rectangular, i.e. abrupt profile of the layer sequences along growth direction is assumed. In order to illustrate the diffraction pattern in a simple way, we use the fact that the real space and the reciprocal space are related to each other via a Fourier transformation. For the intensities the kinematical approximation is used.

In a one-dimensional approximation the period D is represented by a function $f(z)$ which describes the periodic scattering centers by the atomic form factors (which are different in materials A and B). In kinematical approximation the scattering intensity is proportional to the $FF^*(k)$ where $F(k)$ is the Fourier transform of $f(z)$. $f(z)$ is described as a sum of two functions $f_A(z)$ and $f_B(z)$. In real space, these two functions are now presented as multiplications, convolutions and summations according to Fig. 6.32. The slit functions with slits z_1 and z_2 are multiplied with a Dirac comb (a_A, a_B) and are convoluted with δ-function Dirac combs of separation D. The contribution from the two components A and B are added and are finally multiplied with a rectangular function $(N \cdot D)$.

The diffracted intensity is obtained by Fourier transformation of these functions and given by $[F_A(k) + F_B(k)]^2$, where F_A, F_B are the Fourier transforms of $f_A(z)$, $f_B(z)$. For sufficiently large difference between a_A and a_B a diffraction pattern results which is shown in Fig. 6.32.

The main SL maxima appear close to $2\pi/a_A$ and $2\pi/a_B$ accompanied by satellites, the width of which is given by $(4\pi/N \cdot D)$. The separation between two subsequent maxima is given by $2\pi/D$ and the width of the envelope function depends on $4\pi/d_A$ and $4\pi/d_B$; n_0 denotes the diffraction order [6.125]. The angular position of the zeroth order superlattice peak in such a strained-layer situation corresponds to a lattice constant determined

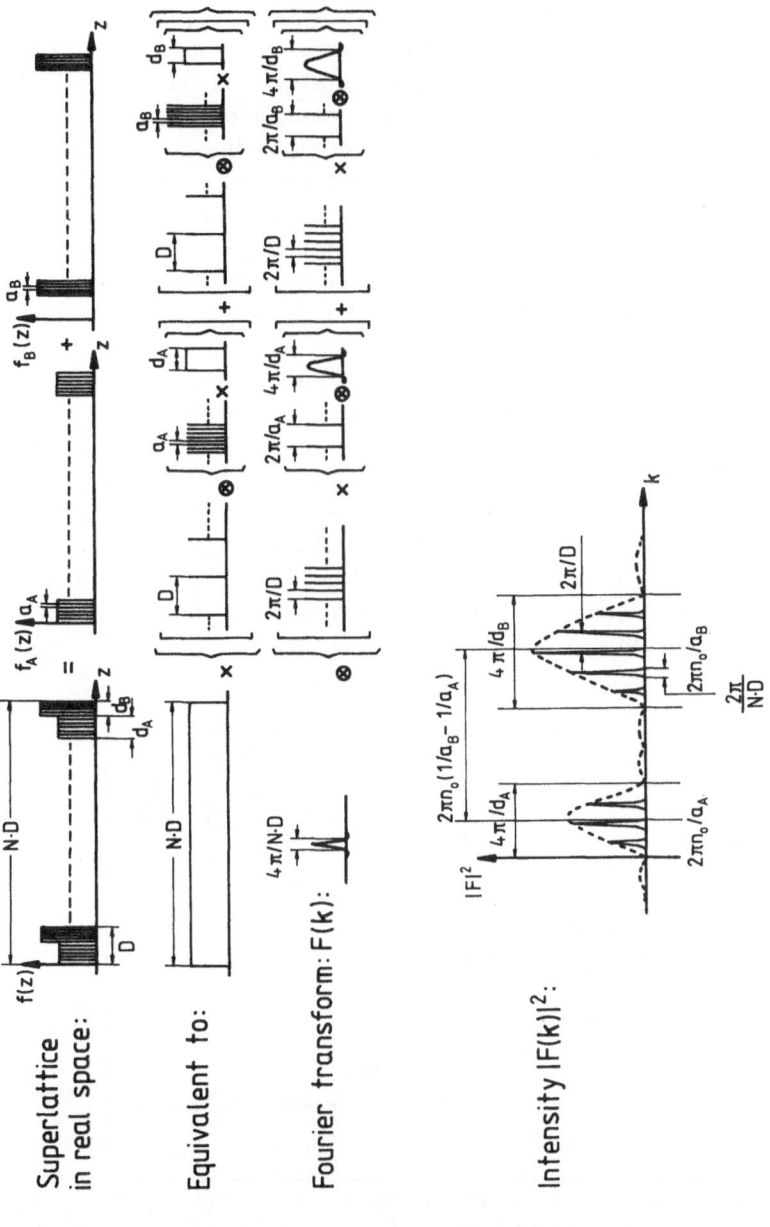

Fig. 6.32. Schematic description of the diffraction profile of a superlattice in terms of a Fourier analysis: It is assumed that the SL is perfect and consisting of N periods of double layers with period $D \,(= d_A + \delta d_B)$ with lattice constants along z-direction a_A and a_B and two different scattering factors [6.125]

by $(d_1 a_A + d_2 a_B)/(d_1 + d_2)$ where d_1 and d_2 are the layer thicknesses of materials A and B in one period and a_A, a_B represent lattice constants in growth direction. Of course, the zero-order SL peak must not be the one with the highest intensity. For growth along [100] direction, and tetragonal distortion Nakashima [6.126] has derived a procedure for finding its position (see also [6.127]): For the ($h00$) oriented substrate the angular distance from a SLS peak to the substrate peak is plotted vs. h^2 for several reflections (hkl). Only the zero-order peak shows a straight line through the origin.

An experimental example for this situation is shown in Fig. 6.33. It shows a $\omega - 2\Theta$-scan of the (004) Bragg diffraction of a Si/SiGe superlattice grown on a SiGe buffer ($x = 0.25$) of 2000 Å thickness on top of a (001) oriented Si wafer. These data illustrate the structures which are expected in a strained layer superlattice. The lattice constant mismatch between Si and $Si_{0.5}Ge_{0.5}$ is about 2%. Since the Si content of the buffer is 25%, the Si layers of the SL and the SiGe layers are strained symmetrically: the Si layers are under biaxial tension and the SiGe layers under biaxial compression. Both the Si-layers of the superlattice and the SiGe layers give rise to their own sub-satellite structures. The indexing in this Figure should not be confused with that of the entire superlattice stack, i.e. the actual zero order peak in the sense of the above discussion appears at an angle of about 68.790°.

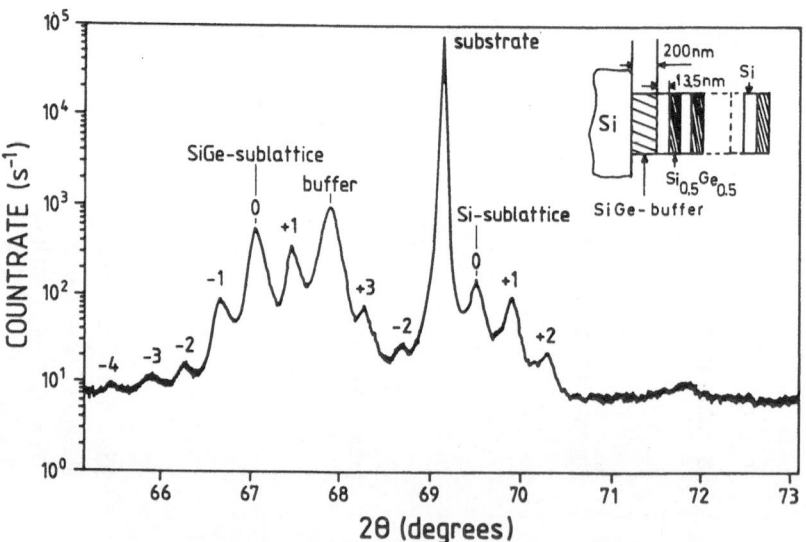

Fig. 6.33. Si/SiGe SL deposited on a SiGe buffer ($D = 270$ Å, $d_A = 135$ Å, $d_B = 135$ Å), 10 periods. Main peak corresponds to the Si substrate, the buffer peak is due to the SiGe alloy and 2 SL systems, one compressive (biaxially) strained SiGe sublattice and one tensile strained Si sublattice of the superlattice, taken in a $\omega - 2\Theta$ scan

This Si/SiGe superlattice is an example for a non-perfect strained-multi-layer structure with quite a limited number of satellite peaks observable (see Sect. 6.6.4).

As shown in Sect. 6.5.1 also the rocking-curves of simple heterostructures exhibit fringes corresponding the layer thicknesses involved. Macrander et al. [6.167] applied Fourier transformation for extracting the thicknesses of a InP/InGaAsP/InP double heterostructure.

At this point we would like to emphasize the importance of the X-Ray optics used for the evaluation of full width of half maxima of diffracted X-Ray peaks, since such quantities are quite often used as a first indication of the structural quality of epitaxial layers and superlattices. In the following we compare two $\omega - 2\Theta$ scans of a PbTe/EuTe superlattice grown on BaF$_2$ substrate along the [111] direction, one recorded with DCD optics and the second one with Triple-Axis Diffractometry (TAD) optics. In both cases a Philips MRD materials research diffractometer employing CuK$_{\alpha 1}$ radiation with a four-crystal Bartels monochromator (set for the (220) Ge reflection mode) in the primary beam was used. For the DCD optics the detector had an opening angle acceptance of 2 degrees. For the TAD optics a channel-cut two-reflection Ge (220) analyser crystal was used in the secondary beam which results in a beam divergence of 12 arcsec and in a high resolution reciprocal space probe as compared to the extent of the RELP's of interest. In Fig. 6.34 the $\omega - 2\Theta$ diffraction curve of the symmetrical (222) Bragg reflection of the PbTe/EuTe SL using DCD optics is shown.

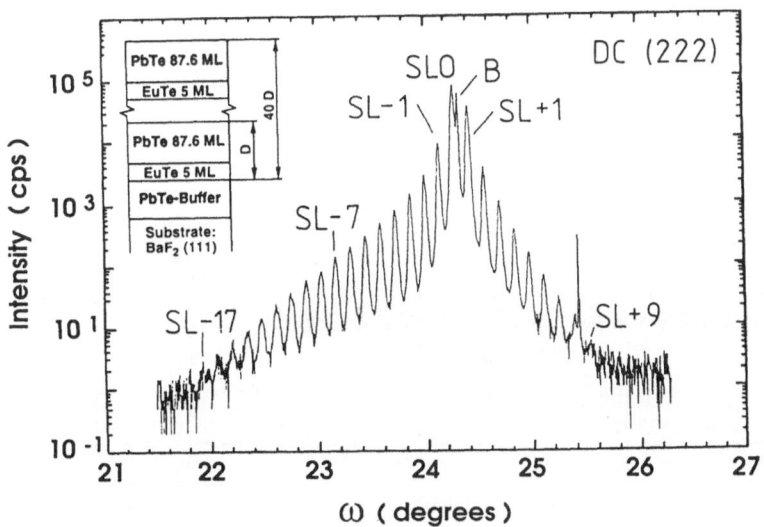

Fig. 6.34. Double-crystal rocking-curve ($\omega-2\Theta$ scan) for the (222) Bragg reflection of a PbTe/EuTe SL (insert with the structural parameters). B denotes the PbTe buffer, SL-17 to SL+8 denote the superlattice satellite reflections [6.128]

The insert shows the nominal structural parameters of the sample. In the double-crystal rocking-curve besides the BaF_2 substrate and the PbTe buffer (B) peaks a large number of superlattice (SL) satellites (SL-17 to SL8) are clearly resolved and few more appear in the background noise. The *FWHM* of the main diffraction peaks are 26 arcsec for the substrate, 42 arcsec for the PbTe buffer, and 102 arcsec for the zero order superlattice peak. These *FWHM* values deduced from the DC rocking-curve are strongly influenced by the limited DCD resolution. In the following Fig. 6.35 the corresponding $\omega - 2\Theta$ triple axis diffractogram, i.e. using the analyser crystal is shown, which clearly demonstrates the much higher instrumental resolution in comparison to the DC rocking-curve.

However, because of the lower X-Ray intensity at the detector a smaller number of satellites can be observed. The *FWHM*s of the diffraction peaks correspond now much better to the real broadening along the $\omega - 2\Theta$ direction, i.e. the growth direction (i.e. 8 arcsec for the BaF_2 substrate, 17 arcsec for the PbTe buffer and 47 arcsec for the SL0 peak). In this Figure also a simulation of the (222) Bragg diffraction curve based on dynamical scattering theory is shown.

An example of a diffraction from a perfect GaAs/AlAs structure is shown in Fig. 6.36. It consists of 710 Å thick GaAs layers and 107 Å thick AlAs layers with 50 periods and shows the (002) diffraction.

The separation of satellite peaks corresponds to a period of 817 Å. One recognises that satellite extrema up to the order $i = \pm 32$ are observable. The envelope of the satellite intensities oscillates due to the final thickness

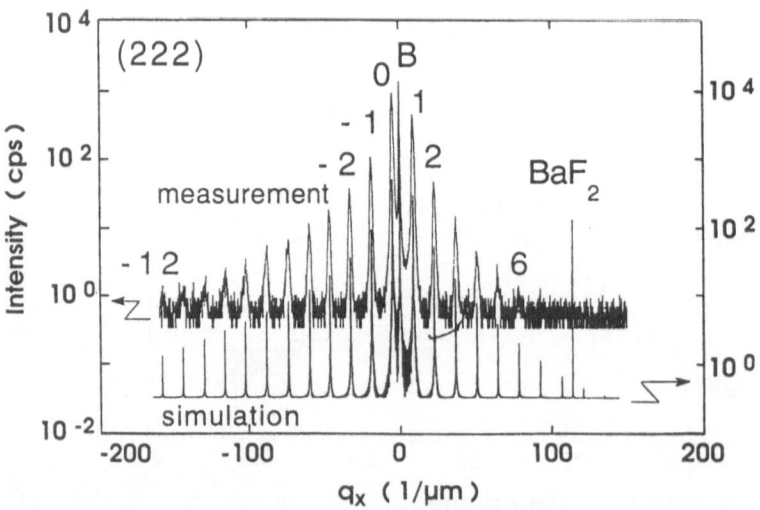

Fig. 6.35. (222) Bragg reflection triple axis rocking-curve ($\omega - 2\Theta$ scan) of the PbTe/EuTe SL shown in Fig. 6.34 with simulation based on dynamical scattering theory [6.128]

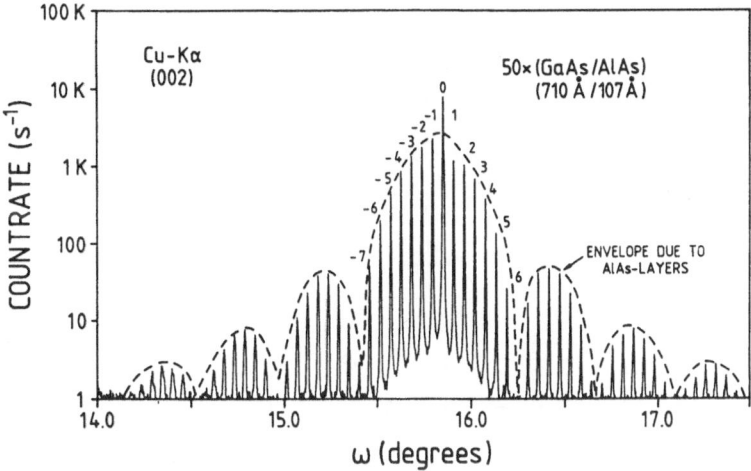

Fig. 6.36. High resolution X-Ray (002) diffraction of a GaAs/AlAs, ($d_A = 710\,\text{Å}$, $d_B = 107\,\text{Å}$) SL with 50 periods showing satellite peaks up to $i = 32$ in an $\omega - 2\theta$ scan. The modulation of the intensities due to an envelope function results from the 107 Å thick AlAs layers since their (002) structure factor is much stronger than that of the GaAs layers. The finite thickness fringes of the AlAs layers cause a spacing approximately eight times as large as that of the superlattice peaks [6.13]

fringes caused by the 107 Å thick AlAs layers which are observable strongly in the (002) Bragg peak since in this case the AlAs structure factor is much stronger than the corresponding GaAs one. The envelope of the superlattice peaks exhibits minima with a spacing approximately eight times as large as that of the satellite peaks.

In strained layer superlattices deposited on a substrate whose lattice constant deviates considerably from the mean SL constant, the intensity distribution turns out to be asymmetric with respect to the $i = 0$ SL peak. Interfacial strain as e.g. caused by introducing monolayers with appreciably different bond lengths than those found on the average in the SL, lead to quite substantial intensity enhancements of higher order satellite peaks as shown in the subsequent Figures.

The extremely high sensitivity of the envelope function to small fluctuations within the layer sequence of the interfaces has been exploited by Vandenberg et al. [6.129, 6.130] to study the presence of interfacial strain at heteroepitaxial interfaces.

At GaInAs/InP interfaces the strain already results from the different group V atoms at both sides of the interfaces without interdiffusion.

The net interfacial strains are caused by the different bond length in arsenic or phosphorous containing compounds. It turns out that X-Ray diffraction is particularly useful to investigate these interfacial strains as shown in Fig. 6.37 [6.131]. For a layer sequence InP/GaInAs/InP the interfaces can

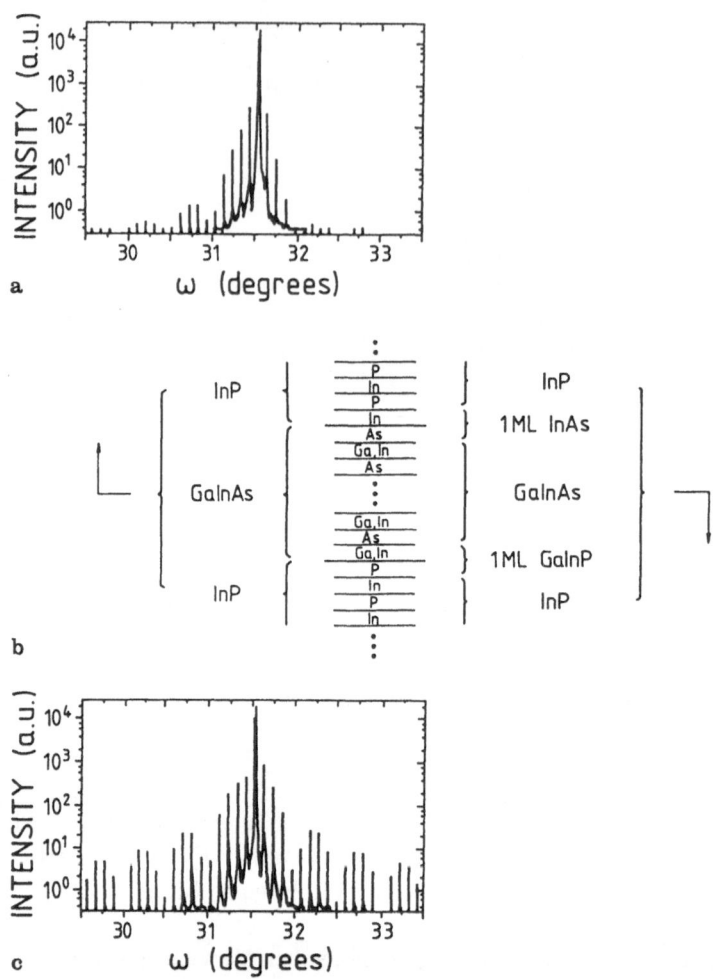

Fig. 6.37. Simulated diffraction patterns in the vicinity of the (004) reflection of a 20 period superlattice consisting of 10 nm thick $Ga_{0.46}In_{0.54}As$ and 40 nm thick InP layers on InP(001) substrate (**a**), same structure, but including additional InAs and GaInP monolayers (**c**) as indicated in (**b**) [6.131]

be defined either as indicated in the left hand part of the central panel of Fig. 6.37 or as shown in the right hand part where at the interfaces one monolayer of InAs or GaInP is introduced. In the simulated X-Ray diffrac- tograms the two cases lead to completely different patterns. The monolayers of InAs and GaInP are strongly lattice mismatched to both InP and the lattice matched GaInAs. Consequently, a similar situation is encountered as already discussed by Tapfer et al. [6.93] (see Fig. 6.26) for Si/Ge/Si. InAs, having a larger lattice constant than InP is under biaxial compressive strain whereas the GaInP layer is under biaxial tensile strain. However, above the

GaInP layer, the subsequent GaInAs layers have the correct lattice constant in comparison to InP, but with slightly shifted atomic positions. Therefore the wave fields from the InP layers below and from the GaInAs layers above the GaInP monolayer are shifted in phase with respect to each other and a similar effect is encountered by the InAs monolayer. Figs. 6.37a,c show the dramatic consequences of the two monolayers on the X-Ray diffractogram.

Vandenberg et al. [6.130] were able to show that high resolution X-Ray diffraction is capable of quantitatively identifying these strains, based on a comparison of experimental data of a $D = 534\,\text{Å}$ SL with 10 doublelayers of InGaAs/InP (Fig. 6.38). Therefore the satellite reflection (up to $i = \pm 34$) are closely spaced. The envelope of the intensities critically depends on strains through the relative shift of atomic positions in the repeated unit cells. In a computer simulation of the diffracted intensities one takes the interfacial strains across the InGaAs/InP into account by incorporating a negative strain producing monolayer at the InP to InGaAs interface and a positively strained monolayer in the InGaAs to InP interface. The interface spacing on one side is $1.4261\,\text{Å}$ and on the other side $1.5347\,\text{Å}$. The total number of layers is $N_{\text{InP}} = 310$ and $N_{\text{InGaAs}} = 54$, i.e. it is important that d_A is quite different from d_B in order to produce the envelope intensity variation and consequently different macroscopic net strains on both sides of the interfaces, ($\epsilon_\perp^+ = +4.6\%$, $\epsilon_\perp^- = -4.6\%$).

The formation of interfacial InAs-, InAsP-, GaInP- or GaInAsP layers in nearly lattice matched InGaAs/InP structures is also very likely due to an exchange of group V elements during the gas switching procedure in Metal

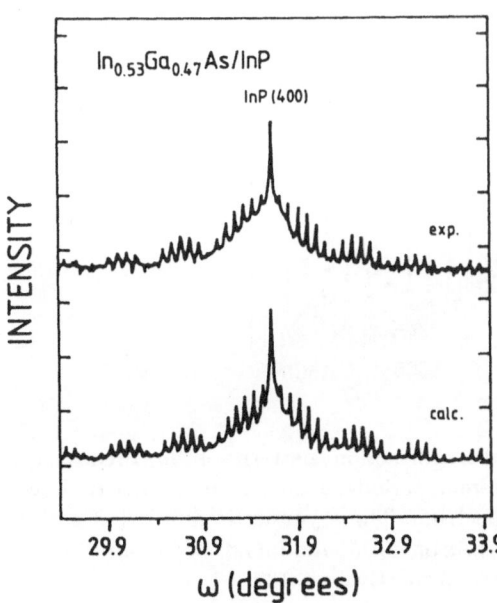

Fig. 6.38. Diffraction scan of a $\text{In}_{0.53}\text{Ga}_{0.47}\text{As/InP}$ superlattice ($D = 534\,\text{Å}$, ten periods). In the best fit $N_{\text{InGaAs}} = 54$ monolayers are assumed with one strained monolayer at each interface. The simulation includes interfacial strain in the closely lattice matched $\text{In}_x\text{Ga}_{1-x}\text{As/InP}$ SL (ϵ^+, ϵ^-) and small linear decrease of the number of InP monolayers N_{InP} from 314 to 306 [6.130]

Organic Vapour Phase Epitaxy (MOVPE) or due thermal interdiffusion. X-Ray diffraction is a valuable tool in identifying interlayers in the sequence of the whole stack of layers. Figure 6.39 shows an X-Ray diffractogram of a multilayer structure designed for an optical switch application in the 1.55 μm range [6.132]. This structure consists of an InP substrate, an InP buffer layer (0.2 μm) and a multilayer structure with a period of 10 nm (InGaAs/InP) double layers, and of a second multilayer structure with a period of 105 nm (which involves the first one (InGaAs/InP) and an additional 55 nm InP layer). On top an InP cap layer is deposited. The experimentally observed diffraction pattern is quite intricate with apparently different periods on the low and

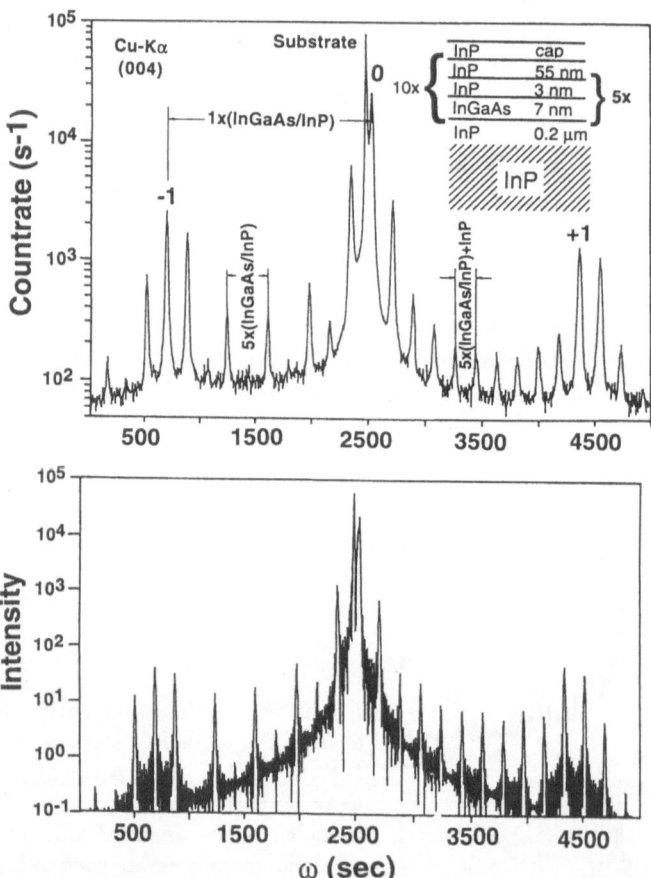

Fig. 6.39. Upper panel: experimental diffraction pattern of an InGaAs/InP MQW-layer structure with three different periods as shown in the insert. Lower panel: best fit to the experimental spectrum. The asymmetric diffraction pattern can only be simulated by introducing InAs or InAsP and InGaP or InGaAsP monolayers at the lower and upper interfaces, respectively [6.132]

high angle side of the substrate peak. The different periods in the structure are clearly observable:

The SL peaks $i = \pm 1$ represent the period of the 10 nm thick double layers. On the low angle side of the $i = 0$ SL peak, the total thickness of the InP/InGaAs SL ($d_{total} = 50$ nm) causes intensity fringes to appear. Finally on the high angle side of the main peak the satellite peaks of the other SL appear with a period $D = (5 * 10 + 55)$ nm $= 105$ nm. This characteristic envelope function allows for an unambiguous determination of the interfacial strain distribution along the growth direction as shown by model calculations based on the dynamical theory. A calculated pattern is shown in the lower part of Fig. 6.39. The calculations clearly show that the experimental pattern can only be simulated when strain-creating interfacial layers are introduced, which is a well known phenomenon in this material system. Figure 6.40 shows

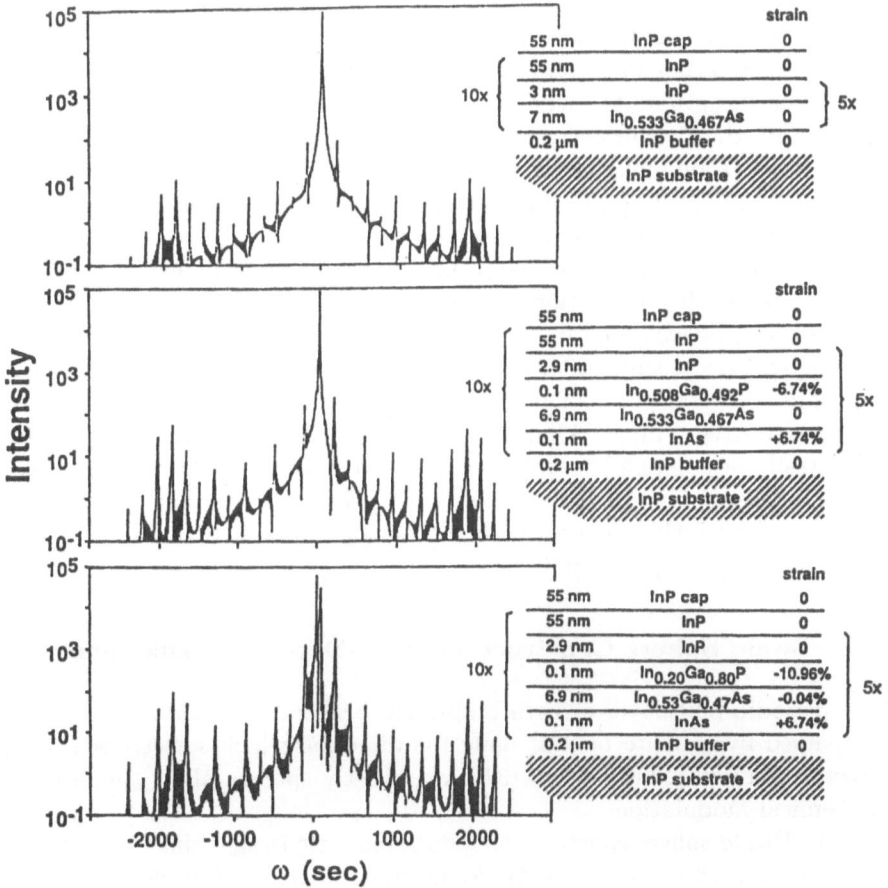

Fig. 6.40. Simulation procedure of Fig. 6.39: upper panel: without any strain; central part: including symmetric interfacial monolayers and lower panel: with asymmetric interfacial monolayers (see text) [6.132]

the different stages of the simulation. Firstly, the structure is simulated as de-signed, without any strain. The pattern is completely symmetric and the main satellite peaks being much too low in intensity (upper panel). The intensity of the satellite peaks can be increased symmetrically by introducing interfacial monolayers with height of same magnitude but opposite sign at both the InP-to-InGaAs and InGaAs-to-InP heterointerfaces (central panel). In the next step (lower panel) the envelope function is influenced by introducing inter-facial monolayers of asymmetric strain. In order to get similarity with the experimental spectrum the nature of the interfacial strain must be compres-sive at the InP-to-InGaAs interfaces and tensile at the InP-to-InGaAs ones. This situation can be verified by introducing a material whose lattice constant is smaller than that of InP between InGaAs/InP and a material with a larger lattice constant between InP/InGaAs. A negatively mismatched GaInAsP or GaInP monolayer at GaInAs-to-InP and a positively mismatched InAs or InAsP monolayer at InP-to-InGaAs fulfill these requirements. In the latter case the best fit is obtained with the highest strain possible ($\epsilon_\perp^+ = +6.74\%$), i.e. with pure InAs monolayers. At the upper interface a higher strain value is needed for an optimal fit, i.e. $\epsilon_\perp^+ = -10\%$. Such high strain can be generated by lowering the In concentration. Thus from X-Ray analysis the interfacial strain distribution along growth direction was deduced. Interfacial layers are also needed in order to fit ellipsometric data on such samples [6.133]. Quite recently, similar measurements were performed on GaInP/GaAs superlattices on GaAs substrate using the (002) diffraction [6.134]. The example demon-strates how much information in particular also on subtle epitaxial growth processes can be extruded from HRXRD.

The numerical simulation of rocking-curves is becoming more and more important. Several companies (Philips, Bede, Siemens, etc) offer program packages. Herres et al. [6.135] have described a program Simulat which calcu-lates strain parameters based on Segmüller-Murakami [6.99], Fewster's [6.87] version of the dynamical X-Ray diffraction theory for reflection and convolves the theoretical rocking-curves with the monochromator function for a com-parison with the experimental data.

6.6.2 Ewald Sphere Construction of SL-Diffraction Diagrams

In a periodic multilayer structure all reciprocal lattice points (hkl) are ac-companied by satellite points, along the direction which corresponds to the growth direction [6.60, 6.136, 6.137], i.e., strictly spoken, along the direction of chemical modulation.

The Ewald sphere construction for a symmetric Bragg reflection (002) of a superlattice is shown in Fig. 6.41. Along the direction of wavevector transfer, i.e. $\mathbf{G}_{hkl}$ the satellite peaks are present. These are broadened and elongated along the perpendicular direction (i.e. parallel to the surface plane) due to the presence of disorder, i.e. misfit dislocations. The broadening along the

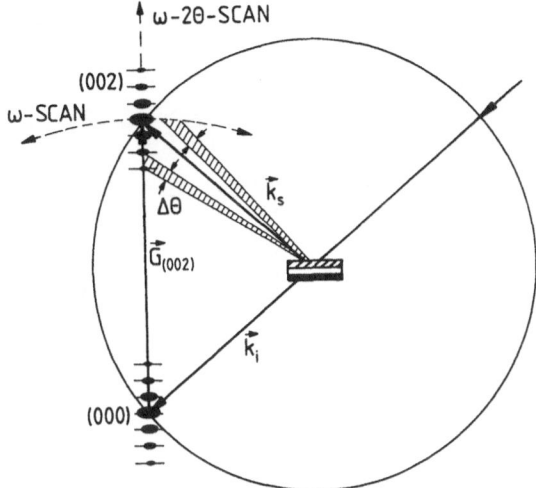

Fig. 6.41. Ewald sphere construction for (002) Bragg diffraction of a SL grown along [001] direction showing satellite reciprocal lattice points accompanying the main (hkl) sites. In an $\omega - 2\Theta$-scan the satellite structure is probed along the direction of **G** whereas in a typical rocking-curve an ω-scan (dashed curve) is performed with open slits in front of the detector corresponding to a certain angular spread around the scattered wavevector $\mathbf{k}_s$

growth direction results from the finite number of lattice planes contributing to the interference pattern. The extent of broadening and elongation can be measured by properly choosing both the scan mode and the Bragg diffraction angle. The procedure is outlined in the following Figures where diffracted intensities vs. angles 2θ or ω are plotted together with the corresponding Ewald sphere construction.

As an example a rather imperfect PbTe/PbSnTe superlattice with a period of 285 Å is chosen. Due to the BaF$_2$ substrate, which is lattice mismatched (about 4%) and in addition not of high crystalline perfection, the superlattice is distorted by misfit dislocations and mosaic structure which contribute to broadening of the reciprocal lattice points. Nevertheless in the symmetrical (222) $\omega - 2\Theta$ scan (Fig. 6.42) a satellite structure is clearly resolved, since for this scan mode the wavevector transfer is along the growth direction.

However, in an ω-scan of the same (222) Bragg reflection of the sample, the $i = \pm 2i$ satellites are barely visible and the $i = \pm 1$ ones are strongly broadened. The buffer peak and the $i = 0$ SL peak have merged together (Fig. 6.43).

In the reciprocal lattice, as shown in Figs. 6.42, 6.43 these facts can be explained with the help of the Ewald sphere considering the different scans in the $\omega - 2\Theta$ (Fig. 6.42) as well as in the ω mode (Fig. 6.43). Provided that the small angle grain boundaries are mainly oriented parallel to the growth

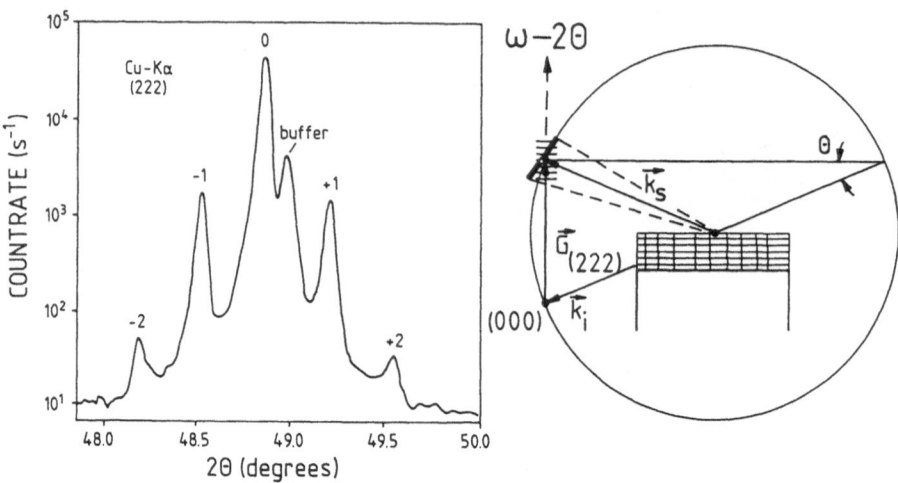

Fig. 6.42. An ω-scan of a PbTe/Pb$_{1-x}$Sn$_x$Te SL deposited on a Pb$_{1-x}$Sn$_x$Te buffer ($x = 0.18$, period $D = 285$ Å) of a (222) Bragg diffraction. The broadening of the reciprocal lattice points in direction perpendicular to the scan mode is due to misfit dislocations

direction the broadening of the SL satellites will not become effective in a scan mode along the **G**-direction for a symmetrical reflection (Fig. 6.42). However, for the ω-scan the SL is probed along the $\mathbf{k}_S$ direction as indicated in Fig. 6.43, upper panel, and several SL-points contribute to the registered Bragg diffraction. Choosing scattering geometry with $\mathbf{k}_S$ approximately parallel to the growth direction, i.e. for that case a (264) Bragg diffraction for which $2\Theta = 126.6°$ and $\omega = 40°$ ($2\Theta - \omega \cong 90°$) produces in the rocking-curve a well resolved satellite SL-structure in an ω-scan (Fig. 6.43, central panel). For that diffraction, the broadening of the reciprocal lattice points parallel to the superlattice planes does not contribute to the observed diffraction pattern. The peak broadening is also influenced by interdiffusion in addition to the effects mentioned earlier.

This information on the extent of the reciprocal lattices points in various directions in reciprocal space are confirmed by choosing a scattering geometry which probes the extension of the reciprocal lattice points along a proper direction which is perpendicular to the growth direction. This is the case for a (062) Bragg reflection with an ω angle of 92.3° and a 2θ of 97.5° ($2\theta - \omega \cong 0°$). In the ω-scan no satellite at all is visible and in the corresponding Ewald sphere plot, the contribution from all satellites are recorded simultaneously for all ω-positions (Fig. 6.43, lower panel). In all the diagrams illustrating the ω-scans in the reciprocal lattice, the finite dispersion, i.e. the angular range of scattered wavevectors $\mathbf{k}_s$ is indicated by dashed lines. This example illustrates the importance of the proper selection of diffraction conditions. The choice of the (264) and (062) Bragg peaks is optimal for the investigation of layers

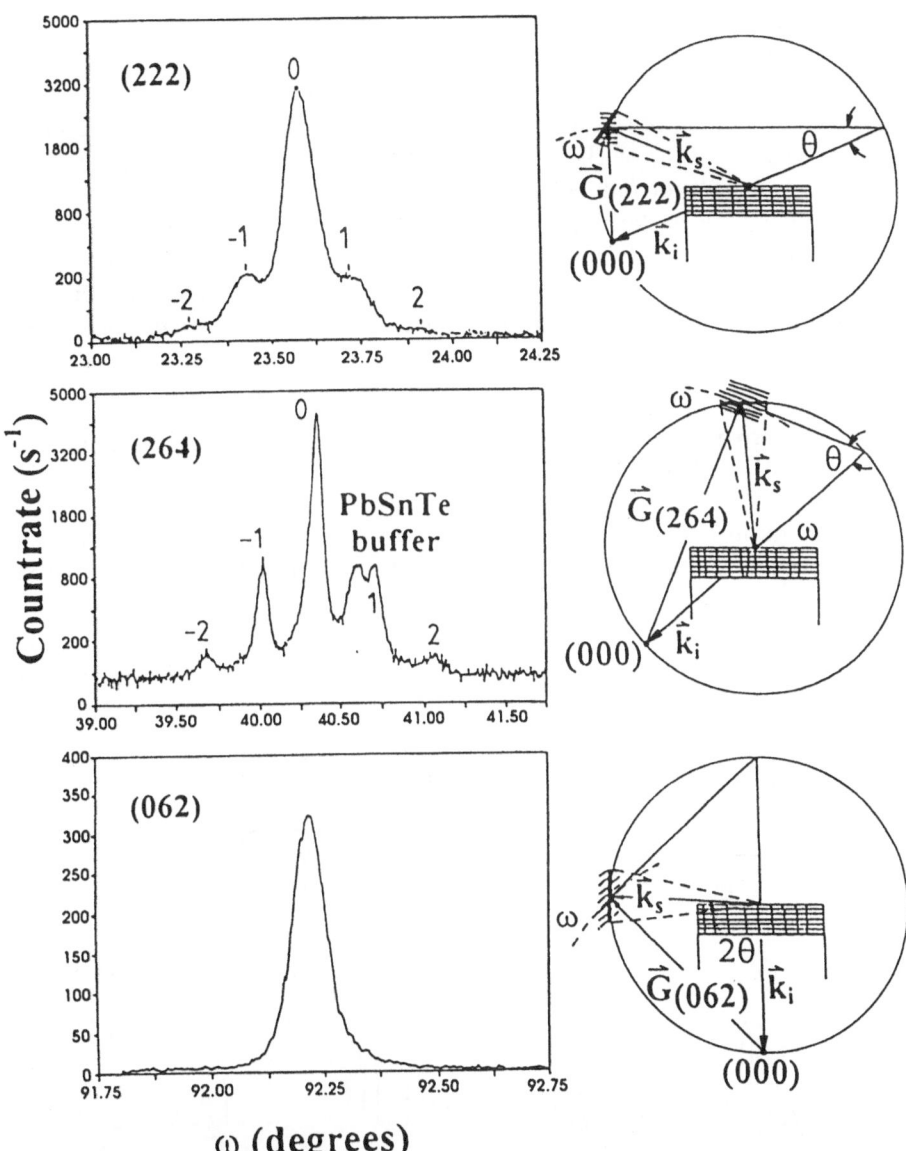

Fig. 6.43. As Fig. 6.42 but for ω-scans for (222), (264) and (062) diffractions. In the symmetric (222) case the satellites are barely resolved since the rocking-curve is taken with wide open slits in front of the detector during the scan. For the (264) diffraction the SL satellites are well resolved (see reciprocal lattice for explanation) whereas in the (062) geometry all satellites contribute at the same time to the reduced intensity for various ω angles

grown in the [111]-direction. In general, the resulting shape of the rocking-curve depends both on the amount of mosaicity and the (hkl) of the reciprocal space probe.

The broadening of the scattering intensity thus immediately yields information on the extent of the reciprocal lattice points parallel to the growth direction which is not influenced by the finite-layer thicknesses.

Therefore, by performing measurements with several Bragg diffractions one can uniquely get information on the orientation dependence of the broadening with respect to the growth direction. Such data are necessary for a determination of the real structure (mosaic structure due to small angle grain boundaries, misfit and threading dislocations) of epitaxial films.

6.6.3 Interpretation of the Fine Structure in X-Ray Diffraction Profiles of SL's

In highly perfect SL's, between the satellite peaks, additional extrema can be observed as shown in Fig. 6.44. A (004) Bragg diffraction of a Si/SiGe SL

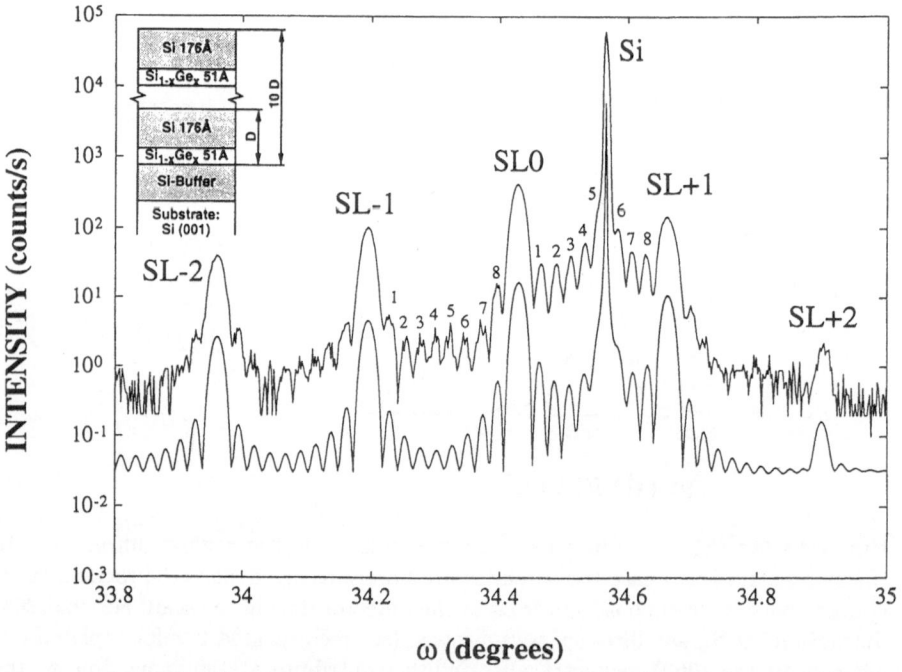

Fig. 6.44. HRXRD of Si/SiGe superlattice deposited on (001) oriented Si. SL period $D = 227$ Å with 10 double layers corresponding to a total thickness of 2270 Å. The number of secondary maxima in-between the main SL satellite peaks is $10 - 2 = 8$

on a Si buffer and thus highly strained with a total thickness of 2270 Å is shown. Expanding the ω-scale additional maxima appear between the main satellites. The SL period for the spacing for the $i = 1, 2, 3, \ldots$ SL satellites is found to be 227 Å, i.e. there are 10 double layers of SiGe. The evaluation of the distance between the fringes accompanying the SL satellites yields a total thickness of about 2270 Å. We would like to point out that between two subsequent satellites there are $N - 2$ side maxima $(10 - 2 = 8)$ which follows from classical diffraction physics for a diffraction grating consisting of N slits.

The intensity of the subsidiary maxima between the main satellite peaks turns out to be extremely sensitive to irregularities in the period. As Powell et al. [6.138] have shown for a Si/Ge SL, a small dispersion in the SL period in the order of 2% causes quite drastic changes in the pattern of the subsidiary maxima, whereas the width of the main satellites is still unaffected.

6.6.4 Imperfect MQW's and Superlattices

Real MQW and SL samples exhibit a number of imperfections which result from the growth procedure itself like interdiffusion, compositional and thickness fluctuations. In the following the consequences of these imperfections on the diffraction profiles are discussed.

6.6.4.1 Interdiffusion in MQW's and SL-Systems.

From the intensity of the satellite peaks of X-Ray diffractograms already Fleming et al. [6.83, 6.139, 6.140] have determined the abruptness of the transition region between GaAs and AlAs layers in $(GaAs)_n(AlAs)_m$ superlattices. Later this technique has been applied by Arch et al. [6.141] for a study of interdiffusion in HgTe-CdTe superlattices and recently by Hogg et al. [6.142] for $Cd_{1-x}Mn_xTe/CdTe$ multilayers. If a rectangular modulation of the chemical composition across the interfaces is assumed without change of the SL-period D, the compositional modulation $c(z)$ is described by a Fourier series:

$$c(z) = c_0 \left[1 + \sum_m Q_m \cos \left(\frac{2\pi m z}{D} \right) \right] \tag{6.56}$$

Q_m is the amplitude of the m-th harmonic. For abrupt interfaces only the odd Fourier components are nonzero, whereas for an arbitrary profile all components have to be considered. During growth or during an annealing procedure the coefficients Q_m vary with time according to the diffusion equation:

$$Q_m(t) = Q_m(0) \exp \left[- \left(\frac{2\pi m z}{D} \right)^2 D(T) \, t \right] \tag{6.57}$$

Depending on growth direction and crystal structure, the influence of the concentration profile $c(z)$ on the X-Ray scattering amplitudes $S(k)$ has to be calculated. In a kinematical approximation then the intensities of the SL peaks are readily calculated and the relation between the intensity of the satellite peak number $\pm m$ and the Fourier coefficients Q_m is given by

$$\frac{I_{\pm m}}{I_0} = \frac{Q_m^2}{4} \left(\frac{2\pi\epsilon\rho_\pm}{m\frac{2\pi}{D}} \pm \eta \right)^2 \tag{6.58}$$

where

$$\rho\pm = la^* \pm \frac{m2\pi}{D} \qquad h = \frac{4\Delta f}{S(00l)}$$

(for growth along the [001] direction). $S(00l)$ is the average structure factor of the $A_{1-x}B_xC$ lattice; ϵ denotes the amplitude of the interplane spacing modulation. $\Delta f = f_A - f_B$ takes into account the modulation of the scattering due to the difference in chemical composition and a^* is an average lattice parameter in reciprocal space.

Thus the intensity of the SL peaks in an X-Ray diffractogram has two contributions:

a) one originating from the modulation of scattering amplitudes due to the differences in chemical composition

b) the second originating from the modulation of interplanar spacing (see also Fig. 6.32).

Using intensities on both sides of the central peak (which yields I_0 as intensity) both Q_m and ϵ can be determined. From the knowledge of the Q_m's the concentration profile is established.

Quite often, especially for large x in $AC/A_{1-x}B_xC$ superlattices, the interdiffusion process is *composition dependent* and the Fick's law according to

$$\frac{\partial c}{\partial t} = D\frac{\partial^2 c}{\partial z^2} + \frac{\partial D}{\partial c}\left(\frac{\partial c}{\partial z}\right)^2 \tag{6.59}$$

has to be used. For the consequences, especially in short period superlattices, we refer to Fleming et al. [6.83], and Mc Whan [6.139, 6.140].

Several authors have investigated the limits of the applicability of the procedure just outlined. Recently, Hogg et al. [6.142] have shown that for large differences in lattice constants of the constituent materials, i.e. large values of strain, the diffusion constant derived from the satellite intensities using the method of Fleming et al. [6.83] breaks down. As a breakdown criterion Hogg et al. [6.142] have suggested:

$$\frac{\lambda}{2\cos\theta}\left(\frac{1}{d_w} + \frac{1}{d_b}\right) \geq \frac{a_w^\perp - a_b^\perp}{a_b^\perp}\tan\theta \tag{6.60}$$

i.e. when the strain is sufficiently large to cause a splitting of the well (w) and barrier (b) diffraction patterns, e.g. like that shown in Fig. 6.33 or schematically in Fig. 6.32.

However, there is another effect which has to be considered especially in SL structures where the interdiffusion coefficient is quite large. The layers close to the substrate are for a longer period of time at growth temperature than those close to the final surface. Consequently, already during growth the layers close to the substrate experience more interdiffusion than those which are close to the growing surface.

Thus the diffusion equation has to be solved with a boundary condition at the surface where the diffusion current equals the flux of component B (in a $AC/A_{1-x}B_xC$ SL grown by MBE) for a sticking coefficient of one [6.143]. The diffusion equation

$$\frac{\delta c}{\delta t} = \frac{\delta}{\delta z}\left(D\frac{\delta c}{\delta t}\right) \tag{6.61}$$

is solved in discretised steps in $i\Delta z$ $(i = 1, 2, \ldots, N)$ equal to the plane distances (monolayer by monolayer) and in $j\Delta t$ $(j = 1, 2, \ldots, N)$

$$\frac{c(1, j+1) - c(i, j)}{\Delta t} = D\frac{c(i+1, j) - 2c(i, j) + c(i-1, j)}{(\Delta z)^2} \tag{6.62}$$

In the simulation Δz corresponds to the lattice plane distances (here d_{111}, but identical considerations hold for d_{100}) growth direction, the temporal step Δt is related to the growth rate v by $\Delta t = d/(vn_{tot})$, where $n_{tot} = d/d_{111}$. With these expressions, the equation above can be rewritten

$$c(1, j+1) = c(i, j) + D\frac{d}{vd_{111}}(c(i+1, j) - 2c(i, j) + c(i-1, j)) \tag{6.63}$$

A unique solution is only possible for special boundary conditions.

For the two-dimensional growth process two spatial boundary conditions must be considered: At the buffer-to-substrate interface the diffusion current vanishes, whereas at the top layer the diffusion current equals the incoming flux I_{beam} from the source, if a sticking coefficient of 1 is assumed. Thus the boundary conditions are

$$\frac{\delta c}{\delta z} = 0(z = 0) \tag{6.64}$$
$$c(1, j) = c(2, j) \tag{6.65}$$
$$\frac{\delta c}{\delta z} = I_{beam}(z = vt) \tag{6.66}$$
$$c(j, j) = I_{beam}(j)\Delta t \tag{6.67}$$

The Eqs. (6.66, 6.67) reflect the moving boundary $i = j = 1, 2, \ldots, N$ during the growth.

This procedure was used for the analysis of rocking-curves of PbTe/ PbMnTe superlattices [6.143]. The X-Ray diffraction data were compared with calculated ones, based on the Mn-profile which was obtained from the numerical solution of the diffusion equation. For large interdiffusion it is difficult to obtain from the damping of the satellite intensities alone reliable information on the diffusion constant and the method just described should be used. Furthermore one has to consider the influence of interface roughness on the satellite intensities as well.

An interesting example for nonlinear interdiffusion in HgCdTe/CdTe multilayers was performed by Kim et al. [6.144] who demonstrated the importance of relaxation of systems like SL's which are far from thermodynamic equilibrium.

In summary, interdiffusion does not affect the periodicity but decreases the contrast between the layers, and thus results in reduced satellite peak intensities particularly for the higher order satellites.

6.6.4.2 Imperfect Superlattices: Period, Thickness, Composition Fluctuations.

Fewster [6.145] has outlined a procedure for determining variations in period D, in interface roughness, and in interface grading from SL diffraction patterns.

Period variations are changes ΔD of D, i.e. of the sum of barrier and well width with depth of the sample, but averaged laterally over a coherently diffracting volume. Interface roughness originates from both lateral changes of well and barrier widths as well as vertical variations of these quantities. This interface roughness can be either random or correlated both laterally and vertically as was discussed by Savage et al. [6.146], Phang et al. [6.147], Holý et al. [6.148]. The separation of correlated and uncorrelated interface roughness phenomena is possible by measuring both the Bragg diffracted intensities as well as the diffuse scattered ones.

In addition, in real semiconductor superlattices consisting of binary and ternary materials (e.g. GaAs/Ga$_{1-x}$Al$_x$As) grading will occur which causes variations of the composition x across the interface, both vertically as well as laterally.

Since in uninterrupted MBE growth processes the growth front usually extends over about three monolayers in vertical direction, imperfect completion of layers during growth is a standard phenomenon.

Fewster has shown [6.149] that the higher order satellites broaden progressively if there are variations in the periods. For satellite peaks close to the central Bragg peak, $\cos\theta$ is nearly constant, D may be simplified to:

$$D = \frac{(L_i - L_j)\lambda}{\Delta\theta\, 2\cos(\theta)} \tag{6.68}$$

where $\Delta\theta$ is the angular distance between two satellites L_i, L_j. From the differentiation of this equation follows the relation between the change in period ΔD and the satellite broadening $\delta(\Delta\theta)$ (difference in angle within one satellite):

$$\delta(\Delta D) = \frac{(L_i - L_j)\lambda}{2\cos(\theta)} \frac{1}{(\Delta\theta)^2} \delta(\Delta\theta) \tag{6.69}$$

If grading occurs at heterointerfaces, both the lattice parameters and the scattering factors change, which have to be included in model calculations. In Fig. 6.33 the substrate and the buffer layer have different chemical compositions and thus different lattice constants. The superlattice itself has a mean lattice constant, which depends on the lattice constants of the constituents A, B and their layer widths d_A, d_B. From the angular separation between the $i = 0$ superlattice peak, the average mismatch can be roughly calculated [6.137]:

$$\left(\frac{\Delta a}{a}\right)_\perp = \frac{-\Delta\theta(1-\nu)}{\cos^2\phi(\tan\theta_B + \tan\phi)(1+\nu)} \tag{6.70}$$

where ν is the Poisson ratio, and ϕ the angle between the diffracting lattice planes and the surface plane. For small total thicknesses of the SL ($< 0.5\,\mu m$) the average $i = 0$ SL peak is shifted to the substrate peak and only a dynamical simulation program yields proper results.

For the description of imperfect superlattices which exhibit composition gradients computer simulations are presented in the following. As an example the (004) diffractogram of an ideal 10 period $In_{0.511}Ga_{0.489}As$ /InP MQW structure is shown in Fig. 6.45a. Figure 6.45b shows the effect of a linear

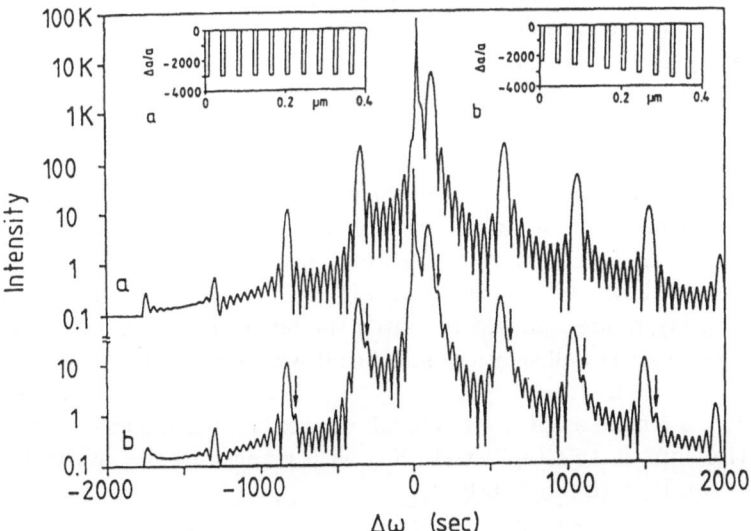

Fig. 6.45a. Calculated (004) Bragg diffractogram for a 10 period $In_{0.511}Ga_{0.489}As$/InP MQW structure on InP(001) with $d_1 = 10\,nm$ and $d_2 = 30\,nm$, **b** Effect of a gradient in In concentration from 0.516 to 0.507 on the diffractogram. Arrows indicate the secondary maxima which would lead to broadening and/or splitting of the experimental diffractogram

gradient in the composition on the diffractogram. The In gradient from 0.516 to 0.507 causes only minute changes in the intensity of the main satellite peaks, whereas the envelope function of the secondary has changed which manifests itself as an enhancement of the $i = 1$ secondary maximum (see arrows). In a measurement such a gradient will appear as a splitting and/or broadening of the satellite peaks. For a statistical fluctuation of the In content within the limits $x = 0.516$ to 0.507, the resulting diffractogram is hardly distinguishable from that of the perfect one. Several authors have discussed and experimentally investigated such effects [6.150–6.155]. Also for the investigation of imperfect SL structures reciprocal space mapping offers some advantages. With DCD scans it is in general difficult to separate strain and composition gradients in SLS structures from each other. Using reciprocal space maps around symmetrical and asymmetrical Bragg reflections, such a distinction can be easily made both from the asymmetry of the iso-intensity contours and from the maximum intensity positions with respect to the substrate reciprocal lattice points.

6.6.5 Strained-Layer Superlattices: Tilt, Terracing and Mosaic Spread

In general the mean superlattice lattice constant is different from that of the substrate and also different from that of the intermediate buffer layer. In the reciprocal lattice each substrate point is accompanied by a buffer point and a set due to the strained-layer superlattice. The relative position of the points depends on the superlattice period D and the strains present parallel and perpendicular to growth direction.

Neumann et al. [6.121] have studied the effects of terracing in GaAs/ GaAs$_{1-x}$Sb$_x$ SL's. In Fig. 6.46a, a schematic presentation of a terraced SL is given, where the angle α denotes the terrace angle, i.e. the angle between the direction of modulation and the normal of the constituent lattice planes. This means that the chemical modulation direction is not parallel to the lattice planes of the SL film. In addition, the SL can also be tilted with respect to the substrate as shown schematically in Fig. 6.46b. There it is assumed that the GaAsSb layers are strained to match the lattice constants of GaAs both along the terraces as well as at the interfacial steps (completely coherent interfaces).

In Fig. 6.46b, a second angle 2β is defined, where β represents the average tilt of the entire superlattice (epitaxial film) with respect to the substrate. The value of β can be estimated to be

$$\beta \approx \frac{\alpha(a_{x'} - a_0)}{a_0} \approx \alpha \left(1 + \frac{2C_{12}}{C_{11}} \right) \frac{\Delta a}{a_0} \tag{6.71}$$

where a_0 and $a_{x'}$ are the lattice parameters at the step and directly above the step in the previous layer, respectively. Δa is the difference in lattice

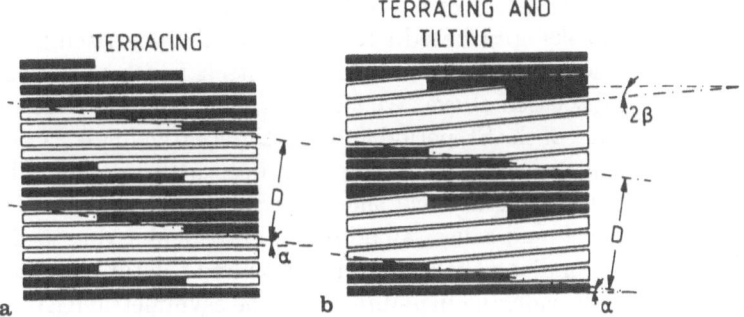

Fig. 6.46. Scheme of a terraced superlattice without (**a**) and with additional (**b**) tilting relative to the substrate [6.121]

constants between the ternary compound and the binary one. $a_{x'}$ is given by:

$$a_{x'} = a_x + \frac{2C_{12}(x)}{C_{11}(x)}\Delta a \tag{6.72}$$

Thereby the value of $a_{x'}$ was calculated by the Poisson expansion for a cubic cell compressed along two of the cube edges. With the assumption $C_{11} \approx 2C_{12}$ which is approximately valid for many zinc-blende semiconductors it follows for the tilt angle of a completely coherent system:

$$\beta \approx 2\alpha\epsilon \tag{6.73}$$

where $\epsilon = \Delta a/a_0$ is the misfit strain. This model describes the tilting mechanism in the regime of small strains i.e. without formation of misfit dislocations in the stack of the SL layer. Neumann et al. have found that for a GaAs substrate, miscut by 2°, and 25 periods of a $GaAs_{1-x}Sb_x/GaAs$ SL with equal thickness and $D = 428\,\text{Å}$ the terracing angle α is 1.5° and the tilt angle β is 0.08°. These experiments were performed with a triple-axis spectrometer which is a convenient tool to separate the effects resulting from the terraced SL from the additional tilt.

Following a presentation given by Holý et al. [6.148, 6.156, 6.157] we summarise the effects of strains and terracing, and mosaic spread in a schematic presentation in Fig. 6.47 for a strained-layer SL grown along a [001]-direction. In Fig. 6.47 for clarity only the reciprocal lattice points (004), (224), and ($\overline{2}\overline{2}4$) for the SL are presented. The azimuth is along a [1$\overline{1}$0] direction.

The position of the strained layer SL peak labeled 0, i.e. the zeroth-order satellite peak depends on the mean strains parallel or perpendicular to the growth plane. The separation of the subsequent satellites numbered $(-i, -i+1, \ldots, -1, 0, +1, \ldots, i-1, i)$ is determined by $2\pi/D$, D being the superlattice period in the growth direction (in the case of Fig. 6.47 the [001]-direction). In a system without any tilt these satellites are arranged exactly in the [001]-direction, the picture for the asymmetric (224) and ($\overline{2}\overline{2}4$) reflexes being completely symmetric with respect to the (004) reflexes. If terracing occurs, the superlattice peaks are rotated about the zeroth-order satellite by the terracing angle α and the zeroth-order satellite itself is rotated about (000) by the angle β as discussed above. We show schematically the effect of mosaicity, of the fluctuation of the SL period and of interface roughness as well as of random deformation on the shape of the reciprocal lattice points. The effect of mosaicity leads to a symmetrical smearing of the reciprocal lattice points according to Fig. 6.47 and causes an additional broadening due to the mosaic block shape function along the $\omega - 2\theta$ direction and the tilts between the mosaic blocks cause a broadening in ω-direction. The asymmetric RELP's are all elongated along the ω-circumference, the further away from (000) the larger the broadening. Interface roughness leaves the contours of constant intensity untilted. The SL0 peak is unaffected whereas with increasing satellite index the broadening perpendicular to the growth direction becomes larger.

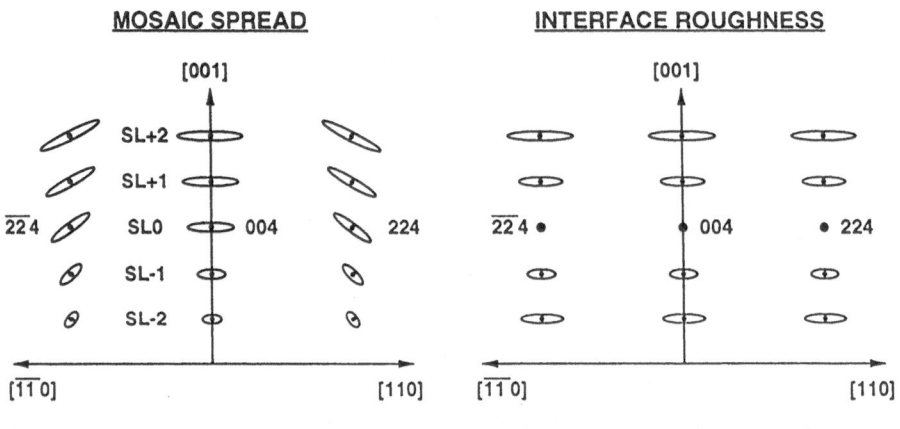

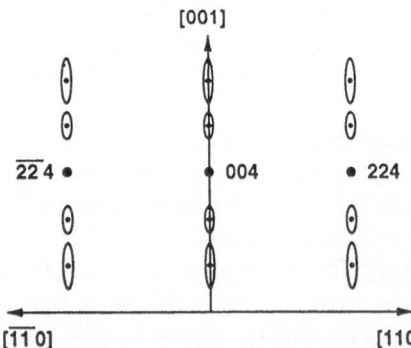

Fig. 6.47. Scheme of unstrained-layer SL of different orders on substrate without tilt in reciprocal space. Influence of mosaic spread, of interface roughness, and of fluctuations in period are shown schematically part) [6.148,6.156], see also Fig. 6.46

Fluctuations in the superlattice period manifest themselves in an elongation of the SL satellite peaks along growth direction, and again the SL0 satellite remains unaffected (Fig. 6.47).

6.7 Scans in the Reciprocal Lattice

So far only scans along one distinct direction in reciprocal space have been discussed. With the triple-axis spectrometers, nowadays available, maps of a reciprocal lattice spot are measured in order to determine independently Bragg plane tilts from asymmetry effects and substrate curvature induced broadening from film mosaicity (see Sect. 6.1.4).

This technique was applied to study intensity contour plots, along a (440) Bragg diffraction from a single QW InGaAs (160 Å) embedded in an InP barrier (500 Å cap layer, 2500 Å InP buffer deposited by MBE on InP:Fe substrate tilted 2° off) [6.42]. In Fig. 6.48 the intensity contours show $i = +4 \ldots -3$ satellites accompanying the strain-split InP peak in the center of the (440) diffraction. The crystal face is oriented in the [100] direction, the [010] direction lies within the surface plane. The intensity distribution (corresponding to 4 orders of magnitude) is recorded over a distance within the reciprocal lattice which corresponds to an ω-scan of 2.26°. Due to tilt of the growth plane by 2° off the [100] direction towards the [110] direction the line connecting the satellite intensity extrema is inclined by a small angle with respect to the vertical axis. For the intensity profile along the [100] and [110] direction it turns out that the position of higher order satellites yields quite precise information on lattice mismatch between the cap layer and the InGaAs quantum well, as well as on surface roughness and on interface roughness.

Another example of the efficient use of contour plots is shown in Fig. 6.49 which was obtained from a DCD together with a position-sensitive detector (PSD) system [6.12] for an $In_{0.2}Ga_{0.8}As/GaAs$ strained-layer SL for the (224) reflection.

The intensity contour map yields much more information than the ω-scan which just resolves the substrate peak and the main SL peak but no satellites. The other projection in the $\Delta 2\theta$ axis which corresponds to an $\omega - 2\theta$ scan reveals the satellite structure, but information of the tilt of the SL with respect to the substrate is lost.

A direct comparison of the information content of a rocking-curve of a ZnSe epitaxial layer on top of a GaAs substrate with a mapping of the reciprocal lattice is given in Fig. 6.50.

The main advantage of the triple-axis mode is the fact that the intensity contour plot gives much more information. The asymmetric broadening of the layer peaks is interpreted as being caused by mosaicity of the ZnSe layer in comparison to the much better perfection of the GaAs substrate. In the follow-

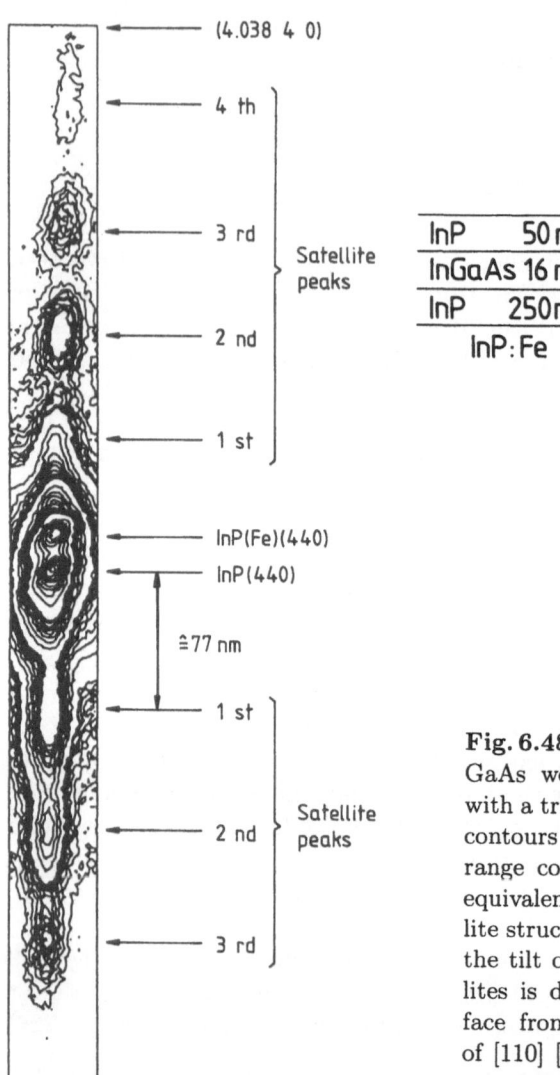

InP   50 nm
InGaAs 16 nm
InP   250nm
InP:Fe

Fig. 6.48. Intensity contour plot of In-GaAs well embedded in InP obtained with a triple-axis spectrometer. Intensity contours are shown over 4 decades. Scan range corresponds to $\pm 0.07a^*$ which is equivalent to an ω-scan of $2.26°$. Satellite structure is due to a period of $770\,\text{Å}$, the tilt of the line connecting the satellites is due to the offset of the crystal face from [110] by $2°$ in the direction of [110] [6.42]. Thicknesses in the insert are approximate values from the crystal growth parameters

ing Fig. 6.51a a direct comparison of a rocking-curve on a PbTe/EuTe SL and a reciprocal space mapping is presented. The SL consists of 40 double-layers of 93 ML of PbTe ($a_0 = 6.462\,\text{Å}$) and 5 ML of EuTe ($a_0 = 6.598\,\text{Å}$) which are deposited on a $4.1\,\mu$m PbTe buffer on a (111)-oriented BaF$_2$-substrate ($a_0 = 6.200\,\text{Å}$). In the symmetric (222) Bragg reflection, superimposed on a broad background, $i = -18$ satellites are observed. The width of the SL maxima increases from 127" for $i = 0$ to 212" for $i = -15$.

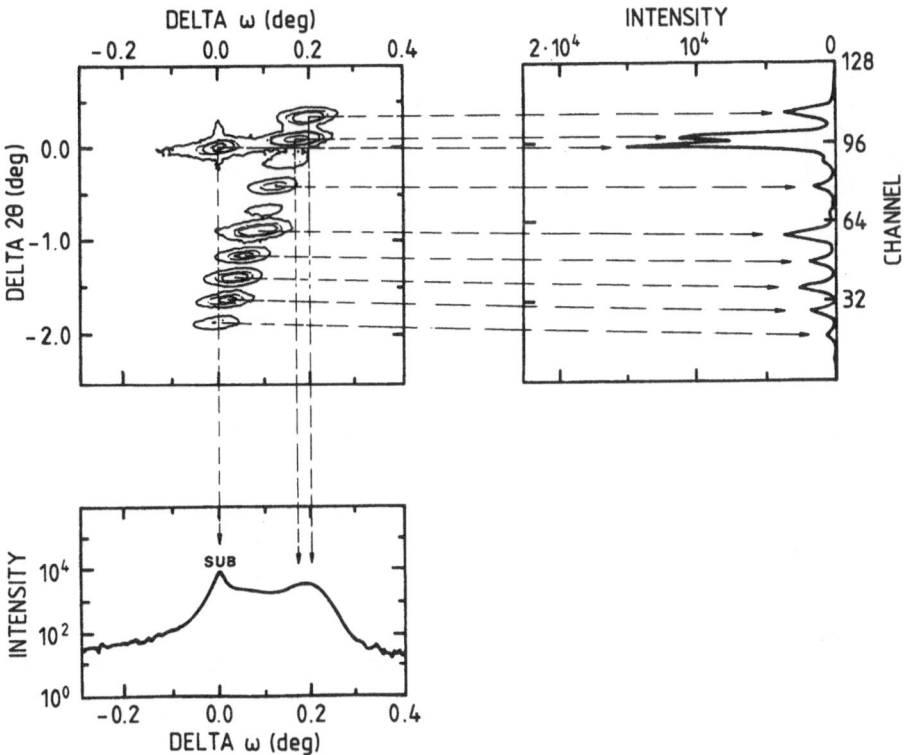

Fig. 6.49. Reciprocal space intensity contours of a $GaAs/In_{0.2}Ga_{0.8}As$ SL ($d_1 = 12\,nm$, $d_2 = 26\,nm$) deposited on an $In_{0.1}Ga_{0.9}As$ buffer ($d = 200\,nm$) on GaAs ((224) reflection). Upper left corner: contours obtained with conventional DCD using a position sensitive detector; upper right: integrated intensity along 2θ and lower left: integrated intensity along ω with identification along lines with arrows [6.12]

According to Fewster [6.9, 6.10] the use of a triple-axis diffractometer, with an analyser which uses more than one reflection, offers the following advantages in the analysis of multilayer structures:

– Strain and strain gradients can be separated from structural imperfections such as tilts and mosaicity: in the reciprocal space maps the intensity distribution along the strain influences the $q_{\perp}$-direction, i. e. the one along the ω-2θ-scan.
– Mosaic spread or bending is observed along the $q_{\parallel}$-direction in the reciprocal space maps, i. e. along the ω-scan.

In superlattices, the half width of the zero order ($i = 0$) peak along the $q_{\parallel}$-direction yields the lateral correlation length ξ. Any additional broadening of the $i = \pm 1$ and higher satellite peaks along the $q_{\parallel}$-direction is associated

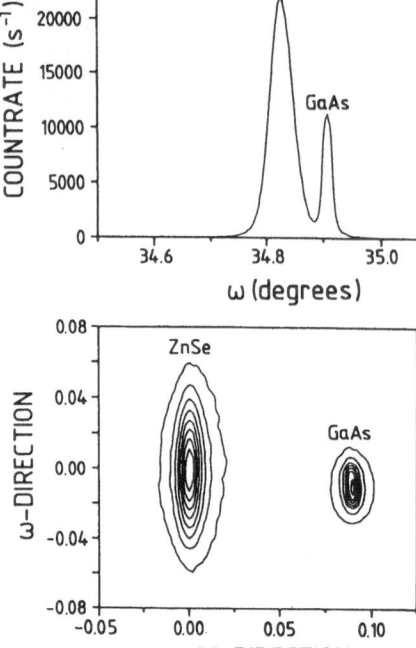

Fig. 6.50. Comparison of a rocking-curve of the (004) Bragg peak of an epitaxial ZnSe layer on top of a GaAs substrate with a iso-intensity contour plot obtained with the 4+1+2 spectrometer. From the $\omega - 2\Theta$ scan the lattice mismatch can be deduced, whereas during the ω-scan the broad structure of the ZnSe diffraction results from mosaicity [6.13]

with superlattice imperfections. Using the Scherrer equation

$$\xi = \frac{k\lambda}{2W\cos\theta} \tag{6.74}$$

where k is a constant (~ 1) which is characteristic for the shape, Θ the Bragg angle and W the width of the diffraction peak, the broader $i > 0$ SL peaks can be analysed to yield a correlation length ξ which characterises a length scale in the interfaces of the SL's. Such an analysis, performed by Fewster [6.9] on short period GaAs/AlAs (16.7 Å/16.7 Å) superlattices gave a correlation length ξ of the order of 400 to 800 Å, in good agreement with results of scanning tunneling microscopy. Hence, triple-crystal diffractometry is particularly useful for the investigation of imperfect superlattices and heterostructures. In strained-layer superlattices, beyond their critical thickness, partial strain relaxation occurs. Using reciprocal space mapping of asymmetric reflections, the complete strain status can be obtained: i.e. the strain components parallel and perpendicular to the surface, and therefore the lattice parameters. Furthermore, also information on the strain variation as a function of depth into the crystal is accessible. In addition, also information on mosaicity and tilting is obtained. Multicrystal-multireflection diffractometry can therefore determine residual strains and mosaic structure or bending in real heterostructure

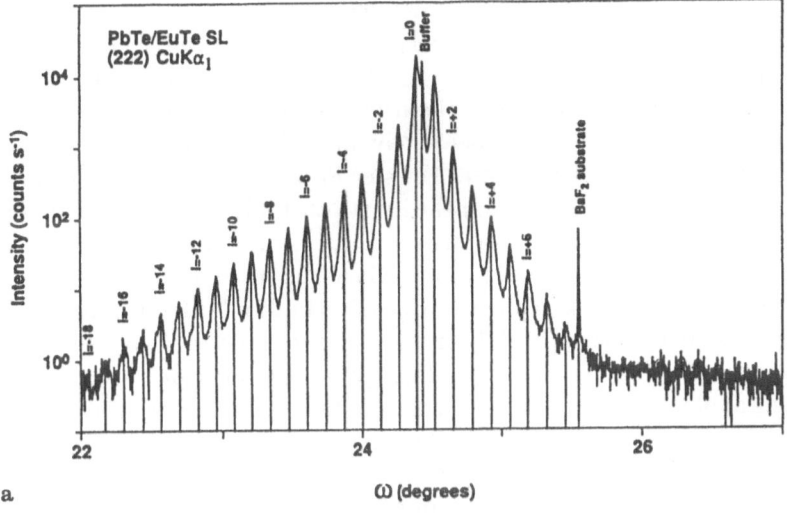

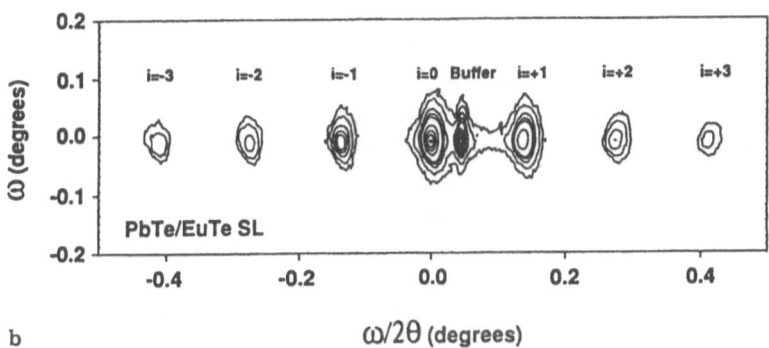

Fig. 6.51a. Rocking-curve of a PbTe/EuTe superlattice, one period consisting of 93 monolayers of PbTe and 5 monolayers of EuTe, 40 periods on a $4.1\,\mu m$ thick PbTe buffer deposited on (111) BaF$_2$ substrate, **b** triple-axis reciprocal lattice scan intensity isocontour plot of the same structure showing broadening in the ω-direction which corresponds to the $q_\parallel$-direction. Intensities span the region from 5 to 5000 in 10 steps

systems. Such information is not accessible from conventional rocking-curves obtained with the DCD.

As an example we show in Fig. 6.51b the reciprocal map for a (222) diffraction of the same PbTe/EuTe SL structure which was shown in Fig. 6.51a. Along the ω-2Θ-direction ($q_\perp$-direction) the strain variations and thickness fluctuations of the SL along the growth direction can be derived. In the ω-direction ($q_\parallel$-direction) it is apparent that already the PbTe buffer layer peak exhibits some mosaic spread which increases considerably for the higher SL satellites. For the SL width along $q_\parallel$, we deduce an interface coherence length of about 700 Å. In order to extract the relevant information from the triple axis diffractograms it is necessary to calculate iso-intensity contour plots around reciprocal lattice points and to perform the necessary projections. For such imperfect structures the usual fitting procedure based on the Takagi-Taupin equations which label for dynamical scattering in nearly perfect layer structures is not adequate. Fewster [6.158] has recently attempted to formulate a dynamical diffraction theory for partially relaxed semiconductor layer structures which contain interfacial defects and has compared his calculation which experimental data on InGaAs/GaAs and on $Si/Si_{1-x}Ge_x$.

The combination of low temperature equipment and a triple-axis X-Ray spectrometer is particularly useful for the study of structural phase transitions. $Pb_{1-x}Ge_xTe$ exhibits a structural instability towards the formation of a rhombohedral phase at low temperatures from the high temperature rock-salt structure. Fig. 6.52 (top) shows the consequences of this phase transition for scattered X-Ray intensities in a reciprocal lattice scan. Below T_c, any of the equivalent $\langle 111 \rangle$-directions transforms into a rhombohedral c-axis. The occurrence of this splitting can be used as a manifestation for the phase transition. For a compound with a Ge content of 6% (film thickness: 2600 nm) a lattice temperature of 20 K is well below T_c. If a PbGeTe layer of the same Ge-content with a thickness of 500 Å is sandwiched between PbTe layers, the occurrence of the structural phase transition is inhibited, apparently by strains exerted on the PbGeTe film (Fig. 6.52 (bottom)). The films are deposited on (111) oriented BaF_2 substrates. The corresponding reciprocal lattice scan, showing iso-intensity contours around a PbGeTe reflection and around a PbTe reflection does not exhibit any evidence for a splitting of the PbGeTe peak even at $T = 20$ K [6.159]. For the Bragg diffractions, a pseudo-orthorhombic notation is used with the cubic [111] orientation parallel to the [001] orthorhombic direction and the orthorhombic [100] and [010] axes parallel to the [211] and [011] axes, respectively.

Another example for a reciprocal space map on a 2 μm AlInAs layer on an InP substrate is shown in Fig. 6.53 for a (004) reflection as obtained with the HRMCMRD (see Fig. 6.9) [6.58, 6.9, 6.10]. The diffraction space probe has a width of 10 arcsec and is presented by the parallelogram A'–B'. The parallel lines A–B denote the finite width of the Ewald sphere due to the wavelength spread leaving the monochromator. Without analyser crystals the detector acceptance angle is several degrees and corresponds to a sector which is longer than A–B. Measurements taken without an analyser (DCS, Bartels-type) and with the 3-diffraction analyser are compared in the right part of Fig. 6.53. From the DCS results using a wide open detector the intensities of layer and substrate are nearly identical. An $\omega - 2\Theta$ scan parallel C–D, however, reveals quite different intensities: smaller and broader for the layer and a larger peak intensity for the substrate.

From a series of scans parallel to the line C–D the diffraction map can be obtained and a detailed analysis of the two-dimensional intensity distribution

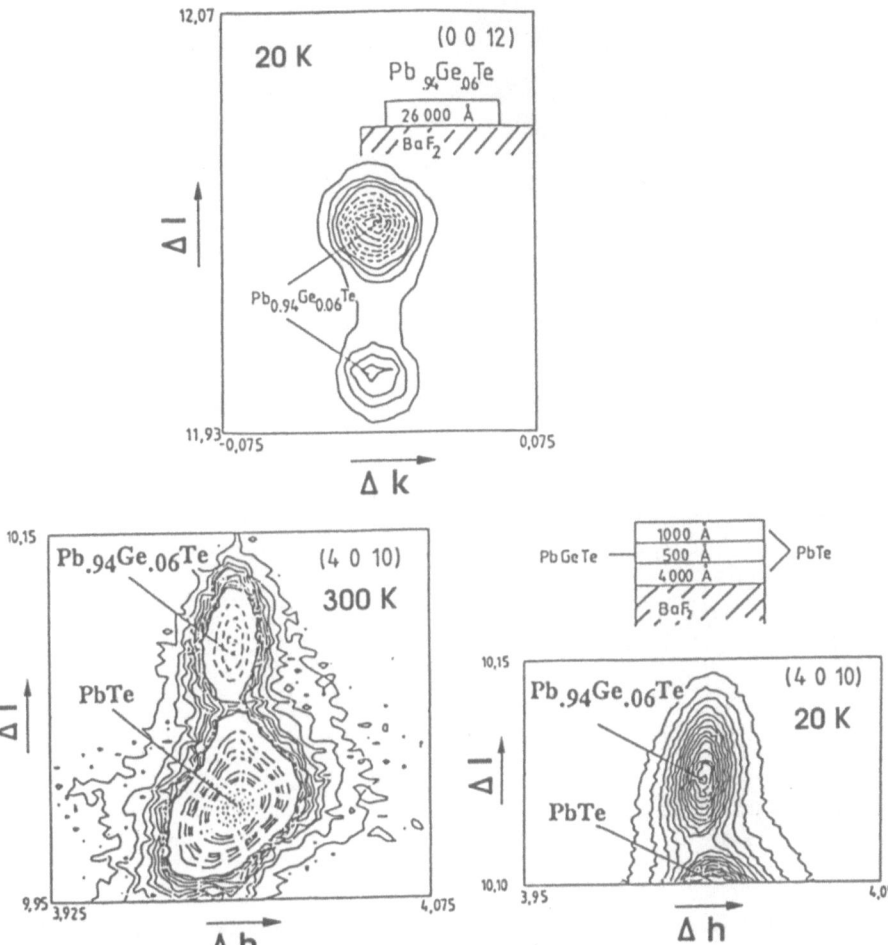

Fig. 6.52. Isocontour plots of PbGeTe layers deposited on a BaF₂ substrate with and without PbTe buffer and cap layers. top: at $T = 20\,K$, below the cubic - rhombohedral phase transitions, splitting of (0012) peak occurs (rhombohedral notation). bottom: with PbTe layers the phase transition is inhibited, probably due to mismatch strains [6.159]

yields information on strain, sample curvature, and eventually on mosaic structure. The excellent signal to noise ratio makes this technique most useful for the study of thin epitaxial layers and superlattices. Diffraction space maps in the vicinity of an asymmetric reflection of a superlattice can be used to determine partial relaxation in strained layer SL's, the strain variation as a function of the z-coordinate (growth direction) as well as any tilting. In combination with topography, where the film is placed behind the analyser crystal, a correlation of diffraction space mapping and topographs which ex-

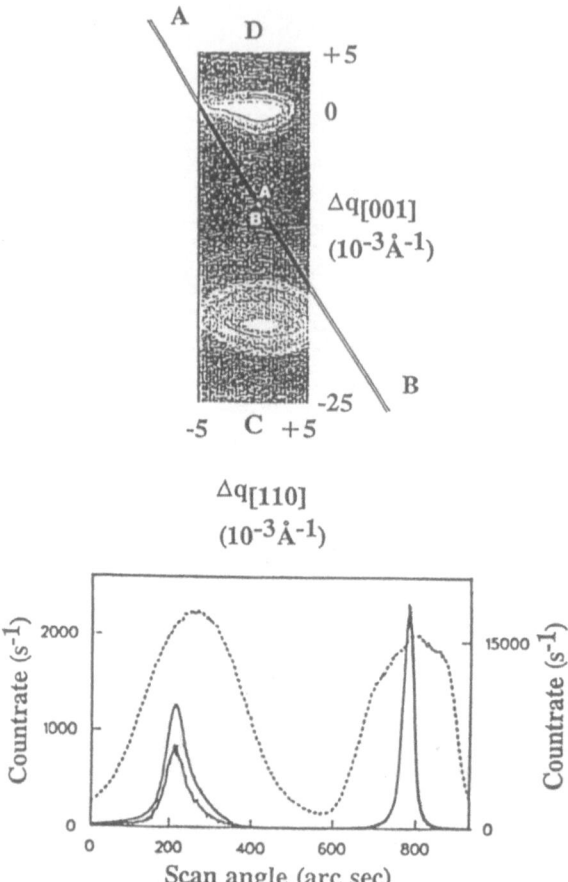

Fig. 6.53. Left panel: reciprocal space map of a (004) reflection from an AlInAs layer ($2\,\mu$m) deposited on an InP substrate. The parallel lines A, B represent the finite width of Ewald sphere due to finite wavelength spread leaving the monochromator crystals. Direction C–D corresponds to a radial direction from (000) along d^*, accessible with an $\omega - 2\omega'$ scan. The $4+1$ crystal monochromator with open detector accepts all intensities between lines A–B. In the HRMCMRD the probe which is swept is indicated by the small parallelogram A–B. Right panel: for the case shown above, different diffraction profiles: dashed lines correspond to ω-scan with $4+1$ crystal monochromator the solid line is obtained with the HRMCMRD ($\omega-2\Theta$ scan)(lhs intensity scale); dotted line: sum of the HRMCMRD scans parallel to C–D [6.9]

hibit intensity modulation due to strain fields of dislocations is possible (see Fig. 6.9). This combination technique has been applied by Keir et al. [6.160] for a study of the structural properties of $Hg_{1-x}Cd_xTe/CdTe$ on GaAs.

In the following we demonstrate the usefulness of reciprocal space mapping for the assessment of the structural properties of two short period

$Si_{0.6}Ge_{0.4}$ structures grown on either single step or step-graded SiGe buffers. Reciprocal space mapping of both symmetric (004) and asymmetric (224) Bragg diffraction peaks is used to establish the status of strain relaxation. The structural characteristics of substrate, buffer, and superlattice influence the positions of the reciprocal lattice points (RELP) and the shape of iso-intensity contours around them. We show that these two- dimensional reciprocal lattice maps yield a wealth of structural information, usually much more than the conventional rocking-curves [6.161]. Both SL samples were grown by molecular beam epitaxy on (001) oriented non-miscut Si substrates. For sample A, a 20 nm thick Si layer was deposited at $T = 550\,°C$, followed by the 20 nm thick single step $Si_{0.60}Ge_{0.4}$ buffer deposited at $450\,°C$. The short period SL $Si_{0.60}Ge_{0.4}$ was grown at $350\,°C$ (145 periods). For sample B, a 100 nm thick Si layer was grown on top of the substrate followed by the step-graded buffer (B1), in which the Ge content was increased stepwise by 3% per 50 nm up to a total thickness of 700 nm., i.e. to a nominal Ge content of 40%. During the buffer growth, the temperature was decreased continuously from $600\,°C$ to $520\,°C$. Subsequently, a 550 nm thick $Si_{0.60}Ge_{0.4}$ alloy buffer layer (B2) was grown at $500\,°C$. Prior to the growth of the SL a monolayer of antimony was deposited as a surfactant. For the thick step-graded buffer, the relaxation mechanism causes a complete relaxation of the individual slices, of which the buffer is composed, and thus the corresponding RELP's finally lie along the [224] direction which connects (000) with the (224) substrate RELP. In such a case the superlattice with the proper mean composition (with respect to the top buffer layer) can grow virtually unaffected by the substrate. The in-plane lattice constants of the SL and the top buffer layer coincide and within the SL layers the values of biaxial compression and dilation are unaffected by the substrate.

In Fig. 6.54 contours of constant scattered intensity around the (004) and (224) RELP's of sample A are shown, which were derived from a series of $\omega - 2\Theta$ scans with ω-offsets using the proper transformation from angular space into reciprocal space [6.161]. The asymmetry around the substrate (004) RELP is an artifact (analyser streak caused by the finite size of the reciprocal space probe, which is defined by the X-Ray optics used). Qualitatively, the strain situation is determined by a partially relaxed buffer layer, on which pseudomorphic growth of the superlattice with respect to the single step buffer occurs as indicated by the position of the zero order superlattice peak. For sample B, the reciprocal space maps are shown in Fig. 6.55. From both RELP's around (004) and (224) it follows that all portions of the step-graded buffer (B1) are fully relaxed, because the intensity contours are symmetric around the $q_\parallel[224]$ direction, which is not entirely the case for the zero order intensity contours. The maximum intensity lies along the $[11\bar{1}]$ direction away from the $q_\parallel[224]$ direction. The region of the SiGe alloy with constant Ge content (B2) yields intensity contours which overlap in their positions with the zero order contours of the superlattice. Because of its low intensity, the

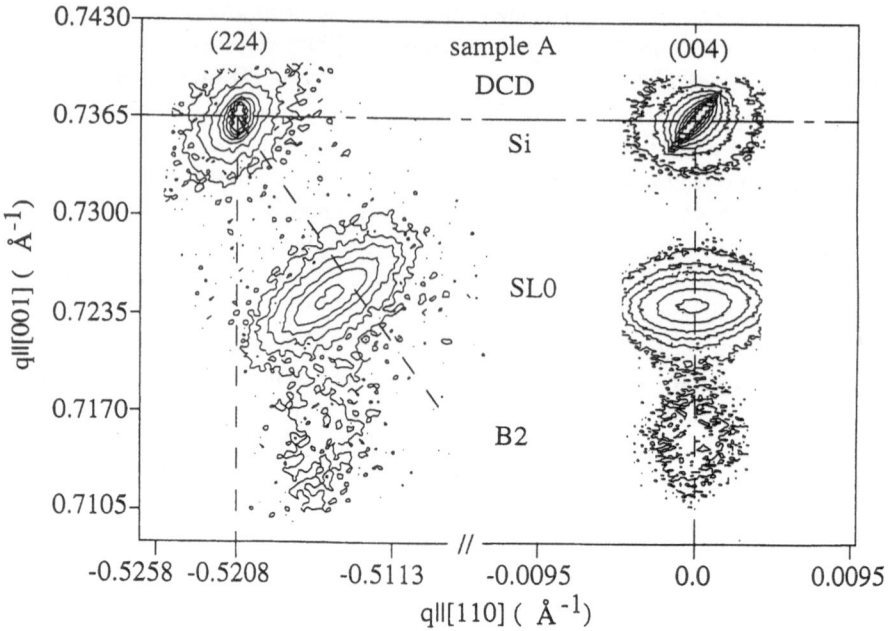

Fig. 6.54. Measured (224) and (004) reciprocal space maps of sample A. The iso-intensity contours correspond to 0.5, 1, 2, 10, 20, 2000 cps in the (224) map and to 1, 2, 10, 20, 100, 2000 cps in the (004) map. The (224) buffer RELP lies below the center of the SL0 RELP along the [001] growth direction [6.161]

zero order peak was not detected in the (224) map. From the (004) reciprocal space map it follows that the tilt between superlattice, buffer and substrate is negligible.

From the reciprocal space maps, $FWHM$'s of the intensity distribution parallel [001] and perpendicular [110] to the growth direction are deduced and for sample A values of 400 and 1150 arcsec are found, respectively, for the zero order peak of the superlattice. For sample B, the corresponding values are 350 and 1070 arcsec, respectively, which do not change for first order (SL-1) intensity contour within the experimental accuracy of ± 15 arcsec.

From Figs. 6.54 and 6.55 it is obvious what potential is offered for immediate identification of the strain status of short-period superlattices by reciprocal space maps. The symmetry of the intensity contours around the zero order peaks indicates the absence of large scale strain gradients along the growth direction, and a rather statistical distribution of the mosaic blocks and a constant strain status within the superlattice stack.

If partial relaxation occurs on SLS grown on alloy buffers, the evaluation of the strain status is complicated when a deviation from Vegard's law is present, which is the case e.g. for SiGe alloys. This deviation has to be

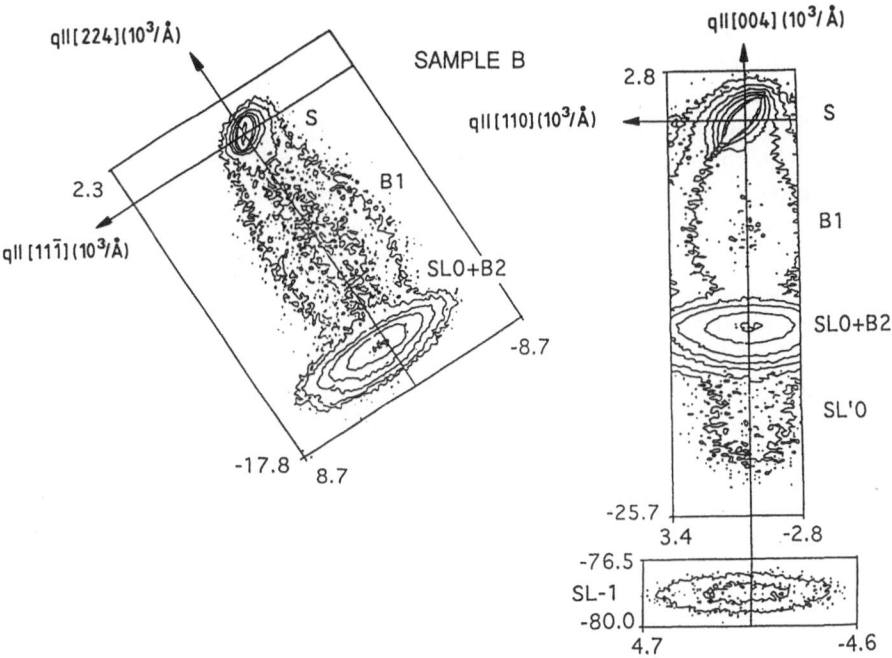

Fig. 6.55. Measured reciprocal space maps of sample B. The iso-intensity contours correspond to 1, 2, 5, 10, 20, 50, 1000 cps in the (224) map, to 1, 2 cps in the (004) map of the SL-1 and to 1, 2, 5, 10, 20, 50, 100 cps in the other (004) map, respectively. In the (224) map, the B1 buffer layers RELP's are symmetric around the [224] direction indicating full relaxation; the shift of the maximum intensity of the SL0 peak away from the $q \parallel$ [224] direction indicates residual in-plane strain SL of -0.33% [6.161]

considered for the evaluation of strain status deduced from the shift of the reciprocal lattice points during the relaxation process.

The distribution of the diffusely scattered intensity around reciprocal lattice points results from structural imperfections like microscopic and macroscopic strain gradients, and statistically distributed lattice plane tilts, usually referred to as mosaicity.

Recently, the correlation function of the random deformation field due to these structural defects was calculated for short period Si_9Ge_6 superlattices. The Fourier transformation of the two-dimensional reciprocal space distribution of the scattered intensity equals the correlation function multiplied by the reflectivity of the perfect structure, i.e. it is obtained from the reciprocal space maps without any assumption on the defect structure in the sample. From the correlation function one obtains directly the region which scatters X-rays coherently, i.e. along growth direction as well as laterally [6.163]. Reciprocal space mapping has been also used quite extensively to study partially relaxed layers of ZnSe grown by MBE on (001) oriented GaAs substrates [6.164]. From measurements of reciprocal space maps around different asymmetric and symmetric RELP's a depth gradient not only in strain but also in mosaicity was derived by the authors.

In the following we show data on pseudomorphic Si/SiGe MQW structures which exhibit considerably smaller $FWHM$'s both along and perpendicular to growth direction. The p-type modulation doped $Si/Si_{1-x}Ge_x$ MQW structures were grown by MBE on (001) oriented Si substrates [6.165]. The MQW sample, on which X-Ray data are shown in Fig. 6.56, consisted of 10 repeats of 52 Å thick $Si_{0.77}Ge_{0.23}$ wells and 175 Å thick Si barriers. A Si cap layer of 440 Å was deposited on top.

In Fig. 6.56 in the insert a DCD rocking-curve of the (004) Bragg reflection is shown. Apart from the substrate peak several superlattice peaks are identified. However this sample shows subsidiary intensity maxima between the main SL maxima. Since the number of periods is $N = 10$, the number of

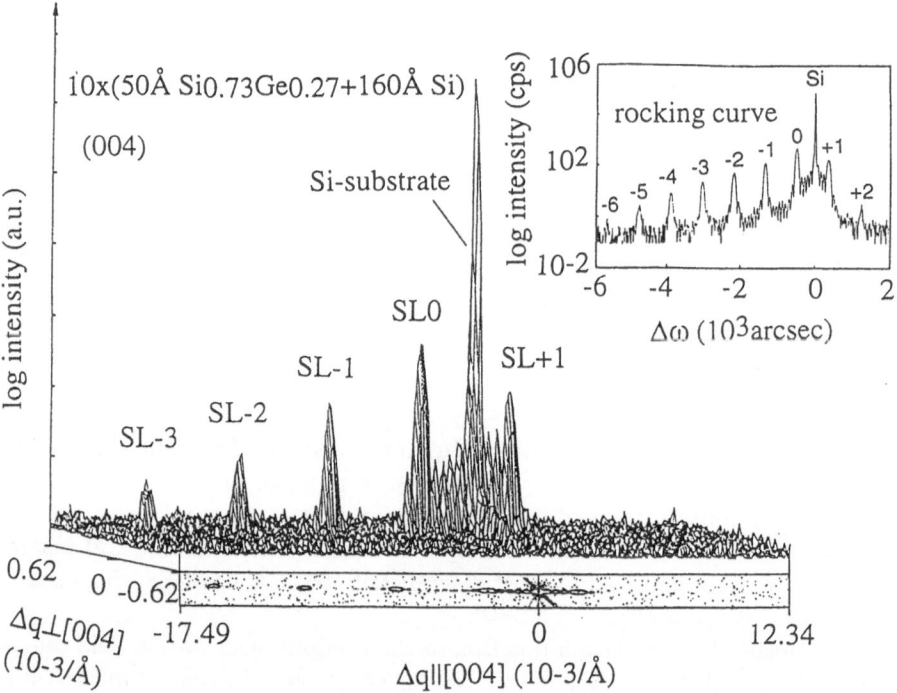

Fig. 6.56. Triple-axis reciprocal space map of a pseudomorphic Si/SiGe MQW structure on Si(001) in the vicinity of (004) substrate reflection along the directions q∥[004] and perpendicular to it. Three- and two-dimensional contour plots of constant scattered intensity (in a logarithmic scale) are shown, transformed to reciprocal space coordinates Δq relative to the substrate. The insert exhibits a DCD rocking-curve measured over the same region in reciprocal space with the Si substrate peak and several orders of superlattice peaks ($-6\ldots2$) [6.165]

subsidiary maxima has to be $N - 2$, which is in this case $10 - 2 = 8$. The subsidiary maxima can be identified not only between SL0 and SL+1, but also between the SL0 and SL-1, SL-1 and SL-2 and even in between SL-2 and SL-3. In Fig. 6.56 also a reciprocal space map around the (004) Bragg reflection for the substrate and the SL+1, SL0, SL-1,-2,-3 peaks is shown, which demonstrates the different *FWHM*'s of the contours of equal intensity along the [004]-direction and perpendicular to it (i.e., the [010]-direction). The intensity in the reciprocal space map is also shown in a three-dimensional representation. In Fig. 6.57 the central part of 6.56 is enlarged for clarity for the demonstration of the subsidiary intensity maxima in between the SL0 and SL+1 superlattice (004) reflections for the same sample. The *FWHM* of the SL extrema along growth direction is 78 arcsec, which is due to the finite thickness of the entire MWQ stack. However, perpendicular to growth direction the *FWHM* is only 13 arcsec for the zero order peak, which indicates the extremely high crystalline perfection and the absence of any appreciable mosaicity. This fact is not astonishing since the analysis of the (004) and the (224) reciprocal space maps yield that the mean in-plane lattice constant

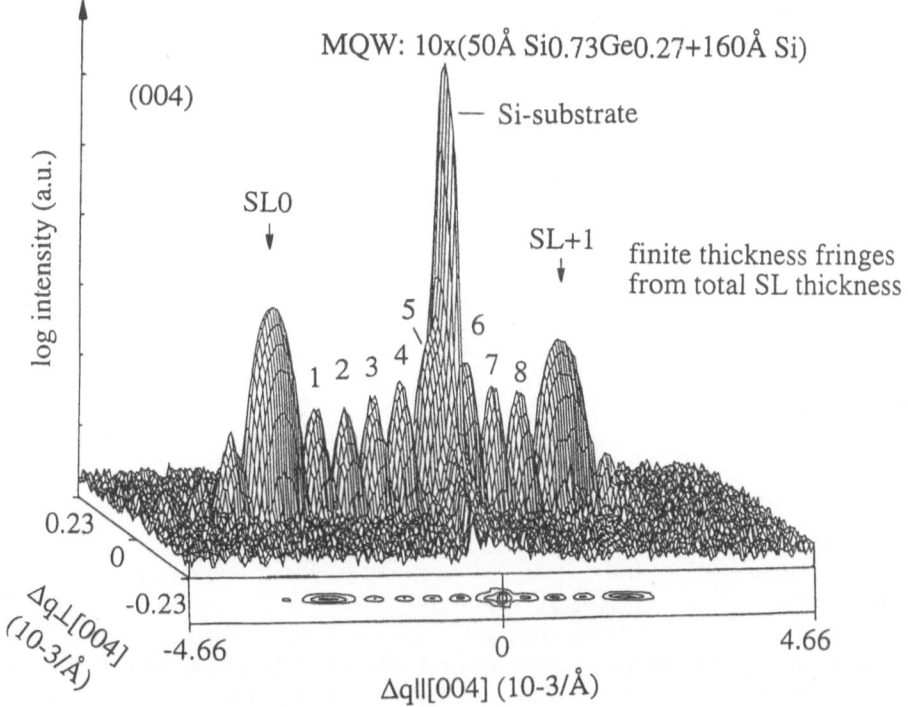

Fig. 6.57. Detail of Fig. 6.56 showing 8 clearly resolved finite thickness fringes inbetween the SL peaks SL0 and SL+1, indicating high structural perfection. The *FWHM* of the SL reciprocal lattice points in the $\mathbf{q}_{\perp}$ direction are remarkably small [6.165]

of the MQW constituent layers is exactly equal to the unstrained Si lattice constant.

For superlattices grown along a [111]-direction like e.g. the one shown in Fig. 6.58 usually reciprocal space maps are recorded around a symmetric (222) and an asymmetric (264) reflection. The data stem from a PbTe (87.6 monolayers)/EuTe (5 monolayer) superlattice grown on a PbTe buffer, deposited on a (111) BaF_2 substrate. The iso-intensity contours in the vicinity of the buffer (B) and the superlattice peaks (SL0, SL $\pm$ 1, SL $\pm$ 2, SL $\pm$ 3) are shown. For these two (111) and (264) reflections the superlattice peaks lie all along the [111]-direction normal to the sample surface and both measurements were performed in the same [1$\bar{1}$0] azimuth orientation of the sample.

A theoretical analysis based on the kinematical diffraction theory of diffuse X-Ray scattering from multilayers exhibiting interface roughness as well as mosaic structure was given by Holý [6.168]. With this method iso-intensitiy

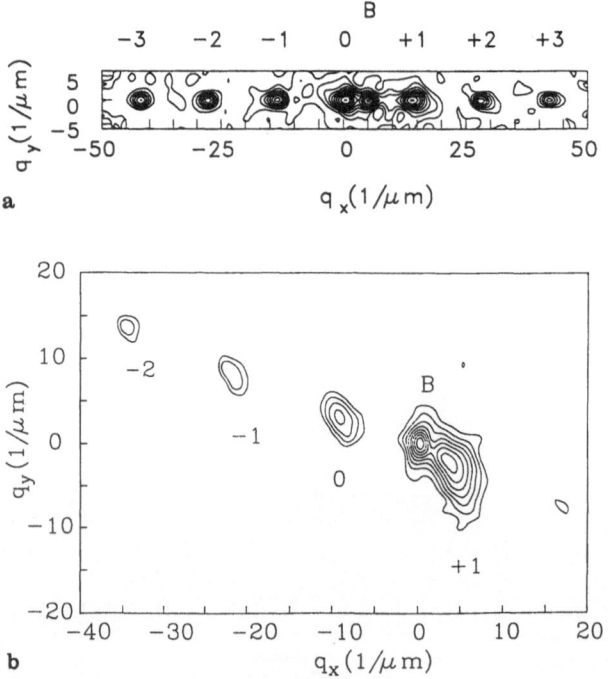

Fig. 6.58. Reciprocal space maps around symmetric (111) (**a**) and asymmetric (264) (**b**) reciprocal lattice points of a PbTe (87.6 monolayers)/ EuTe (5 monolayers) superlattice grown on a PbTe buffer deposited on (111) BaF_2 substrate. The ω and $\omega - 2\theta$ scan directions are transformed to the reciprocal space axes q_x (parallel to $q_{[222]}$) and q_y (perpendicular to $q_{[222]}$) for (**a**), and parallel to $q_{[264]}$ and perpendicular to $q_{[264]}$ for (**b**) [6.128]

contours and various satellite reflections in reciprocal space maps as well as DCD rocking-curves are simulated. The analysis yields interface roughness, the correlation of interface roughness from different interfaces, and independently the mosaic spread.

6.8 New Developments

6.8.1 Analysis of Quantum Wire Structures Using HRXRD

Laterally structured semiconductors (quantum wires and quantum dots) are in the focus of current interest. Since these are periodic structures with large coherence length and typical dimensions of the order of 1000 Å, X-Ray diffraction can be used to get information on their structural quality.

Recently Tapfer et al. [6.169, 6.170] have studied deep mesa etched AlGaAs/GaAs quantum wire structures and obtained both the quantum-wire period as well as the quantum wire width. In the angular range close to a Bragg diffraction peak interference of X-Ray waves scattered by the quantum-wires occurs. For the analysis a combination of X-Ray diffraction- and a multiple-slit Fraunhofer model has been employed. The normalised reflectivity as a function of the deviation $\Delta\theta$ from the Bragg angle is given by

$$R(\Delta\theta) = (|Y_h|DI)^2 \tag{6.75}$$

where Y_h is the Fourier coefficient of the polarisability for the h-th Bragg reflection and

$$D = \frac{\sin\beta}{\beta} \tag{6.76}$$

the diffraction term with

$$\beta = \frac{\pi d_w \sin(2\theta_B)\Delta\theta}{\lambda \cos\theta_e} \tag{6.77}$$

where d_w is the quantum-wire width, θ_e is the angle between reflected X-Ray beam and the crystal surface. I is the interference term

$$I = \frac{\sin(N\alpha)}{N\sin\alpha} \tag{6.78}$$

with

$$\alpha = \frac{\pi L_p \sin(2\theta_B)\Delta\theta}{\lambda \cos\theta_e} \tag{6.79}$$

where N is the total number of irradiated quantum wires and L_p the QW period. From these equations the QW width d_w and period L_p are given by

$$d_w = \frac{\pm m\lambda \cos\theta_e}{\Delta\theta_{\min}\sin(2\theta_B)} \tag{6.80}$$

$$L_p = \frac{\pm n\lambda \cos\theta_e}{\Delta\theta_{\max}\sin(2\theta_B)}. \tag{6.81}$$

There is the possibility that the maxima of I coincide with the minima of D. When $L_p = (n/m)d_w$ the n-th order QW peaks are suppressed. This allows determination of not only the QW period but also the QW width d_w. An example of such a measurement is given in Fig. 6.59 with inserts explaining the scattering geometry. In reciprocal space the rods have dimensions Å^{-1}, the inverse height of the mesa-etched structures and they are a distance $1/L_p$

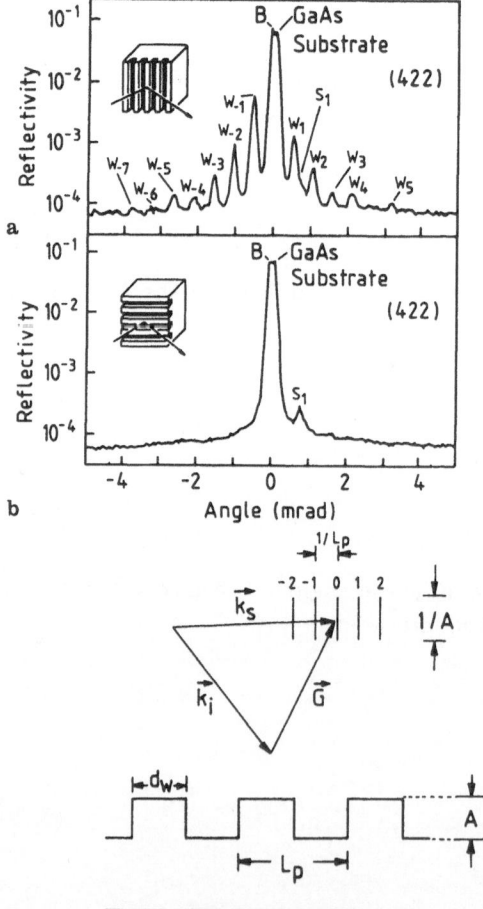

Fig. 6.59. Diffraction pattern of a GaAs/Al$_{0.36}$Ga$_{0.64}$AS multi quantum well wire structure (**a**) perpendicular and (**b**) parallel to the scattering plane as indicated by the inserts [6.171]. The satellite peaks W_i are due to the periodic array of the quantum wires. Lower part shows the corresponding scattering scheme in reciprocal space [6.166]

apart. The accuracy of the determination of the quantum wire period L_p is about 30 Å.

Figure 6.59 shows experimental X-Ray diffraction pattern of a MQW $Al_{0.36}Ga_{0.64}As/GaAs$ deposited on a $Al_{0.36}Ga_{0.64}As$ buffer layer on top of a GaAs substrate for an asymmetric (422) reflection. The satellites designed W_i (with i up to ± 6) are due to the lateral periodicity caused by the quantum wires and their separation yields a period of 282 nm and a quantum wire width of 60 nm. B denotes the diffraction peak of the buffer and S_1 the first order SL peak from the chemical modulation along the growth direction. In addition, Tapfer et al. deduced from the comparison of the SL peak positions in the two scattering geometries an in-plane lattice strain normal to the quantum wires.

Macrander and Slusky [6.166] have investigated InP substrates with a sawtooth corrugation and overgrown epilayers of InGaAsP on these substrates. They found satellite structures, if the corrugations were oriented perpendicular to the diffraction plane. For the corrugation orientation parallel to the diffraction plane no evidence for satellite peaks was found.

Recently, a diffraction model based on a semi-dynamical diffraction theory for the calculation of diffraction pattern of corrugated surfaces was presented [6.172]. This model is useful for the determination of the shape (rectangular, trapezoid, ...) i.e. of the geometrical parameters of the surface corrugation from the observed diffractograms. The analysis of symmetric and asymmetric diffractograms is necessary for this purpose, the latter should be recorded in glancing exit reflection geometry, since this is the most sensitive one to surface corrugation. Apparently, two-dimensional reciprocal space maps offer similar advantages for the analysis of periodic corrugated surfaces as well as for periodic semiconductor wires or dots fabricated by deep mesa etching, as they do for the analysis of epilayers and heterostructures. Gailhanou et al. [6.173], van der Sluis et al. [6.174] and Holý et al. [6.175] have recently used this technique for the study of corrugated GaAs, InP, Si substrates as well as InAs/GaAs quantum wires. Gailhanou has pointed out that surface gratings act simultaneously as reflection and transmission gratings.

Fukui and Saito [6.176] have in principle made similar observations on fractional-layer superlattices $(AlAs)_{0.5}/(GaAs)_{0.5}$ grown on vicinal (001) surfaces which exhibited satellite peaks in X-Ray diffraction due to the lateral compositional modulation (AlAs rods–GaAs rods). Structural studies on InAs microclusters (quantum dots) were also performed by Brandt et al. [6.177]. The InAs microclusters were prepared by deposition of fractional monolayers of InAs on terraced (001) GaAs surfaces and subsequent overgrowth of InAs. An asymmetry of the angular position of the satellite peaks (± 1) was observed when the sample was rotated around the surface normal.

In the following reciprocal space maps of the diffraction pattern of reactive ion etched 150 nm and 175 nm wide GaAs/AlAs periodic quantum wires and quantum dots, fabricated by electron beam lithography and $SiCl_4/O_2$

reactive ion etching, are presented [6.178]. The GaAs/AlAs wires and dots were realized by nanostructuring a 30 period AlAs-GaAs multiquantum well grown on a 15 nm thick GaAs buffer. The nominally 8 nm thick GaAs wells are separated by nominally 12 nm AlAs barriers resulting in a total thickness of 600 nm. The MQW was capped by a 20 nm GaAs layer. Beneath the GaAs buffer 25 periods of a 5 ML/5 ML short period AlAs-GaAs SL with a total thickness of approximately 75 nm was grown on the GaAs substrate with a 80 nm buffer. The lateral macroperiodicity of the wire and dot arrays gives rise to additional intensity maxima in the diffraction pattern along q_x-direction perpendicular to the growth direction. In Fig. 6.60a reciprocal space maps around the (004) reciprocal lattice point (RELP) of an unstructured (as-grown) GaAs/AlAs-reference sample are shown. "S" denotes the GaAs-substrate peak, SL_0 and SL_1 the zero and first-order MQW peak, respectively. "A" is a symbol for an artifact, the analyser streak. Both the substrate and SL_0 peak are elongated along the Ewald sphere intersecting the growth direction with the Bragg angle Θ_B. Thickness fringes inbetween the MQW peaks SL_0 and SL_1 indicate the good crystalline quality of the system. Their spacing corresponds to the total thickness of the superlattice of approximately 640 nm. In Fig. 6.60b the diffraction pattern of the periodic wire array is shown. Wire satellites accompanying the SL_0 peak and the first-order MQW peak SL_1 are observed. The wire period determined from the spacing of the satellites along the q_x direction is 303 nm. The inset in Fig. 6.60b defines the diffraction geometry, the arrow is the normal to the diffraction plane, which is defined by the incident and diffracted (004) X-Ray

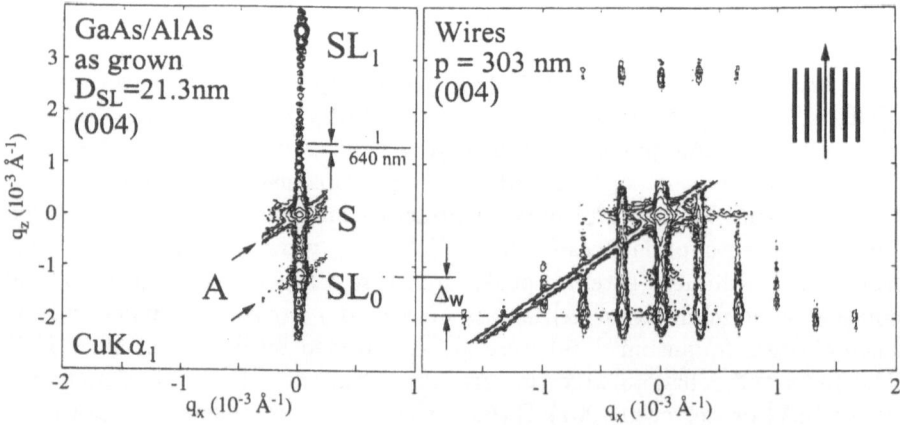

Fig. 6.60. Reciprocal space maps of a MQW GaAs/AlAs reference sample (**a**) and a MQW GaAs/AlAs wire array (**b**) on GaAs(001) around the GaAs (004) reflection. The levels of the isointensity contours are varying between 1.2 to 18000 counts/s in (**a**) and between 1.5 and 15000 counts/s in (**b**)

wavevectors. The half width of the satellite peaks beside the GaAs substrate and buffer peak is much larger than that of the actual GaAs/AlAs dot fringes indicating a corrugation of the GaAs buffer.

In Fig. 6.61 maps for the periodic dot array are shown. The sample was oriented with the [110]-direction perpendicular to the diffraction plane (q_x-direction coincides with [1$\bar{1}$0], (a) and with [100] perpendicular to the diffraction plane (q_x-direction coincides with [010], (b). Clearly dot satellites are observable around the SL_0 satellite RELP. The satellite spacing decreases by a factor of $1/\sqrt{2}$ when the diffraction plane is rotated by 45° out of the principal direction of the dot array Fig. 6.61b. After a further rotation of 45° the original spacing is observed again. From the position of the MQW wire satellites in Fig. 6.60b it is concluded that the mean lattice constant along growth direction is larger than in the unpatterned reference sample Fig. 6.60a, probably due to oxidizing of the AlAs layers upon etching.

Schuller et al. [6.179] investigated GaAs/AlAs short period superlattices which were grown with individual layer thicknesses which were different from integral numbers of atomic planes. This can be viewed as the introduction of controlled interfacial roughness modulation, which caused additional satellite structure. The non-integral but periodic modulation of the AlAs/GaAs SL structure can be interpreted as causing a splitting of the satellites. Schuller et al. interpreted this finding as evidence that controlled interface roughness induces changes in peak positions and not merely changes in peak intensities

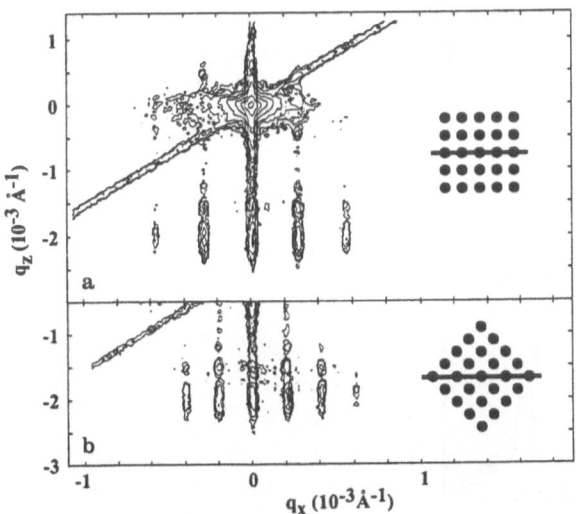

Fig. 6.61. Reciprocal space maps of a periodic MQW GaAs/AlAs dot array for $\phi = 0°$ (**a**) and $\phi = 45°$ (**b**) on GaAs(001). Due to the larger dot period in the incidence plane of the X-Rays in geometry B the satellite spacing in the reciprocal plane is reduced by a factor of $1/\sqrt{2}$

and linewidth, which is the case for random interface roughness. Fullerton et al. [6.154] derived an expression based on kinematical diffraction theory that includes random, continuous, and discrete fluctuations from the average superlattice structure.

Miceli et al. [6.180] measured and modeled both the low angle reflectivity as well as the Bragg scattering for (001) ErAs/GaAs (Fig. 6.62). X-Ray scattering was performed using Mo-K_{α_1} radiation from an 18 kW rotating anode generator and a triple-axis spectrometer with Ge(111) monochromator and analyser crystals. They analysed their data in the kinematical approximation including discrete interface fluctuations and the influence of diffuse scattering. They were able to fit both the data for the specular reflectivity and in the region of the (002) Bragg reflection peak with one single set of data. From their data, the authors concluded that the ErAs films grow with pinholes for 1 or 2 atomic layers of coverage and evolve into continuous films by 5 atomic layers coverage. It is claimed that the extended range reflectivity method gives structural details at the sub-monolayer scale and that the growth morphology of epitaxial layers can consequently be followed throughout the entire growth. Similar experiments were performed by Baribeau [6.181] on nonideal Si/SiGe superlattices. From the comparison of the conventional large angle high resolution X-Ray diffraction patterns on the same sample with the X-Ray reflectivities the interface roughness parameters, thickness fluctuations, and partial strain relaxation could be determined unambiguously.

In particular the X-Ray spectra of thin pseudomorphic epilayers exhibit binomial fluctuations for small numbers of atomic layers. This is characteristic for layer by layer growth. If the number of atomic layers increases Gaussian

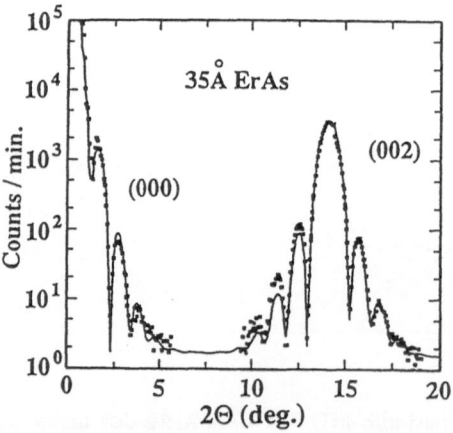

Fig. 6.62. Extended range specular reflectivity for 200 Å a-Si/35 Å ErAs/ GaAs(001). The whole spectrum is fitted by one single set of parameters (solid line) both for the specular reflection (000) as well as for the (002) Bragg reflection [6.180].

interface fluctuations occur. For sufficiently thick layers, when lattice relaxation occurs, the intensity distribution indicates exponential fluctuations. The line shape of the intensity distribution along the $q_\parallel$-direction (in the intensity contour plots: ω-direction) yields information on the lateral disorder correlations. As already shown by Sinha et al. [6.182] the intensity distribution along q$_\parallel$ contains two components. The first one is delta-like and the second one is a broad background from diffuse scattering which can even exhibit side peaks. The diffuse scattering originates from lateral fluctuations, usually mosaic-like, at least for the thicker films. The delta-peak stems from flatness over a correlation range of several μm. The intensity of the delta-peak decreases with increasing film thickness, whereas the diffuse scattering intensity increases, especially with the occurrence of misfit dislocations. Miceli has shown, that the lineshapes in $q_\parallel$ geometry change considerably as $q_\perp$ is varied from a grazing angle condition to Bragg angle geometry.

Recently, Yasuami et al. [6.183] investigated by diffuse X-Ray scattering the sublattice ordering among group III atoms in $In_{0.5}Ga_{0.5}P$ and in $In_{0.5}Al_{0.5}P$. They determined the Warren-Cowley [6.184] short-range parameters and found excellent agreement with structure models from high resolution transmission electron microscopy.

6.8.2 Real Time X-Ray Diffraction

In-situ HRXRD has been recently applied by Tsuchiya et al. [6.185] for growth monitoring of InGaAs on InP using X-Ray diffraction data as an input for a feedback for the control of the metal organic sources. For that purpose the MOVPE reactor was equipped with Be-windows for the incident radiation, which passed through a Bartels monochromator and for the scattered radiation. A precise control of the lattice mismatch of the growing InGaAs layer was achieved.

Fast structural X-Ray diffraction methods have become possible through the use of synchrotron sources, angular dispersive methods and position-sensitive detectors. Using these advanced methods, problems like the strain relaxation dynamics in epitaxial layers can be addressed as shown by Lowe et al. [6.186]. In a recent study on SiGe films grown on Si substrates in a metastable region close to the critical thickness for misfit dislocation production, evidence was found for the coexistence of different strain states. During rapid thermal treatment the change of the strain status can be followed with a temporal resolution of about 1 s. The results indicate that the interface relaxation process is not dominated by the kinetics of single misfit dislocations but is rather a cooperative phenomenon.

6.9 Grazing-Incidence X-Ray Techniques

So far, the application of different X-Ray diffraction techniques for the characterisation of advanced epitaxial layers has been described. Measurements of the lattice mismatch and — in the case of alloy compounds — composition of epitaxial semiconductor structures are nowadays well established. In recent years ultrathin and thin layers, meaning a few tens to a few hundred Å, as well as multilayer structures are of increasing importance in science and technology. Although high resolution X-Ray techniques are now available, measurements on such structures are complicated because the relatively large penetration depth of X-Rays (typically some μm) leads to a poor surface layer to substrate scattering ratio. Nevertheless, for the past 10 years surface X-ray diffraction has become an important tool to solve various problems in surface and interface science. This development is strongly connected with the rapidly increasing number of high intensity synchrotron X-Ray sources because the diffracted waves are usually of very low intensity. Different techniques like crystal truncation rod analysis, two-dimensional crystallography, three-dimensional structure analysis and the evanescent wave method have been successfully applied. Previous reviews to this field of scattering techniques in surface science were made by Feidenhans'l [6.187], Fuoss and Brennan [6.188] and Robinson and Tweet [6.189]. The surface sensitivity of all widely used X-Ray techniques (diffraction, fluorescence analysis, X-Ray absorption, topography, ...) can be considerably enhanced when the X-Ray beam meets the surface at a glancing angle (a few milliradians) below the critical angle for which total external reflection occurs and an evanescent wave propagating parallel to the surface is created. In the last few years there has been a remarkable increase of the number of experimental and theoretical reports concerning grazing-incidence techniques for probing surfaces, surface layers and internal interfaces (see e.g. [6.190, 6.191]).

In the next subsections a discussion of the applications and capabilities of these techniques is presented. First we shall briefly review the basic concepts of the propagation of X-Ray radiation crossing interfaces. In particular, we shall consider refraction and total reflection of an incident wave at glancing angle. Next the methods of Grazing-Incidence Diffraction (GID) will be discussed.

Figure 6.63 illustrates a typical glancing-angle geometry. Let us consider a plane wave

$$\mathbf{E} = \mathbf{E_0} \exp[i\,(\mathbf{k} \cdot \mathbf{r} - \omega t)] \tag{6.82}$$

with wavevector $\mathbf{k}_i$ impinging on a sharp interface which separates two different media (e.g. a single-crystal surface in vacuum). The angle of incidence ϑ_i is measured from the plane of the surface. Under suitable conditions, which will be explained below, the incoming wave splits into a reflected (wavevector $\mathbf{k}_r$) and a refracted one (not shown) and also a diffracted wave (wavevector $\mathbf{k}_s$)

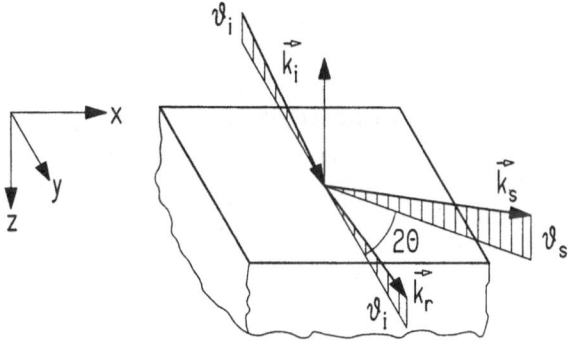

Fig. 6.63. Grazing-incidence geometry. The X-Rays meet the surface at an angle ϑ_i; the beam is diffracted through an angle 2θ and leaves the surface at an angle ϑ_s. 2θ is the angle between the projection of the specularly reflected X-Ray on the surface and the projection of the diffracted one

in the case of obeying the Bragg condition for a lattice plane nearly perpendicular to the surface. The situation for the reflected and refracted wave $\mathbf{k}_t$ is shown separately in Fig. 6.64. For simplicity all angles are assumed to be small (usually $\vartheta < 1°$) so that the small-angle approximation $[\sin(\vartheta) \approx \vartheta]$ can be used.

6.10 Reflection of X-Rays at Grazing Incidence

Refraction and reflection are well-known optical phenomena and are described by Snell's law and the Fresnel equations (see also Chapter 5 on FIR spectroscopy in this textbook) [6.193]. The index of refraction is the fundamental quantity for the description of wave propagation in any media. In the X-Ray range it can be written as [6.194, 6.77]:

$$n = 1 - \delta - \mathrm{i}\,\beta \tag{6.83}$$

with the dispersive term

$$\delta = \frac{N_a r_e \lambda^2}{2\pi}\, \rho\, \frac{Z + f'}{A} \tag{6.84}$$

and the absorptive term

$$\beta = \frac{N_a r_e \lambda^2}{2\pi}\, \frac{\rho f''}{A} = \frac{\mu \lambda}{4\pi} \tag{6.85}$$

[N_a : Avogadro's number, r_e : classical electron radius, λ : wavelength of X-Rays, ρ : mass density, Z : atomic number, f', f'' : real (dispersion) and imaginary (absorption) part of the dispersion corrections [6.195], respectively, A : mass number, μ : linear absorption coefficient [6.196]]

For a multielement specimen in (6.84) and (6.85) the sum is to be taken over the weighted fractions of each element within the compound, respectively. The real part of n, $1 - \delta$, is connected to the phase-lag of the propagating wave, the imaginary part, β, corresponds to the decrease of the wave amplitude. δ and β are small positive quantities of order 10^{-5} to 10^{-7} for X-Ray wavelengths at about 1.5 Å (Tab. 6.2). Consequently, the refractive index is slightly less than 1 and the transmitted wave will be refracted from the normal (Fig. 6.64) according to Snell's law :

$$n_1 \cos \vartheta_i = n_2 \cos \vartheta_t \tag{6.86}$$

It is widely known that the effect of refraction must to be taken into account for precise lattice constant determination from measured Bragg angles [6.34]. On the other hand the deviation from Bragg's law has been used to determine the refractive index of crystals for X-Rays [6.77]. It was Compton [6.197] who first pointed out as early as 1923 that since the refractive index of matter is less than unity for X-Rays it ought to be possible to obtain external total reflection from a smooth surface. He verified his prediction experimentally determining the critical angle ϑ_c for the tungsten L line ($\lambda = 1.28$ Å) being totally reflected from a glass surface. Neglecting absorption ($\beta = 0$), the X-Ray critical angle ϑ_c is given by

$$\cos \vartheta_c = n_2 \tag{6.87}$$

which leads, on expansion of the cosine for small angles, to

$$\vartheta_c \approx \sqrt{2\delta_2} \sim \sqrt{\rho(Z + f')}\lambda \tag{6.88}$$

Values of δ, β and ϑ_c (calculated for CuK$_\alpha$ radiation ($\lambda = 1.54$ Å)) for three different materials are listed in Table 6.2. External total reflection typically takes place for glancing angles below 0.7° for X-Ray wavelengths at about 1.5 Å (CuK$_\alpha$ radiation).

It is well known from optical theory [6.193] that if the second medium is absorbing, Snell's law (6.86) is only true in a generalised form with a complex

Table 6.2. Comparison of the calculated dispersive term δ, the absorptive term β and the critical angle ϑ_c ($\lambda = 1.54$ Å)

	δ	β	ϑ_c
Si	$7.47 \cdot 10^{-6}$	$0.18 \cdot 10^{-6}$	0.22°
GaAs	$14.79 \cdot 10^{-6}$	$0.46 \cdot 10^{-6}$	0.31°
Au	$49.78 \cdot 10^{-6}$	$4.59 \cdot 10^{-6}$	0.57°

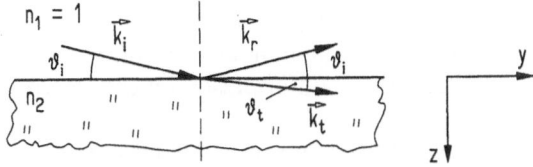

Fig. 6.64. Reflection and refraction of X-Rays incident upon a plane boundary between vacuum ($n_1 = 1$) and a medium whose refractive index n_2 is for X-Rays always less then 1

angle of refraction. This means that the planes of constant amplitude in the medium differ from the planes of constant phase. The refracted wave is conveniently called an inhomogeneous "plane" wave. From (6.86), reformulated in terms of ϑ_c such that

$$\vartheta_t^2 = \vartheta_i^2 - \vartheta_c^2, \tag{6.89}$$

we note that the internal angle ϑ_t is obviously always imaginary for $\vartheta_i < \vartheta_c$. This means that the amplitude of the transmitted wave field drops off exponentially

$$E_t(\mathbf{r}') \propto \exp[i \cdot \mathrm{Re}(\mathbf{k_t}) \cdot \mathbf{r}'] \exp[-z/z_0] \tag{6.90}$$

with a damping constant z_0 [6.198]

$$z_0 = \frac{\sqrt{2}}{k} \left[\sqrt{(\vartheta_i^2 - 2\delta_2)^2 + 4\beta_2^2} - (\vartheta_i^2 - 2\delta_2) \right]^{-\frac{1}{2}} \tag{6.91}$$

as the wave penetrates the less dense medium propagating parallel to the surface (y-direction in Fig. 6.64). Figure 6.65 shows the damping constant z_0 as a function of glancing angle ϑ_i. Note that for angles below ϑ_c the amplitude of the so-called evanescent wave decays at a short distance into the solid

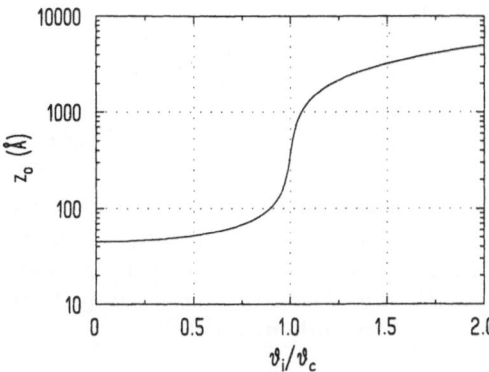

Fig. 6.65. Damping constant z_0 (1/e depth of the transmitted wave amplitude) for CuK$_\alpha$ radiation in GaAs as a function of glancing angles ϑ_i. The critical angle is $\vartheta_c = 0.31°$

(typically 50 Å). This fact makes grazing-incidence X-Ray techniques surface sensitive since the signal originates from a short depth below the surface. On the other hand the penetration of the wave into the bulk can be controlled by increasing the angle of incidence providing precise information on the electron density depth profile.

In order to derive expressions for the reflected and transmitted wave fields one has to apply the boundary conditions found from Maxwell's equations like in optical theory [6.193]. This leads to the Fresnel equations (see also Chapter on FIR spectroscopy) defining the amplitude reflection and transmission coefficients r and t, which are given by

$$r = \frac{\vartheta_i - \vartheta_t}{\vartheta_i + \vartheta_t} \tag{6.92}$$

$$t = \frac{2\vartheta_i}{\vartheta_i + \vartheta_t} \tag{6.93}$$

In the small angle limit considered here, the reflectance $R = r \cdot r^*$ and the transmittance $T = t \cdot t^*$ are independent of polarisation of the incident wave. A detailed discussion of R and T for X-Ray wavelengths is given by Parrat [6.199] and by Vineyard [6.200]. A theoretical reflection curve for GaAs is shown in Fig. 6.66. For angles well below ϑ_c, R becomes nearly unity. When ϑ_i exceeds the critical angle ϑ_c, R drops off rapidly as ϑ^{-4}. But there is obviously no sharp limit of total reflection. The reflectivity is significantly reduced in a range close to ϑ_c. This is due to photoelectric absorption of GaAs which was included in the calculation. It has first been verified by Prins [6.201] and Kiessig [6.202] in a series of experiments, that when X-Ray wavelengths approach the absorption edge of an element, then the sharpness of the limit of total reflection is much affected by the absorption. Their results gave for the first time definite evidence of the existence of anomalous dispersion of X-Rays. Measurements of reflectivity, carried out with synchrotron radiation, have recently been used for the determination of the energy dependence of the

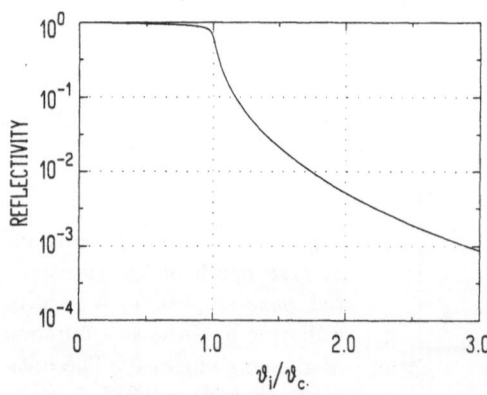

Fig. 6.66. Calculated reflectivity of GaAs (CuK$_\alpha$ radiation) as a function of glancing angles ϑ_i where the surface is perfectly smooth. ϑ_c is 0.31°

anomalous scattering factor f' (6.84) [6.203]. The knowledge of this quantity is quite important for the interpretation of many X-Ray experiments. We have to keep in mind that the critical angle as well as the shape of the Fresnel reflectivity curve (Fig. 6.66) depends on some characteristic material properties only, such as electron density, atomic form factor and absorption coefficient. They are independent of the crystalline structure or the orientation of crystallites on (or in) the surface. However, the measured reflectivity from a real surface departs from the predicted Fresnel reflectivity because in reality a surface is never ideally flat [6.199]. This problem will be discussed in the next Section.

6.11 Specular Reflection and Non-Specular Scattering of X-Rays from Layered Structures

Let us now consider a thin film on a substrate (Fig. 6.67). When the incident angle ϑ_i of the incoming X-Ray wave exceeds ϑ_c for the layer, the reflectivity will show oscillations as a function of ϑ_i due to interferences of waves reflected from the top surface and waves reflected from the interface (Fig. 6.68).

These phase-sensitive structures are geometric resonances known in optics as Fabry-Perot interferences. The angular spacing of the intensity maxima of specularly reflected X-Rays was first measured by Kiessig [6.204] from which he determined the thickness of thin films. For the situation depicted in Figs. 6.67, 6.68, respectively, $n_2 < n_3$, the thickness d is related to the maxima positions ϑ_m by

$$\frac{1}{2}(2m + 1)\lambda = 2d \cdot \sqrt{\vartheta_m^2 - \vartheta_c^2} \qquad (6.94)$$

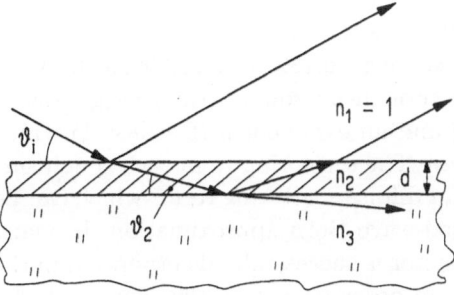

Fig. 6.67. X-Ray reflection at grazing incidence from a layer of thickness d on a substrate ($n_2 < n_3$). The incident beam is partially transmitted. This beam is reflected then back from the layer/substrate interface. The resulting path-length difference determines the interference pattern of the two beams

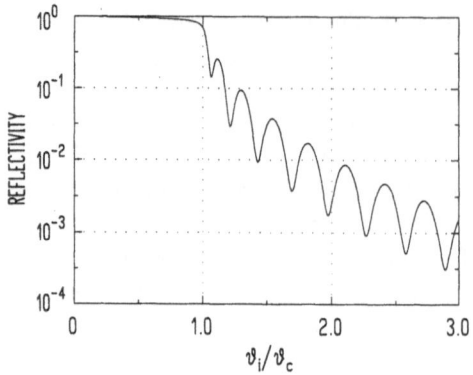

Fig. 6.68. Calculated reflectivity for a 42 nm thick GaAs layer on silicon (CuK_α radiation) assuming ideally abrupt interface and a flat substrate

where m is an integer, λ is the wavelength of the incident radiation. This leads to a linear relationship if the square of the angular position of the maxima, ϑ_m^2, is plotted versus $(m + 1/2)^2$. The slope gives the layer thickness d and intercepts the critical angle for the layer. While the oscillations in the reflectivity profile result from the thickness of the specimen, the amplitude of the oscillations depends upon the contrast at both interfaces, that is the difference of the dispersive term δ (6.84) at the film-to-substrate interface. Consequently, the greater the refractive index difference at the two interfaces, the more pronounced will be the oscillations. For layered structures the analysis of the reflected intensity requires a calculation of the fringe pattern based for example on the recursion formulae given by Parrat [6.199] following from the Fresnel formalism of optical reflection and refraction in a multilayer with smooth interfaces. But in reality, the surface and the interfaces are not atomically sharp, therefore the calculated reflectivity on the basis of the simple model of ideal layers is often in substantial disagreement with the measured data. The interfacial roughness and its correlation from interface to interface in multilayer thin films devices influences also their novel optical, electrical, magnetic, mechanical and superconducting properties. Therefore the knowledge and control of the quality of the interfaces is of both practical and fundamental interest. In several theoretical approaches concerning the X-Ray scattering from non-ideal interfaces (e.g. [6.182, 6.205, 6.206, 6.207]) the rough surface is characterized by the root mean square (rms) roughness σ, the height-height correlation length ξ and an exponent h ($0 < h < 1$) which determines the texture of the roughness. The scattering from such rough surfaces is split into specular reflection and diffuse scattering terms which can be calculated on the basis of the distorted-wave Born approximation. In many cases a Debye-Waller type roughness factor is successfully incorporated in the Fresnel reflectivity model calculations in order to deduce the rms roughness σ. In the method applied by Vidal and Vincent [6.208] the Fresnel amplitude coefficients r_{Fj} at each interface j is corrected by

$$r_j = r_{Fj} \cdot \exp\left(-2k_j^2\, \vartheta_{ji}\vartheta_{(j+1)i}\, \sigma_j^2\right) \tag{6.95}$$

where σ_j is the rms value of roughness for layer j. The surface roughness determines the decay near ϑ_c while interfacial roughness leads, for layered structures, to progressive damping of the oscillations with increasing angle ϑ_i. But from reflectivity experiments no information on the in-plane correlation length ξ can be obtained. This incoherent contribution affects the scattering of X-Rays into non-specular directions which are observed in a wide angular range around the speculary reflected beam. A typical feature of the diffusely scattered intensity is a maximum if either the angle of incidence or the detection angle equals the critical angle of total reflection. Yoneda [6.209] was the first who observed this anomalous surface reflection for glancing angles. Warren and Clark [6.210] and Guentert [6.211] interpreted the effect in terms of small-angle scattering of surface or interface irregularities. The diffuse scattering technique has proved to be a useful tool for the determination of surface quality and surface contaminants on the atomic scale. In combination with measurements of the reflectivity it is possible to separate the roughness of external and internal interfaces and its correlation in multilayer films as well as to investigate the fractal dimension of the interface [6.206, 6.212, 6.175, 6.213, 6.214].

Furthermore it has been demonstrated that the nonspecular scattering is affected in particular by the vertical roughness correlation [6.215]. These theoretical results have been used for the analysis of specular and nonspecular X-Ray reflectivity experiments on a 20 period MBE-grown AlAs/GaAs SL performed whith a synchrotron source using various scans and reciprocal space mapping around (000).

Some of the principles of specular reflection given in the previous sections will now be illustrated. The characterisation of thin films by means of X-Ray reflectivity measurements under grazing incidence conditions is not restricted to single-crystal samples like in the case of diffraction techniques. One can obtain information on electron density and thickness of single-crystal, polycrystal as well as amorphous films. But also surface and interface roughness on an atomic scale can be analysed. While the reflectivity profile of a single-layer sample often exhibits simple oscillations which can easily be evaluated, in case of multilayer structures the interpretation is not that straightforward. Because of recent improvements of deposition techniques multilayer structures of stratified and periodic media have been produced, which found a wide interest in basic research as well as in applications to the design of electronic and optoelectronic devices, soft X-Ray mirror structures, etc.. X-Ray scattering at small angle has proved to be useful for their characterisation [6.216, 6.217]. A typical situation of a semiconductor superlattice with large period thickness is shown in Fig. 6.69. The sample was nominally 10 * (60 nm GaAs, 20 nm AlAs) grown by MBE on a GaAs substrate. Beyond the critical angle at 0.31°,

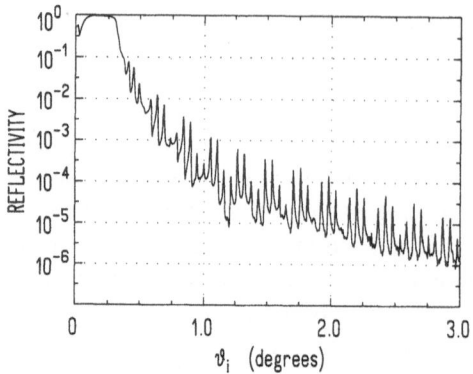

Fig. 6.69. Grazing incidence X-ray reflectivity of a 10 period GaAs/AlAs ($d_A = 60\,\text{nm}$, $d_B = 200\,\text{nm}$) superlattice grown on (100) GaAs (The sample was grown by A. Förster, ISI, KFA Jülich, the reflectivity measurement was carried out by U. Klemradt, IFF, KFA Jülich)

which corresponds to the value for GaAs, the reflectivity curve exhibits a series of peaks in reflection. A full analysis of the data requires a calculation of the reflected intensity and adjustment of the parameters until a good fit is obtained. For this purpose least-squares curve fitting procedures [6.218] and Fourier analysis [6.219] have been employed.

An good example for the interface characterisation by means of X-Ray total reflection was given by Krol et al. [6.220]. For thin epitaxial $In_xGa_{1-x}As$ layers ($x = 0.53$ and 0.60) on InP and GaAs substrates they determined the roughness of the interfaces, and epilayer thickness by fitting the soft X-Ray reflectivities (used X-Ray energies $550\ldots700$ eV), assuming a model with uncorrelated interfacial roughness. The surface roughness parameters of all investigated samples were always smaller than interfacial roughness and did not depend on the type of the substrate or presence of stress in the epilayer. They concluded that the surface quality of the substrate and the MBE growth conditions influence more strongly the morphologic structure than strain or lattice mismatch. Slijkerman et al. [6.221] demonstrated that grazing-incidence X-Ray reflection technique is sensitive enough to detect even a delta-doping profile at a depth of a few nanometers below the surface. They characterised samples with a very narrow distribution of Sb dopant atoms capped, after deposition on a Si (001) crystal, with an ultrathin (a few nm) Si overlayer by analysing the reflectivity profiles. A schematic representation of the sample geometry is shown in Fig. 6.70. In the right-hand part, the depth profile of the electron density used in their model calculations is illustrated. The parameters they used to adjust their data (Fig. 6.71) indicate that the Sb doping profile drops off abruptly towards the substrate and more smoothly towards the surface with a $1/e$ decay length of $1.01 \pm 0.37\,\text{nm}$.

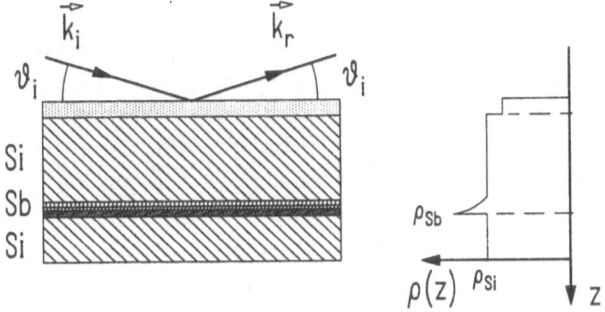

Fig. 6.70. Schematic illustration of a reflectance experiment from a buried delta-doping layer of Sb on Si. The corresponding density depth profile is given in the right-hand part [6.221]

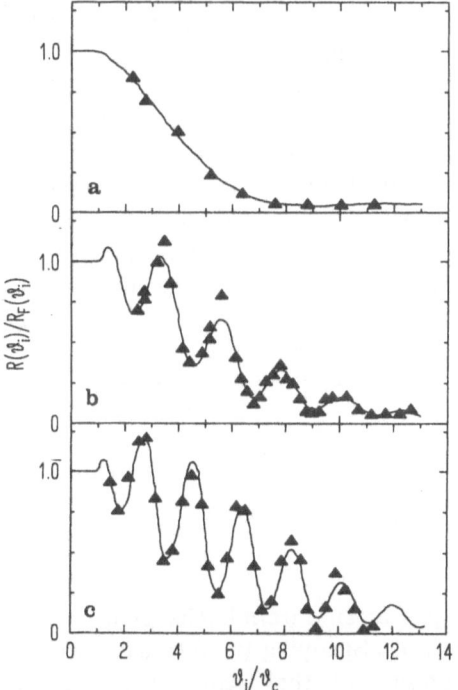

Fig. 6.71. Measured normalised reflectivities $R(\vartheta_i)/R_F(\vartheta_i)$ as a function glancing angle ϑ_i/ϑ_c. **a** Reflectivity data for a sample without Sb. The solid line was simulated for a model with only oxide on top of the sample surface, **b** the reflectivity curve for a Sb delta-doping layer on Si with c-Si cap. The calculated profile (solid line) to the measured data is given by the density profile with exponential Sb decay towards the surface in Fig. 6.70, **c** reflectivity data and fit obtained from a Sb doping profile with an a-Si overlayer [6.221]

Grazing X-Ray reflection was also applied by Baribeau [6.222] to study monolayer-thick buried Ge layers on Si. He measured and analysed strong oscillations, arising from the reflected X-Rays at the Ge/Si interface and the surface of the Si capping layer, for Ge film thicknesses down to one monolayer. The X-Ray data which are in agreement with additional transmission electron microscope studies of the samples, suggest a transition from a two-dimensional to three-dimensional growth mode at Ge coverage of about 6 monolayers. Similar results were obtained from the analysis of X-Ray reflectivity curves from various short-period $(Si_mGe_n)_p$ superlattices [6.223]. Sharp interfaces (half a monolayer wide) were found only for samples grown on Si with $n < 4$. Structures with larger n had rough interfaces owing to three-dimensional growth phenomena. The great sensitivity of X-Ray reflectometry to thickness variations and interfacial roughness makes this technique a powerful tool also for the characterisation of non-ideal multilayer structures. Recently it was demonstrated [6.181], how X-Ray reflectometry can complement double-crystal diffractometry analysis of non-ideal Si/Si_{1-x} superlattices that contain thickness fluctuations or in which partial strain relaxation is present.

These examples clearly demonstrate the capacity of this X-Ray technique for nondestructive charaterisation of thin films and interfaces. In addition it has been successfully applied to study a wide variety of interesting problems in surface and interface physics and chemistry and also in epitaxial growth. Examples include: average composition determination in multilayer structures [6.224], oxidation of metal and semiconductor surfaces [6.42], structure of Langmuir-Blodgett films [6.225], capillary waves on the surface of liquids [6.226], liquid organic monolayers on water and spreading of polymer micro-droplets on solid surfaces [6.227]. Combined X-Ray fluorescence and reflectivity measurements turned out to be very promising for depth profile element analysis of surface and layer structures [6.228, 6.229, 6.217].

6.12 Grazing-Incidence X-Ray Diffraction

Marra et al. [6.230] demonstrated for the first time that X-Rays striking a sample at a glancing angle can be Bragg scattered from lattice planes normal to the surface of ordered-layered structures. Studying the interface structure of epitaxial Al films on a GaAs single crystal, they achieved an considerably enhanced surface sensitivity under external total reflection conditions. Because of the improved signal-to-noise ratio they were able to measure the lattice spacing parallel to the interface in the Al layer as a function of layer thickness from 1000 down to 35 atomic layers (202 nm to 7.5 nm). Since this first experimental report grazing-incidence X-Ray scattering techniques with simultaneous total external reflection have been rapidly developed and successfully applied to study different aspects of surface and interfaces structures. A general survey of the current state of art is given by the recent conference proceedings [6.190, 6.191]. But the first theoretical concept dealing with this subject has already been treated in the early paper by Farwig and Schürmann [6.231]. Meanwhile the grazing-incidence X-Ray scattering theory has been studied in detail by Vineyard [6.200], Dietrich and Wagner [6.232, 6.233], Afanas'ev and Melkonyan [6.234] and Cowan [6.235]. These approaches range from semikinematic to dynamic scattering theory. The experimental progress in this new and promising research area are stimulated by advanced surface- and thin-film-preparation techniques in combination with the availability of X-Ray facilities like high intensity synchrotron sources. But also laboratory high-resolution X-Ray diffractometers have proved to be useful. The ideas of glancing angles reflectivity and surface diffraction have also been extended to other forms of radiation, particularly thermal neutrons [6.236, 6.237].

A typical scattering geometry is sketched in Fig. 6.72. The X-Ray beam strikes the sample surface at an angle ϑ_i $(< \vartheta_c)$ and is specular reflected as described in the previous sections. Bragg diffraction occurs from planes perpendicular to the surface. The scattered X-rays leave the surface at an angle 2θ about the surface normal at a takeoff angle ϑ_s (typically equal to ϑ_i). Conventional Bragg diffraction monitors the structural properties perpendicular to the interface or close to it, in Grazing-Incidence Diffraction (GID) the diffraction vector $\mathbf{k}_\parallel$ is parallel or nearly parallel to the surface. The small perpendicular momentum transfer $\mathbf{k}_z$ (6.98) depending on ϑ_i and ϑ_s has to be carefully considered [6.238, 6.237], too. In many cases the samples of interest exhibit a certain mosaic structure; therefore a semikinematic approach is adequate for the description of the scattering phenomena. The concept is known as "Distorted Wave Born Approximation" (DWBA). The

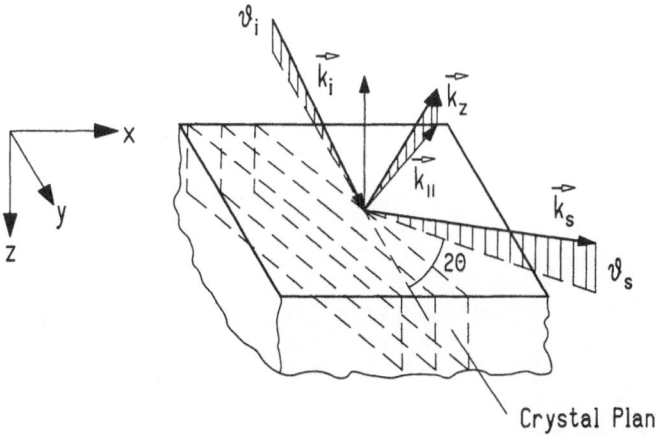

Fig. 6.72. Diffraction geometry under grazing incidence conditions showing incident ($\mathbf{k}_i$), scattered ($\mathbf{k}_s$) wavevectors as well as the components of the scattering vector. The speculary reflected beam is not shown

DWBA was first conceived by Vineyard [6.200] who treated only one grazing angle. Later this approach was extended to an arbitrary scattering geometry by Dietrich and Wagner [6.232, 6.233] and put into practice by Dosch et al. [6.238]. Within the framework of DWBA, the evanescent wave field inside the medium ("distorted wave") instead of the wave field in vacuum is considered to illuminate the crystal. The scattering intensity near the conditions of total external reflection is then given by [6.239]

$$I(k_\parallel, k_z) \sim |t_i|^2 \cdot S(k_\parallel, k_z) \cdot |t_s|^2 \tag{6.96}$$

with the structure factor $S(k_\parallel, k_z)$ for the momentum transfer components $k_\parallel$ parallel and k_z perpendicular to the surface. Furthermore, the Fresnel transmission coefficients $t_{i,s}$ are associated with the angles $\vartheta_{i,s}$

$$t_{i,s} = \frac{2\vartheta_{i,s}}{\vartheta_{i,s} + (\vartheta_{i,s}^2 - \vartheta_c^2)^{\frac{1}{2}}} \tag{6.97}$$

which are related to the amplitude of transmitted wave of the incident and diffracted beams. The ϑ-dependence of $t_{i,s}$ features an interesting phenomenon for $\vartheta_i = \vartheta_c$ (Fig. 6.73).

Here, the amplitude inside the medium is twice that of the incident wave (neglecting absorption) because of the phase matching of the two external waves that form a standing wave in front of the surface. It should be noted that grazing incidence (ϑ_i) and grazing exit (ϑ_s) are equivalent in case of

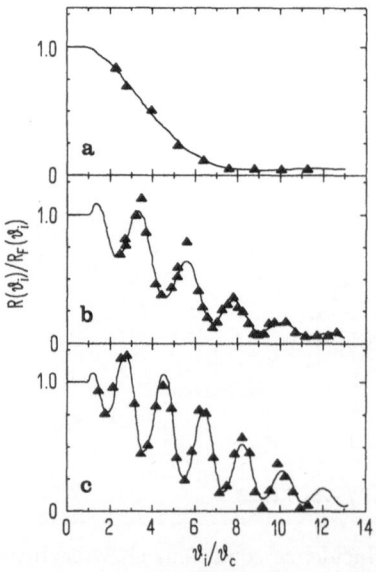

Fig. 6.73. Magnitude of the Fresnel transmission coefficient as a function of ϑ_i/ϑ_c

evanescent X-Ray waves. This interesting property has been experimentally verified by Becker et al. [6.240] detecting absorption (external incidence) and emission (internal excitation in the vicinity of a surface) of X-Ray waves in the total reflection region. Taking into account the refraction correction, k_z is given inside the solid by [6.233]:

$$k_z = k_{iz} - k_{sz} = \frac{2\pi}{\lambda} \left[(\sin^2 \vartheta_i - 2\delta - i2\beta)^{\frac{1}{2}} + (\sin^2 \vartheta_s - 2\delta - i2\beta)^{\frac{1}{2}} \right] \quad (6.98)$$

Whenever both angles ϑ_i and ϑ_s are less than ϑ_c ($\sin^2 \vartheta_c = 2\delta$), $\mathrm{Re}(k_z)$ vanishes in the case of no absorption, although the vacuum value may be unequal zero. The consequences for Bragg scattering are that $\vartheta_{i,s}$ can be controlled independently within the total reflection regime without affecting the Bragg condition which has to be fulfilled inside the crystal. The associated scattering depth Λ for which the scattered intensity is reduced by $1/e$ is defined as [6.238]:

$$\Lambda \equiv |\mathrm{Im}(k_z)|^{-1} \quad (6.99)$$

This length has to be distinguished from the penetration depth z_0 (6.91) of the evanescent wave which depends only on the the incidence angle α_i. Because of (6.98) Λ is symmetrically dependent on ϑ_i and ϑ_s. Fixing one of the angles, ϑ_i or ϑ_s, at a value below ϑ_c the other one can be chosen freely, providing a variation of Λ between a minimum value of typically $20 \ldots 50 \, \text{Å}$ and a maximum which is independent on the scanning angle (Fig. 6.74). The situation is qualitatively different when the fixed angle ϑ_i in Fig. 6.74,

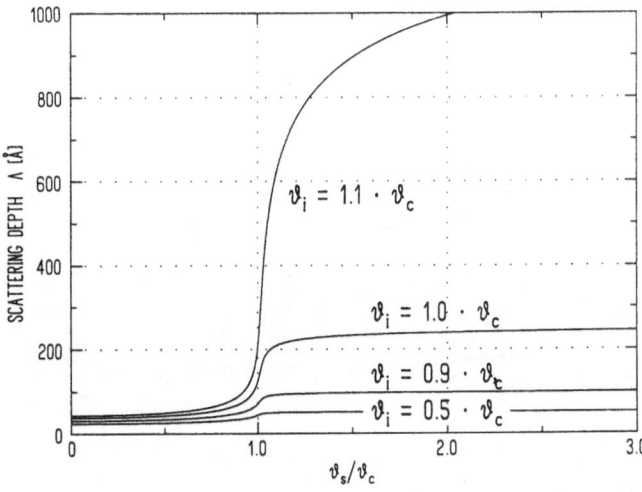

Fig. 6.74. Calculated scattering depth Λ for GaAs as a function of ϑ_s/ϑ_c for different incident angles ϑ_i/ϑ_c

exceeds ϑ_c. Although for $\vartheta_s/\vartheta_c \leq 1$ there still exists an evanescent wave ("grazing exit diffraction") in this case the upper limit of Λ is determined by photoabsorption and increases continuously with ϑ_s.

The behavior of Λ in the vicinity of total reflection has to be carefully considered when discussing surface Bragg scattering [6.238]. Figure 6.75 shows the results for the measured intensity distributions of the (222) reflection from a Fe_3Al ($\bar{1}10$) surface for different incident and exit angles [6.238]. The Bragg maximum is displaced from its bulk position ($\vartheta_{i,s} = 0$) by an amount comparable to ϑ_c. These results can be qualitatively understood as follows. At small angles Λ is also small and so is the effective number of layers participating in the diffraction, thereby reducing the scattered intensity. The maximum in the diffracted intensity appears near $\vartheta_{i,s} \approx \vartheta_c$ because at larger angles $Re(k_z)$ becomes finite and the Bragg condition is no longer obeyed. This effect overcompensates the increase of the scattering depth Λ. Similar scattering studies were carried out at highly perfect single crystals in order to get data for a detailed comparison with dynamical scattering theory (see e.g. [6.241, 6.242]). Because of the properties of X-Ray waves outside and inside a perfect crystal, the calculated profiles of the specular reflected and diffracted beam show interesting features which are not described by the kinematical approximation [6.234, 6.235].

High angular resolution measurements were achieved with synchrotron radiation revealing some of these features. Although effects of natural oxide layers [6.243] and of a tilt angle between the scattering lattice planes and surface normal has been carefully considered in the interpretation of the experimental data systematic deviations are still observed [6.241]. Recently

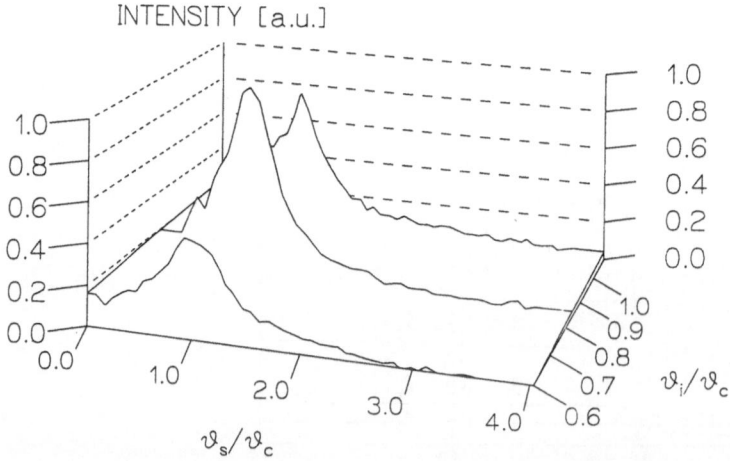

Fig. 6.75. Intensity profiles of an in-plane Bragg reflection from Fe_3Al (100) as a function of the take off angle ϑ_s/ϑ_c at three different glancing angles ϑ_i/ϑ_c [6.238]

Stepanov [6.244] proposed an explanation of the disagreements giving corrections for the angular dependence of the parameter α which describes the deviation from the Bragg condition. He also outlined ideas for further experimental tests which should to be taken on very clean crystal surfaces utilizing back diffraction conditions ($\Theta_B \approx 90°$).

The high angular resolution which is required in probing near surface profiles leads to very low intensity of the diffracted waves, so this kind of measurements are commonly performed with synchrotron radiation. A different diffraction scheme, for which a conventional X-Ray source is sufficient, has been proposed by Golovin and Imamov [6.245]. Instead of measuring the angular dependence of the intensity of the diffracted beam the whole diffracted cone is recorded with an open detector for varied angle of incidence of a well collimated X-Ray beam. Rahn and Pietsch [6.246, 6.247] demonstrated the application of the so-called "integral mode" for the characterisation of nanometer heterostructure layers. In their experimental setup they measured both the GID and the grazing-incidence reflectivity simultaneously. From the analysis of the profiles it was possible to separate the contributions of crystalline and amorphous part of the layer to the signals.

In order to determine both crystallite size and strains parallel to the interface also "$\Theta - 2\Theta$" scans are performed as in conventional diffraction. The direction of the scattering vector $\mathbf{k}_\parallel$ remains fixed, but the magnitude changes.

The interpretation of the data is straightforward and the structural properties are obtained from the formalism described in the previous passages of this Chapter. Segmüller et al. [6.3] applied this GID technique to characterise very different kinds of epitaxial films with thicknesses down to a few atomic layers. Although with their scattering arrangement depth profiling cannot be done, nor are the intensity calculation feasible, the measurements performed with a laboratory rotating anode X-Ray generator demonstrate the great application potential of GID in laboratory environment for thin film characterisation. Complementary results are obtained by symmetric Bragg diffraction along the direction parallel to the surface normal. In thin epitaxial InAs layers on GaAs (001) substrates Munekata et al. [6.248] detected for example by GID two domains with different strain distribution (Fig. 6.76). Conventional Bragg scattering only gives the domain with the relatively low strain.

Taking advantage of synchrotron sources, which provide high intensity in combination with small beam divergence, complex dynamic surface processes of scientific and technological importance can be characterised by high resolution surface scattering. The first *in-situ* grazing-incidence X-Ray scattering studies of a chemical vapor deposition process were presented by Fuoss et al. [6.249]. The authors directly observed the Organometallic Vapour-Phase Epitaxial (OMVPE) growth of ZnSe on GaAs(001). While the characteristic p(2×4) reconstruction was monitored from a clean GaAs surface, after an

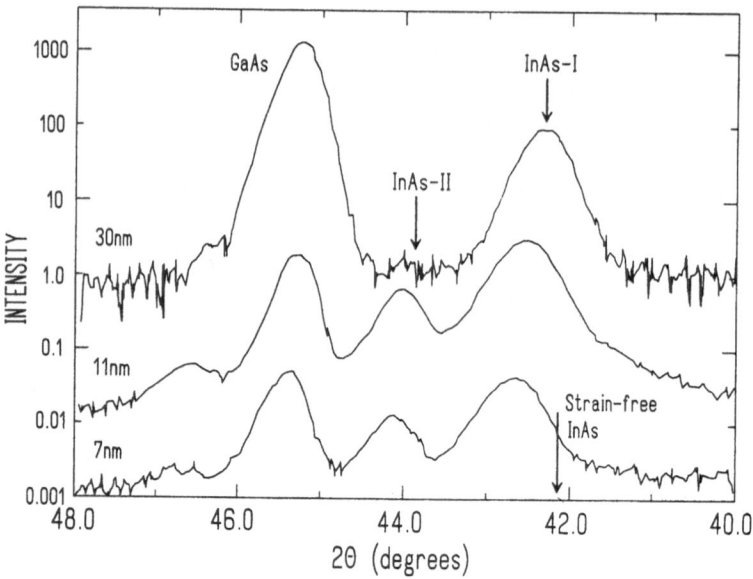

Fig. 6.76. (220) reflections of thin epitaxial InAs layers on GaAs (001) of different thicknesses obtained by X-Ray grazing-incidence diffraction. Two InAs domains with different strains are evident: a weekly strained InAs-I and a strongly strained InAs-II [6.248]

initial transient, a very well defined (2×1) ZnSe reconstruction was found independent of growth conditions. In a further experiment, they observed cyclic changes in X-Ray reflectivity of growing ZnSe films during alternating source epitaxy (Fig. 6.77). These oscillations are believed to characterise the kinetics of the decomposition of the selenium source compound [6.250]. Further interesting applications of grazing-incidence scattering for the atomic scale characterisation of OMVPE growth have recently been reported by Kisker et al. [6.251] and Lamelas et al. [6.252]. Several surface reconstructions were unambiguously determined for the growth of GaAs under OMVPE conditions. The nucleation of islands, their growth and finally, coalescence was observed by monitoring the surface sensitive crystal truncation rods. In addition, the average island spacing, its temperature dependence, and the anisotropy of the islands shape has been estimated from the diffuse scattering near the truncation rods. These are a particularly interesting example of the capacity of X-Ray techniques for surface studies so far thought to be the domain of electron diffraction techniques (e.g. RHEED, LEED).

In this short review only the basic aspects of X-Ray surface scattering techniques could be summarised. Its importance for nondestructive surface sensitive characterisation of epitaxial thin films has been demonstrated at several examples. Since X-Ray reflection does not depend on strain distribution

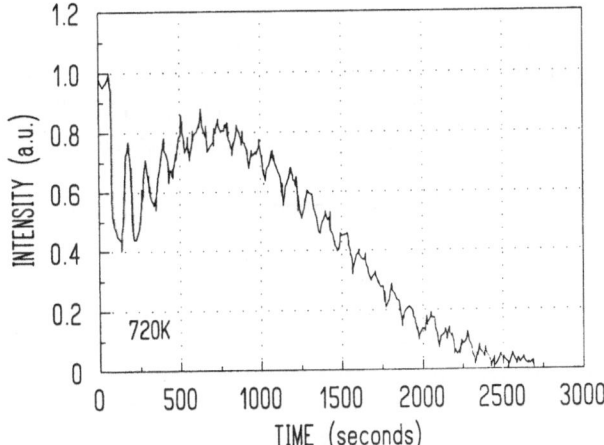

Fig. 6.77. Intensity oscillations from the specularly reflected X-Ray beam during alternating source epitaxial growth of ZnSe at 450 °C. For this study the film grew while the gas flow of the Se and Zn source was abruptly switched on and off [6.250]

or defects within a microstructure it allows the study also of heavily-defected structures or even amorphous materials. The reflectivity data can be used to determine the thickness of layers, the depth profiles of the electron density as well as interface and surface roughness. Simultaneous measurements of the X-Ray fluorescence which is excited by the standing wave formed at total reflection give information on the compositional properties of thin films and surfaces. This combination appears to be a promising new technique for non-destructive near-surface analysis. In case of crystalline samples Bragg diffraction under total external reflection conditions can be studied. Taking advantage of the reduced penetration depth crystallite size, strain, lateral lattice mismatch as well as mosaic spread in very thin epitaxial layers are determined. Although the available high intense synchrotron sources offer the suitable experimental conditions to perform all kind of grazing incidence measurements with variable X-Ray energies, conventional X-Ray sources have also proved to be applicable. For a more extensive treatment to this subject, we refer to the numerous original papers and recent reviews.

6.13 Summary

High-resolution X-Ray diffraction has become a standard tool in many laboratories where emphasis is on the growth of epitaxial layers and layer systems. Double-crystal diffractometry has been recently in widespread use and the introduction of commercially available instrumentation, especially the four-crystal monochromator has opened a wide field of applications. Through its non-destructive nature it is ideal for the determination of composition of heteroepitaxial layers, of their thicknesses, of strains and of interdiffusion at heterointerfaces. Through careful analysis strain relaxation in layers with thicknesses beyond the critical thickness for misfit dislocation can also be obtained. Lattice mismatch can be measured with an accuracy of $1.4 \cdot 10^{-5}$ (!). The tilt of epilayers with respect to their substrates can be determined as well.

The instrumentation has been further developed and nowadays triple-axis spectrometers or double-crystal diffractometers equipped with position-sensitive detectors allow one to perform a mapping of the reciprocal space. These maps are particulary advantageous for the analysis of not so perfect heterosystems or superlattices or of epilayers which show even some mosaicity. Further recent developments are the combination of high resolution X-Ray diffraction with X-Ray topography which is useful for producing maps of lattice strains with lateral resolutions down to 1 micrometer. Standard semiconductor wafers (thicknesses of 0.5 mm) can be used for this technique in transmission.

Structuring heteroepitaxial layers laterally has become quite important in order to realise 1-dimensional or 0-dimensional electronic systems. X-Ray diffraction techniques can be succesfully employed in order to determine the lateral periodicity of quantum wires or dots.

The use of synchrotron sources has permitted a real-time analysis of strain relaxation in heteroepitaxial layers and *in-situ* analysis during growth.

Grazing incidence diffraction has become a versatile technique for the characterisation of epitaxial layers during the past ten years. Since in this technique the diffraction vector is parallel to the surface, this method is quite sensitive to all properties which have to be probed in regions close to the surface or interface. In particular strains, crystallographic orientation and crystallite size, or domains within very thin epitaxial films can be determined by this technique. With rotating anodes or synchrotron radiation sources even monolayers are accessible.

6.14 Concluding Remarks

Due to the vast interest in epitaxial thin films (not necessarily semiconducting layers) there has been a rapid development in several X-Ray diffraction methods. Since the films are usually deposited on rather thick substrates with conventional laboratory instrumentation only the Bragg case reflection is accessible. Double crystal diffraction has become a general tool, usual in conjunction with several epitaxial techniques for immediate feedback of structural qualities of heterostructures, multilayers and superlattices for the crystal grower.

The use of an analyser crystal between the sample and the detector has become popular. Reciprocal space maps provide much more information than rocking-curve modelling alone. Whereas for the latter extensive software is generally available for the former such kind of analysis programs are right now under development. Since the reciprocal space maps provide information on the films both in the direction of the surface normal as well as parallel to it, it is forseeable that triple-crystal techniques will become as important as double-crystal techniques are nowadays. Together with further developments in grazing incidence techniques which yield some new complementary information, X-Ray techniques provide a wealth of structural information on heteroepitaxial systems which range from crystallography, also for layers covered by others, as well as surface and interface morphology.

New directions are:

- *in-situ* growth control using Bragg diffraction of growing films, for a feedback to influence the growth process directly. This technique has so far been realised for MOVPE using DCD (with a Bartels type of monochromator [6.185]).
- Observation of *in-situ* growth with grazing incidence techniques relying on high intensity X-Ray sources such as a synchrotron [6.250].
- Observation of *in-situ* strain relaxation in Si/Ge growth, again using synchrotron radiation [6.186].
- extended range specular reflectivity [6.180] which starts at grazing angle incidence and extends to Bragg scattering at higher angles in a single experiment (using a rotating anode generator as a source). This technique seems to provide unique information on specular and diffuse scattering which combined in the analysis yields information on atomically discrete interface fluctuations.

In our opinion it is just a matter of time until also *in-situ* X-Ray diffraction will be used in conjunction with molecular beam growth technology. *In-situ* X-Ray scattering provides important additional information which is not easily or not at all obtainable with *in-situ* reflection high energy electron diffraction which is so far used as a standard probe in MBE growth:

- *in-situ* control of layer uniformity, chemical composition, and thickness.
- *in-situ* control of interdiffusion,
- *in-situ* control of structural, chemical, and morphological changes of buried layers due to subsequent growth process of further layers.

Whereas the use of synchrotron radiation tremendeously increases the sensitivity of X-Ray scattering due to monolayers and surfaces, standard laboratory instrumentation provides a wealth of the type of information outlined above. Due to its non-destructive nature the role of X-Ray diffraction will likely even increase as an invaluable method for probing the structural properties of heterostructures in near future.

References

1.1 B. A. Joyce: Rep. Prog. Physics **48**, 1637 (1985)

1.2 D. L. Partin: in *Semiconductors and Semimetals*, ed. T.P. Pearsall (Academic Press N.Y.) **33**, 331 (1991)

1.3 R. L. Gunshor, L. A. Kolodziejski, A. V. Nurmikko, N. Otsuka: in *Semiconductors and Semimetals*, ed. T. P. Pearsall, (Academic Press N.Y.) **33**, 337 (1991)

1.4 A. Y. Cho: J. Crystal Growth **111**, 1 (1991)

1.5 M. Razeghi: *The MOCVD Challenge*, (Adam Hilger, Bristol 1989)

1.6 E. Kasper, J. C. Bean: *Silicon Molecular Beam Epitaxy*, (CRC Press Florida 1988)

1.7 E. Kasper, F. Schäffler: in *Semiconductors and Semimetals*, ed. T. P. Pearsall (Academic Press N.Y.) **33**, 223 (1991)

1.8 M. A. Herman, H. Sitter: *Molecular Beam Epitaxy*, (Springer Berlin, Göttingen, Heidelberg 1989)

1.9 M. Illegems: in *Technology and Physics of Molecular Beam Epitaxy*, ed. E. M. C. Parker (Plenum Press N.Y.), p. 83 (1985)

1.10 P. K. Larsen and P. J. Dobson: in *Reflection High Energy Electron Diffraction and Reflection Electron Imaging of Surfaces* (NATO ARW, Plenum Press, N.Y. 1988)

1.11 D. E. Aspnes: in *Handbook on Semiconductors*, Vol.2, Ed. M. Balkanski (North-Holland Publishing Company 1980), p. 109

1.12 D. E. Aspnes: Proc. SPIE Aachen 1991, **1361**, p. 551

1.13 W. Richter: Phil. Trans. R. Soc. Lond. **A344**, 453 (1993)

1.14 S. M. Koch, O. Acher, F. Omnes, M. Defour, B. Drevillon, M. Razeghi: J. Appl. Phys. **69**, 3 (1991)

1.15 P. F. Fewster: Applied Surface Science **50**, 9 (1991)

1.16 A. Segmüller, I. C. Noyen, V. S. Speriosa: Progress in Crystal Growth and Characterization **18**, 21 (1989)

1.17 B. K. Tanner: in *Analysis of Microelectronic Materials and Devices*, eds. M. Grasserbauer and H. W. Werner (J. Wiley and Sons, Chichester and New York 1991), p. 609

1.18 A. Segmüller: J. Vac. Sci. Technology **A9**, 2447 (1991)

1.19 E. Betzig and J. Trautman: Science **257**, 189 (1992)

1.20 E. L. Buckland, P. J. Moyer and M. A. Paesler: J. Appl. Phys. **73**, 1018 (1993)

1.21 M. Henzler: Surface Science **152/153**, 963 (1985)

1.22 M. P. Seah: Vaccum **34**, 453 (1984)

1.23 M. Grasserbauer and H. W. Werner (eds.): *Analysis of Microelectronic Materials and Devices*, (J. Wiley, Chichester and New York 1991)

1.24 A. Benninghoven, F. G. Lüdemann, H. Werner: *Secondary Ion Mass Spectroscopy*, (J. Wiley, New York 1987)

1.25 S. T. Picraux, B. L. Doyle, J. Y. Tsao: in *Semiconductors and Semimetals*, ed. T. P. Pearsall (Academic Press N.Y.) **33**, 139 (1991)

1.26 W. F. van der Weg and F. H. P. M. Habraken: in *Analysis of Microelectronic Materials and Devices*, eds. M. Grasserbauer and H. W. Werner (J. Wiley and Sons, Chichester and New York), p. 563 (1991)

1.27 H. Cerva and H. Oppholzer: Prog. Crystal Growth and Characterisation **20**, 189 (1990)

1.28 H. Cerva: Appl. Surface Science **50**, 19 (1991)

1.29 F. A. Ponce, G. B. Anderson, J. M. Ballignall: Surface Science **168**, 564 (1986)

1.30 L. Reimer: *Transmission Electron Microscopy*, (Springer Series in Optical Science **36**, 1989)

1.31 D. E. Newbury, D. C. Joy, P. Ecklin, C. E. Fiori, J. I. Goldstein: *Advanced Scanning Electron Microscopy and X-Ray Analysis* (Plenum Press, N.Y. 1986)

1.32 *Scanning Tunneling Microscopy*, I, II, and III
H.-J. Güntherodt, R. Wiesendanger (eds.) (Springer Series in Surface Science, 1992)

1.33 R. C. Newman: in *Growth and Characterisation of Semiconductors*, R. A. Stradling and P. C. Klipstein (eds.) (Adam Hilger, Bristol), p. 119 (1990)

1.34 R. A. Stradling and P. C. Klipstein: *Growth and Characterization of Semiconductors* (Adam Hilger, Bristol 1990)

1.35 T. Dumelow, T. J. Parker, S. R. P. Smith, D. R. Tiley: Surf. Sci. Rep. **17**, 151 (1993)

1.36 M. Cardona and L. Ley: in *Photoemission in Solids* I and II, Topics in Applied Physics **26** and **27** (1978)

1.37 G. V. Marr: *Handbook on Synchrotron Radiation* (North Holland, Amsterdam) **2** (1987)

1.38 *Principles, Applications, Techniques of EXAFS, SEXAFS and XANES*, C. C. Koningsberger and R. Prius (eds.) (Wiley, New York 1988)

1.39 E. C. Lightowlers: in *Growth and Characterisation of Semiconductors*, R. A. Stradling and P. C. Klipstein (eds.) (Adam Hilger, Bristol), p. 135 (1990)

1.40 M. A. Herman, J. Christen, D. Bimberg: J. Appl. Phys. **70**, R1 (1991)

1.41 F. H. Pollak: in *Proceedings of the Society of Photo-Optical Instrumentation Engineers, San Jose*, D. E. Aspnes, S. So, R. F. Potter (eds.), SPIE, Bellingham (1981)

1.42 M. Cardona and G. Güntherodt: *Light Scattering in Solids* II to V, Topics in Applied Physics, **50** (1982), **51** (1982), **54** (1984), **66** (1989), (Springer, Berlin, Heidelberg, New York)

1.43 J. B. Theeten: Surface Science **96**, 275 (1980)

1.44 HRXRD Conf. Proc., J. Phys. D **26**, 4A (1993)

1.45 Z. J. Radzimski, B. I. Jiang, G. A. Rozgonyi, T. P. Humphreys, N. Hamaguchi, S. M. Bedair: J. Appl. Phys. **64**, 2328 (1988)

1.46 R. Köhler, Proc. HRXRD Workshop, Aigen 1992, Appl. Phys. **A58**, 149 (1994)

394   References

2.1 M. A. Herman, H. Sitter: *Molecular Beam Epitaxy* (Springer Berlin, Göttingen, Heidelberg 1989)

2.2 K. Ploog: *Angewandte Chemie* **100**, 611 (1988)

2.3 B. A. Joyce: Rep. Prog. Physics **48**, 1637 (1985)

2.4 A. Y. Cho: J. Crystal Growth **111**, 1 (1991)

2.5 A. Lopez-Otero: Thin Solid Films **49**, 1 (1978)

2.6 G. H. Olsen, T. J. Zamerowski: Prog. Cryst. Growth Characterisation **2**, 309 (1979)

2.7 G. B. Stringfellow: *Metalorganic Vapour Phase Epitaxy* (North Holland, Amsterdam 1986)

2.8 M. Razeghi: *The MOCVD Challenge*, Vol. 1, (Adam Hilger, Bristol 1989)

2.9 E. Kuphal, D. Fritzsche: J. Electr. Mat. **12**, 743 (1983)

2.10 M. G. Astles: *Liquid Phase Epitaxial Growth of III-V Compound Semiconductor Materials and Their Device Applications*, (Hilger, Bristol 1990)

2.11 D. Fotiadis, M. Boekholt, K. F. Jensen, W. Richter: J. Crystal Growth **100**, 577 (1990)

2.12 F. Durst, A. Melling, J. H. Whitelaw: *Principle and Practice of Laser Doppler Anemometry* (Academic Press, London, UK 1976)

2.13 R. J. Schodl: Fluid Eng. **102**, 412 (1980)

2.14 L. Stock, W. Richter: J. Crystal Growth **77**, 144 (1986)

2.15 W. Richter: Festkörperprobleme XXVI, Advances in Solid State Physics **26**, 335 (1986)

2.16 K. F. Jensen: J. Cryst. Growth **98**, 148 (1989)

2.17 K. C. Chiu, F. Rosenberger: Int. J. Heat and Mass Transfer **30**, 1645 (1987)

2.18 A. Compaan, H. J. Trodahl: Phys. Rev. **29**, 793 (1984)

2.19 A. Weber: *Raman Spectroscopy of Gases and Liquids* (Springer Berlin, 1979), p. 123

2.20 T. O. Sedgwick, J. E. Smith, R. Ghcz, M. E. Cowher: J. Crystal Growth **31**, 264 (1975)

2.21 T. O. Sedgwick, J. E. Smith: J. Electrochem. Soc. **123**, 254 (1976)

2.22 J. E. Smith, T. O. Sedgwick: Thin Solid Films **40**, 1 (1977)

2.23 G. A Hebner, K. P. Killeen, R. M. Bielfeld: J. Crystal Growth **98**, 293 (1989)

2.24 V. M. Donnelly, R. F. Karlicek: J. Appl. Phys. **53**, 6399 (1982)

2.25 M. L. Fischer, R. Lückerath, P. Balk, W. Richter: Chemtronics **3**, 156 (1988)

2.26 K. P. Killeen: Appl. Phys. Lett. **61**, 1864 (1992)

2.27 J. E. Butler, N. Bottka, R. S. Sillmon, D. K. Gaskill: J. Crystal Growth **77**, 163 (1986)

2.28 P. Ho, W. G. Breiland: J. Appl. Phys. **63**, 5184 (1988)

2.29 W. G. Breiland, P. Ho, M. E. Coltrin: J. Appl. Phys. **60**, 1505 (1986)

2.30 W. G. Breiland, M. E. Coltrin, P. Ho: J. Appl. Phys. **59**, 3267 (1986)

2.31 P. Ho, W. G. Breiland: Appl. Phys. Lett. **44**, 51 (1984)

2.32 P. Ho, W. G. Breiland: Appl. Phys. Lett. **43**, 125 (1983)

2.33 S. Ishizaka, J. Simpson, J. O. Williams: Chemtronics **1**, 175 (1986)

2.34 K. J. Mackey, D. C. Rodway, P. C. Smith, A. W. Vere: Chemtronics **5**, 85 (1991)

2.35 W. G. Breiland, M. J. Kushner: Appl. Phys. Lett. **42**, 395 (1983)

2.36 M. Koppitz, W. Richter, R. Bahnen, M. Heyen: in Springer Series in Chemical Physics **39**, Laser Processing and Diagnostics, ed. by D. Bäuerle, Springer Verlag, Berlin, p. 530 (1984)

2.37 Y. Monteil, R. Favre, P. Raffin, J. Bouix, M. Vaille, P. Gibart: J. Crystal Growth **93**, 159 (1988)

2.38 S. A. J. Druet, J. P. E. Taran: Prog. Quant. Electr. **7**, 1 (1981)

2.39 R. Brakel, F. W. Schneider: in *Advances in Spectroscopy* Vol. 15, eds. R. J. H. Clark, R. E. Hester, Wiley, Chichester 1988

2.40 W. Richter, P. Kurpas, R. Lückerath, M. Motzkus: J. Crystal Growth **107**, 13 (1991)

2.41 M. Alden, A. L. Schawlow, S. Svanberg, W. Wendt, P. L. Zhang: Optics Letters **9**, 211 (1984)

2.42 International Colloquium on Optogalvanic Spectroscopy and its Application, J. de Physique Colloque **C7**, 44 (1983)

2.43 G. Leyendecker, J. Doppelbauer, D. Bäuerle, P. Geittner, H. Lydtin: Appl. Phys. **A30**, 237 (1983)

2.44 W. Richter, P. Kurpas, R. Lückerath, M. Motzkus: J. Crystal Growth **107**, 13 (1991)

2.45 M. E Coltrin, R. J. Kee, J. A. Miller: J. Electrochem. Soc. **131**, 425 (1984)

2.46 R. Lückerath, P. Tommack, A. Hertling, H. J. Koss, P. Balk, K. F. Jensen, W. Richter: J. Crystal Growth **93**, 151 (1988)

2.47 T. J. Mountziaris, K. F. Jensen: J. Electrochem. Soc. **138**, 2426 (1991)

2.48 M. P. Seak, W. A. Dench: Surf. Interf. Anal. **1**, 2 (1979)

2.49 D. E. Aspnes, A. A. Studna: Phys. Rev. **B27**, 985 (1983)

2.50 M. Cardona, F. H. Pollak, K. L. Shaklee: J. Phys. Soc. Jap. **21**, 89 (1966)

2.51 R. M. A. Azzam: Optics Communications **19**, 122 (1976)

2.52 H. Wormeester, D. J. Wentink, P. L. deBoeij, C. M. J. Wijers, A. van Silfhout: Phys. Rev. B **47**, 12663 (1993)

2.53 D. E. Aspnes, J. P. Harbison, A. A. Studna, L. T. Folrez: J. Vac. Sci. Technol. **A6**, 1327 (1988)

2.54 D. E. Aspnes: J. Vac. Sci. Technol. **B3**, 1498 (1985)

2.55 P. Chiaradia, G. Chiarotti, F. Ciccacci, R. Memeo, S. Nannarone, P. Sassaroli, S. Selci: Surf. Sci. **99**, 70 (1980)

2.56 P. Chirardia, A. Skrebtii, C. P. Goletti, Wang-Jian, R. del Sole: Surface Science **85**, 497 (1993)

2.57 C. P. Goletti, P. Chirardia, Wang-Jian, G. Chiarotti: Solid State Commun. **84**, 421 (1992)

2.58 D. K. Biegelsen, R. D. Bringans, J. E. Northrup, A. Schwartz: Phys. Rev. **B 41**, 5701 (1990)

2.59 S. N. Jasperson, S. E. Schnatterly: Rev. Sci. Instr. **40**, 761 (1969)

2.60 A. A. Studna, D. E. Aspnes, L. T. Florez, J. P. Harbison and R. Ryan: J. Vac. Sci. Technol. **A7**, 3291 (1989)

2.61 D. E. Aspnes, Y. C. Chang, A. A. Studna, L. T. Florez, H. H. Farrell, J. P. Harbison: Phys. Rev. Lett. **64**, 192 (1990)

2.62 L. Däweritz, R. Hey: Surf. Sci. **236**, 15 (1990)

2.63 S. M. Koch, O. Acher, F. Omnes, M. Defour, B. Drevillon, M. Razeghi: J. Appl. Phys. **69**, 1389 (1991)

2.64 T. Yasuda, D. E. Aspnes, D. R. Lee, C. H. Bjorkman and G. Lucovsky: J. Vac. Sci. Technol. **A12**, 1152 (1994)

2.65 D. E. Aspnes, J. P. Harbison, A. A. Studna, L. T. Florez: Phys. Rev. **B59**, 1687 (1987)

2.66 J. J. Harbison, D. E. Aspnes, A. A. Studna, L. T. Florez and M. K. Kelly: Appl. Phys. Lett. **52**, 2046 (1988)

2.67 F. Briones, D. Golmayo, L. Gonzalez, J. L. de Miguel: J. Appl. Phys. Japan **24**, L478 (1985)

2.68 J. E. Epler, T. A. Jung, H. P. Schweizer, Appl. Phys. Lett. **62**, 143 (1993)

2.69 B. Y. Maa, P. D. Dapkus, P. Chen, A. Madhukar: Appl. Phys. Lett. **62**, 2551 (1993)

2.70 E. Colas, D. E. Aspnes, R. Bhat, A. A. Studna, J. P. Harbison, L. T. Florez, M. A. Koza, V. G. Keramidas: J. Cryst. Growth **107**, 47 (1991)

2.71 I. Kamiya, H. Tanaka, D. E. Aspnes, L. T. Florez, E. Colas, J. P. Harbison, R. Bhat: Appl. Phys. Lett **60**, 1238 (1992)

2.72 I. Kamiya, D. E. Aspnes, H. Tanaka, L. T. Florez, E. Colas, J. P. Harbison, R. Bhat: Applied Surface Science **60/61**, 534 (1992)

2.73 I. Kamiya, D. E. Aspnes, H. Tanaka, L. T. Florez, J. P. Harbison, R. Bhat: Phys. Rev. Lett. **68**, 627 (1992)

2.74 Y.-Ch. Chang, D. E. Aspnes: J. Vac. Sc. Technol. **8**, 896 (1990)

2.75 D. W. Kisker, G. B. Stephenson, P. H. Fuoss, F. J. Lamelas, S. Brennan, P. Imperatori: J. Crystal Growth **124**, 1 (1992)

2.76 K. Ploska, W. Richter, F. Reinhardt, J. Jönson, J. Rumberg, M. Zorn, MRS Proceedings **334**, 155 (1993)

2.77 M. Wassermeier, I. Kamiya, D. E. Aspnes, L. T. Florez, J. P. Harbison, P. M Petroff: J. Vac. Sci. Technol. **B9**, 2263 (1991)

2.78 I. Kamiya, D. E. Aspnes, H. Tanaka, L. T. Florez, J. P. Harbison, R. Bhat: J. Vac. Sci. Technol. **B10**, 1716 (1992)

2.79 L. Samuelson, K. Deppert, S. Jeppesen, J. Jönnson, G. Paulsson, P. Schmidt: Crystal Properties and Preparation **32**, 338 (1991)

2.80 L. Samuelson, K. Deppert, S. Jeppesen, J. Jönsson, G. Paulsson, P. Schmidt: J. Cryst. Growth **107**, 68 (1991)

2.81 O. Acher, S. M. Koch, F. Omnes, M. Defour, B. Drevillon, M. Razeghi: *Condensed Systems of Low Dimensionality*, ed. by J. L. Beeby et al., (Plenum Press, New York 1991)

2.82 F. Reinhardt, W. Richter, A. B. Müller, D. Gutsche, P. Kurpas, K. Ploska, K. C. Rose, M. Zorn: J. Vac. Sci. Technol. **B11**, 1427 (1993)

2.83 W. Richter: Phil. Trans. R. Soc. London A **344**, 453 (1993)

2.84 P. Kurpas, J. Jönssson, W. Richter, D. Gutsche, M. Pristovsek, M. Zorn: J. Cryst. Growth **145**, 35 (1994)

2.85 M. Zorn, J. Jönsson, A. Krost, W. Richter, J.-T. Zettler, K. Ploska, F. Reinhardt: J. Cryst. Growth **145**, 53 (1994)

2.86 K. Ploska, J.-T. Zettler, J. Jönsson, F. Reinhardt, M. Zorn, J. Rumberg, M. Pristovsek, W. Richter: J. Cryst. Growth **145**, 44 (1994)

2.87 F. Reinhardt, J. Jönsson, M. Zorn, W. Richter, K. Ploska, J. Rumberg, P. Kurpas: J. Vac. Sci. Technol. **B12**, 2541 (1994)

2.88 S. M. Scholz, A. B. Müller, W. Richter, D. R. T. Zahn, D. I. Westwood, D. A. Woolf, R. H Williams: J. Vac. Sci. Technol. **B10**, 1710 (1992)

2.89 N. Kobayashi, Y. Horikoshi: Jpn. J. Appl. Phys. **28**, L1880 (1989)

2.90 N. Kobayashi, Y. Horikoshi: Jpn. J. Appl. Phys. **30**, L319 (1991)

2.91 T. Makimoto, Y. Yamauchi, N. Kobayashi, Y. Horikoshi: Jpn. J. Appl. Phys. **29**, L645 (1990)

2.92 N. Kobayashi, Y Horikoshi: Jpn. J. Appl. Phys. **29**, L702 (1990)

2.93 N. Kobayashi, Y. Horikoshi: Jpn. J. Appl. Phys. **30**, L1443 (1991)

2.94 K. Hingerl, D. E. Aspnes, I. Kamiya, L. T. Florez: Appl. Phys. Lett. **63**, 885 (1993)

2.95 H. Ibach, D. L. Mills: *Electron Energy Loss Spectroscopy and Surface Vibrations* (Academic Press, New York 1982)

2.96 A. Förster: Ph.D. Thesis RWTH Aachen (1988)

2.97 Y. J. Chabal: Surf. Sci. Reports **8**, 211 (1988)

2.98 R. G. Greenler: J. Chem. Phys. **44**, 310 (1966)

2.99 H. Ibach: Surf. Sci. **66**, 56 (1977)

2.100 M. A. Chesters, A. B. Horn, E. J. C. Kellar, S. F. Parker, R. Raval, in: *Mechanisms of Reaction of Organometallic Compounds with Surfaces*, eds. D. J. Cole-Hamilton, J. O. Williams, NATO ASI Series, Series:B, Physics **198**, Plenum (1989)

2.101 J. D. E. McIntyre, D. E. Aspnes: Surf. Sci. **24**, 417 (1971)

2.102 D. S. Buhaenko, S. M. Francis, P. A. Goulding, M. E. Pemble: J. Cryst. Growth **97**, 591 (1989)

2.103 H. Patel, M. E. Pemble: J. Phys. IV, Colloq. **1**, 167 (1991)

2.104 N. Bloembergen, R. K. Chang, S. S. Jha, C. H. Lee: Phys. Rev. **174**, 813 (1968)

2.105 H. W. K. Tom, T. F. Heinz, Y. R. Shen: Phys. Rev. Lett. **51**, 1983 (1983)

2.106 T. F. Heinz, M. M. T. Loy., W. A. Thompson: Phys. Rev. Lett. **54**, 63 (1985)

2.107 H. W. K Tom., G. D. Aumiller: Phys. Rev. **B33**, 8818 (1986)

2.108 J. F. McGilp , Y. Yeh: Solid State Commun. **59**, 91 (1986)

2.109 J. F. McGilp: Semicond. Sci. Technol. **2**, 102 (1987)

2.110 J. F. McGilp: J. Vac. Sci. Technol. **A5**, 1442 (1987)

2.111 J. F. McGilp: J. of Phys.: Condens. Matter **2**, 7985 (1990)

2.112 D. Guidotti, T. A. Driscoll, H. J. Gerritsen: Solid State Commun. **46**, 337 (1983)

2.113 T. Stehlin, M. Feller, P. Guyot-Sionnest, Y. R. Shen: Opt. Letts. **13**, 389 (1988)

2.114 R. W. J. Hollering, A. J. Hoeven, J. M. Lenssinck: J. Vac. Sci. Technol. **A8**, 3194 (1989)

2.115 M. E. Pemble, D. S. Buhaenko, S. M. Franas, P. A. Goulding, J. T. Allen: J. Cryst. Growth **107**, 37 (1991)

2.116 J. M. Olson, A Kibbler: J. Cryst. Growth **77**, 182 (1986)

2.117 D. J. Robbins, A. J. Pidduck, A. G. Cullis, N. G. Chew, R. W. Hardeman, D. B Gasson, C. Pickering, A. C. Daw, M. Johnson, R. Jones: J. Cryst. Growth **81**, 421 (1987)

2.118 C. Pickering, D. J. Robbins, I. M. Young, J. L. Glasper, M. Johnson, R. Jones: Mater. Res. Soc. Symp. Proc. **94**, 173 (1987)

2.119 A. J. Pidduck, D. J. Robbins, A. G. Cullis, D. B. Glasson, J. L. Glasper: J. Electrochem. Soc. **136**, 3083 (1989)

2.120 A. J. Pidduck, D. J. Robbins, D. B. Glasson, C. Pickering, J. L. Glasper: J. Electrochem. Soc. **136**, 3088 (1989)

2.121 E. L. Church, H. A Jenkinson, J. M. Zavada: Opt. Eng. **18**, 125 (1979)

2.122 G. W. Smith, A. J. Pidduck, C. R. Whitehouse, J. L. Glasper, A. M. Keir, C. Pickering: Appl. Phys. Lett. **59**, 3282 (1991)

2.123 W. Tsang, R. F. Hampson: J. Phys. Chem. Ref. Data **15**, 1087 (1986)

2.124 M. E. Coltrin, R. J. Kee, J. A. Miller: J. Electrochem. Soc. **133**, 1206 (1986)

2.125 M. Tirowidjo, R. Pollard: J. Crystal Growth **93**, 108 (1988)

3.1 J. Jamin: Ann. de chim. et phys. **29**, 263 (1850)

3.2 P. Drude: Ann. d. Phys. Chem. (Leipzig) **34**, 489 (1888)

3.3 D. E. Aspnes, A. A. Studna: Phys. Rev. **B27**, 985 (1983)

3.4 A. Röseler: *Infrared Spectroscopic Ellipsometry* (Akademie Verlag, Berlin, Germany 1990)

3.5 R. M. A. Azzam, N. M. Bashara: *Ellipsometry and Polarized Light* (North-Holland Publishing Company 1977)

3.6 F. L. McCrackin, E. Passaglia, R. R. Stromberg, H. L. Steinberg: J. of Res. Nat. Bur. of Stan. **67A**, 363 (1963)

3.7 D. E. Aspnes: J. Opt. Soc. Am. **64**, 812 (1974)

3.8 D. E. Aspnes, A. A. Studna: Appl. Opt. **14**, 220 (1975)

3.9 P. S. Hauge, F. H. Dill: IBM J. Res. Dev. **17**, 472 (1973)

3.10 S. N. Jasperson, S. E. Schnatterly: Rev. Sci. Instr. **40**, 761 (1969)

3.11 S. N. Jasperson, D. K. Burge, R. C. O'Handley: Surf. Sci. **37**, 548 (1973)

3.12 J. C. Kemp: *Polarized Light and its Interaction with Modulating Devices* (Hinds Int. Inc. 1987)

3.13 H. Mueller: J. Opt. Soc. Am. **38**, 661 (1948)

3.14 C. Pickering, D. J. Robbins, I. M. Young, J. L. Glasper, M. Johnson, R. Jones: Mat. Res. Soc. Symp. Proc. **94**, 173 (1987)

3.15 J. F. Nye: *Physical Properties of Crystals* (Clarendon Press, Oxford 1985)

3.16 M. Born: *Optik* (Springer, Berlin 1985) 2nd edition

3.17 see for example: Hamamatsu Technical Notes TN-106-03 *Super Quiet Xenon Lamps*, August 1984

3.18 E. Hecht, A. Zajac: *Optics* (Addison-Wesley 1974) chapter 8

3.19 D. E. Aspnes: J. Opt. Soc. Am. **64**, 639 (1974)

3.20 N. V. Nguyen, B. S. Pudliner, Ilsin An, R. W. Collins: J. Opt. Soc. Am. **A8**, 919 (1991)

3.21 O. Acher, E. Bigan, B. Drevillon: Rev. Sci. Instr. **60**, 65 (1989)

3.22 B. Drevillon: Thin Solid Films **163**, 157 (1988)

3.23 F. Ferrieu, D. Dutatre: J. Appl. Phys. **68**, 5810 (1990)

3.24 J. Brehmer, O. Hunderi, K. Fanping, T. Skauli, E. Wold: Appl. Optics **31**, 471 (1992)

3.25 K.-L. Barth, D. Böhme, K. Kamaras, F. Keilmann, M. Cardona: Thin Solid Films **234**, 314 (1993)

3.26 R. L. Johnson, J. Barth, M. Cardona, D. Fuchs, A. M. Bradshaw: Rev. Sci. Instr. **60**, 2209 (1989)

3.27 Proceedings of the first conference on spectroscopic ellipsometry, ICSE Paris 1993 in: Thin Solid Films **233/234** (1993)

3.28 R. M. A. Azzam, K. A. Giardina, A. G. Lopez: Optical Engineering **30**, 1583 (1991) and references therein

3.29 P. S. Hauge: Surf. Sci. **96**, 108 (1980)

3.30 J. A. Woollam, P. G. Snyder: J. Appl. Phys. **62**, 4867 (1987)

3.31 R. Calvani, R. Caponi, F. Cisternino: Optics Communication **54**, 63 (1985)

3.32 F. Bassani, G. Pastori Parravicini: *Electronic States and Optical Transitions in Solids* (Pergamon Press, Oxford, 1975)

3.33 M. Cardona: in *Modulation Spectroscopy*, Suppl. **11** of Solid State Physics, edited by F. Seitz, D. Turnbull, and H. Ehrenreich (Academic Press, New York 1969)

3.34 P. Lautenschlager, Ph.D. Thesis, University of Stuttgart 1987

3.35 J. R. Chelikowski, M. L. Cohen: Phys. Rev. **14**, 556 (1976)

3.36 H. Arwin, D. E. Aspnes: J. Vac. Sci. Technol. **A2**, 1316 (1984)

3.37 Landolt-Börnstein, New series, Vol. 15b, Springer, Berlin 1985

3.38 J. H. Weaver, C. Krafka, D. W. Lynch, and E. E. Koch: in *Physik Daten: Optical Properties of Metals*, Nr. 18-1 (Fachinformationszentrum Energie, Physik, Mathematik GmbH, Karlsruhe, 1981), pp. 239 ff.

3.39 *Handbook of Optical Constants of Solids*, edited by E. D. Palik (Academic Press, New York 1985)

3.40 D. E. Aspnes: in *Handbook on Semiconductors*, Vol. 2 ed. M. Balkanski (North Holland, Amsterdam 1980)

3.41 C. C. Kim, J. W. Garland, H. Abad, P. M. Raccah: Phys. Rev. **B45**, 11749 (1992)

3.42 K. Cho: in *Excitons*, edited by K. Cho, Topics in Current Physics (Springer, Berlin, Heidelberg, New York 1979)

3.43 D. E. Aspnes, R. P. H. Chang: Mat. Res. Soc. Symp. Proc. **29**, 217 (1984)

3.44 G. E. Jellison: Optical Materials **1**, 41 (1992)

3.45 D. E. Aspnes, G. P. Schwartz, G. J. Gualtieri, A. A. Studna, B. Schwartz: J. Electrochem. Soc. **128**, 590 (1981)

3.46 L. Viña, M. Cardona: Phys. Rev. **B29**, 6739 (1984)

3.47 B. Drevillon, C. Godet, S. Kumar: Appl. Phys. Lett. **50**, 1651 (1987)

3.48 S. Logothetidis, G. Kiriakidis: J. Appl. Phys. **64**, 2389 (1988)

3.49 L. Viña, S. Logothetidis, M. Cardona: Phys. Rev. **B30**, 1979 (1984)

3.50 D. E. Aspnes, C. E. Bouldin, E. A. Stern: in *Proc. 17th Internat. Conf. Physics of Semiconductors*, ed. J. D. Chadi and W. A. Harrison (Springer, New York 1985), p. 841.

3.51 D. E. Aspnes: *Spectroscopic Ellipsometry of Solids*, in B. O. Seraphin (ed.) *Optical Properties of Solids* (North Holland, Amsterdam 1976), chapter 15

3.52 M. Cardona, D. L. Greenaway: Phys. Rev. **A133**, 1685 (1964)

3.53 U. Rossow, U. Frotscher, N. Esser, U. Resch, Th. Müller, W. Richter, D. A. Woolf, R. H. Williams: Appl. Surf. Sci. **63**, 35 (1993)

3.54 R. Strümpler, H. Lüth: Thin Solid Films **177**, 287 (1989)

3.55 B. J. Schäfer, A. Förster, M. Londschien, A. Tulke, K. Werner, M. Kamp, H. Heinecke, M. Weyers, H. Lüth, P. Balk: Surf. Sci. **204**, 485 (1988)

3.56 O. Hunderi: J. Phys. **F5**, 2214 (1975)

3.57 O. Hunderi, R. Ryberg: J. Phys. **F4**, 2096 (1974)

3.58 D. Y. Smith, E. Shiles, M. Inokuti: in *Handbook of Optical Constants of Solids*, edited by E. D. Palik (Academic, New York, 1985), Table XII

3.59 D. L. Greenaway, G. Harbeke, F. Bassani, E. Tosatti: Phys. Rev. **178**, 1340 (1969)

3.60 H. Arwin, D. E. Aspnes: Thin Solid Films **138**, 193 (1986)

3.61 J. Wagner, P. Lautenschlager: J. Appl. Phys. **59**, 2044 (1986)

3.62 U. Schmid, J. Humliček, F. Lukes, M. Cardona, H. Presting, H. Kibbel, E. Kasper, K. Eberl, W. Wegscheider, G. Abstreiter: Phys. Rev. **B45**, 6793 (1992)

3.63 S. Ninomiya, S. Adachi: Jpn. J. Appl. Phys. **33**, 2479 (1994)

3.64 S. Zollner, C. Lin, E. Schönherr, A. Böhringer, M. Cardona: J. Appl. Phys. **66**, 383 (1989)

3.65 M. Garriga, P. Lautenschlager, M. Cardona, K. Ploog: Solid State Commun. **61**, 157 (1987)

3.66 D. E. Aspnes, S. M. Kelso, R. A. Logan, R. Bhat: J. Appl. Phys. **60**, 754 (1986)

3.67 H. Burkhard, H. W. Dinges, E. Kuphal: J. Appl. Phys. **53**, 655 (1982)

3.68 S. Logothetidis, J. Petalas, M. Cardona, T. D. Moustakas: Proc. of the EMRS, Strasbourg, France, May 24–27, 1994 in print

3.69 S. Adachi, S. Ozaki: Jpn. J. Appl. Phys. **32**, 4446 (1993)

3.70 M. von der Emde, U. Rossow, G. Kudlek, A. Hoffmann, A. Krost, W. Richter, S. Morley, A. C. Wright, J. O. Williams, D. R. T. Zahn: Proc. of the ICFSI-4 Jülich, 14-18 June 1993, p. 684
eds. B. Lengeler, H. Lüth, W. Mönch, J. Pollmann (World Scientific Press Singapore, New Jersey, London, Hong Kong 1994)

3.71 S. Logothetidis, L. Viña, M. Cardona: Phys. Rev. **B31**, 947 (1985)

3.72 S. Adachi, T. Taguchi: Phys. Rev. **B43**, 9569 (1991)

3.73 M. Aven, D. T. F. Marple, B. Segall: J. Appl. Phys. **32**, 2261 (1961)

3.74 J. L. Freeouf: Phys. Rev. **B7**, 3810 (1973)

3.75 S. Ozaki, S. Adachi: Jpn. J. Appl. Phys. **32**, 5008 (1993)

3.76 M. Cardona, M. Weinstein, G. A. Wolff: Phys. Rev. **140**, A633 (1965)

3.77 S. Adachi: J. Appl. Phys. **68**, 1198 (1990)

3.78 K. Sato, S. Adachi: J. Appl. Phys. **73**, 926 (1993)

3.79 S. Logothetidis, M. Cardona, P. Lautenschlager, M. Garriga: Phys. Rev. **B34**, 2458 (1986)

3.80 S. Adachi, T. Kimura, N. Suzuki: J. Appl. Phys. **74**, 3435 (1993)

3.81 W. H. Weber, J. T. Remillard, J. R. McBride: Phys. Rev. **B46**, 15085 (1992)

3.82 N. Suzuki, S. Adachi: Jpn. J. Appl. Phys. **33**, 193 (1994)

3.83 R. Matz, H. Lüth: Appl. Phys. **18**, 123 (1979)

3.84 M. Altwein, H. Finkenrath, C. Konak, J. Stuke, G. Zimmerer: Phys. Stat. Sol. **29**, 203 (1968)

3.85 S. Logothetidis, P. Lautenschlager, M. Cardona: Phys. Rev. **B33**, 1110 (1986)

3.86 H. J. Trodahl, L. Viña: Phys. Rev. **B27**, 6498 (1983)

3.87 F. Ferrieu, C. Viguier, A. Cros, A. Humbert, O. Thomas, R. Madar, J. P. Senateur: Solid State Commun. **62**, 455 (1987)

3.88 F. Ferrieu, J. H. Lecat: Thin Solid Films **164**, 43 (1988)

3.89 C. Viguier, A. Cros, A. Humbert, C. Ferrieu, O. Thomas, R. Madar, J. P. Senateur: Solid State Commun. **60**, 923 (1986)

3.90 J. R. Jimenez, Z.-C. Wu, L. J. Schowalter, B. D. Hunt, R. W. Fathauer, P. J. Grunthaner, T. L. Lin: J. Appl. Phys. **66**, 2738 (1989)

3.91 H.-W. Chen, J.-T. Lue: J. Appl. Phys. **59**, 2165 (1986)

3.92 P. Chindaudom, K. Vedam: Thin Solid Films **234**, 439 (1993)

3.93 D. E. Aspnes, B. Schwartz, A. A. Studna, L. Derick, L. A. Koszi: J. Appl. Phys. **48**, 3510 (1977)

3.94 I. H. Malitson: J. Opt. Soc. Am. **55**, 1205 (1965)

3.95 L. Pajasova: Czech. J. Phys. **B19**, 1265 (1969)

3.96 H. J. Mattausch, D. E. Aspnes: Phys. Rev. **B23**, 1896 (1981)

3.97 J. Barth, R. L. Johnson, M. Cardona, D. Fuchs, A. M. Bradshaw: Phys. Rev. **B41**, 3291 (1990)

3.98 S. Y. Kim, K. Vedam: Thin Solid Films **166**, 325 (1988)

3.99 G. M. Hale, M. R. Querry: Appl. Opt. **12**, 555 (1973)

3.100 D. E. Aspnes, S. M. Kelso: SPIE Proc. **452**, 79 (1983)

3.101 D. E. Aspnes, H. J. Stocker: J. Vac. Sci. Technol. **21**, 413 (1982)

3.102 B. Drevillon, E. Bertran, P. Alnot, J. Olivier, M. Razeghi: J. Appl. Phys. **60**, 3512 (1986)

3.103 H. W. Dinges, H. Burkhard, R. Lösch, H. Nickel, W. Schlapp: Appl. Surf. Sci. **54**, 477 (1992)

3.104 S. Imai, S. Adachi: Jpn. J. Appl. Phys. **32**, 3860 (1993)

3.105 S. Ozaki, S. Adachi: J. Appl. Phys. **75**, 7470 (1994)

3.106 S. Ozaki, S. Adachi: Jpn. J. Appl. Phys. **32**, 2620 (1993)

3.107 S. Adachi, T. Kimura: Jpn. J. Appl. Phys. **32**, 3496 (1993)

3.108 L. Viña, C. Umbach, M. Cardona, L. Vodopyanov: Phys. Rev. **29**, 6752 (1984)

3.109 P. Lautenschlager, S. Logothetidis, L. Viña, M. Cardona: Phys. Rev. **B32**, 3811 (1985)

3.110 S. M. Kelso, D. E. Aspnes, M. A. Pollack, R. E. Nahory: Phys. Rev. **B26**, 6669 (1982)

3.111 J. Humliček, E. Schmidt, L. Bočánek, M. Garriga, M. Cardona: Solid State Commun. **73**, 127 (1990)

3.112 S. Adachi, H. Kato, A. Moki, K. Ohtsuka: J. Appl. Phys. **75**, 478 (1994)

3.113 D. E. Aspnes, B. Schwartz, A. A. Studna, L. Derick, L. A. Koszi: J. Appl. Phys. **48**, 3510 (1977)

3.114 W. H. Press, B. P. Flannery, S. A. Teukolsky, W. T. Vetterling: *Numerical Recipes in C: The Art of Scientific Computing* Cambridge University Press, 1988

3.115 J. C. Maxwell-Garnett: Phil. Tr. of the Roy. Soc. of London **203**, 385 (1904) and **205A**, 237 (1906)

3.116 H. Looyenga: Physica **31**, 401 (1965)

3.117 D. A. G. Bruggeman: Ann. Phys. (Leipzig) **24**, 636 (1935)

3.118 D. E. Aspnes: SPIE Proc. **452**, 60 (1983)

3.119 D. Bergman: Phys. Rep. **C43**, 377 (1978)

3.120 D. Bergman: *Bulk physical properties of composite media, Les methodes de l'homogeneisation*, Edition Eyrolles (1985)

3.121 W. Theiß, Ph.D. Thesis, RWTH Aachen 1989

3.122 S. Logothetidis, H. M. Polatoglou, S. Ves: Solid State Commun. **68**, 1075 (1988)

3.123 Y. Cong, R. W. Collins, R. Messier, K. Vedam, G. F. Epps, H. Windischmann: J. Vac. Sci. Techn. **A9**, 1123 (1991)

3.124 S. M. Sze: *VLSI Technology* (McGraw Hill 1988)

3.125 P. Lautenschlager, P. B. Allen, M. Cardona: Phys. Rev. **B31**, 2163 (1985)

3.126 U. Rossow, J. Wagner, W. Richter: unpublished

3.127 R. W. Collins, J. M. Cavese: J. Appl. Phys. **60**, 4169 (1986)

3.128 J. Stuke, G. Zimmerer: Phys. Stat. Sol. **B49**, 513 (1972)

3.129 M. Erman, J. B. Theeten, P. Chambon, S. M. Kelso, D. E. Aspnes: J. Appl. Phys. **56**, 2664 (1984)

3.130 R. E. Williams: *GaAs Processing Technology* (Artech House Inc. 1984)

3.131 M. Erman, J. B. Theeten, P. Frijlink, S. Gaillard, F. J. Hia, C. Alibert: J. Appl. Phys. **56**, 3241 (1984)

3.132 U. Rossow, A. Krost, T. Werninghaus, K. Schatke, W. Richter, A. Hase, H. Künzel, H. Roehle: Thin Solid Films **233**, 180 (1993)

3.133 M. A. Haase, J. Qiu, J. M. DePuydt, H. Cheng: Appl. Phys. Lett. **59**, 1272 (1991)

3.134 S. Yamaga: Jpn. J. Appl. Phys. **30**, 437 (1991)

3.135 D. R. T. Zahn, Ch. Maierhofer, A. Winter, M. Reckzügel, R. Srama, U. Rossow, A. Thomas, K. Horn, W. Richter: Appl. Surf. Sci. **56-58**, 684 (1992)

3.136 D. R. T. Zahn, G. Kudlek, U. Rossow, A. Hoffmann, I. Broser, W. Richter: Adv. Mat. for Opt. and Electr. **3**, 11 (1994)

3.137 T. Matsuno, H. Masato, A. Ryoji, K. Inoue: Proceedings of *Int. Symp. on Gallium arsenide and related compounds*, Karuizawa (Japan), Sept. 1992, T. Ikegami, F. Hasegawa, Y. Takeda eds., Inst. Phys. Conf. Ser. **129**, 729 (1993) Institute of Physics, Bristol (UK)

3.138 A. Adams, E. O'Reilly: Physics World **5**(10), 43 (1992)

3.139 J. Kircher, W. Böhringer, W. Dietrich, H. Hirt, P. Etchegoin, M. Cardona: Rev. Sci. Instr. **63**, 3733 (1992)

3.140 L. T. Canham: Appl. Phys. Lett. **57**, 1046 (1990)

3.141 A. Bsiesy, J. C. Vial, F. Gaspard, R. Herino, M. Ligeon, F. Muller, R. Romestain, A. Wasiela, A. Halimaoui, G. Bomchil: Surf. Sci. **254**, 195 (1991)

3.142 R. T. Carline, C. Pickering, N. S. Garawal, D. Lancefield, L. K. Howard, M. T. Emeny: SPIE proc. **1678**, 285 (1992)

3.143 S. Luryi, E. Suhir: Appl. Phys. Lett. **49**, 140 (1986)

3.144 H. Münder, C. Andrzejak, M. G. Berger, T. Eickhoff, H. Lüth, W. Theiß, U. Rossow, W. Richter, R. Herino, M. Ligeon: Appl. Surf. Sci. **56-58**, 6 (1992)

3.145 V. Lehmann, H. Cerva, U. Gösele: Mat. Res. Soc. Symp. Proc. **256**, 3 (1992) *Light Emission from Silicon*, eds. S. S. Iyer, R. T. Collins, L. T. Canham

3.146 H. Münder, M. G. Berger, H. Lüth, U. Rossow, U. Frotscher, W. Richter, M. Ligeon, R. Herino: Appl. Surf. Sci. **63**, 57 (1993)

3.147 C. Pickering, M. I. J. Beale, D. J. Robbins, P. J. Pearson, R. Greef: J. Phys. **C17**, 6535 (1984)

3.148 H. Münder, M. G. Berger, U. Rossow, U. Frotscher, W. Richter, R. Herino, M. Ligeon: Appl. Surf. Sci. **63**, 57 (1993)

3.149 U. Rossow, H. Münder, M. Thönissen, W. Theiß: J. of Luminescence **57**, 205 (1993)

3.150 P. Roca i Cabarrocas, Satyendra Kumar, B. Drevillon: J. Appl. Phys. **66**, 3236 (1989)

3.151 V. Chu, M. Fang, B. Drevillon: J. Appl. Phys. **69**, 3363 (1991)

3.152 E. A. Irene: Thin Solid Films **233**, 96 (1993)

3.153 E. D. Palik, V. M. Bermudez, O. J. Glembocki: J. Electrochem. Soc. **132**, 871 (1985)

3.154 Y. Demay, D. Arnoult, J. P. Gailliard, P. Medina: J. Vac. Sci. Technol. **A5**, 3139 (1987)

3.155 Y. Demay, J. P. Gailliard, P. Medina: J. Cryst. Growth **81**, 97 (1987)

3.156 G. Laurence, F. Hottier, J. Hallais: J. Cryst. Growth **55**, 198 (1981)

3.157 D. E. Aspnes, W. E. Quinn, M. C. Tamargo, M. A. A. Pudensi, S. A. Schwarz, M. J. S. P. Brasil, R. E. Nahory, S. Gregory: Appl. Phys. Lett. **60**, 1244 (1992)

3.158 R. I. G. Uhrberg, R. D. Bringans, M. A. Olmstead, R. Z. Bachrach, J. E. Northrup: Phys. Rev. **B35**, 3945 (1987)

3.159 S. P. Kowalczyk, D. L. Miller, J. R. Waldrop, P. G. Newman, R. W. Grant: J. Vac. Sci. Technol. **19**, 255 (1981)

3.160 P. Etienne, P. Alnot, J. F. Rochette, J. Massies: J. Vac. Sci. Technol. **B4**, 1301 (1986)

3.161 R. W. Bernstein, A. Borg, H. Husby, B.-O. Fimland, A. P. Grande, J. K. Grepstadt: Appl. Surf. Sci. **56-58**, 74 (1992)

3.162 U. Resch, N. Esser, I. Raptis, J. Waßerfall, A. Förster, D. I. Westwood, W. Richter: Surf. Sci. **269/270**, 797 (1992)

3.163 H. Wilhelm, W. Richter, U. Rossow, D. R. T. Zahn, D. A. Woolf, D. I. Westwood, R. H. Williams: Surf. Sci. **251/252**, 556 (1991)

3.164 R. I. G. Uhrberg, R. D. Bringans, M. A. Olmstead, R. Z. Bachrach: Phys. Rev. **B35**, 3945 (1987)

3.165 D. A. Woolf UWC Cardiff, A. Förster KFA Jülich: private communication

3.166 T. H. Shen, C. C. Matthai: Surf. Sci. **287/288**, 672 (1993)

3.167 S. A. Alterowitz, P. G. Snyder, K. G. Merkel, J. A. Woollam, D. C. Radulescu, L. F. Eastman: J. Appl. Phys. **63**, 5081 (1988)

3.168 J. L. Freeouf: Appl. Phys. Lett. **53**, 2426 (1988)

3.169 M. Erman, C. Alibert, J. B. Theeten, P. Frijlink, B. Catte: J. Appl. Phys. **63**, 465 (1988)

3.170 M. Erman, Doctorat d'Etat thesis, University Pierre and Marie Curie, Paris, 1986

3.171 P. Apell, O. Hunderi: Optical Properties of Superlattices, in *Handbook of Optical Constants of Solids* II (Academic Press 1990)

3.172 K. Vedam, P. J. McMarr, J. Narayan: Appl. Phys. Lett. **47**, 339 (1985)

3.173 K. Vedam, S. So: Surf. Sci. **29**, 379 (1972)

3.174 J. A. Woollam, B. Johs, W. A. McGahan, P. G. Snyder, J. Hale, H. W. Yao: MRS Symp. Proc. **324**, 15 (1994)

3.175 J . C. C. Fan, J. M. Poate (eds.), *Heteroepitaxy on Silicon*, MRS Symp. Proc. **67**; J. C. C. Fan, J. M. Phillips, B. Y. Tsaur (eds.), *Heteroepitaxy on Silicon II*, MRS Symp. Proc. **91**

3.176 D. E. Aspnes: J. Opt. Soc. Am. **A10**, 976 (1993)

3.177 F. H. P. M. Habraken, O. L. J. Gijzeman, G. A. Bootsma: Surf. Sci. **96**, 482 (1980)

3.178 D. E. Aspnes: J. Vac. Sci. Techn. **B3**, 1498 (1985)

3.179 P. Etchegoin, M. Cardona: Thin Solid Films **233**, 137 (1993)

3.180 U. Resch-Esser, N. Blick, N. Esser, Th. Werninghaus, U. Rossow, W. Richter: Proc. of the ICFSI-4 Jülich, 14-18 June 1993, p. 321

eds. B. Lengeler, H. Lüth, W. Mönch, J. Pollmann, World Scientific Press, Singapore, New Jersey, London, Hong Kong 1994

3.181 U. Rossow, Ph.D. Thesis, TU Berlin 1993

4.1 W. Hayes, R. Loudon: *Light Scattering in Solids*, (J. Wiley and Sons, New York 1978)

4.2 W. Richter: in *Springer Tracts in Modern Physics* **78**, Resonant Raman Scattering in Semiconductors, ed. by G. Höhler, (Springer, Berlin, Heidelberg, New York 1976)

4.3 M. Cardona: in *Topics in Applied Physics* **50**, Light Scattering in Solids II, ed. by M. Cardona and G. Güntherodt, (Springer, Berlin, Heidelberg, New York 1982), p. 19

4.4 G. Abstreiter, M. Cardona, A. Pinczuk: in *Topics in Applied Physics* **54**, Light Scattering in Solids IV, ed. by M. Cardona and G. Güntherodt, (Springer, Berlin, Heidelberg, New York 1984), p. 5

4.5 B. Jusserand, M. Cardona: in *Topics in Applied Physics* **66**, Light Scattering in Solids V, ed. by M. Cardona and G. Güntherodt, (Springer, Berlin, Heidelberg, New York 1989) p. 49

4.6 J. Sapriel, B. Djafari Rouhani: Surf. Science Rep. **10**, 189 (1989)

4.7 R. Loudon: Proc. Royal Soc. **A275**, 218 (1963)

4.8 A. Pinczuk, E. Burstein: in *Topics in Applied Physics* **8**, Light Scattering in Solids, ed. by M. Cardona and G. Güntherodt, (Springer, Berlin, Heidelberg, New York 1975), p. 23

4.9 E. O. Kane: Phys. Rev. **178**, 1368 (1969)

4.10 M. Born, K. Huang: *Dynamical Theory of Crystal Lattices,* (Oxford: Clarendon Press, 1954)

4.11 P. Brüesch: in *Springer Series in Solid-State Sciences* **34**, Phonons: Theory and Experiments I, ed. by M. Cardona, P. Fulde, H. J. Queisser (Springer, Berlin, Heidelberg, New York 1982), p. 117

4.12 H. Fröhlich: Adv. Phys. **3**, 325 (1954)

4.13 G. F. Koster, J. O. Dimmock, R. G. Wheeler, H. Statz: *Properties of the thirtytwo point groups*, (M.I.T. Press, Cambridge (Mass.) 1963)

4.14 R. Loudon: Adv. Phys. **13**, 423 (1964)

4.15 E. Hecht, A. Zajac: *Optics* (Addison-Wesley Publ. Comp. 1974)

4.16 R. K. Chang, M. B. Long: in [4.3], p. 179

4.17 J. C. Tsang: in *Topics in Applied Physics* **66**, Light Scattering in Solids V, ed. by M. Cardona and G. Güntherodt, Springer, Berlin, Heidelberg, New York (1989) p. 233

4.18 R. B. Bilhorn, J. V. Sweedler, P. M. Epperson, M. B. Denton: Appl. Spectroscopy **41**, 1114 (1987)

4.19 A. Krost, W. Richter and D. R. T. Zahn: Appl. Surf. Science, **56–58**, 691 (1992)

4.20 V. Wagner, D. Drews, N. Esser, D. R. T. Zahn, J. Geurts, W. Richter: J. of Appl. Phys. **75**, 7330 (1994)

4.21 D. W. Pohl, W. Denk, M. Lanz: Appl. Phys. Lett. **44**, 651 (1984)

4.22 E. L. Buckland, P. J. Moyer, M. A. Paesler: J. Appl. Phys. **73**, 1018 (1993)

4.23 H. F. Hess, E. Betzig, T. D. Harris, L. N. Pfeiffer, K. West: Science **264**, 1740 (1994)

4.24 M. Cardona, L. Ley: in *Topics in Applied Physics* **26**, Photoemission in Solids, ed. by M. Cardona and L. Ley, (Springer, Berlin, Heidelberg, New York 1978), p. 1

4.25 Y. Shirakawa, H. Kukimoto: J. Appl. Phys. **51**, 2014 (1980)

4.26 M. Aven, D. T. F. Marple, B. Segall: J. Appl. Phys. Suppl. **32**, 2261 (1961)

4.27 D. J. Olego, K. Shahzad, J. Petruzzello, D. Cammack: Phys. Rev. **B36**, 7674 (1987)

4.28 M. A. Chesters: Reflection-Absorption Infrared Spectroscopy of Adsorbates on Metal Surfaces, in *Analytical Applications of Spectroscopy*, ed. C. S. Creaser, A. M. C. Davies, Royal Society of Chemistry, London (1988)

4.29 P. Skeath, C. Y. Su, W. A. Harrison, I. Lindau, W. Spicer: Phys. Rev. **B27**, 6246 (1983)

4.30 F. Schäffler, R. Ludeke, A. Taleb-Ibrahimi, G. Hughes, D. Rieger: J. Vac. Sci. Technol. **B5**, 1048 (1987)

4.31 R. D. Bringans: Critical Reviews in Solid State and Material Sciences **17**, 353 (1992)

4.32 W. Richter, N. Esser, A. Kelnberger, M. Köpp: Solid State Commun. **84**, 165 (1992)

4.33 M. Hünermann, J. Geurts, W. Richter: Phys. Rev. Lett. **66**, 640 (1991)

4.34 H. Wilhelm, W. Richter, U. Rossow, D. R. T. Zahn, D. A. Woolf, D. I. Westwood, R. H. Williams: Surface Science **251/252**, 556 (1991)

4.35 R. B. Doak, D. B. Nguyen: Phys. Rev. **B41**, 3578 (1989)

4.36 N. Esser, M. Köpp, P. Haier, W. Richter: Journal of Electron Spectroscopy and Related Phenomena **64/65**, 85 (1993)

4.37 A. V. Nurmikko, R. L. Gunshor, L. A. Kolodziejski: IEEE J. Quantum Electr. **QE-22** 1785 (1986)

4.38 C. B. Duke, A. Paton, W. K. Ford, A. Kahn, J. Carelli: Phys. Rev. **B26**, 803 (1982)

4.39 C. Mailhiot, C. B. Duke, D. J. Chadi: Phys. Rev. **B31**, 2213 (1985)

4.40 P. Martensson, G. V. Hansson, M. Lähdeniemi, K. O. Magnusson, S. Wiklund, J. M. Nicholls: Phys. Rev. **B33**, 7399 (1986)

4.41 P. Martensson, R. M. Feenstra: Phys. Rev. **B39**, 7744 (1989)

4.42 M. Hünermann: Ph.D. Thesis, RWTH Aachen, 1991

4.43 N. Esser, M. Reckzügel, R. Srama, U. Resch, D. R. T. Zahn, W. Richter, C. Stephens, M. Hünermann: J. Vac. Sci. Technol. **B8**, 680 (1990)

4.44 N. Esser: Ph.D. Thesis (Berlin University of Technology, 1991)

4.45 W. Pletschen, N. Esser, H. Münder, D. Zahn, J. Geurts, W. Richter: Surf. Science **178**, 140 (1986)

4.46 M. Hünermann, W. Pletschen, U. Resch, U. Rettweiler, W. Richter, J. Geurts, P. Lautenschlager: Surf. Science **189/190**, 322 (1987)

4.47 J. S. Lannin: Phys. Rev. **B15**, 3863 (1977)

4.48 R. N. Zitter: in *The Physics of Semimetals and Narrow-Gap Semiconductors*, ed. by E. L. Carter, R. T. Bate (Pergamon Press, Oxford 1971), p. 285

4.49 R. I. Sharp, E. Warming: J. Phys. **F1**, 570 (1971)

4.50 U. Resch, N. Esser, W. Richter: Surf. Science **251/252**, 621 (1991)

4.51 N. Esser, M. Hünermann, U. Resch, D. Spaltmann, J. Geurts, D. R. T. Zahn, W. Richter, R. H. Williams: Appl. Surf. Science **41/42**, 169 (1989)

4.52 A. B. McLean, R. M. Feenstra, A. Taleb-Ibrahimi, R. Ludeke: Phys. Rev. **B39**, 12925 (1989)

4.53 T. Guo, R. E. Atkinson, W. K. Ford: Phys. Rev. **B41**, 5138 (1990)

4.54 A. G. Milnes, D. L. Feucht: *Heterojunctions and Metal-Semiconductor Junctions* (Academic Press, New York 1972)

4.55 R. Trommer, M. Cardona: Phys. Rev. **B 17**, 1865 (1978)

4.56 M. Sinyukov, R. Trommer, M. Cardona: Phys. Stat. Sol. (b) **86**, 563 (1978)

4.57 A. Compaan, H. J. Trodahl, Phys. Rev. **B 29**, 793 (1984)

4.58 N. Esser, M. Köpp, P. Haier, W. Richter, J. Electron. Spectrosc. **64/65**, 85 (1993)

4.59 N. Esser, R. Hunger, J. Rumberg, W. Richter, R. Del Sole, A. I. Shkrebtii, Surf. Sci. **307–309**, 1045 (1994)

4.60 W. Mönch, H. Gant: Phys. Rev. Lett. **48**, 512 (1982)

4.61 R. Merlin, A. Pinczuk, W. T. Beard, C. E. E. Wood: J. Vac. Sci. Technol. **21**, 516 (1982)

4.62 H. Brugger, F. Schäffler, G. Abstreiter: Phys. Rev. Lett. **52**, 141 (1984)

4.63 J. G. Brugger: Ph.D. Thesis, (Munich University of Technology, 1987)

4.64 D. R. T. Zahn, Ch. Maierhofer, A. Winter, M. Reckzügel, R. Srama, U. Rossow, A. Thomas, K. Horn, W. Richter: Appl. Surf. Sci. **56-58**, 684 (1992)

4.65 Ch. Maierhofer: Ph.D. Thesis (Technical University of Berlin, 1992)

4.66 K. J. Chang, S. Froyen, M. L. Cohen: Phys. Rev. **B 28**, 4736 (1983)

4.67 *Proc. of the II-VI-Conference Newport 1993*, J. Crystal Growth **138** (1994)

4.68 F. H. Pollak: in *Analytical Raman Spectroscopy*, ed. by J. G. Grasseli and B. J. Bulkin, Chemical Analysis Series, Vol. **114** (1991)

4.69 C. W. Snyder, B. G. Orr, D. Kessler, L. M. Sander: Phys. Rev. Lett. **66** 3032 (1991)

4.70 H. J. van der Merwe: Surf. Science **31**, 198 (1972)

4.71 W. Matthews, A. E. Blakeslee: J. Cryst. Growth **29**, 273 (1975)

4.72 R. People, IEEE J. Quantum Elekctron. **QE-22**,1696 (1986)

4.73 J. F. Nye: *Physical Properties of Crystals, Their Representation by Tensors and Matrices* (Clarendon Press, Oxford 1957)

4.74 E. Anastassakis: in *Physical Problems in Microelectronics*, Proceedings 4th Int. School ISSPPME, Varna (Bulgaria), ed. by J. Kassabov (World Scientific, Singapore (1985), p. 128

4.75 E. Anastassakis: J. Appl. Phys. **68**, 4561 (1990)

4.76 Landolt-Börnstein: *Numerical Data and Functional Relationships in Science and Technology, New. Series, Group III*, Vol. 17a, ed. O. Madelung, (Springer, Berlin, Heidelberg, New York, Tokyo 1983)

4.77 G. Landa, R. Carles, C. Fontaine, E. Bedel, A. Munoz-Yagüe: J. Appl. Phys. **66**, 196 (1988)

4.78 T. Nishioka, Y. Shinoda, Y. Ohmachi: J. Appl. Phys. **57**, 276 (1985)

4.79 K. Brunner, G. Abstreiter, B. O. Kolbesen, H. W. Meul: Appl. Surf. Sci. **39**, 116 (1989)

4.80 G. Abstreiter: Appl. Surf. Sci. **50**, 73 (1991)

4.81 A. Krost, W. Richter, O. Brafman: Appl. Phys. Lett. **56**, 343 (1990)

4.82 O. Brafman, A. Krost, W. Richter: J. Phys. : Condens. Matter **3**, 6203 (1991)

4.83 F. Cerdeira, C. J. Buchenauer, F. H. Pollak, M. Cardona, Phys. Rev. **B 5**, 580 (1972)

4.84 J. C. Tsang, F. H. Dacol, P. Mooney, J. O. Chu, B. S. Meyerson: Appl. Phys. Lett. **62**, 1146 (1993)

4.85 C. Fontaine, H. Benarfa, E. Bedel, A. Munoz-Yague, G. Landa, R. Carles: J. Appl. Phys. **60**, 208 (1986)

4.86 V. Wagner, J. Geurts, M. Eube, J. Woitok: Proc. of the Int. Conf. on Semicond. Interf. 7, Jülich (1993), *Formation of Semiconductor Interfaces*, ed. B. Lengeler, H. Lüth, W. Mönch, J. Pollmann (World Scientific, Singapore 1994), p. 550

4.87 J. Woitok, Ph.D. Thesis, RWTH Aachen (1989)

4.88 R. Beserman, C. Hirlimann, M. Balkanski: Solid State Commun. **20**, 485 (1976)

4.89 B. Jusserand, S. Slempkes: Solid State Commun. **44**, 95 (1984)

4.90 S. Emura, S. Gonda, Y. Matsui, Hayashi: Phys. Rev. **B38**, 3280 (1988)

4.91 X. Wang, X. Zhang: Solid State Commun. **59**, 869 (1986)

4.92 M. A. Renucci, J. B. Renucci, M. Cardona: in *Proc. 2nd Int. Conf. on Light Scattering in Solids*, ed. M. Balkanski (Flammarion, Paris 1971), p. 326

4.93 J. Finders, J. Geurts, A. Kohl, M. Weyers, B. Opitz, O. Kayser, P. Balk: J. Crystal Growth **107**, 151 (1991)

4.94 J. C. Tsang, F. H. Dacol, P. M. Mooney, J. O. Chu, B. S. Meyerson: Appl. Phys. Lett. **62** 1146 (1993)

4.95 P. M. Mooney, F. H. Dacol, J. C. Tsang, J. O. Chu: Appl. Phys. Lett. **62** 2069 (1993)

4.96 R. Schorer, E. Friess, K. Eberl, G. Abstreiter: Phys. Rev. **B44**, 1772 (1991)

4.97 A. Gomyo, T. Suzuki, S. Iijima: Phys. Rev. Lett. **60**, 2645 (1988)

4.98 A. Gomyo, K. Kobayashi, S. Kawata, I. Hino, T. Suzuki: J. Crystal Growth **77**, 367 (1986)

4.99 A. Mascarenhas, S. Kurtz, A. Kibbler, J. M. Olson: Phys. Rev. Lett. **63**, 2108 (1989)

4.100 J. Geurts, J. Finders, O. Kayser, B. Opitz, M. Maassen, R. Westphalen, P. Balk: SPIE Conference Proceedings **1361**, 744 (1991)

4.101 A. Krost, N. Esser, H. Selber, J. Christen, W. Richter, D. Bimberg, L. C. Su, G. B. Stringfellow: J. Vac. Sci. Technol. **B12**, 2558 (1994)

4.102 A. Mascarenhas, S. R. Kurtz, A. Kibbler, J. M. Olson: Phys. Rev. Lett. **63**, 2108 (1989)

4.103 J. C. Tsang, Y. Yokota, R. Matz, G. W. Rubloff: Appl. Phys. Lett. **44**, 430 (1984)

4.104 R. J. Nemanich, R. T. Fulks. , B. L. Stafford, H. A. Vander Plas: Appl. Phys. Lett. **46**, 670 (1985)

4.105 A. Krost, W. Richter, D. R. T. Zahn, K. Hingerl, H. Sitter: Appl. Phys. Lett. **57**, 1981 (1990)

4.106 D. R. T. Zahn, W. Richter, T. Eickhoff, J. Geurts, T. D. Golding, J. H. Dinan, K. J. MacKey, R. H. Williams: Applied Surface Science **41/42**, 497 (1989)

4.107 K. J. MacKey, P. M. G. Allen, W. G. Herrenden-Harker, R. H. Williams: Surf. Sci. **178**, 7 (1986)

4.108 G. P. Schwartz, B. Schwartz, D. Distefano, G. J. Gualtieri, J. E. Griffiths: Appl. Phys. Lett. **34**, 205 (1979)

4.109 R. L. Farrow, R. K. Chang, S. Mroczkowski: Appl. Phys. Lett. **31**, 768 (1977)

4.110 R. L. Farrow, R. K. Chang, S. Mroczkowski, F. H. Pollak: Appl. Phys. Lett. **31**, 768 (1977)

4.111 D. Drews, M. Langer, W. Richter, D. R. T. Zahn: Proc. of the Int. Conf. on Semicond. Interf. 7, Jülich (1993), *Formation of Semicoductor Interfaces*, ed. B. Lengeler, H. Lüth, W. Mönch, J. Pollmann (World Scientific, Singapore 1994), p. 506

4.112 V. Wagner, D. Drews, N. Esser, W. Richter, D. R. T. Zahn, J. Geurts, W. Richter: Proc. of the Int. Conf. on Semicond. Interf. 7, Jülich (1993), *Formation of Semicoductor Interfaces*, ed. B. Lengeler, H. Lüth, W. Mönch, J. Pollmann (World Scientific, Singapore 1994), p. 546

4.113 M. V. Klein: IEEE J. QE-**22**, 1760 (1986)

4.114 B. Jusserand, D. Paquet: in *Semiconductor Heterojunctions and Super-lattices*, ed. by G. Allan, G. Bastard, N. Boccara, M. Lannoo, M. Voos, (Springer, Berlin, Heidelberg 1986), p. 108

4.115 M. Cardona: in *Lectures of Surface Science*, ed. by G. R. Castro, M. Cardona, (Springer, Berlin, Heidelberg, New York 1987), p. 2

4.116 R. Enderlein, D. Suisky, J. Röseler, Phys. Stat. Sol. (b) **165**, 9 (1991)

4.117 A. Fasolino, E. Molinari: Surf. Science **228**, 112 (1990)

4.118 B. Jusserand, F. Alexandre, D. Paquet, G. Le Roux: Appl. Phys. Lett. **47**, 301 (1986)

4.119 J. Geurts et al., Phys. Stat. Sol. (a) **152** (1995)

4.120 J. Finders, J. Geurts, Y. Pusep (to be published)

4.121 D. J. Olego, K. Shahzad, D. A. Cammack, H. Cornelissen: Phys. Rev. **38**, 5554 (1988)

4.122 M. K. Jackson, R. H. Miles, T. C. McGill, J. P. Faurie: Appl. Phys. Lett. **55**, 786 (1989)

4.123 R. Merlin, C. Colvard, M. V. Klein, H. Morkoc, A. Y. Cho, A. C. Gossard: Appl. Phys. Lett. **36**, 43 (1980)

4.124 A. K. Sood, J. Menendez, M. Cardona, K. Ploog, Phys. Rev. Lett. **54**, 2111 (1985)

4.125 R. E. Camley, D. L. Mills: Phys. Rev. **29**, 1695 (1984)

4.126 Akhilesh K. Arora, A. K. Ramdas, M. R. Melloch, N. Otsuka: Phys. Rev. **B36**, 1021 (1987)

4.127 M. V. Klein: in *Light Scattering in Solids*, ed. M. Cardona (Springer, Heidelberg 1975)

4.128 J. Wagner, M. Ramsteiner, H. Seelewind, J. Clarc: J. Appl. Phys. **64**, 802 (1988)

4.129 D. J. Olego, T. Marshall, J. Gaines, and K. Shahzad: Phys. Rev. **B42**, 9067 (1990)

4.130 D. J. Olego, T. Marshall, D. Cammack, K. Shahzad, and J. Petruzzello: Appl. Phys. Lett. **58**, 2654 (1991)

4.131 D. J. Olego, J. Petruzello, T. Marshall, and D. Cammack: Appl. Phys. Lett. **59**, 961 (1991)

4.132 A. A. Gogolin, E. I. Rashba: Solid State Commun. **19**, 1177 (1976)

4.133 A. A. Gogolin, E. I. Rashba: in *Proc. 13th Int. Conf. on the Physics of Semiconductors*, ed. by F. G. Fumi (Tipografia Marves, Rome, 1976), p. 231

4.134 R. Trommer: Ph.D. Thesis, University of Stuttgart, 1977

4.135 D. E. Aspnes: in *Handbook on Semiconductors* **Vol. 2**, ed. M. Balkanski (North Holland Publishing Company 1980)

4.136 J. Menendez, M. Cardona: Phys. Rev. **B31**, 3696 (1985)

4.137 A. Pinczuk, G. Abstreiter: in *Light Scattering in Solids V*, ed. M. Cardona and G. Güntherodt (Springer, Berlin)

4.138 B. B. Varga: Phys. Rev. **A137**, 1896 (1965)

4.139 E. Burstein, A. Pinczuk, S. Iwasa: Phys. Rev. **157**, 611 (1967)

4.140 P. Grosse: *Freie Elektronen in Festkörpern*, (Springer, Berlin, Heidelberg, New York 1979)

4.141 W. Richter, U. Nowak, A. Stahl: Proc. 15th Int. Conf. Physics of Semiconductors, Kyoto, 1980, J. Phys. Soc. Japan **49**, Suppl. A, 703 (1980)

4.142 E. Burstein, A. Pinczuk, S. Buchner: in *Physics of Semiconductors 1978*, ed. by B. L. H. Wilson, The Institute of Physics, London (1979), p. 1231

4.143 U. Nowak, W. Richter, G. Sachs: Phys. stat. sol. (b) **108**, 131 (1981)

4.144 W. Richter, U. Nowak, H. Jürgensen, U. Rössler: Solid State Commun. **67**, 199 (1988)

4.145 B. Boudart, B. Prévot, C. Schwab, Appl. Surf. Sci. **50**, 295 (1991)

4.146 U. Resch, N. Esser, Y. S. Raptis, W. Richter, J. Wasserfall, A. Förster, D. I. Westwood: Surf. Sci. **269/270**, 797 (1992)

4.147 A. Mooradian, A. L. McWhorter: Phys. Rev. Lett. **19**, 849 (1967)

4.148 A. Pinczuk, S. Schmitt-Rink, G. Danan, J. P. Valladares, L. N. Pfeiffer, K. W. West: Phys. Rev. Lett. **63**, 1633 (1989)

4.149 J. M. Worlock. (ed.): *Proc. 7th Conf. Electronic Properties of Two Dimensional Systems*, Surface Sci. **196** (1988)

4.150 T. Ando, A. B. Fowler, F. Stern: Rev. Mod. Phys. **54**, 437 (1982)

4.151 J. Wagner: *Proc. of the SPIE* **1678**, New Jersey (1992), p. 110

4.152 D. Olego, A. Pinczuk, A. C. Gossard, W. Wiegmann: Phys. Rev. **B25**, 7867 (1982)

4.153 A. Nurmikko, A. Pinczuk: in Physics Today **6**, 24 (1993)

4.154 G. Y. Robinson: in *Physics and Chemistry of III-V Semiconductor Interfaces*, ed. C. F. Wilmsen, (Plenum Press, New York 1985), p. 73

4.155 W. Mönch: in *Advances in Solid State Physics*, **XXVI**, ed. P. Grosse (Vieweg, Braunschweig 1986), p. 67

4.156 E. H. Rhoderick, R. H. Williams: *Metal Semiconductor Contacts* (Clarendon, Oxford 1988)

4.157 W. Mönch: Rep. Prog. Phys. **53**, 221 (1990)

4.158 W. Mönch: *Semiconductor Surfaces and Interfaces* (Springer, Berlin, Heidelberg, New York 1993)

4.159 W. Schottky: Zeitschrift f. Physik **118**, 539 (1942)

4.160 L. Ley, M. Cardona, F. H. Pollak: in *Topics in Applied Physics* **27**, Photoemission in Solids II, ed. by M. Cardona and L. Ley (Springer, Berlin, Heidelberg, New York 1978), p. 11

4.161 F. Schäffler, G. Abstreiter: Phys. Rev. B **34** 4017 (1986)

4.162 A. Pinczuk, A. A. Ballman, R. E. Nahory, M. A. Pollack, J. M. Worlock: J. Vac. Sci. Technol. **16** 1168 (1979)

4.163 L. A. Farrow, C. J. Sandroff, M. C. Tamargo: Appl. Phys. Lett. **51**, 1931 (1987)

4.164 R. E. Viturro, J. L. Shaw, C. Mailhiot, L. J. Brillson, N. Tache, J. McKinley, G. Margaritondo, J. M. Woodall, R. D. Kirchner, G. D. Petit, S. L. Wright: Appl. Phys. Lett. **52**, 2052 (1988)

4.165 G. P. Schwartz, G. J. Gualtieri: J. Electrochem. Soc. **133**, 1266 (1986)

4.166 D. J. Olego: J. Vac. Sci. Technol. **B6**, 1193 (1988)

4.167 E. T. Yu, T. C. McGill: Appl. Phys. Lett. **53**, 60 (1988)

4.168 D. J. Olego: Appl. Phys. Lett. **51**, 1422 (1987)

4.169 D. J. Olego: Phys. Rev. **B39**, 12743 (1989)

4.170 W. Franz: Z. Naturforsch. **13a**, 484 (1958)

4.171 L. V. Keldysh: Soviet Phys. JETP **34**, 788 (1958)

4.172 J. G. Gay, J. D. Dow, E. Burstein, A. Pinczuk: in *Light Scattering in Solids*, ed. by M. Balkanski (Flammarion, Paris 1971), p. 33

4.173 W. Richter, R. Zeyher, M. Cardona: Phys. Rev. **B18**, 4312 (1978)

4.174 W. R. Pletschen: Ph.D. Thesis (RWTH Aachen, 1986)

4.175 A. Huijser, J. van Laar, T. L. van Rooy: Surf. Sci. **62**, 472 (1977)

4.176 G. M. Guichard, C. A. Sebenne, C. D. Thualt: J. Vac. Sci. Technol. **16**, 1212 (1979)

4.177 L. J. Brillson, E. Burstein: Phys. Rev. Lett. **27**, 808 (1971)

4.178 H. J. Stolz, G. Abstreiter: J. Vac. Sci. Technol. **19**, 380 (1981)

4.179 M. Mattern-Klosson, H. Lüth: Solid State Commun. **56**, 1001 (1985)

4.180 R. H. Williams, D. R. T. Zahn, N. Esser, W. Richter: J. Vac. Sci. Technol. **B7**, 997 (1989)

4.181 T. Kendelewicz, K. Miyano, R. Cao, J. C. Woicik, I. Lindau, W. E. Spicer: Surf. Sci. **220**, L726 (1989)

4.182 F. Schäffler, R. Ludeke, A. Taleb-Ibrahimi, G. Hughes, D. Rieger: Phys. Rev. **B36**, 1328 (1987)

4.183 G. Annovi, M. -G. Betti, U. del Pennino, C. Mariani: Phys. Rev. **B41**, 11978 (1990)

4.184 C. K. Shih, R. M. Feenstra, P. Martensson: J. Vac. Sci. Technol. **A8**, 3379 (1990)

4.185 A. Kumar, O. P. Katyal: J. Mater. Sci. **24**, 4037 (1989)

4.186 H. Brugger, G. Abstreiter: in *Semiconductor Quantum Well structures and Superlattices*, ed. by K. Ploog and N. T. Linh (Editions de Physique, Les Ulis (1985), p. 209

4.187 H. J. Stolz, G. Abstreiter: J. Phys. Soc. Jpn. **49** Suppl. A, 1101 (1980)

4.188 K. Smit, L. Koenders, W. Mönch: J. Vac. Sci. Technol. **B7**, 888 (1989)

4.189 T. U. Kampen, D. Troost, X. Y. Hou, L. Koenders, W. Mönch: J. Vac. Sci. Technol. **B9**, 2095 (1991)

4.190 C. Trallero-Giner, A. Cantarero, M. Cardona, M. Mora: Phys. Rev. **B 45**, 6601 (1992)

4.191 M. Ramsteiner, J. Wagner, P. Hiesinger, K. Köhler, U. Rössler: J. Appl. Phys. **73**, 5023 (1993)

4.192 J. Geurts: Surf. Sci. Rep. **18**, 1 (1993)

5.1 R. J. Bell: *Introduction to Fourier Transform Spectroscopy* (Academic Press, New York 1972)

5.2 P. R. Griffiths, J. A. de Haseth: *Fourier Transform Infrared Spectrometry* (John Wiley & Sons, New York 1986)

5.3 R. Geick: *Topics in Current Chemistry* **58** (Springer, Berlin 1975), p. 73

5.4 A. G. Marshall, F. R. Verdun: *Fourier Transforms in NMR, Optical, and Mass Spectrometry* (Elsevier, Amsterdam 1990)

5.5 P. Appel, O. Hunderi: in *Optical Properties of Superlattices*, ed. E. D. Palik, Academic Press (Boston, 1991), p. 97

5.6 T. Dumelow, T. J. Parker, S. R. P. Smith, D. R. Tilley, Surface Science Reports **17**, 151 (1993)

5.7 L. D. Landau, E. M. Lifshitz: *Electrodynamics of Continuous Media* (Pergamon, Oxford 1960)

5.8 E. E. Bell: *Encyclopedia of Physics* **XXV/2** ed. by L. Genzel, (Springer, Berlin 1967), p. 1

5.9 J. D. Jackson: *Classical Electrodynamics*, (John Wiley, New York 1975)

5.10 D. Palik: *Handbook of Optical Constants of Solids* II (New York, Academic 1991)

5.11 H. A. Kramers: Estratto dagli Atti del Congresso Internazionale, de Fisici Como **2**, 545 (1927)

5.12 R. de L. Kronig: J. Opt. Soc. Am. **12**, 547 (1926)

5.13 P. Grosse, V. Offermann: Appl. Phys. **A52**, 138 (1991)

5.14 M. G. Sceats, G. C. Morris: Phys. Stat. Sol. (a) **14**, 643 (1972)

5.15 C. W. Peterson, B. W. Knight: J. Opt. Soc. Am. **63**, 1238 (1973)

5.16 B. Harbecke: Appl. Phys. **A40**, 154 (1986)

5.17 B. Harbecke: Appl. Phys. **B39**, 165 (1986)

5.18 Z. Knittel: *Optics of Thin Films* (John Wiley, London 1976)

5.19 P. Grosse: *Freie Elektronen in Festkörpern* (Springer, Berlin 1979)

5.20 A. F. Terzis, X. C. Liu, A. Petrou, B. D. McCombe, M. Dutta, H. Shen, D. D. Smith, M. W. Cole, M. Maysing-Lara, P. G. Newman: J. Appl. Phys. **67**, 2501 (1990)

5.21 D. W. Berreman: Phys. Rev. **130**, 2193 (1963)

5.22 D. W. Berreman: Proc. Intern. Conf. Lattice Dynamics Copenhagen, ed. by R. F. Wallis (Pergamon, Oxford 1963), p. 397

5.23 B. Harbecke, B. Heinz, P. Grosse: Appl. Phys. **A38**, 263 (1985)

5.24 R. Brendel: Appl. Phys. **A50**, 587 (1990)

5.25 M. A. Chesters: in *Analytical Applications of Spectroscopy*, ed. by C. S. Creaser, A. M. C. Davies (Roy. Soc. of Chemistry, London 1988)

5.26 N. J. Harrick: *Internal Reflection Spectroscopy* (Wiley, New York 1967), p. 138

5.27 E. Kretschmann: Z. Physik **241**, 313 (1971)

5.28 A. Otto: Z. Physik **216**, 398 (1968)

5.29 V. M. Agranovich, D. L. Mills (ed.), *Surface Polaritons* (North-Holland, Amsterdam 1982)

5.30 N. Wiener: Act. Math. Stockholm **55**, 117 (1930)

5.31 A. Khintchine: Math. Ann. **109**, 604 (1934)

5.32 J. D. Saalmüller: Ph.D. Thesis, RWTH Aachen (1987)

5.33 K. Krishnan, P. J. Stout, M. Watanabe: *Practical Fourier Transform Infrared Spectroscopy*, ed. by J. R. Ferraro, K. Krishnan, (Academic Press, San Diego 1990), p. 286

5.34 H. R. Chandrasekhar, A. K. Ramdas: Phys. Rev. **B21**, 1511 (1980)

5.35 U. Kreibig, C. v. Fragstein: Z. Physik **224**, 307 (1969)

5.36 A. Petrou, B. D. McCombe: in *Landau Level Spectroscopy*, ed. by G. Landwehr, E. I. Rashba (Elsevier Science Publisher, Amsterdam, 1990), p. 679

5.37 T. Duffield, R. Bhat, M. Koza, D. De Rosa, D. M. Hwang, P. Grabbe, S. J. Allen: Phys. Rev. Lett. **56**, 2724 (1986)

5.38 T. Ando, A. B. Fowler, F. Stern: Rev. Mod. Phys. **54**, 437 (1982)

5.39 Special issue on QW's and superlattices: IEEE J. Quantum Electron. **QE 22** (1986),

5.40 Special issue on QW's and superlattices: IEEE J. Quantum Electron. **QE 24** (1988),

5.41 D. Heitmann: in *Physics and Applications of Quantum Wells and Superlattices*, ed. by E. E. Mendez, K. v. Klitzing, (Plenum Press, New York 1987), p. 317

5.42 D. Heitmann, T. Demel, P. Grambow, K. Ploog: in *Festkörperprobleme/Advances in Solid State Physics* **29**, ed. by U. Rössler, (Vieweg, Braunschweig 1989), p. 285

5.43 D. Heitmann, U. Mackens: Phys. Rev. **B33**, 8269 (1986)

5.44 E. Gornik: in *Landau Level Spectroscopy*, ed. by G. Landwehr, E. I. Rashba, (North Holland, Elsevier, Amsterdam, The Netherlands, 1991), p. 911

5.45 E. Batke: in *Festkörperprobleme/Advances in Solid State Physics* **31**, ed. by U. Rössler (Vieweg, Braunschweig 1992), p. 297

5.46 S. J. Allen Jr. , D. C. Tsui, B. Vinter: Solid State Commun. **20**, 425 (1976)

5.47 T. Ando: Solid State Commun. **21**, 133 (1977)

5.48 W. L. Bloss: J. Appl. Phys. **66**, 3639 (1989)

5.49 M. Helm: in *Intersubband Transitions in Quantum Wells*, ed. by E. Rosencher, (Plenum Press, New York 1992), p. 151

5.50 M. Helm, W. Hilber, T. Fromherz, F. M. Peeters, K. Alavi, R. N. Pathak: Phys. Rev. **B48**, 1601 (1993)

5.51 J. Faist, F. Capasso, D. L. Sivco, C. Sirtori, A. L. Hutchinson, A. Y. Cho: Science **264**, 553 (1994)

5.52 B. F. Levine: J. Appl. Phys. **74**, R1 (1993)

5.53 H. Hertle, G. Schuberth, E. Gornik, G. Abstreiter, F. Schäffler: Appl. Phys. Lett. **59**, 2977 (1991)

5.54 T. Fromherz, E. Koppensteiner, M. Helm, G. Bauer, J. Nützel, G. Abstreiter: Phys. Rev. **B50**, 15073 (1994)

5.55 T. Fromherz, E. Koppensteiner, M. Helm, G. Bauer, J. Nützel, G. Abstreiter: Phys. Rev. **B50**, 15073 (1994)

5.56 P. S. Zory: *Quantum Well Lasers* (Academic Press, Boston 1993)

5.57 S. Yuan, H. Krenn, G. Springholz, G. Bauer, M. Kriechbaum: Appl. Phys. Lett. **62**, 885 (1993)

5.58 S. Yuan, N. Frank, G. Bauer, M. Kriechbaum: Phys. Rev. **B50**, 5286 (1994)

5.59 M. M. Pradhan, R. K. Garg, M. Arora: Infrared Phys. **27**, 25 (1987)

5.60 A. S. Oates, W. Lin: J. Cryst. Growth **89**, 117 (1988)

5.61 F 123-91: ASTM - American Society for Testing and Materials, 1916 Race Street, Philadelphia, PA 19103-1187, USA

5.62 DIN 50 438, part 2 in DIN, Beuth Verlag GmbH, Berlin, Wien, Zürich (1982)

5.63 DIN 50 438, part 1 in DIN, Beuth Verlag GmbH, Berlin, Wien, Zürich (1990)

5.64 A. Baghdadi, N. M. Bullis, M. C. Croarkin, Y. Z. Li, R. I. Scace, R. W. Series, P. Stallhofer, M. Watanabe: J. Electrochem. Soc. **136**, 2015 (1989)

5.65 P. Wagner: Appl. Phys. **A53**, 20 (1991)

5.66 A. Borghesi, M. Geddo, B. Pivac: J. Appl. Phys. **69**, 7251 (1991)

5.67 M. Geddo, B. Pivac, A. Borghesi, A. Stella: Appl. Phys. Lett. **58**, 370 (1991)

5.68 A. V. Annapragada, F. F. Jensen, T. F. Kuech: J. Cryst. Growth **107**, 248 (1991)

5.69 A. K. Ramdas, S. Rodriguez: Rep. Prog. Phys. **44**, 1297 (1981)

5.70 F. Bassani, G. Pastore-Parravicini: *Eletronic States and Optical Transitions in Solids* (Pergamon Press,Oxford, 1975)

5.71 M. Altarelli, F. Bassani: in *Handbook of Semiconductors* Vol. 1, ed. by W. Paul, (North Holland, Amsterdam, 1980), p. 269

5.72 W. Zawadzki: in *Landau Level Spectroscopy*, ed. by G. Landwehr, E. I. Rashba, (North Holland-Elsevier, Amsterdam, 1991), p. 1305

5.73 B. O. Kolbesen: Appl. Phys. Lett. **27**, 353 (1975)

5.74 E. E. Haller, H. Navarro, F. Keilmann: Proc. Int. Conf. Phys. Semicond., Stockholm 1986 (World Scientific, Singapore 1987), p. 837

5.75 C. Jagannath, Z. W. Grabowski, A. K. Ramdas: Solid State Commun. **29**, 355 (1979)

5.76 G. Bastard: Phys. Rev. **B24**, 4714 (1981)

5.77 G. Bastard: *Wave Mechanics Applied to Heterostructures* (Editions de Physique, Les Ulis 1989)

5.78 C. Mailhiot, Yia-Chung Chang, T. C. McGill: Phys. Rev. **B26**, 4449 (1982)

5.79 R. Greene, K. K. Bajaj: Phys. Rev. **B31**, 913 (1985)

5.80 P. Lane, R. L. Greene: Phys. Rev. **B33**, 5871 (1986)

5.81 A. A. Reeder, J. M. Mercy, B. D. McCombe: IEEE J. Quant. Electron. , **QE24**, 1690 (1988)

5.82 T. W. Masselink, Y. C. Chang, M. Morkoc: Phys. Rev. **B28**, 7373 (1983)

5.83 T. W. Masselink, Y. C. Chang, M. Morkoc: Phys. Rev. **B32**, 5190 (1985)

5.84 L. T. Canham: Appl. Phys. Lett. **57**, 1046 (1990)

5.85 A. Halimaoui, C. Oules, G. Bomchil, A. Bsiesy, F. Gaspard, R. Herino, M. Ligeon, F. Muller: Appl. Phys. Lett. **59**, 304 (1991)

5.86 A. Gee: J. Electrochem. Soc. **107**, 787 (1960)

5.87 D. C. Bensahel, L. T. Canham, S. Ossicini. *Optical Properties of Low Dimensional Silicon Structures*, NATO ASI Series E: Applied Sciences **244** (1993)

5.88 S. S. Iyer, L. T. Canham, R. T. Collins: *Light Emission from Silicon*, Proceedings of the MRS Fall Meeting (Boston 1991)

5.89 A. G. Cullis: unpublished

5.90 A. Halimaoui: in *Optical Properties of Low Dimensional Silicon Structures*, ed. by D. C. Bensahel, L. T. Canham, S. Ossicini, (Kluwer Academic Publ., Dordrecht, Boston, London 1993), p. 11

5.91 J. Fricke (ed.): Journal of Non-Crystalline Solids **145** (1992)

5.92 M. Stutzmann, J. Weber, M. S. Brandt, H. D. Fuchs, M. Rosenbauer, P. Deak, A. Höpfner, A. Breitschwerdt: in *Festkörperprobleme/Advances in Solid State Physics* **32**, 179 (1992)

5.93 S. Frohnhoff: in *Berichte des Forschungszentrums Jülich* (Jül-2765), ISSN 0944-2952 (1993)

5.94 H. Münder, M. G. Berger, S. Frohnhoff, M. Thönissen, H. Lüth, W. Theiß, L. Küpper: in *Optical Properties of Low Dimensional Silicon Structures*, ed. by D. C. Bensahel, L. T. Canham, S. Ossicini, (Kluwer Academic Publ., Dordrecht, Boston, London 1993), p. 75

5.95 M. H. Berger, C. Dieker, M. Thönissen, L. Vescan, H. Lüth, H. Münder: J. Phys. D **27**, 1333 (1994)

5.96 K. H. Beckmann: Surf. Sci. **3**, 314 (1965)

5.97 T. Unagami: Jpn. J. Appl. Phys. **19**, 231 (1980)

5.98 Y. Kato, T. Ito, A. Hiraki: Jpn. J. Appl. Phys. **27**, 1406 (1988)

5.99 W. Theiß, P. Grosse, H. Münder, H. Lüth, R. Herino, M. Ligeon: Mat. Res. Soc. Symp. Proc. **238**, 215 (1993)

5.100 W. Theiß, P. Grosse, H. Münder, H. Lüth, R. Herino, M. Ligeon: Applied Surface Science **63**, 240 (1993)

5.101 W. Theiß, in *Festkörperprobleme/Advances in Solid State Physics* **33**, ed. by R. Helbig (Vieweg, Braunschweig, Wiesbaden 1994), p. 149

5.102 J. C. Maxwell Garnett: Philos. Trans. R. Soc. London **203**, 385 (1904)

5.103 D. A. G. Bruggeman: Ann. Phys. (Leipzig) **24**, 636 (1935)

5.104 D. J. Bergman: Phys. Rep. C **43**, 377 (1978)

5.105 D. J. Bergman, D. Stroud: Solid State Phys., **46**, p. 147, eds. H. Ehrenreich and D. Turnbull (Academic Press, Boston 1992),

5.106 M. Evenschor, P. Grosse, W. Theiß: Vibrational Spectroscopy **1**, 173 (1990)

5.107 M. Hornfeck, R. Clasen, W. Theiß: J. Non-Cryst. Solids **145**, 154 (1992)

5.108 H. Münder, C. Andrzejak, M. G. Berger, U. Klemradt, H. Lüth, R. Herino, M. Ligeon: Thin Solid Films **221**, 27 (1992)

5.109 P. Grosse: Vibrational Spectroscopy **1**, 187 (1990)

6.1 A. Segmüller, M. Murakami in: Thin Films from Free Atoms and Particles,

6.2 B.M. Paine: MRS Symp. Proc. **69**, 39 (1986)

6.3 A. Segmüller, I.C. Noyan, V.S. Speriosu: Progr. Crystal Growth and Charact. **18**, 21 (1989)

6.4 A. Segmüller: J.Vac.Sci.Technol. **A9**, 2477 (1991)

6.5 C. Schiller, G. Martin, W.W. v.d. Hoogenhof, J. Corno: Philips J.Res. **47**, 217 (1993)

6.6 B.K. Tanner: Advances in X-Ray Analysis **33**, 1 (1990)

6.7 B.K. Tanner in: Analysis in Microelectronic Materials and Devices, ed. by M. Grasserbauer and H.W. Werner, (J. Wiley, New York), p. 609 (1991)

6.8 M.A.G. Halliwell: Prog. Crystal Growth and Charact. **19**, 249 (1989)

6.9 P.F. Fewster: J. Appl. Cryst. **24**, 178 (1991)

6.10 P.F. Fewster: Appl. Surf. Sci. **50**, 9 (1991)

6.11 P.F. Fewster: Semicond. Sci. Technol. **8**, 1915 (1993)

6.12 T. Picraux, B.L. Doyle, J.Y. Tsao in: Semiconductors and Semimetals, ed. by T.P. Pearsall, (Academic Press, N.Y.), p.139-220 (1991)

6.13 T.W. Ryan, M. Halliwell, S. Bates, I. Bassignana: Materials Research Society, Short Course on Characterisation of Compound Semiconductors by High Resolution X-Ray Diffraction 1990 and 1991

6.14 C.R. Wie: Materials Science and Engineering **R13**, No. 1, (1994)

6.15 E.J. Fantner, K. Lischka (eds.): Proc. High Resolution X-Ray Diffraction Workshop Aigen 1992, Appl. Phys. A **58**, No 3 (1994)

6.16 Proc. Int. Conf. on High Resolution X-Ray Diffraction and Topography, published in J. Phys. D (Appl. Phys.) **26**, No 4A (1993)

6.17 K. Kohra: J. Phys. Soc. Japan **30**, 1136 (1971)

6.18 R. Köhler: Appl. Phys. **A58**, 149 (1994)

6.19 C. Malgrange, D. Ferret: Nuclear Instruments and Methods in Physics Research **A314**, 285 (1992)

6.20 W. L. Bragg: Proc. Cambridge Phil. Soc. **17**, 43 (1913)

6.21 A. Segmüller: Advances in X-Ray Analysis **29**, 353 (1986)

6.22 M. Hart: J. Crystal Growth **55**, 409 (1981)

6.23 B. Davis, W. M. Stempel: Phys. Rev. **17**, 526 (1921)

6.24 W. Ehrenberg, H. Mark: Z. Physik **42**, 807 (1927)

6.25 J. W. M. DuMond: Phys. Rev. **52**, 872 (1937)

6.26 R. Bubakova in: Brümmer and Stephanik, loc. cit. [6.27], 148 (1976)

6.27 O. Brümmer, H. Stephanik (eds.): *Dynamische Interferenztheorie*, (Akad. Verlagsgesellschaft, Leipzig) 1976

6.28 A. Fingerland: in Brümmer and Stephanik, loc. cit. [6.27], 159 (1976)

6.29 K. J. Godwod: in Brümmer and Stephanik, loc. cit. [6.27], 165 (1976)

6.30 M. M. Schwarzschild: Phys. Rev. **32**, 162 (1928)

6.31 H. W. Schnopper: J. Appl. Phys. **36**, 1415 (1965)

6.32 J. H. Beaumont, M. Hart: J. Phys. **E7**, 823 (1974)

6.33 W. J. Bartels: J. Vac. Sci. Technol. **B1**, 338 (1983)

6.34 W. L. Bond: Acta Cryst. **13**, 814 (1960)

6.35 P. van der Sluis: J. Appl. Cryst. **27**, 50 (1994)

6.36 B. K. Tanner, D. K. Bowen: J. Cryst. Growth **126**,1 (1993)

6.37 M. Renninger: Z. Naturforschung **16a**, 1110 (1961)

6.38 K. Kohra: J. Phys. Soc. Japan **17**, 589 (1962)

6.39 M. Renninger: Z. Kristallogr. **99**, 181 (1938)

6.40 M. Renninger: Acta Cryst. **A24**, 143 (1968)

6.41 M. Lefeld-Sosnowska in: Brümmer and Stephanik, loc. cit. [6.27], 148 (1976)

6.42 T. W. Ryan, P. D. Hatton, S. Bates, M. Watt, C. M. Sotomayor Torres, P. A. Claxton, J. S. Roberts: Semicond. Sci. Technol. **2**, 241 (1987)

6.43 E. Koppensteiner, T. Ryan, M. Heuken, J. Söllner: J. Phys. D **26**, A35 (1993)

6.44 P. van der Sluis: J. Phys. D **26**, A188 (1993)

6.45 P. Zaumseil, U. Winter, F. Cembali, M. Servidori, Z. Sorek: Phys. Stat Sol. (a) **100**, 95 (1987)

6.46 R. Thompson, B. L. Doyle: Mat. Res. Proc. EA- **18**, 141 (1988)

6.47 A. McL. Mathieson: Acta Cryst. **A38**, 378 (1982)

6.48 R. L. Thompson, G. J. Collins, B. L. Doyle, J. A. Knapp: J. Appl. Phys. **70**, 4760 (1991)

6.49 T. Picraux T, B. L. Doyle, J. Y. Tsao: Mat. Sci. Technology **33**, 139 (1991)

6.50 N. Itoh, K. Okamoto: J. Appl. Phys. **63**, 1486 (1988)

6.51 M. Renninger: Z. Physik **106**, 141 (1937)

6.52 B. Post: J. Appl. Crystallogr. **8**, 452 (1975)

6.53 B. Post, P. P. Gong, L. Kern, J. Ladell: Acta Crystallogr. **A42**, 178 (1986)

6.54 B. Greenberg, J. Ladell: Appl. Phys. Lett. **50**, 436 (1987)

6.55 S. L. Morelhao, L. P. Cardoso, J. M. Sasaki, M. M. G. de Carvalho: J. Appl. Phys. **70**, 2589 (1991)

6.56 S. L. Morelhao, L. P. Cardoso: J. Appl. Phys. **73**, 4218 (1993)

6.57 Z. G. Pinsker: *Dynamical Scattering of X-Rays in Crystals*, Springer Verlag, Berlin Heidelberg New York 1978

6.58 P. F. Fewster: J. Appl. Cryst. **22**, 64 (1989)

6.59 P. F. Fewster: Electrochem. Soc. Symp. Proc. **89**-5, 278 (1989)

6.60 T. W. Ryan: Ph.D. Thesis, University of Edinburgh, 1986

6.61 C. G. Darwin C. G: Philos. Mag. **27**, 315 and 675 (1914)

6.62 P. P. Ewald: Ann. Physik **54**, 519 (1917)

6.63 M. von Laue: Ergebnisse d. exakt. Naturwiss. **10**, 133 (1931)

6.64 N. Kato, A. R. Lang: Acta Cryst. **12**, 787 (1959)

6.65 B. W. Batterman, G. Hildebrandt: Acta Cryst. **A24**, 150 (1968)

6.66 J. A. Prins: Zeit. f. Physik **63**, 477 (1930)

6.67 W. H. Zachariasen: *Theory of X-Ray Diffractions in Crystals*, (Wiley, New York 1945)

6.68 S. Takagi: Acta Cryst. **15**, 1311 (1962)

6.69 S. Takagi: J. Phys. Soc. Japan **26**, 1239 (1969)

6.70 D. Taupin: Bull. Soc. Fr. Mineral. Crystallogr. **87**, 469 (1964)

6.71 W. J. Bartels, J. Hornstra, D. J. W. Lobeek: Acta Cryst. **A42**, 539 (1986)

6.72 R. Zaus, M. Schuster, H. Göbel, J.-P. Reithmaier: Appl. Surf. Sci. **50**, 92 (1991)

6.73 M. Servidori, F. Cembali, R. Fabri, A. Zani: J. Appl. Cryst. **25**, 46 (1992)

6.74 M. O. Möller: Thesis, University of Würzburg, Germany

6.75 R. Zaus: J. Appl. Cryst. **26**, 801 (1993)

6.76 Y. C. Chen, P. K. Bhattacharya: J. Appl. Phys. **73**, 7389 (1993)

6.77 R. W. James: *The Optical Principles of the Diffraction of X-Rays, The Crystalline State-Vol* II, ed. by L. Bragg (G. Bell and Sons Ltd) (1962)

6.78 P. V. Petrashen: Sov. Phys. Sol. St. **16**, 1417 (1975)

6.79 P. V. Petrashen: Sov. Phys. Sol. St. **17**, 1882 (1976)

6.80 L. Tapfer, K. Ploog: Phys. Rev. **B33**, 5565 (1986)

6.81 L. Tapfer, K. Ploog: Phys. Rev. **B40**, 9802 (1989)

6.82 A. Segmüller, A. E. Blakeslee: J. Appl. Cryst. **6**, 19 (1973)

6.83 R. M. Fleming, D. B. Mc Whan, A. C. Gossard, W. Wiegemann, R. A. Logan: J. Appl. Phys. **51**, 357 (1980)

6.84 V. S. Speriosu, T. Vreeland Jr.: J. Appl. Phys. **56**, 1591 (1984)

6.85 V. S. Speriosu: J. Appl. Phys. **52**, 6094 (1981)

6.86 U. Lienert: Thesis, Technical University of Berlin 1989, unpublished

6.87 P. F. Fewster, C. J. Curling: J. Appl. Phys. **62**, 4154 (1987)

6.88 L. Tapfer, Phys. Scr.: **T25**, 6094 (1989)

6.89 H. Holloway: J. Appl. Phys. **67**, 6229 (1990)

6.90 C. R. Wie, H. M. Kim: J. Appl. Phys. **69**, 6406 and 6412 (1991)

6.91 A. T. Macrander, E. R. Minami, D. W. Berreman: J. Appl. Phys. **60**, 1364 (1986)

6.92 F. Abelès: Ann. de Physique **5**, 596 and 706 (1955)

6.93 L. Tapfer, M. Ospelt, H. von Känel: J. Appl. Phys. **67**, 1298 (1990)

6.94 D. M. Vardanyan, H. M. Manoukyan, H. M. Petrosyan: Acta Cryst. **A 41**, 212 and 218 (1985)

6.95 C. R. Wie: J. Appl. Phys. **65**, 1036 and 2267 (1989)

6.96 A. Caticha: Phys. Rev. **B49**, 33 (1994)

6.97 B. D. Cullity: *Elements of X-ray diffraction*, Addison-Wesley, Reading, Mass., (1956)

6.98 H. Meyerheim: Thesis, Ludwig-Maximilians University of Munich, 1985, unpublished

6.99 A. Segmüller, M. Murakami in: *Analytical Techniques for Thin Films*, ed. by K. N. Tu and R. Rosenberg, (Academic Press, New York 1988)

6.100 J. F. Nye: *Physical Properties of Crystals: Their Representation by Tensors and Matrices*, (Oxford 1957)

6.101 Y. Kawamura, H. Okamoto: J. Appl. Phys. **50**, 4457 (1979)

6.102 J. Hornstra, W. J. Bartels: J. Cryst. Growth **44**, 513 (1978)

6.103 B. Ortner: Advances in X-Ray Analysis **29**, 387 (1986)

6.104 E. Anastassakis: J. Appl. Phys. **68**, 4561 (1990)

6.105 M. Grundmann, U. Lienert, D. Bimberg, A. Fischer-Colbrie, J. N. Miller: Appl. Phys. Lett. **55**, 1765 (1989)

6.106 C. Giannini, L. De Caro, L. Tapfer: Solid State Commun. **91**, 635 (1994)

6.107 B. R. Bennett, J. A. del Alamo: J. Electron. Mater. **20**, 1075 (1991)

6.108 B. R. Bennett, J. A. del Alamo: Materials Research Society Proceedings (MRS, Pittsburgh, PA) **240**, 153 (1992)

6.109 W. J. Bartels, W. Nijman: J. Cryst. Growth **44**, 518 (1978)

6.110 J. W. Matthews in: *Epitaxial Growth B*, ed. by J. W. Matthews, Academic Press, New York 1975, p. 560

6.111 H. J. van der Merwe: J. Appl. Phys. **34**, 117 (1963)

6.112 H. J. van der Merwe, W. A. Jesser: J. Appl. Phys. **64**, 4968 (1988)

6.113 M. A. G. Halliwell, M. H. Lyons, M. J. Hill: J. Crystal Growth **68**, 523 (1984)

6.114 C. R. Wie, T. A. Tombrello, T. Vreeland: J. Appl. Phys. **59**, 3743 (1986)

6.115 B. M. Paine, V. S. Speriosu: J. Appl. Phys. **62**, 1704 (1987)

6.116 S. Bensoussan, C. Malgrange, M. Sauvage-Simkin: J. Appl. Cryst. **20**, 222 (1987)

6.117 H. Nagai: J. Appl. Phys. **43**, 4254 (1972)

6.118 H. Nagai: J. Appl. Phys. **45**, 3789 (1974)

6.119 P. Auvray, M. Baudet, A. Regreny: J. Cryst. Growth **95**, 228 (1989)

6.120 P. Auvray, A. Poudoulec, M. Baudet, B. Guenais, A. Regreny, C. d'Anterroches, J. Massies: Appl. Surf. Sci. **50**, 109 (1991)

6.121 D. A. Neumann, H. Zabel, H. Morkoc: J. Appl. Phys. **64**, 3024 (1988)

6.122 A. Pesek, K. Hingerl, F. Riesz, K. Lischka: Semicond. Sci. Technol. **6**, 705 (1991)

6.123 A. Pesek: Thesis, University of Linz, (1993)

6.124 A. Leiberich, J. Levkoff: Mat. Res. Symp. **159**, 101 (1990)

6.125 M. Quillec, L. Goldstein, G. LeRoux, J. Burgeat, J. Primot: J. Appl. Phys. **55**, 2904 (1984)

6.126 K. Nakashima: J. Appl. Phys. **71**, 1189 (1992)

6.127 P. van der Sluis: Appl. Phys. Lett. **62**, 1898 (1993)

6.128 E. Koppensteiner, G. Springholz, P. Hamberger, G. Bauer: J. Appl. Phys. **74**, 6062 (1993)

6.129 J. M. Vandenberg, M. B. Panish, H. Temkin, R. A. Hamm: Appl. Phys. Lett. **53**, 1920 (1988)

418   References

6.130 J. M. Vandenberg, A. T. Macrander, R. A. Hamm, M. B. Panish: Phys. Rev.
     B44, 3991 (1991)
6.131 R. Meyer, M. Hollfelder, H. Hardtdegen, B. Lengeler, H. Lüth: J. Cryst.
     Growth **124**, 583 (1992)
6.132 A. Krost, J. Böhrer, H. Roehle, G. Bauer: Appl. Phys. Lett. **64**, 469 (1994)
6.133 U. Rossow, A. Krost, T. Werninghaus, K. Schatke, W. Richter, A. Hase, H.
     Künzel, H. Roehle: Thin Solid Films, **233**, 180 (1993)
6.134 X. G. He, M. Erdtmann, R. Williams, S. Kim, M. Razeghi: Appl. Phys.
     Lett., **65**, 2812 (1994)
6.135 N. Herres, G. Bender, G. Neumann: Appl. Surf. Sci. **50**, 97 (1991)
6.136 A. Segmüller, P. Krishna, L. Esaki: J. Appl. Cryst. **10**, 1 (1977)
6.137 J. Kervarac, M. Baudet, J. Caulet, P. Auvray, Y. Y. Emery, A. Regreny: J.
     Appl. Cryst. **17**, 196 (1984)
6.138 A. Powell, R. Kubiak, E. Parker, K. Bowen, M. Polcarova: Mater. Res. Symp.
     Proc. Anaheim, 1991
6.139 D. B. McWhan in: *Synthetic Modulated Structures*, ed. by L. L. Chang and
     B. C. Giessen, (Academic Press, N. Y. 1985)
6.140 D. B. McWhan in: *Physics, Fabrication and Applications of Multilayered
     Structures*, ed. by P. Dhez and C. Weisbuch, (Plenum Press, N. Y.), p. 67
     (1988)
6.141 D. K. Arch, J. P. Faurie, J.-L. Staudenmann, M. Hibbs-Brenner, P. Chow:
     J. Vac. Sci. Technology **A4**, 2101 (1986)
6.142 J. H. C. Hogg, D. Shaw, M. Staudte: Appl. Surf. Sci. **50**, 87 (1991)
6.143 H. Krenn, A. Holzinger, A. Voiticek, G. Bauer, H. Clemens: J. Appl. Phys.
     72, 97 (1992)
6.144 Y. Kim, A. Ourmazd, M. Bode, R. Feldman: Phys. Rev. Lett. **63**, 636 (1989)
6.145 P. F. Fewster: J. Appl. Cryst. **21**, 524 (1988)
6.146 D. E. Savage, J. Kleiner, N. Schimke, Y.-H. Phang, T. Jankowski, J. Jacobs,
     R. Kariotis, M. G. Lagally: J. Appl. Phys. **69**, 1411 (1991)
6.147 Y. H. Phang, D. E. Savage, T. F. Kuech, M. G. Lagally, J. S. Park, K. L.
     Wang: Appl. Phys. Lett **60**, 2986 (1992)
6.148 V. Holý, J. Kubena, I. Ohlidal, K. Ploog: Superlattices and Microstructures
     12, 25 (1992)
6.149 P. F. Fewster: Philips J. Res. **41**, 268 (1986)
6.150 W. J. Bartels in: *Thin Film Growth Techniques for Low-Dimensional Struc-
     tures*, ed by R. F. C. Farrow et al., Nato ASI Series **163**, Plenum Press,
     (1987), p. 441
6.151 S. J. Barnett, G. T. Brown, D. C. Houghton, J. M. Baribeau: Appl. Phys.
     Lett **54**, 1781 (1989)
6.152 L. Hart, M. R. Fahy, R. C. Newman, P. F. Fewster: Appl. Phys. Lett. **62**,
     2218 (1993)
6.153 M. A. Hollanders, B. J. Thijsse: J. Appl. Phys. **70**, 1270 (1991)
6.154 E. E. Fullerton, I. K. Schuller, H. Vanderstraeten, Y. Bruynserade: Phys.
     Rev. **B45**, 9292 (1992)
6.155 P. van der Sluis: Philips J. Research **47**, 203 (1993)
6.156 V. Holý, J. Kubena: Phys. Stat. Sol. (b) **170**, 9 (1992)
6.157 V. Holý, J. Kubena, E. Abramof, K. Lischka, A. Pesek, E. Koppensteiner:
     J. Appl. Phys. **74**, 1736 (1993)

6.158 P. F. Fewster: J. Appl. Cryst. **25**, 714 (1992)

6.159 P. Ofner, H. Krenn, S. Bates, R. A. Cowley, H. Clemens, G. Bauer: *Proc. Int. Conf. Phys. Semicond. Warszawa* (Polish Academy of Sciences Warszawa 1988), p. 483

6.160 A. M. Keir, S. J. Barnett, J. Fiess, T. D. Walsh, M. G. Astles: Appl. Surf. Sci. **50** 103 (1991)

6.161 E. Koppensteiner, P. Hamberger, G. Bauer, A. Pesek, H. Kibbel, H. Presting, E. Kasper: Appl. Phys. Lett. **62**, 1783 (1993)

6.162 E. Koppensteiner, G. Bauer, H. Kibbel, E. Kasper: J. Appl. Phys. **76**, 3489 (1994)

6.163 E. Koppensteiner, P. Hamberger, G. Bauer, V. Holý, E. Kasper: Appl. Phys. Lett. **64**, 172 (1994)

6.164 H. Heinke, M. O. Möller, D. Hommel, G. Landwehr: J. Crystal Growth **135**, 41 (1994)

6.165 G. Bauer, E. Koppensteiner, P. Hamberger, J. Nützel, G. Abstreiter, H. Kibbel, H. Presting, E. Kasper: Acta Physica Polonica A **84**, 475 (1993)

6.166 A. T. Macrander, S. E. G. Slusky: Appl. Phys. Lett. **56**, 443 (1990)

6.167 A. T. Macrander, S. Lau, K. Strege, S. N. G. Chu: Appl. Phys. Lett. **52**, 1985 (1988)

6.168 V. Holý: Appl. Phys. **A58**, 173 (1994)

6.169 L. Tapfer, P. Grambow: Appl. Phys. **A50**, 3 (1990)

6.170 L. Tapfer, G. C. LaRocca, H. Lage, O. Brandt, D. Heitmann, K. Ploog: Appl. Surf. Sci. **60/61**, 517 (1992)

6.171 R. Cingolani, H. Lage, L. Tapfer, H. Kalt, D. Heitmann, K. Ploog: Phys. Rev. Lett. **67**, 891 (1991)

6.172 L. De Caro, P. Sciacovelli, L. Tapfer: Appl. Phys. Lett. **64**, 34 (1994)

6.173 M. Gailhanou, T. Baumbach, U. Marti, P. C. Silva, F. K. Reinhart, M. Ilegems: Appl. Phys. Lett. **62**, 1623 (1993)

6.174 P. van der Sluis, J. J. M. Binnsma, T. van Dongen: Appl. Phys. Lett. **62**, 3186 (1993)

6.175 V. Holý, L. Tapfer, E. Koppensteiner, G. Bauer, H. Lage, O. Brandt, K. Ploog: Appl. Phys. Lett. **63**, 3140 (1993)

6.176 T. Fukui, H. Saito: Appl. Phys. Lett. **50**, 824 (1987)

6.177 O. Brandt, L. Tapfer, K. Ploog, R. Bierwolf, M. Hohenstein, F. Phillip, H. Lage, A. Heberle: Phys. Rev. **B44**, 8043 (1991)

6.178 A. A. Darhuber, E. Koppensteiner, G. Bauer, P. D. Wang, Y.P. Song, C.M. Sotomayor Torres, M.C. Holland: Appl. Phys. Lett. **66**, No. 8, 20 February 1995

6.179 I. K. Schuller, M. Grimsditch, F. Chambers, G. Devane, H. Vanderstraeten, D. Neerinck, J. P. Locquet, Y. Bruynseraede: Phys. Rev. Lett. **65**, 1235 (1990)

6.180 P. F. Miceli in: *Semiconductor Interfaces, Microstructures and Devices: Properties and Applications*, ed. by Z. C. Feng (Adam Hilger IOP Publishing Ltd, Bristol 1992), p. 87

6.181 J. M. Baribeau: J. Appl. Phys. **72**, 4452 (1992)

6.182 S. K. Sinha, E. B. Sirota, S. Garoff, H. B. Stanley: Phys. Rev. **B38**, 2297 (1988)

6.183 S. Yasuami, K. Ohshima, S. Sasaki, M. Ando: J. Appl. Cryst. **25**, 514 (1992)

6.184 J. M. Cowley: J. Appl. Phys. **21**, 24 (1950)

6.185 T. Tsuchiya, T. Taniwatari, K. Uomi, T. Kawano, Y. Ono: *Proc. 4th InP and Related Materials Conference*, Newport, Rhode Island (1992)

6.186 W. Lowe, R. A. MacHarrie, J. C. Bean, L. Peticolas, R. Clarke, W. Dos Passos, C. Brizard, B. Rodricks: Phys. Rev. Lett. **67**, 2513 (1991)

6.187 R. Feidenhans'l: Surf. Sci. Rep. **10**, 105 (1989)

6.188 P. H. Fuoss, S. Brennan: Ann. Rev. Mater. Sci. **20**, 365 (1990)

6.189 I. K. Robinson, D. J. Tweet: Reports on Progress in Phys. **55**, 599 (1992)

6.190 M. Bienfait, J. M. Gay (eds.): *X-Ray and Neutron Scattering from Surfaces and Thin Films*, Colloque de Physique **C7**, Supp. 10, Tome 50, Marseille, France 1989

6.191 H. Zabel, J. K. Robinson (eds.): *Surface X-Ray and Neutron Scattering*, Springer Proceedings in Physics, Vol. **61**, (Springer, Berlin, Heidelberg, New York 1991)

6.192 H. Zabel: Appl. Phys. **A58**, 159 (1994)

6.193 M. Born, E. Wolf: *Principles of Optics* (Pergamon Press, Third Ed.) 1965

6.194 A. H. Compton, S. K. Allison: *X-Rays in Theorie and Experiment*, D. van Nostrand Company, Inc. Princeton, New Jersey, Second Edition, 674 (1935)

6.195 D. T. Cromer, D. Liberman: J. Chem. Phys. **53**, 1891 (1970)

6.196 W. J. Veigele in: *Handbook of Spectroscopy*, Vol. 1, ed. by J. W. Robinson (CRC Press, Cleveland, Ohio 1974)

6.197 A. H. Compton: Philos. Mag. **2**, 897 (1923)

6.198 B. Lengeler in: *Synchrotronstrahlung in der Festkörperphysik*, 18. IFF Ferienkurs, March 16-27, Jülich (1987)

6.199 L. G. Parrat: Phys. Rev. **95**, 359 (1954)

6.200 G. H. Vineyard: Phys. Rev. **B26**, 4146 (1982)

6.201 J. A. Prins: Z. Physik **47**, 4791 (1928)

6.202 H. Kiessig: Ann. Phys. **10**, 715 (1931)

6.203 F. Stanglmeier: Report JÜL-2346, Januar 1990, ISSN 0366-0885

6.204 H. Kiessig: Ann. Phys. **10**, 769 (1931)

6.205 D. G. Stearns: J. Appl. Phys. **71**, 4286 (1992)

6.206 W. Weber, B. Lengeler: Phys. Rev. **B46**, 7953 (1992)

6.207 V. Holý, J. Kubena, I. Ohlodal, K. Lischka, W. Plotz: Phys. Rev. **B47**, 15896 (1993)

6.208 B. Vidal, P. Vincent: Appl. Optics **23**, 1794 (1984)

6.209 Y. Yoneda: Phys. Rev. **131**, 2010 (1963)

6.210 B. E. Warren, J. S. Clarke: J. Appl. Phys. **36**, 324 (1965)

6.211 O. J. Guentert: Phys. Rev. **138**, A732 (1965)

6.212 D. E. Savage, N. Schimke, Y.-H. Phang, M. G. Lagally: J. Appl. Phys. **71**, 3283 (1992)

6.213 Y.-H. Phang, R. Cariotis, D. E. Savage, M. G. Lagally, J. Appl. Phys. **72**, 4627 (1992)

6.214 D. Bar, W. Press, R. Jebasinski, S. Mantel: Phys. Rev. **B47**, 4385 (1993)

6.215 V. Holý, T. Baumbach: Phys. Rev. **B49**, 10668 (1994)

6.216 L. L. Chang, A. Segmüller, L. Esaki: Appl. Phys. Lett. **28**, 39 (1976)

6.217 A. R. Powell, J. Bradler, C. R. Thomas, R. A. Kubiak, K. D. Bowen, M. Wormington, J. M. Hudson: Mater. Res. Symp. Proc. (Boston), 1991

6.218 W. Spirkl: J. Appl. Phys. **74**, 1776 (1993)

6.219 K. Sakurai, A. Iida: Jpn. J. Appl. Phys. **31**, L113 (1992)

6.220 A. Król, H. Resat H, C. J. Sher, S. C. Woronick, W. Ng, Y. H. Kao, T. L. Cole, A. K. Green, W. Lowe-Ma, V. Rehn: J. Appl. Phys. **69**, 949 (1990)

6.221 W. F. J. Slijkerman, J. M. Gay, P. M. Zagwijn, J. F. van der Veen, J. E. Macdonald, A. A. Williams, D. J. Gravesteijn, G. F. A. van de Walle: J. Appl. Phys. **68**, 5105 (1990)

6.222 J. M. Baribeau: Appl. Phys. Lett. **57**, 1748 (1990)

6.223 J. M. Baribeau: J. Appl. Phys. **70**, 5710 (1991)

6.224 P. F. Miceli, D. A. Neumann, H. Zabel: Appl. Phys. Lett. **48**, 24 (1986)

6.225 M. Pomerantz: Thin Solid Films **152**, 165 (1987)

6.226 A. Braslau, P. S. Pershan, G. Swislow, B. M. Ocko, J. Als-Nielsen: Phys. Rev. **A38**, 2457 (1988)

6.227 J. J. Benattar, J. Daillant, L. Bosio, L. Leger: Colloque de Physique **C7**, Tome 50, Suppl. 10, ed. by M. Bienfait, J. M. Gay, Marseille, France (1989)

6.228 A. Król, C. J. Sher, Y. H. Kao: Phys. Rev. **B38**, 8579 (1988)

6.229 U. Weisbrod, R. Gutschke, J. Knoth, H. Schwenke: Appl. Phys. **A53**, 449 (1991)

6.230 W. C. Marra, P. Eisenberger, A. Y. Cho: J. Appl. Phys. **50**, 6927 (1979)

6.231 P. Farwig, H. W. Schürmann: Z. Physik **204**, 489 (1967)

6.232 S. Dietrich, H. Wagner: Phys. Rev. Lett. **51**, 1469 (1983)

6.233 S. Dietrich, H. Wagner: Z. Phys. B -Condensed Matter **56**, 207 (1984)

6.234 A. M. Afanas'ev, M. K. Melkonyan: Acta. Cryst. **A39**, 207 (1983)

6.235 P. L. Cowan: Phys. Rev. **B32**, 5437 (1985)

6.236 H. Zabel: in *Festkörperprobleme XXX - Advances in Solid State* Physics, ed. by U. Rössler (Vieweg Braunschweig, Wiesbaden 1990), p. 197

6.237 H. Dosch: *Critical Phenomena at Surfaces and Interfaces, Springer Tracts in Modern Physics*, **126**, (Springer, Berlin, Heidelberg, New York 1992)

6.238 H. Dosch, B. W. Batterman, D. C. Wack: Phys. Rev. Lett. **56**, 1144 (1986)

6.239 H. Dosch: Phys. Rev. **B35**, 2137 (1987)

6.240 R. S. Becker, J. A. Golovchenko, J. R. Patel: Phys. Rev. Lett. **50**, 153 (1983)

6.241 N. Bernhard N, E. Burkel, G. Gompper, H. Metzger, J. Peisl, H. Wagner, G. Wallner: Z. Phys. B - Condensed Matter **69**, 3031 (1987)

6.242 T. Jach, P. L. Cowan, Q. Shen, M. J. Bedzyk: Phys. Rev. **B39**, 5739 (1989)

6.243 P. A. Aleksandrov, A. M. Afanasiev, M. K. Melkonyan, S. A. Stepanov: Phys. Stat. Sol.(a) **81**, 47 (1984)

6.244 S. A. Stepanov: Phys. Stat. Sol. (a) **126**, K15 (1991)

6.245 A. L. Golovin, R. M. Imamov: Phys. Stat. Sol. (a) **77**, K91 (1983)

6.246 H. Rhan, U. Pietsch: Phys. Stat. Sol. (a) **107**, K93 (1988)

6.247 H. Rhan, U. Pietsch: Z. Phys. B - Condensed Matter **80**, 347 (1990)

6.248 H. Munekata, A. Segmüller, L. L. Chang: Appl. Phys. Lett. **51**, 587 (1987)

6.249 P. H. Fuoss, D. W. Kisker, S. Brennan, J. L. Kahn, G. Renaud, K. L. Tokuda in: *Heteroepitaxial Approaches in Semiconductors - Lattice Mismatch and its Consequences*, ed. by A. T. Macrander and T. J. Drummond, p. 159 (The Electrochemical Society, Pennington NJ) (1989)

6.250 D. W Kisker, P. H. Fuoss, S. Brennan, G. Renaud, K. L. Tokuda, J. L. Kahn: J. Cryst. Growth **101**, 42 (1990)

6.251 D. W Kisker, G. B. Stephenson, P. H. Fuoss, F. J. Lamelas, S. Brennan, P. Imperatori: J. Cryst. Growth **124**, 1 (1992)

6.252 F. J. Lamelas, P. H. Fuoss, D. W. Kisker, G. B. Stephenson, P. Imperatori, S. Brennan: Phys. Rev. **B49**, 1957 (1994)

Subject Index

Ergebnisse der Mathematik und ihrer Grenzgebiete, 3. Folge

A Series of Modern Surveys in Mathematics

Ed.-in-chief: **R. Remmert.** Eds.: **E. Bombieri, S. Feferman, M. Gromov, H. W. Lenstra, P-L. Lions, W. Schmid, J-P. Serre, J. Tits**

Volume 1: **A. Fröhlich**

Galois Module Structure of Algebraic Integers

1983. X, 262 pp. ISBN 3-540-11920-5

Volume 2: **W. Fulton**

Intersection Theory

1984. XI, 470 pp. ISBN 3-540-12176-5

Volume 3: **J. C. Jantzen**

Einhüllende Algebren halbeinfacher Lie-Algebren

1983. V, 298 pp. ISBN 3-540-12178-1

Volume 4: **W. Barth, C. Peters. A van de Ven**

Compact Complex Surfaces

1984. X, 304 pp. ISBN 3-540-12172-2

Volume 5: **K. Strebel**

Quadratic Differentials

1984. XII, 184 pp. 74 figs. ISBN 3-540-13035-7

Volume 6: **M. J. Beeson**

Foundations of Constructive Mathematics

Metamathematical Studies
1985. XXIII, 466 pp. ISBN 3-540-12173-0

Volume 8: **R. Mañé**

Ergodic Theory and Differentiable Dynamics

Translated from the Portuguese by Silvio Levy
1987. XII, 317 pp. 32 figs. ISBN 3-540-15278-4

Volume 9: **M. Gromov**

Partial Differential Relations

1986. IX, 363 pp. ISBN 3-540-12177-4

Volume 10: **A. L. Besse**

Einstein Manifolds

1986. XII, 510 pp. 22 figs. ISBN 3-540-15279-2

Volume 11: **M. D. Fried, M. Jarden**

Field Arithmetic

1986. XVII, 458 pp. ISBN 3-540-16640-8

Volume 12: **J. Bochnak, M. Coste, M.-F. Roy**

Géométrie algébrique réelle

1987. X, 373 pp. 44 figs. ISBN 3-540-16951-2

Volume 13: **E. Freitag, R. Kiehl**

Etale Cohomology and the Weil Conjecture

With an Historical Introduction by J. A. Dieudonné
1987. XVIII, 317 pp. ISBN 3-540-12175-7

Volume 14: **M. R. Goresky, R. D. MacPherson**

Stratified Morse Theory

1988. XIV, 272 pp. 84 figs. ISBN 3-540-17300-5

Volume 15: **T. Oda**

Convex Bodies and Algebraic Geometry

An Introduction to the Theory of Toric Varieties
1987. VIII, 212 pp. 42 figs. ISBN 3-540-17600-4

Volume 16: **G. van der Geer**

Hilbert Modular Surfaces

1988. IX, 291 pp. ISBN 3-540-17601-2

■ ■ ■ ■ ■ ■ ■ ■ ■ ■

Springer-Verlag, Postfach 31 13 40, D-10643 Berlin, Fax 0 30 / 82 07 - 3 01 / 4 48, e-mail: orders@springer.de

■ ■ ■ ■ ■ ■ ■ ■ ■ ■

Springer

Springer-Verlag, Postfach 31 13 40, D-10643 Berlin, Fax 0 30 / 82 07 - 3 01 / 4 48, e-mail: orders@springer.de   BA95.10.18